Theory of Stochastic Objects

Probability, Stochastic Processes and Inference

CHAPMAN & HALL/CRC
Texts in Statistical Science Series

Statistical Theory: A Concise Introduction
F. Abramovich and Y. Ritov

Practical Multivariate Analysis, Fifth Edition
A. Afifi, S. May, and V.A. Clark

Practical Statistics for Medical Research
D.G. Altman

Interpreting Data: A First Course in Statistics
A.J.B. Anderson

Introduction to Probability with R
K. Baclawski

Linear Algebra and Matrix Analysis for Statistics
S. Banerjee and A. Roy

Modern Data Science with R
B. S. Baumer, D. T. Kaplan, and N. J. Horton

Mathematical Statistics: Basic Ideas and Selected Topics, Volume I, Second Edition
P. J. Bickel and K. A. Doksum

Mathematical Statistics: Basic Ideas and Selected Topics, Volume II
P. J. Bickel and K. A. Doksum

Analysis of Categorical Data with R
C. R. Bilder and T. M. Loughin

Statistical Methods for SPC and TQM
D. Bissell

Introduction to Probability
J. K. Blitzstein and J. Hwang

Bayesian Methods for Data Analysis, Third Edition
B.P. Carlin and T.A. Louis

Statistics in Research and Development, Second Edition
R. Caulcutt

The Analysis of Time Series: An Introduction, Sixth Edition
C. Chatfield

Introduction to Multivariate Analysis
C. Chatfield and A.J. Collins

Problem Solving: A Statistician's Guide, Second Edition
C. Chatfield

Statistics for Technology: A Course in Applied Statistics, Third Edition
C. Chatfield

Analysis of Variance, Design, and Regression: Linear Modeling for Unbalanced Data, Second Edition
R. Christensen

Bayesian Ideas and Data Analysis: An Introduction for Scientists and Statisticians
R. Christensen, W. Johnson, A. Branscum, and T.E. Hanson

Modelling Binary Data, Second Edition
D. Collett

Modelling Survival Data in Medical Research, Third Edition
D. Collett

Introduction to Statistical Methods for Clinical Trials
T.D. Cook and D.L. DeMets

Applied Statistics: Principles and Examples
D.R. Cox and E.J. Snell

Multivariate Survival Analysis and Competing Risks
M. Crowder

Statistical Analysis of Reliability Data
M.J. Crowder, A.C. Kimber, T.J. Sweeting, and R.L. Smith

An Introduction to Generalized Linear Models, Third Edition
A.J. Dobson and A.G. Barnett

Nonlinear Time Series: Theory, Methods, and Applications with R Examples
R. Douc, E. Moulines, and D.S. Stoffer

Introduction to Optimization Methods and Their Applications in Statistics
B.S. Everitt

Extending the Linear Model with R: Generalized Linear, Mixed Effects and Nonparametric Regression Models, Second Edition
J.J. Faraway

Linear Models with R, Second Edition
J.J. Faraway

A Course in Large Sample Theory
T.S. Ferguson

Multivariate Statistics: A Practical Approach
B. Flury and H. Riedwyl

Readings in Decision Analysis
S. French

Discrete Data Analysis with R: Visualization and Modeling Techniques for Categorical and Count Data
M. Friendly and D. Meyer

Markov Chain Monte Carlo: Stochastic Simulation for Bayesian Inference, Second Edition
D. Gamerman and H.F. Lopes

Bayesian Data Analysis, Third Edition
A. Gelman, J.B. Carlin, H.S. Stern, D.B. Dunson, A. Vehtari, and D.B. Rubin

Multivariate Analysis of Variance and Repeated Measures: A Practical Approach for Behavioural Scientists
D.J. Hand and C.C. Taylor

Practical Longitudinal Data Analysis
D.J. Hand and M. Crowder

Linear Models and the Relevant Distributions and Matrix Algebra
D.A. Harville

Logistic Regression Models
J.M. Hilbe

Richly Parameterized Linear Models: Additive, Time Series, and Spatial Models Using Random Effects
J.S. Hodges

Statistics for Epidemiology
N.P. Jewell

Stochastic Processes: An Introduction, Third Edition
P.W. Jones and P. Smith

The Theory of Linear Models
B. Jørgensen

Pragmatics of Uncertainty
J.B. Kadane

Principles of Uncertainty
J.B. Kadane

Graphics for Statistics and Data Analysis with R
K.J. Keen

Mathematical Statistics
K. Knight

Introduction to Functional Data Analysis
P. Kokoszka and M. Reimherr

Introduction to Multivariate Analysis: Linear and Nonlinear Modeling
S. Konishi

Nonparametric Methods in Statistics with SAS Applications
O. Korosteleva

Modeling and Analysis of Stochastic Systems, Third Edition
V.G. Kulkarni

Exercises and Solutions in Biostatistical Theory
L.L. Kupper, B.H. Neelon, and S.M. O'Brien

Exercises and Solutions in Statistical Theory
L.L. Kupper, B.H. Neelon, and S.M. O'Brien

Design and Analysis of Experiments with R
J. Lawson

Design and Analysis of Experiments with SAS
J. Lawson

A Course in Categorical Data Analysis
T. Leonard

Statistics for Accountants
S. Letchford

Introduction to the Theory of Statistical Inference
H. Liero and S. Zwanzig

Statistical Theory, Fourth Edition
B.W. Lindgren

Stationary Stochastic Processes: Theory and Applications
G. Lindgren

Statistics for Finance
E. Lindström, H. Madsen, and J. N. Nielsen

The BUGS Book: A Practical Introduction to Bayesian Analysis
D. Lunn, C. Jackson, N. Best, A. Thomas, and D. Spiegelhalter

Introduction to General and Generalized Linear Models
H. Madsen and P. Thyregod

Time Series Analysis
H. Madsen

Pólya Urn Models
H. Mahmoud

Randomization, Bootstrap and Monte Carlo Methods in Biology, Third Edition
B.F.J. Manly

Statistical Regression and Classification: From Linear Models to Machine Learning
N. Matloff

Introduction to Randomized Controlled Clinical Trials, Second Edition
J.N.S. Matthews

Statistical Rethinking: A Bayesian Course with Examples in R and Stan
R. McElreath

Statistical Methods in Agriculture and Experimental Biology, Second Edition
R. Mead, R.N. Curnow, and A.M. Hasted

Statistics in Engineering: A Practical Approach
A.V. Metcalfe

Theory of Stochastic Objects: Probability, Stochastic Processes and Inference
A.C. Micheas

Statistical Inference: An Integrated Approach, Second Edition
H. S. Migon, D. Gamerman, and F. Louzada

Beyond ANOVA: Basics of Applied Statistics
R.G. Miller, Jr.

A Primer on Linear Models
J.F. Monahan

Stochastic Processes: From Applications to Theory
P.D Moral and S. Penev

Applied Stochastic Modelling, Second Edition
B.J.T. Morgan

Elements of Simulation
B.J.T. Morgan

Probability: Methods and Measurement
A. O'Hagan

Introduction to Statistical Limit Theory
A.M. Polansky

Applied Bayesian Forecasting and Time Series Analysis
A. Pole, M. West, and J. Harrison

Statistics in Research and Development, Time Series: Modeling, Computation, and Inference
R. Prado and M. West

Essentials of Probability Theory for Statisticians
M.A. Proschan and P.A. Shaw

Introduction to Statistical Process Control
P. Qiu

Sampling Methodologies with Applications
P.S.R.S. Rao

A First Course in Linear Model Theory
N. Ravishanker and D.K. Dey

Essential Statistics, Fourth Edition
D.A.G. Rees

Stochastic Modeling and Mathematical Statistics: A Text for Statisticians and Quantitative Scientists
F.J. Samaniego

Statistical Methods for Spatial Data Analysis
O. Schabenberger and C.A. Gotway

Bayesian Networks: With Examples in R
M. Scutari and J.-B. Denis

Large Sample Methods in Statistics
P.K. Sen and J. da Motta Singer

Introduction to Statistical Methods for Financial Models
T. A. Severini

Spatio-Temporal Methods in Environmental Epidemiology
G. Shaddick and J.V. Zidek

Decision Analysis: A Bayesian Approach
J.Q. Smith

Analysis of Failure and Survival Data
P. J. Smith

Applied Statistics: Handbook of GENSTAT Analyses
E.J. Snell and H. Simpson

Applied Nonparametric Statistical Methods, Fourth Edition
P. Sprent and N.C. Smeeton

Data Driven Statistical Methods
P. Sprent

Generalized Linear Mixed Models: Modern Concepts, Methods and Applications
W. W. Stroup

Survival Analysis Using S: Analysis of Time-to-Event Data
M. Tableman and J.S. Kim

Applied Categorical and Count Data Analysis
W. Tang, H. He, and X.M. Tu

Elementary Applications of Probability Theory, Second Edition
H.C. Tuckwell

Introduction to Statistical Inference and Its Applications with R
M.W. Trosset

Understanding Advanced Statistical Methods
P.H. Westfall and K.S.S. Henning

Statistical Process Control: Theory and Practice, Third Edition
G.B. Wetherill and D.W. Brown

Generalized Additive Models: An Introduction with R, Second Edition
S. Wood

Epidemiology: Study Design and Data Analysis, Third Edition
M. Woodward

Practical Data Analysis for Designed Experiments
B.S. Yandell

Texts in Statistical Science

Theory of Stochastic Objects

Probability, Stochastic Processes and Inference

Athanasios Christou Micheas

Department of Statistics, University of Missouri, USA

CRC Press is an imprint of the
Taylor & Francis Group an Informa business
A CHAPMAN & HALL BOOK

CRC Press
Taylor & Francis Group
6000 Broken Sound Parkway NW, Suite 300
Boca Raton, FL 33487-2742

First issued in paperback 2021

Version Date: 20171219

ISBN 13: 978-1-4665-1520-8 (hbk)
ISBN 13: 978-1-03-224288-0 (pbk)

DOI: 10.1201/9781315156705

Library of Congress Cataloging-in-Publication Data

Names: Micheas, Athanasios Christou, author.
Title: Theory of stochastic objects : probability, stochastic processes and inference / by Athanasios Christou Micheas.
Description: Boca Raton, Florida : CRC Press, [2018] | Includes bibliographical references and index.
Identifiers: LCCN 2017043053| ISBN 9781466515208 (hardback) | ISBN 9781315156705 (e-book)
Subjects: LCSH: Point processes. | Stochastic processes.
Classification: LCC QA274.42 .M53 2018 | DDC 519.2/3--dc23
LC record available at https://lccn.loc.gov/2017043053

Visit the Taylor & Francis Web site at
http://www.taylorandfrancis.com

and the CRC Press Web site at
http://www.crcpress.com

Publisher's Note
The publisher has gone to great lengths to ensure the quality of this reprint but points out that some imperfections in the original copies may

To my family

Contents

Preface

Random variables and random vectors have been well defined and studied for over a century. Subsequently, in the history of statistical science, researchers began considering collections of points together, which gave birth to point process theory and more recently, to random set theory. This was mainly motivated due to advances in technology and the types of data that experimenters began investigating, which in turn led to the creation and investigation of advanced statistical methods able to handle such data.

In this book we take the reader on a journey through some of the most essential topics in mathematics and statistics, constantly building on previous concepts, making the transition from elementary statistical inference to the advanced probabilistic treatment more natural and concrete. Our central focus is defining and exploring the concept of a random quantity or object in different contexts, where depending on the data under consideration, "random objects" are described using random variables, vectors or matrices, stochastic processes, integrals and differential equations, or point processes and random sets.

This view of random objects has not been adequately investigated and presented in mathematics and statistics textbooks that are out there since they have mostly concentrated on specific parts of the aforementioned concepts. This is one of the reasons why I undertake the task of writing a textbook that would present the knowledge in a concrete way, through examples and exercises, which is sorely needed in understanding statistical inference, probability theory and stochastic processes. This approach will help the instructor of these topics to engage the students through problem sets and present the theory and applications involved in a way that they will appreciate.

Since this monumental task cannot be accomplished in a single textbook, the theoretical and modeling topics considered have been organized in two texts; this text is concerned with rudimentary to advanced theoretical aspects of random objects based on random variables, including statistical inference, probability theory and stochastic processes. The modeling of these objects and their applications to real life data is presented in the text *Theory and Modeling of Stochastic Objects: Point Processes and Random Sets* (forthcoming, hereafter referred to as *TMSO-PPRS*). The latter stochastic objects are a natural extension of random variables and vectors and we can think of the *TMSO-PPRS* text as a natural continuation of the theory presented herein.

In particular, we present a comprehensive account of topics in statistics in a way that can be a natural extension of a more traditional graduate course in probability theory. This is especially true for Chapters 1 and 2, which is a feature that has been lacking from available texts in probability theory. Another distinguishing feature of this text is that we have included an amazing amount of material. More precisely, one would need to use at least one book on real analysis, one in measure and/or probability theory, one in stochastic processes, and at least one on statistics to capture just the expository material that has gone into this text.

Being a teacher and mentor to undergraduate and graduate students, I have seen their attempts to comprehend new material from rudimentary to advanced mathematical and statistical concepts. I have also witnessed their struggles with essential topics in statistics, such as defining a probability space for a random variable, which is one of the most important constructs in statistics. This book attempts to introduce these concepts in a novel way making it more accessible to students and researchers through examples. This approach is lacking in most textbooks/monographs that one can use to teach students.

Instructors and researchers in academia often find themselves complementing material from several books in order to provide a spherical overview of the topics of a class. This book is the result of my efforts over the years to provide comprehensive and compact accounts of topics I had to teach to undergraduate and graduate students.

Therefore, the book is targeted toward students at the master's and Ph.D. levels, as well as academicians in the mathematics and statistics disciplines. Although the concepts will be built from the master's level up, the book addresses advanced readers in the later chapters. When used as a textbook, prior knowledge of probability or measure theory is welcomed but not necessary.

In particular, Chapters 1 and 2 can be used for several courses on statistical inference with minor additions for any proofs the instructor chooses to further illustrate. In these chapters we summarize over a century and a half of development in mathematical statistics. Depending on the level of the course, the instructor can select specific exercises to supplement the text, in order to provide a better understanding and more depth into the concepts under consideration. For example, using selectively the material and exercises from Chapters 1, 2, 4 and 5, I have taught several sequences on statistical inference at the University of Missouri (MU), including Stat7750/60 and 4750/60 (statistical inference course at the undergraduate and master's level), Stat8710/20 (intermediate statistical inference course at the Ph.D. level) and Stat9710/20 (advanced inference for Ph.D. students).

At the master's level, it is recommended that the instructor omits advanced topics from Chapter 2, including most of the decision-theoretic topics and the corresponding proofs of the relevant results. Basic theorems and their proofs, such as the Bayes or the factorization theorem, should be presented to the students in detail. The proofs of such results are included as exercises, and the instructor can use the solution manual in order to choose what they deem appropriate to illustrate to the students.

For example, when teaching a statistical inference course for Ph.D. students, all concepts presented in Chapters 1 and 2 should be introduced, as well as topics on asymptotics from Chapter 5. However, certain proofs might be beyond the level of an intermediate statistical inference course for Ph.D. students. For example, when it comes to introducing evaluation of point estimators, we may omit the explicit proof of all parts of the important remark 2.12 and simply present the material, or the compactness results in Chapter 5, and focus only on the central limit theorems or Slutsky and Cramér theorems.

For an advanced course on statistical inference at the Ph.D. level, one would omit most of the rudimentary results of Chapter 1, and focus on topics from Chapter 2 (inference), Chapter 4 (e.g., characteristic functions), and Chapter 5 (asymptotics), including all the important proofs of the theorems and remarks presented in the text. Once again, the instructor can find the solution manual invaluable in this case, since it will allow them to select the topics they want to present along with concrete proofs.

Chapters 3-5 can be used to introduce measure theoretic probability to mathematics and statistics graduate students. Some of the proofs should be skipped since it would take more than one semester to go through all the material. More precisely, over the past decade when I taught the advanced probability theory course Stat9810 at MU, I had to omit most of the measure theoretic proofs and be quite selective in the material for a one-semester course. For example, important theorems and their proofs, like Fubini, Kolmogorov 0-1 Law, Radon-Nikodym or Kolmogorov Three Series, should be illustrated to the students in detail.

In contrast, one may skip the proofs of the theoretical development of the Carathodory extension theorem, or omit thc proofs of the decomposition theorems (Chapter 3) and the compactness theorems of Chapter 5. Of course, most of the important results in measure and probability theory and their proofs are still there for the inquisitive student and researcher who needs to go deeper. These chapters are fairly comprehensive and self-contained, which is important for Ph.D. students that have not had an advanced real analysis course.

Chapter 6 is a fairly comprehensive account of stochastic processes in discrete time and in particular Markov chains. This material has been used to teach an introductory course on stochastic processes to both undergraduate and master's students (Stat4850/7850), as well as Ph.D.-level students in one semester (Stat 9820, a continuation of Stat9810). Note that most of the development and exposition of discrete Markov chains and processes does not require heavy measure theory as presented in Chapters 6 and 7, therefore making it accessible to a wide variety of students, including undergraduates. A good working knowledge of matrix algebra is required in this case, which is a requirement for the undergraduate and graduate students when they take this course. In particular, the instructor simply needs to explain in a rudimentary way "transition probability measures," e.g., replace it with the notion of transition probabilities and matrices, and then the material can be presented to the students in a non-measure theoretic way.

The material in Chapter 7 has been used to teach stochastic processes in continuous time to Ph.D. (Stat 9820) and advanced master's level students, including topics from Chapter 6, as mentioned above. The instructor can supplement materials from other chapters as they see fit in order to build the mathematical foundations of the concepts presented as needed. For example, in the beginning of the class we may conduct a mini review of probability theory and Markov chains before jumping into continuous time stochastic processes.

As you begin reading, several features that help with the learning process should immediately draw your attention; each chapter begins with basic illustrations and ends with a more advanced treatment of the topic at hand. We are exploring and reconciling, when feasible, both the frequentist and Bayesian approaches to the topics considered. In addition, recent developments in statistics are presented or referenced in the text and summary of each chapter.

Proofs for most of the theorems, lemmas and remarks presented in each chapter are given in the text or are requested as exercises, with the exception of the rudimentary Chapters 1 and 2, where the proofs are requested as exercises only. Proofs and additional information on the topics discussed can be found in the books or journal papers referenced at the summary section of each chapter. Of course, the interested reader can find proofs to selected exercises in the supplementary online material for the book (sec website below).

The theorems and results presented in the text can range from easy to complicated, and therefore, we usually follow them with an illustrative remark or example to explain the new concept. To further help in our understanding of the material and for quick reference, various topics and complements from mathematics and statistics are included in an appendix.

The MATLAB® code used for the examples presented along with solutions to exercises and other material, such as errata, can be found at the book website https://www.crcpress.com/9781466515208.

There are many people that have contributed, in their own way, to the creation of this book. I am grateful to the faculty members of the Department of Statistics at the University of Missouri, USA, for their constructive interactions and discussions over the years. In particular, special thanks go to my friends and colleagues Christopher Wikle, Scott Holan, Stamatis Dostoglou and Joe Cavanaugh (University of

Iowa), and my friend and mentor Konstantinos Zografos from the Department of Mathematics, University of Ioannina, Greece. Lastly, my academic advisor, Distinguished Professor of Statistics Dipak Dey, Department of Statistics, University of Connecticut, USA, has been an inspiration to me over the years.

I am grateful to Professors Stamatis Dostoglou, Department of Mathematics, University of Missouri, USA, Georg Lindgren, Department of Mathematical Statistics, Centre for Mathematical Sciences, Lund, Sweden, and an anonymous reviewer, for their invaluable comments and suggestions regarding earlier versions of the manuscript. Special thanks go to my friend and colleague Distinguished Professor Noel Cressie, School of Mathematics and Applied Statistics, University of Wollongong, Australia, for his support and encouragement over the years as I was working on the manuscript, as well as for his advice regarding all aspects of the book, including its title.

Additional thanks go to the hundreds of students for their undivided attention while they had to take classes from me on these topics and have helped me better myself through the teaching process. In particular, special thanks goes to all my graduate students, especially to Jiaxun Chen and Alex Oard. I am also grateful to Rob Calver, Michele Dimont, Becky Condit and the friends at Chapman-Hall/CRC for their patience while the manuscript was composed and for their help with the copy edit process.

Above all, my appreciation and love to my family, my daughters Vaso, Evi and Christina, my wife Lada and my father Christos, for their unconditional love and understanding.

I apologize in advance for any typos or errors in the text and I would be grateful for any comments, suggestions or corrections the kind reader would like to bring to my attention.

Sakis Micheas

December 2017

List of Figures

List of Tables

List of Tables

List of Abbreviations

a.s.	Almost Surely
BCT	Bounded Convergence Theorem
Be	Beta function
Càdlàg	Continue à Droite, Limites à Gauche (RCLL)
cdf	Cumulative Distribution Function
cf	Characteristic Function
CI	Confidence Interval
CLT	Central Limit Theorem
CR-LB	Cramer-Rao Lower Bound
DCT	Dominated Convergence Theorem
ev.	Eventually
HBM	Hierarchical Bayesian Model
HPD	Highest Posterior Density
iid	Independent and Identically Distributed
i.o.	Infinitely Often
LHS	Left-Hand Side
MAP	Maximum *a Posteriori*
MCMC	Markov Chain Monte Carlo
MCT	Monotone Convergence Theorem
MLE	Maximum Likelihood Estimator
MG	Martingale
mgf	Moment Generating Function
PEL	Posterior Expected Loss
pgf	Probability Generating Function

PP	Point Process
RHS	Right-Hand Side
RCLL	Right Continuous Left Limits
SLLN	Strong Law of Large Numbers
sMG	Sub-Martingale
SMG	Super-Martingale
TMSO-PPRS	Theory and Modeling of Stochastic Objects: Point Processes and Random Sets
UMVUE	Uniformly Minimum Variance Unbiased Estimator
WLLN	Weak Law of Large Numbers
wlog	Without Loss of Generality
w.p.	With Probability

List of Symbols

$Q \ll \mu$	Q is absolutely continuous with respect to μ
$\mathcal{B}(\mathcal{X})$	Borel sets of the space $\mathcal{X}$
$D^{\Psi}_{[0,+\infty)}$	Càdlàg space: Ψ-valued functions defined on $[0, +\infty)$ with RCLL
$\mathcal{M}$ or $\mathcal{M}(\mu^*)$	Carathéodory measurable sets
$C^n_x = \frac{n!}{(n-x)!x!}$	Combination: the number of ways we can select x objects out of n objects
$C^n_{x_1,x_2,\ldots,x_k} = \frac{n!}{x_1!x_2!\ldots x_k!}$	Multinomial Coefficient, $\sum\limits_{i=1}^{k} x_i = n$, $x_i = 0, 1, \ldots, n$, $i = 1, 2, \ldots, k$
$C^{\mathcal{R}}_{[0,+\infty)}$	Continuous $\mathcal{R}$-valued functions defined on $[0, +\infty)$
$\downarrow$	Decreasing Sequence Converging to
$X \overset{d}{=} Y$	X and Y have the same distribution
$X \sim f_X, F_X, Q_X$	X has density f_X or cdf F_X or distribution Q_X
$x! = 1 * 2 * \ldots * (x-1) * x$	Factorial of an integer x
$\mathbb{F}$	Closed subsets of $\mathcal{R}^p$
$\forall$	For every (or for all)
$\exists$	There exist(s)
$\Rightarrow$	Implies
$\Leftrightarrow$	If and only if
$\uparrow$	Increasing Sequence Converging to
$I(x \in A) = I_A(x)$	Indicator function of the set A, 1 if $x \in A$, 0 if $x \notin A$
μ_p	Lebesgue measure on $\mathcal{R}^p$
$\omega_1 \vee \omega_2$	$\max\{\omega_1, \omega_2\}$
$\omega_1 \wedge \omega_2$	$\min\{\omega_1, \omega_2\}$

$a_n = o(b_n)$	$\frac{a_n}{b_n} \to 0$ as $n \to \infty$
1:1	One-to-one (function)
$\left[\frac{dQ}{d\mu}\right]$	Radon-Nikodym derivative of Q with respect to μ
$\mathcal{R}$, $\mathcal{R}^p$	Real numbers in 1 and p dimensions
$\overline{\mathcal{R}} = \mathcal{R} \cup \{-\infty\} \cup \{+\infty\}$	Extended real line
$\mathcal{R}^+$, $\mathcal{R}_0^+$	$\{x \in \mathcal{R} : x > 0\}$, $\{x \in \mathcal{R} : x \geq 0\}$
$Q \perp \mu$	Q and μ are mutually singular measures
$[P]$, $[\mu]$	With respect to measure P or μ
$\mathcal{Z}$	Integers, $\{\ldots, -2, -1, 0, 1, 2, \ldots\}$
$\overline{\mathcal{Z}} = \mathcal{Z} \cup \{-\infty\} \cup \{+\infty\}$	Extended integers
$\mathcal{Z}^+$, $\mathcal{Z}_0^+$, $\overline{\mathcal{Z}^+}$	$\{1, 2, \ldots\}$, $\{0, 1, 2, \ldots\}$, $\{0, 1, 2, \ldots, +\infty\}$

List of Distribution Notations

Beta	Beta distribution
Binom	Binomial distribution
Cauchy	Cauchy distribution
Dirichlet	Dirichlet distribution
DUnif	Discrete Uniform distribution
Exp	Exponential distribution
F	F distribution
Gamma	Gamma distribution
Geo	Geometric distribution
HyperGeo	Hyper Geometric distribution
InvGamma	Inverse Gamma distribution
W_p^{-1}	Inverse Wishart distribution
$\mathcal{N}$	Univariate Normal distribution
Multi	Multinomial distribution
$\mathcal{N}_p$	Multivariate Normal distribution
NB	Negative Binomial distribution
Poisson	Poisson distribution
t_n	Student's t distribution
Unif	Uniform distribution
W_p	Wishart distribution
χ_n^2	Chi-square distribution

Chapter 1

Rudimentary Models and Simulation Methods

1.1 Introduction

Statistical modeling involves building mathematical equations (models) that try to describe a natural phenomenon. The model used typically depends on some unknown parameters and then statistical inference attempts to infer about the values of these parameters in order to create the model that would best describe the phenomenon. We think of the phenomenon as an experiment conducted and what we observe as a possible outcome (or realization). The collection of all outcomes is called the sample space, denoted by a set Ω, and any subset of Ω is called an event of the experiment. An event of Ω that cannot be further decomposed in terms of other events is called a simple event.

A phenomenon characteristic is some deterministic function $X(\omega)$ that depends on $\omega \in \Omega$ and hence it can potentially take different values for different outcomes ω. We call this characteristic a random quantity or object. Notice that this non-deterministic view of the world incorporates deterministic phenomena as special cases. The experiment can be anything, from flipping a coin to observing radar reflectivities of a severe storm system as it propagates over an area. The characteristic of interest in these cases could be whether we observe tails in the toss of the coin or the amount of rainfall in the area in the next ten minutes.

If we conduct an experiment many times while keeping all parameters the same, then the observed values of the characteristic of interest becomes our random sample often referred to as the data. The main goal in statistical inference is to utilize random sample values in order to infer about the true value of the parameters used in the model. In general, the models we fit are used to accomplish one or more of the following:

1) Fit the data well, so that if the experiment is conducted in the future, the model proposed describes the values of characteristics well.

2) Capture the evolution of a characteristic (a process). This is the case where the experiment characteristics of interest evolve over time.

3) Allow us to adequately predict (forecast) the characteristic in future observation times.

Depending on the type of data we observe we require models for random variables, vectors and matrices, random integrals and derivatives, random point processes, or even random sets and fields. We use the term random object to refer to each of these random quantities and our goal is to study each of these types of random objects.

Example 1.1 (Coin toss) To formulate these ideas, consider an experiment where we flip a coin, with the possible outcomes described using the simple events $\{Heads\}$ or $\{Tails\}$ and the sample space is the set $\Omega_0 = \{Heads, Tails\}$. A characteristic of interest in this experiment has to be based on these simple events; for example, we might be interested in the outcome of the toss. To obtain a random sample, we repeat the experiment and record the outcome each time, e.g., $\omega_i \in \Omega_0$, $i = 1, 2, \ldots, n$, denotes the outcome of the ith toss. Hence we can represent the random sample as a vector $\omega_0 = (\omega_1, \omega_2, \ldots, \omega_n)$ of $0s$ and $1s$. Now if we consider the experiment where we toss the coin n times, then its sample space Ω_1 consists of all n-dimensional sequences of $0s$ and $1s$ and thus a random sample of size n from Ω_0 is a single possible outcome of this experiment. Therefore, the sample space for the n-toss experiment can be described using the Cartesian product $\Omega_1 = \Omega_0 \times \cdots \times \Omega_0 = \underset{i=1}{\overset{n}{\times}} \Omega_0$.

1.2 Rudimentary Probability

Probability is the essential concept required to help us build from rudimentary to advanced statistical models. Let 2^Ω be the collection of all events (subsets) of Ω.

Definition 1.1 Probability

Probability is a set function $P : 2^\Omega \to [0, 1]$ that satisfies
(i) $P(\Omega) = 1$,
(ii) $P(A) \geq 0$, $\forall A \in 2^\Omega$, and
(iii) for any sequence of disjoint events $\{A_i\}_{i=1}^{+\infty}$, we have

$$P\left(\bigcup_{i=1}^{+\infty} A_i\right) = \sum_{i=1}^{+\infty} P(A_i).$$

The collection 2^Ω can be so large that it makes it hard to assign probabilities to all of its elements (events). Therefore, when it comes to defining probability (measures) we consider a smaller collection of sets than 2^Ω, called σ-fields (see definition 3.2).

Remark 1.1 (Probability arithmetic) Using the properties of definition 1.1, we can build a basic arithmetic over events. Let $A, B \in 2^{\Omega}$.

1. $P(A \cup B) = P(A) + P(B) - P(A \cap B)$.

2. $P(A) + P(A^c) = 1$, where $A^c = \Omega \setminus A$, the complement of the set A.

3. If $A \subseteq B$ then $P(A) \leq P(B)$.

4. $P(\emptyset) = 0$, where $\emptyset$ the empty set.

The following basic definitions will be used throughout the rest of the book.

Definition 1.2 Rudimentary definitions

Let P be defined as in definition 1.1.

1. Almost surely If $P(A) = 1$, then we say that A holds almost surely (a.s.) or almost everywhere (a.e.) or with probability (w.p.) 1.

2. Independence Events A and B are called independent if and only if $P(A \cap B) = P(A)P(B)$.

3. Conditional probability The conditional probability of A given B, denoted by $P(A|B)$, is defined by $P(A|B) = P(A \cap B)/P(B)$, provided that $P(B) > 0$. Note that if A and B are independent then $P(A|B) = P(A)$.

From a Bayesian viewpoint conditional probability is important since it leads to Bayes' theorem.

Theorem 1.1 (Bayes' theorem) For any events $A, B \in 2^{\Omega}$, with $P(B) > 0$, we have $P(A|B) = P(B|A)P(A)/P(B)$.

Next we collect the definitions of the most basic random objects.

Definition 1.3 Random variable or vector

A random variable or vector X is a set function from some sample space Ω into some space $\mathcal{X}$. We refer to $\mathcal{X}$ as the set of values of X.

Although X is always defined as a function $X(\omega)$ of simple events $\omega \in \Omega$, as we will see later we typically ignore this underlying structure and work directly with the values of the random variable. Note that for random vectors the space $\mathcal{X}$ is a subset of the p-dimensional real line $\mathcal{R}^p$. For now we collect the rudimentary definitions required in order to define basic statistical models.

1.2.1 Probability Distributions

In order to introduce randomness, we need to assign probabilities to the range of values entertained by a random object, namely, define $P(X = x)$, $\forall x \in \mathcal{X}$. We consider first aggregated probabilities over values of X.

Definition 1.4 Cumulative distribution function

The cumulative distribution function (cdf) of a random variable X is defined as the function $F_X(x) = P(X \leq x) = P(\{\omega \in \Omega : X(\omega) \leq x\})$, $x \in \mathcal{X}$.

A first distinction between random variables is now possible based on the latter definition.

Definition 1.5 Discrete or continuous random variable

A random variable X is called continuous if its cdf $F_X(x)$ is a continuous function, otherwise, it is called discrete.

The following theorem provides the necessary and sufficient conditions for a function to be a cdf.

Theorem 1.2 (Cdf requirements) The function $F_X(x)$ is a cdf if and only if the following conditions hold:
(i) $F_X(-\infty) = \lim\limits_{x \to -\infty} F_X(x) = 0$ and $F_X(+\infty) = \lim\limits_{x \to +\infty} F_X(x) = 1$, and
(ii) $F_X(x)$ is non-decreasing, and
(ii) $F_X(x)$ is right-continuous, i.e., $\lim\limits_{\varepsilon \to 0} F_X(x + \varepsilon) = F_X(x)$, $\forall x \in \mathcal{X}$.

The generalization of the latter definition to random vectors is immediate.

Remark 1.2 (Inclusion probability) The cdf $F_X(x)$ can be viewed as a set function, since it provides the probability of the event $\{\omega : X(\omega) \in (-\infty, x]\}$. For random objects being vectors of random variables, namely, $\mathbf{X} = (X_1, X_2, \ldots, X_p)$, we define the joint cdf similarly by

$$F_{\mathbf{X}}(\mathbf{x}) = P(X_1 \leq x_1, X_2 \leq x_2, \ldots, X_p \leq x_p) = P\left(\mathbf{X} \in A(\mathbf{x})\right),$$

$\mathbf{x} = (x_1, x_2, \ldots, x_p)$, $x_i \in \mathcal{X}_i$, $i = 1, 2, \ldots, p$, where $A(\mathbf{x}) = \{\mathbf{y} \in \mathcal{R}^p : y_i \leq x_i, i = 1, 2, \ldots, p\}$ and hence we can think of $F_{\mathbf{X}}(\mathbf{x})$ as the inclusion probability of the singleton $\mathbf{X}$ in the set $A(\mathbf{x})$. Consequently, if X is any type of random object defined on some space $\mathcal{X}$, we can unify the definition of the cdf by defining $F_X(A) = P(X \subseteq A)$, for all $A \subseteq \mathcal{X}$. For example, if X is a random closed set, a set valued random object taking values in $\mathcal{X} = \mathbb{F}$, the collection of all closed subsets of $\mathcal{R}^p$, then the cdf of X is defined as the containment functional $F_X(A) = P(X \subset A)$, for all $A \in \mathcal{X}$. For more details on this construction see Molchanov (2005, p. 22). We discuss these types of random objects in the *TMSO-PPRS* text.

We now turn to defining "point probabilities" of random variables. Naturally, we begin with the discrete case.

Definition 1.6 Probability mass function

The function $f_X(x) = P(X = x) = P(\{\omega \in \Omega : X(\omega) = x\})$, $x \in \mathcal{X}$, is called the probability mass function (pmf) or density of a discrete random object X.

The pmf satisfies: (i) $f_X(x) \geq 0, \forall x \in \mathcal{X}$, and (ii) $\sum\limits_{x \in \mathcal{X}} f_X(x) = 1$. Note that for X discrete $F_X(x) = \sum\limits_{t \leq x} f(t)$, whereas, for X continuous $F_X(x)$ is a continuous function. The definition of the density in the continuous case needs to be modified since $P(X = x) = 0$. Indeed, using remark 1.1.3 we see that $\{X = x\} \subseteq \{x - \varepsilon < X \leq x\}$, $\forall \varepsilon > 0$, which leads to

$$P(X = x) \leq P(x - \varepsilon < X \leq x) = F_X(x) - F_X(x - \varepsilon),$$

and therefore

$$0 \leq P(X = x) \leq \lim_{\varepsilon \downarrow 0} [F_X(x) - F_X(x - \varepsilon)] = 0,$$

by the continuity of F_X.

Definition 1.7 Probability density function

The function $f_X(x)$, $x \in \mathcal{X}$, is called the probability density function (pdf) or density of a continuous random object X, if it satisfies $F_X(x) = \int\limits_{-\infty}^{x} f_X(t)dt$, $\forall x \in \mathcal{X}$.

The pdf is such that: (i) $f_X(x) \geq 0, \forall x \in \mathcal{X}$, and (ii) $\int\limits_{\mathcal{X}} f_X(x)dx = 1$. The support of X is the subset $\mathcal{S} \subset \mathcal{X}$ for which $f_X(x) > 0$, $x \in \mathcal{S}$. In both the discrete and continuous case $f_X(x)$ is called the probability distribution or density of the random object X. The rigorous definition of the density of a random object will be obtained via the Radon-Nikodym theorem (see theorem 3.20).

The definition of independence allows us to write the joint distribution for a random sample from some model $f_X(x)$. We collect the rudimentary definition for the bivariate case.

Definition 1.8 Independence of random variables

Let $f(x, y)$ be the joint density of the random vector (X, Y) and $f_X(x)$ and $f_Y(y)$ the (marginal) densities of X and Y, respectively. We say that X and Y are independent if and only if $f(x, y) = f_X(x)f_Y(y)$, for all $x \in \mathcal{X}$ and $y \in \mathcal{Y}$.

Next, we collect some consequences of the definitions we have seen thus far, including the definition and properties of conditional densities.

Remark 1.3 (Conditional densities) For any random variables X and Y and event A we have the following:

1. The conditional cdf of a random variable X given an event A is defined by

$$F(x|A) = P(X \leq x|A) = P(\{X \leq x\} \cap A)/P(A),$$

$x \in \mathcal{X}$, provided that $P(A) > 0$.

2. Given the joint density $f(x, y)$ of X and Y we define the conditional density of X given $Y = y$ by $f(x|y) = f(x, y)/f_Y(y)$, $x \in \mathcal{X}$, when $f_Y(y) > 0$, and $f(x|y) = 0$, otherwise. If X and Y are independent then $f(x|y) = f(x)$.

3. If X is continuous then by definition and the fundamental theorem of calculus, we have $f(x) = \frac{dF(x)}{dx}$ and $f(x|A) = \frac{dF(x|A)}{dx}$. Moreover, for continuous X and Y the bivariate fundamental theorem of calculus yields $f(x, y) = \frac{\partial^2}{\partial x \partial y} F(x, y)$, at continuity points of $f(x, y)$.

4. It can be shown that $P(A) = \int\limits_{\mathcal{X}} P(A|X = x) f(x) dx$, where $P(A|X = x) f(x) = P(A \cap \{X = x\})$.

When we observe a random sample from some distribution, i.e., n independent and identically distributed (iid) observations (or realizations) of the random object X, say, $x_1, \dots, x_n$, then owing to the independence property of the random sample we can easily obtain the joint distribution of the data and utilize it to obtain estimates of the parameters of the distribution of X. This joint distribution is an essential element of the classical approach in statistics.

Definition 1.9 Likelihood function

The likelihood function for a parameter θ of a random sample of size n, $x_1, \dots, x_n$, from some distribution $f(x|\theta)$ of a random object X is the joint distribution of the sample defined as

$$L(\theta|x_1, \dots, x_n) = \prod_{i=1}^{n} f(x_i|\theta).$$

An appeal to Bayes' Theorem for the events $\{\omega \in \Omega : X(\omega) = x\}$ and $\{\omega \in \Omega : Y(\omega) = y\}$ yields an important quantity for Bayesian statistics, the posterior distribution of $Y|X$.

Definition 1.10 Posterior distribution

The posterior (or *a posteriori*) distribution of a random object Y given the random object X with prior (or *a priori*) distribution $f(x)$ is defined by

$$\pi(y|x) = f(x|y) f(y)/f(x),$$

for all $x \in \mathcal{X}$ such that $f(x) > 0$.

1.2.2 Expectation

Every probability distribution has features that reveal important characteristics of a random object, including central tendency and variability. In order to quantify these features we require the concept of expectation.

Definition 1.11 Expectation

Let X be a random variable with density $f(x)$, $x \in \mathcal{X}$ and let g be a real-valued function with domain $\mathcal{X}$. The expected value (or mean) of $g(X)$, denoted by $E(g(X))$, is given by

$$E(g(X)) = \sum_{x \in \mathcal{X}} g(x)f(x),$$

if X is discrete and

$$E(g(X)) = \int_{\mathcal{X}} g(x)f(x)dx,$$

if X is continuous. When $f(x)$ is replaced by the conditional distribution of X given $Y = y$, for some random variable Y, then the definition above corresponds to the conditional expectation of $g(X)$ given $Y = y$, denoted by $E(g(X)|Y = y)$.

The extension of the definition of expectation to random vectors and random matrices is straightforward as we see below.

Remark 1.4 (Expectation definition extensions) Suppose that the random object $\mathbf{X}$ is a random vector in $\mathcal{R}^p$. When the function $g(.) = [g_1(.), \ldots, g_q(.)]^T$ is $\mathcal{R}^q$–valued, the definition of the mean is straightforward, namely, $E(g(\mathbf{X})) = [Eg_1(\mathbf{X}), \ldots, Eg_q(\mathbf{X})]^T$. If $X = [(X_{ij})]$, is a random matrix then we can easily extend the original definition above by writing $EX = [(EX_{ij})]$. Unfortunately, this is not the case when it comes to more complicated random objects, like random sets. There are several definitions that have been proposed over the years that coincide under certain conditions, but there is no unique way of defining the expectation for random sets. We study these random objects in more detail in the *TMSO-PPRS* text.

Several important characteristics about a random vector can be described using moments and functions of moments.

Definition 1.12 Moments and functions of moments

Let $\mathbf{X}$ and $\mathbf{Y}$ be random vectors in $\mathcal{R}^p$. The kth-order non-central moment is defined by $\mu_k = E(\underbrace{\mathbf{X}\mathbf{X}^T\mathbf{X}\ldots\mathbf{X}^T\mathbf{X}}_{k-times})$ and the variance-covariance matrix of $\mathbf{X}$ is defined as $Var(\mathbf{X}) = Cov(\mathbf{X}, \mathbf{X}) = E(\mathbf{X} - \mu)(\mathbf{X} - \mu)^T = \mu_2 - \mu\mu^T$, where $\mu = \mu_1$ is the mean vector of $\mathbf{X}$. The covariance between the two random vectors $\mathbf{X}$ and $\mathbf{Y}$ is defined as $Cov(\mathbf{X}, \mathbf{Y}) = E(\mathbf{X} - \mu_x)(\mathbf{Y} - \mu_y)^T$, where $\mu_x = E\mathbf{X}$ and $\mu_y = E\mathbf{Y}$. The Pearson correlation between two random variables X and Y is defined as

$$Corr(X, Y) = \frac{Cov(X, Y)}{\sqrt{Var(X)Var(Y)}}.$$

Example 1.2 (Exponential family) Consider the q-parameter exponential family in its canonical form with pdf given by

$$f(\mathbf{x}|\boldsymbol{\theta}) = h(\mathbf{x}) \exp\{\boldsymbol{\theta}^T \mathbf{T}(\mathbf{x}) - A(\boldsymbol{\theta})\}, \tag{1.1}$$

where $\boldsymbol{\theta} = [\theta_1, \ldots, \theta_q]^T \in \Theta \subseteq \mathcal{R}^q$, the parameters of the model, $\mathbf{T}(\mathbf{x}) = [T_1(\mathbf{x}), \ldots, T_q(\mathbf{x})]^T$ and $h(\mathbf{x}) \geq 0$, with $\mathbf{x} \in \mathcal{R}^p$. First note that

$$\int_{\mathcal{R}^p} h(\mathbf{x}) \exp\{\boldsymbol{\theta}^T \mathbf{T}(\mathbf{x})\} d\mathbf{x} = \exp\{A(\boldsymbol{\theta})\}$$

since $f(\mathbf{x}|\boldsymbol{\theta})$ is a density so that differentiating both sides with respect to $\boldsymbol{\theta}$ yields

$$\int_{\mathcal{R}^p} \mathbf{T}(\mathbf{x}) h(\mathbf{x}) \exp\{\boldsymbol{\theta}^T \mathbf{T}(\mathbf{x})\} d\mathbf{x} = \exp\{A(\boldsymbol{\theta})\} \nabla A(\boldsymbol{\theta}) \tag{1.2}$$

and hence

$$E\mathbf{T}(\mathbf{x}) = \nabla A(\boldsymbol{\theta}), \tag{1.3}$$

where ∇ denotes the gradient, i.e., $\nabla A(\boldsymbol{\theta}) = \left[\frac{\partial}{\partial \theta_1} A(\boldsymbol{\theta}), \ldots, \frac{\partial}{\partial \theta_q} A(\boldsymbol{\theta})\right]^T$. Now taking the gradient on both sides of (1.2) we obtain

$$\int_{\mathcal{R}^p} \mathbf{T}(\mathbf{x}) \mathbf{T}(\mathbf{x})^T h(\mathbf{x}) \exp\{\boldsymbol{\theta}^T \mathbf{T}(\mathbf{x})\} d\mathbf{x} = \exp\{A(\boldsymbol{\theta})\} \nabla A(\boldsymbol{\theta}) \nabla A(\boldsymbol{\theta})^T + \exp\{A(\boldsymbol{\theta})\} \nabla^2 A(\boldsymbol{\theta})$$

and hence

$$E(\mathbf{T}(\mathbf{x}) \mathbf{T}(\mathbf{x})^T) = \nabla A(\boldsymbol{\theta}) \nabla A(\boldsymbol{\theta})^T + \nabla^2 A(\boldsymbol{\theta}), \tag{1.4}$$

where $\nabla^2 A(\boldsymbol{\theta}) = \left[\left(\frac{\partial^2}{\partial \theta_i \theta_j} A(\boldsymbol{\theta})\right)\right]$ the Hessian matrix. Using (1.3) and (1.4) we obtain

$$Cov(\mathbf{T}(\mathbf{x}), \mathbf{T}(\mathbf{x})) = E(\mathbf{T}(\mathbf{x}) \mathbf{T}(\mathbf{x})^T) - E\mathbf{T}(\mathbf{x}) E\mathbf{T}(\mathbf{x})^T = \nabla^2 A(\boldsymbol{\theta}).$$

This family is called minimal when neither of the vectors $\mathbf{T}$ and $\boldsymbol{\theta}$ satisfy a linear constraint, i.e., there do not exist constants $\mathbf{a}$ such that $\mathbf{a}^T \mathbf{T} = 0$ and constants $\mathbf{b}$ such that $\mathbf{b}^T \boldsymbol{\theta} = 0$. In addition, if the parameter space contains a q-dimensional rectangle then the family is said to be of full rank.

Expectations of certain functions of a random object can be very useful in obtaining features of the corresponding density, as we see in the following definitions.

Definition 1.13 Moment generating functions

Let $\mathbf{X}$ be a random vector in $\mathcal{R}^p$. The moment generating function (mgf) $m_{\mathbf{X}}(\mathbf{t})$ of $\mathbf{X}$ is defined as $m_{\mathbf{X}}(\mathbf{t}) = Ee^{\mathbf{X}^T \mathbf{t}}$. The characteristic function (cf) of $\mathbf{X}$ is defined as $\varphi_{\mathbf{X}}(\mathbf{t}) = Ee^{i\mathbf{X}^T \mathbf{t}}$, where i is the imaginary unit.

The following remark shows us how we can obtain moments from the cf of a random vector.

Remark 1.5 (Obtaining moments from the cf) Note that the cf exists for all values of $\mathbf{t}$ but that is not the case for the mgf in general. Both quantities provide non-central moments for random vectors upon taking the gradient repeatedly with respect to the vector $\mathbf{t}$ and setting $\mathbf{t}$ to the zero vector. In particular, we can show that $E(\mathbf{X}\mathbf{X}^T\mathbf{X}\ldots\mathbf{X}\mathbf{X}^T) = i^{-k}\nabla^k\varphi_{\mathbf{X}}(\mathbf{t})|_{\mathbf{t}=\mathbf{0}}$, where $\nabla^k f(\mathbf{x}) = \underbrace{\nabla\nabla\ldots\nabla}_{k-times} f(\mathbf{x})$.

For discrete random variables we have the following.

Definition 1.14 Probability generating function

Let X be a discrete random variable in $\mathcal{X}$. The probability generating function (pgf) of X is defined as $\rho_X(a) = E(a^X) = \sum\limits_{x\in\mathcal{X}} P(X = x)a^x$, for $|a| \leq 1$. Note that $\rho_X(a) = m_X(\ln a)$.

1.2.3 Mixtures of Distributions

Mixture models are useful in modeling the distribution of a random object when there is indication that there are sub-populations within the main population.

Definition 1.15 Mixture

Let Y be a random variable with density $p(y)$ and values in $\mathcal{Y}$. The p-mixture distribution of a random object X is the density defined as

$$f(x) = \sum_{y\in\mathcal{Y}} p(y)f(x|y),$$

if Y is discrete and

$$f(x) = \int_{\mathcal{Y}} p(y)f(x|y)dy,$$

if Y continuous, where $f(x|y)$ is the conditional distribution of $x|y$. We say that $f(x|y)$ is the mixture component for the given y and $p(y)$ are the component probabilities.

Example 1.3 (Mixture of exponential family components) Consider exponential family components (recall example 1.2) of the form

$$f_j(\mathbf{x}|\boldsymbol{\theta}_j) = h(\mathbf{x})\exp\{\boldsymbol{\theta}_j^T\mathbf{T}(\mathbf{x}) - A(\boldsymbol{\theta}_j)\},$$

where $\boldsymbol{\theta}_j = [\theta_{j1},\ldots,\theta_{jq}]^T \in \Theta_j$, the parameters of the model, $\mathbf{T}(\mathbf{x}) = [T_1(\mathbf{x}),\ldots,T_q(\mathbf{x})]^T$ and $h(\mathbf{x}) \geq 0$. Let $\mathbf{p} = [p_1,\ldots,p_m]^T \in \mathcal{S}_m = \{\mathbf{p} : p_j \geq 0$ and $\sum\limits_{j=1}^{m} p_j = 1\}$, where $\mathcal{S}_m$ is the unit simplex and assume that we treat $\mathbf{p}$ as the component probabilities. We define the m-component mixture of q-parameter exponential family components as follows:

$$f(\mathbf{x}|\boldsymbol{\theta}) = \sum_{j=1}^{m} p_j f_j(\mathbf{x}|\boldsymbol{\theta}_j) = \sum_{j=1}^{m} p_j h(\mathbf{x})\exp\{\boldsymbol{\theta}_j^T\mathbf{T}(\mathbf{x}) - A(\boldsymbol{\theta}_j)\},$$

where $\boldsymbol{\theta} = [\boldsymbol{\theta}_{1:m}, \mathbf{p}] \in \Theta = \mathop{\times}\limits_{j=1}^{m}\Theta_j \times \mathcal{S}_m$, all the parameters of the model and $\boldsymbol{\theta}_{1:m} = \{\boldsymbol{\theta}_j\}_{j=1}^m$.

Example 1.4 (Data augmentation for mixture models) Suppose that $\mathbf{p} = [p_1, \ldots, p_m] \in \mathcal{S}_m$, the component probabilities of a mixture model with bivariate normal components, where $\mathcal{S}_m$ is the unit simplex and let

$$\varphi_j(\mathbf{x}|\boldsymbol{\mu}_j, \boldsymbol{\Sigma}_j) = |2\pi\boldsymbol{\Sigma}_j|^{-1/2} \exp\{-.5(\mathbf{x} - \boldsymbol{\mu}_j)^T \boldsymbol{\Sigma}_j^{-1}(\mathbf{x} - \boldsymbol{\mu}_j)\}, \tag{1.5}$$

the density of a $\mathcal{N}_2(\boldsymbol{\mu}_j, \boldsymbol{\Sigma}_j)$, $j = 1, 2, \ldots, m$. For a datum $\mathbf{x}$ the mixture pdf is given by

$$f(\mathbf{x}|\boldsymbol{\theta}) = \sum_{j=1}^{m} p_j\varphi_j(\mathbf{x}|\boldsymbol{\mu}_j, \boldsymbol{\Sigma}_j) = \sum_{j=1}^{m} p_j|2\pi\boldsymbol{\Sigma}_j|^{-1/2} \exp\{(\mathbf{x} - \boldsymbol{\mu}_j)^T \boldsymbol{\Sigma}_j^{-1}(\mathbf{x} - \boldsymbol{\mu}_j)\}, \tag{1.6}$$

with $\boldsymbol{\theta}$ denoting all the parameters of the model, $\mathbf{p}$, $\boldsymbol{\mu}_j$ and $\boldsymbol{\Sigma}_j$, $j = 1, 2, \ldots, m$. Now assume that $\mathbf{x}_1, \ldots, \mathbf{x}_n$ is a random sample from $f(\mathbf{x}|\boldsymbol{\theta})$. The likelihood of the sample is given by

$$L(\boldsymbol{\theta}|\mathbf{x}_1, \ldots, \mathbf{x}_n) = \prod_{i=1}^{n} \sum_{j=1}^{m} p_j|2\pi\boldsymbol{\Sigma}_j|^{-1/2} \exp\{(\mathbf{x}_i - \boldsymbol{\mu}_j)^T \boldsymbol{\Sigma}_j^{-1}(\mathbf{x}_i - \boldsymbol{\mu}_j)\},$$

which is an intractable form, that is, both classical and Bayesian methods of estimation for $\boldsymbol{\theta}$ cannot handle the product of the sum, in order to provide closed form solutions for estimators of $\boldsymbol{\theta}$.

However, any mixture model can be thought of as a missing or incomplete data model as follows; for each datum $\mathbf{x}_i$ we define the latent variable vector $\mathbf{z}_i = (z_{i1}, \ldots, z_{im})$ (known as the data augmentation or membership indicator vector), which indicates to which component $\mathbf{x}_i$ belongs; that is, $z_{ij} = 1$, if the ith observation comes from the jth component, 0 otherwise, with $\sum_{j=1}^{m} z_{ij} = 1$, for each $i = 1, 2, \ldots, n$, with the vectors $\mathbf{z}_i$ assumed to be independent.

Let $\mathcal{M}_z = \{\mathbf{z} = (z_1, \ldots, z_m) : z_j \in \{0, 1\}$ and $\sum_{j=1}^{m} z_j = 1\}$, the space of all values of $\mathbf{z}_i$, $i = 1, 2, \ldots, n$. Further assume that $P(z_{ij} = 1|\boldsymbol{\theta}) = p_j$, so that the joint distribution of the vector $\mathbf{z}_i$ is multinomial on one trial, that is $f(\mathbf{z}_i|\boldsymbol{\theta}) = p_1^{z_{i1}} \ldots p_m^{z_{im}}$, $z_{ij} = 0$ or 1, $\sum_{j=1}^{m} z_{ij} = 1$ and $f(\mathbf{x}_i|z_{ij} = 1, \boldsymbol{\theta}) = \varphi_j(\mathbf{x}_i|\boldsymbol{\mu}_j, \boldsymbol{\Sigma}_j)$, for $i = 1, 2, \ldots, n$, $j = 1, \ldots, m$. Then the completed data $(\mathbf{x}_1, \mathbf{z}_1), \ldots, (\mathbf{x}_n, \mathbf{z}_n)$ has a joint distribution (augmented likelihood) given by

$$f(\mathbf{x}_{1:n}, \mathbf{z}_{1:n}|\boldsymbol{\theta}) = \prod_{i=1}^{n} \prod_{j=1}^{m} p_j^{z_{ij}} \left(\varphi_j(\mathbf{x}_i|\boldsymbol{\mu}_j, \boldsymbol{\Sigma}_j)\right)^{z_{ij}}. \tag{1.7}$$

Notice that upon summing the joint over the $\mathbf{z}_i$ we obtain the original form of the mixture model. Indeed

$$\begin{aligned} f(\mathbf{x}_i|\boldsymbol{\theta}) &= \sum_{\mathbf{z}_i} f(\mathbf{x}_i, \mathbf{z}_i|\boldsymbol{\theta}) = \sum_{\mathbf{z}_i \in \mathcal{M}_z} \prod_{j=1}^{m} p_j^{z_{ij}} \left(\varphi_j(\mathbf{x}_i|\boldsymbol{\mu}_j, \boldsymbol{\Sigma}_j)\right)^{z_{ij}} \\ &= p_1\varphi_1(\mathbf{x}_i|\boldsymbol{\mu}_1, \boldsymbol{\Sigma}_1) + \cdots + p_m\varphi_m(\mathbf{x}_i|\boldsymbol{\mu}_m, \boldsymbol{\Sigma}_m). \end{aligned}$$

The mixture model with $\sum_{j=1}^{m} z_{ij} = 1$ yields what is known as the additive mixture model with data augmentation. There are situations where we might want to allow

$\sum_{j=1}^{m} z_{ij} \geq 1$ and in this case the mixture model is known as a multiplicative mixture model with data augmentation.

1.2.4 Transformations of Random Vectors

Suppose that $\mathbf{X} \sim f(\mathbf{x}|\boldsymbol{\theta})$ and let $\mathbf{Y} = \mathbf{g}(\mathbf{X})$, where $\mathbf{X} \in \mathcal{X}$ and $\mathbf{Y} \in \mathcal{Y} = \mathbf{g}(\mathcal{X})$, are p-dimensional, with $Y_j = g_j(X_j)$, $j = 1, 2, \ldots, p$. We are interested in the distribution of the random vector $\mathbf{Y}$. First note that

$$f_{\mathbf{Y}}(\mathbf{y}) = P(\mathbf{Y} = \mathbf{y}) = P(\{\omega : \mathbf{g}(\mathbf{X}(\omega)) = \mathbf{y}\}),$$

which provides a generic approach to transformations of random vectors. There are three basic approaches that follow below.

The cdf approach

This general approach considers the definition of the joint cdf of $\mathbf{Y}$ given by

$$F_{\mathbf{Y}}(\mathbf{y}) = P(Y_1 \leq y_1, \ldots, Y_p \leq y_p) = P(g_1(X_1) \leq y_1, \ldots, g_p(X_p) \leq y_p)$$

which is a probability involving the random vector $\mathbf{X}$.

The pdf approach

Let $J(\mathbf{y}) = \left|\frac{\partial \mathbf{x}}{\partial \mathbf{y}}\right| = \left\|\left[\left(\frac{\partial x_j}{\partial y_i}\right)\right]\right\| = \left\|\left[\left(\frac{\partial g_j^{-1}(y_j)}{\partial y_i}\right)\right]\right\|$, the Jacobian of the transformation $\mathbf{X} \longmapsto \mathbf{Y}$. Then the joint of the random vector $\mathbf{Y}$ is given by

$$f_{\mathbf{Y}}(\mathbf{y}) = f_{\mathbf{X}}(g_1^{-1}(Y_1), \ldots, g_p^{-1}(Y_p))|J|, \ \mathbf{y} \in \mathbf{Y},$$

with $|J|$ the absolute value of the Jacobian.

The mgf or cf approach

This method computes the mgf or cf of the transformed variables, namely, $m_{\mathbf{Y}}(\mathbf{t}) = E\left(e^{\mathbf{Y}^T\mathbf{t}}\right) = E\left(e^{\mathbf{g}(\mathbf{X})^T\mathbf{t}}\right)$ and $\varphi_{\mathbf{Y}}(\mathbf{t}) = E\left(e^{i\mathbf{Y}^T\mathbf{t}}\right) = E\left(e^{i\mathbf{g}(\mathbf{X})^T\mathbf{t}}\right)$. A direct application of this approach yields the distribution of sums of independent random variables X_i, $i = 1, 2, \ldots, n$. Indeed, letting $Y = X_1 + \cdots + X_n$, with X_i having cf $\varphi_{X_i}(t)$ and utilizing the independence assumption we obtain

$$\varphi_Y(t_i) = E\left(e^{itY}\right) = E\left(e^{it(X_1+\cdots+X_n)}\right) = E\left(e^{itX_1}\right)\ldots E\left(e^{itX_n}\right) = \prod_{i=1}^{n} \varphi_{X_i}(t_i).$$

1.3 The Bayesian Approach

Now we discuss some of the basic elements required to perform a Bayesian analysis based on the posterior distribution. Prior choice is one of the most crucial tasks in any Bayesian analysis and we begin with the concept of a conjugate prior.

1.3.1 Conjugacy

Conjugacy is the most commonly used approach when selecting a prior distribution for a model parameter.

Definition 1.16 Conjugacy

A family of prior distributions Π is said to be a conjugate family for a class of density functions $\mathcal{D}$, if $\pi(\theta|x)$ is in the class Π for all $f \in \mathcal{D}$ and $\pi \in \Pi$.

Several classic examples employing conjugacy are presented next.

Example 1.5 (Conjugacy for exponential family) Suppose that a random vector $\mathbf{X}$ is distributed according to the exponential family (recall example 1.2) and $\boldsymbol{\theta}$ is assumed to have prior distribution

$$\pi(\boldsymbol{\theta}|\boldsymbol{\xi}, a) = \exp\{\boldsymbol{\xi}^T\boldsymbol{\theta} - aA(\boldsymbol{\theta}) - k(\boldsymbol{\xi}, a)\},\ \boldsymbol{\theta} \in \Theta,$$

with

$$\exp\{k(\boldsymbol{\xi}, a)\} = \int_{\Theta} \exp\{\boldsymbol{\xi}^T\boldsymbol{\theta} - aA(\boldsymbol{\theta})\}d\boldsymbol{\theta}.$$

We show that this is the conjugate prior for the exponential family. The posterior distribution can be written as

$$\pi(\boldsymbol{\theta}|\mathbf{x}, \boldsymbol{\xi}, a) = \exp\{[\boldsymbol{\xi} + \mathbf{T}(\mathbf{x})]^T \boldsymbol{\theta} - (a+1)A(\boldsymbol{\theta}) - B(\mathbf{x}, \boldsymbol{\xi}, a)\},$$

where

$$\begin{aligned}\exp\{B(\mathbf{x}, \boldsymbol{\xi}, a)\} &= \int_{\Theta} \exp\{[\boldsymbol{\xi} + \mathbf{T}(\mathbf{x})]^T \boldsymbol{\theta} - (a+1)A(\boldsymbol{\theta})\}d\boldsymbol{\theta} \\ &= \exp\{k(\boldsymbol{\xi} + \mathbf{T}(\mathbf{x}), a+1)\},\end{aligned}$$

so that

$$\pi(\boldsymbol{\theta}|\mathbf{x}, \boldsymbol{\xi}, a) = \exp\left\{[a\boldsymbol{\xi} + \mathbf{T}(\mathbf{x})]^T \boldsymbol{\theta} - (a+1)A(\boldsymbol{\theta}) - k(\boldsymbol{\xi} + \mathbf{T}(\mathbf{x}), a+1)\right\},$$

and hence the posterior is of the same type as the prior distribution.

Example 1.6 (Conjugate priors for multivariate normal models) Let $\mathbf{x}_{1:n} = \{\mathbf{x}_1, \dots, \mathbf{x}_n\}$ be observations from a $\mathcal{N}_p(\boldsymbol{\mu}, \Sigma)$, $\boldsymbol{\mu} \in \mathcal{R}^p, \Sigma > 0, n > p$ and consider the prior $\boldsymbol{\mu}|\boldsymbol{\xi}, A \sim \mathcal{N}_p(\boldsymbol{\xi}, A)$. Assume that all other parameters are fixed. The posterior distribution of $\boldsymbol{\mu}$ given the data $\mathbf{x}_{1:n}$ is obtained from the full posterior distribution by keeping the terms that involve $\boldsymbol{\mu}$ as follows

$$\pi(\boldsymbol{\mu}|\mathbf{x}_{1:n}, \Sigma, \boldsymbol{\xi}, A) \propto \exp\left\{-\frac{1}{2}\left[-2\left(\Sigma^{-1}\sum_{i=1}^{n}\mathbf{x}_i + A^{-1}\boldsymbol{\xi}\right)^T \boldsymbol{\mu} + \boldsymbol{\mu}^T(\Sigma^{-1}+A^{-1})\boldsymbol{\mu}\right]\right\},$$

and letting $V = (\Sigma^{-1}+A^{-1})^{-1} > 0$ and $\boldsymbol{\lambda} = V\left(\Sigma^{-1}\sum_{i=1}^{n}\mathbf{x}_i + A^{-1}\boldsymbol{\xi}\right)$ we obtain

$$\pi(\boldsymbol{\mu}|\mathbf{x}_{1:n}, \Sigma, \boldsymbol{\xi}, A) \propto \exp\left\{-\frac{1}{2}(\boldsymbol{\mu} - \boldsymbol{\lambda})^T V^{-1}(\boldsymbol{\mu} - \boldsymbol{\lambda})\right\},$$

and hence $\boldsymbol{\mu}|\mathbf{x}_{1:n}, \Sigma, \boldsymbol{\xi}, A \sim \mathcal{N}_p(\boldsymbol{\lambda}, V)$, which makes the multivariate normal the conjugate prior when we estimate a normal mean $\boldsymbol{\mu}$.

Now turn to the estimation of Σ keeping all other parameters fixed and assume that $\Sigma|\Psi, m \sim W_p^{-1}(\Psi, m)$, $\Psi > 0$, an inverted Wishart distribution with density

$$\pi(\Sigma|\Psi, m) = \frac{1}{2^{mp/2}\Gamma_p(m/2)}|\Psi|^{m/2}|\Sigma|^{-(m+p+1)/2}\exp\left\{-\frac{1}{2}tr(\Psi\Sigma^{-1})\right\},$$

where

$$\Gamma_p(m/2) = \pi^{p(p-1)/4} \prod_{i=1}^{p} \Gamma(m/2 - (i-1)/2), \tag{1.8}$$

the multivariate gamma function. The posterior distribution is obtained from the full posterior distribution by keeping the terms involving Σ as follows

$$\pi(\Sigma|\Psi, m, \boldsymbol{\mu}, \mathbf{x}_{1:n}) \propto |\Sigma|^{-((m+n)+p+1)/2} \exp\left\{-\frac{1}{2}tr\left((H+\Psi)\Sigma^{-1}\right)\right\},$$

and hence $\Sigma|\Psi, m, \boldsymbol{\mu}, \mathbf{x}_{1:n} \sim W_p^{-1}(H+\Psi, m+n)$, where $H = \sum_{i=1}^{n}(\mathbf{x}_i - \boldsymbol{\mu})(\mathbf{x}_i - \boldsymbol{\mu})^T$, so that the inverted Wishart is the conjugate prior for the covariance matrix Σ.

When $\boldsymbol{\mu}$ and Σ are both unknown we consider the joint prior on $\boldsymbol{\mu}$ and Σ via a conditioning argument (letting $A = K^{-1}\Sigma$), as the product of a $\mathcal{N}_p(\boldsymbol{\xi}, K^{-1}\Sigma)$ density with an $W_p^{-1}(\Psi, m)$ density, for some constant K, namely,

$$\pi(\boldsymbol{\mu}, \Sigma|\boldsymbol{\xi}, \Psi, m, K) = \left|\frac{2\pi}{n}\Sigma\right|^{-1/2} \exp\left\{-\frac{1}{2}(\boldsymbol{\mu}-\boldsymbol{\xi})^T\left(\frac{1}{K}\Sigma\right)^{-1}(\boldsymbol{\mu}-\boldsymbol{\xi})\right\} \tag{1.9}$$

$$\frac{|\Psi|^{m/2}|\Sigma|^{-(m+p+1)/2}\exp\left\{-\frac{1}{2}tr(\Psi\Sigma^{-1})\right\}}{2^{mp/2}\Gamma_p(m/2)}.$$

Letting $C = \sum_{i=1}^{n}(\mathbf{x}_i - \overline{\mathbf{x}})(\mathbf{x}_i - \overline{\mathbf{x}})^T$, the full posterior distribution of $\boldsymbol{\mu}, \Sigma|\boldsymbol{\xi}, \Psi, m, K, \mathbf{x}_{1:n}$ can be written as

$$\begin{aligned}\pi(\boldsymbol{\mu}, \Sigma|\boldsymbol{\xi}, \Psi, m, K, \mathbf{x}_{1:n}) \propto\; & |\Sigma|^{-((n-1+m)+p+1)/2} \exp\{-\frac{1}{2}tr\left([\Psi + C]\Sigma^{-1}\right) \\ & -\frac{1}{2}n(\overline{\mathbf{x}}-\boldsymbol{\mu})^T\Sigma^{-1}(\overline{\mathbf{x}}-\boldsymbol{\mu}) - \frac{K}{2}(\boldsymbol{\mu}-\boldsymbol{\xi})^T\Sigma^{-1}(\boldsymbol{\mu}-\boldsymbol{\xi})\}.\end{aligned}$$

Consider the part of the exponent that involves $\boldsymbol{\mu}$ and write the quadratic forms involved as

$$\begin{aligned}& n(\overline{\mathbf{x}}-\boldsymbol{\mu})^T\Sigma^{-1}(\overline{\mathbf{x}}-\boldsymbol{\mu}) + K(\boldsymbol{\mu}-\boldsymbol{\xi})^T\Sigma^{-1}(\boldsymbol{\mu}-\boldsymbol{\xi}) = \\ & \quad (n+K)\left[\boldsymbol{\mu}^T\Sigma^{-1}\boldsymbol{\mu} - 2\boldsymbol{\mu}^T\Sigma^{-1}\left(\frac{n}{n+K}\overline{\mathbf{x}} + \frac{K}{n+K}\boldsymbol{\xi}\right)\right] + n\overline{\mathbf{x}}^T\Sigma^{-1}\overline{\mathbf{x}} + K\boldsymbol{\xi}^T\Sigma^{-1}\boldsymbol{\xi},\end{aligned}$$

and letting $\boldsymbol{\nu} = \frac{n}{n+K}\overline{\mathbf{x}} + \frac{K}{n+K}\boldsymbol{\xi}$, with $\boldsymbol{\nu}\boldsymbol{\nu}^T = \frac{1}{n+K}(n^2\overline{\mathbf{x}}\overline{\mathbf{x}}^T + K^2\boldsymbol{\xi}\boldsymbol{\xi}^T + 2nK\overline{\mathbf{x}}\boldsymbol{\xi}^T)$, we can write

$$\begin{aligned}& n(\overline{\mathbf{x}}-\boldsymbol{\mu})^T\Sigma^{-1}(\overline{\mathbf{x}}-\boldsymbol{\mu}) + K(\boldsymbol{\mu}-\boldsymbol{\xi})^T\Sigma^{-1}(\boldsymbol{\mu}-\boldsymbol{\xi}) = \\ & \quad (n+K)(\boldsymbol{\mu}-\boldsymbol{\nu})^T\Sigma^{-1}(\boldsymbol{\mu}-\boldsymbol{\nu}) + tr\left[\left(K\boldsymbol{\xi}\boldsymbol{\xi}^T + n\overline{\mathbf{x}}\overline{\mathbf{x}}^T - \boldsymbol{\nu}\boldsymbol{\nu}^T\right)\Sigma^{-1}\right],\end{aligned}$$

so that the posterior distribution is given by

$$\begin{aligned}\pi(\boldsymbol{\mu}, \Sigma|\boldsymbol{\xi}, \Psi, m, K, \mathbf{x}_{1:n}) \propto\; & |\Sigma|^{-((n-1+m)+p+1)/2} \exp\{-\frac{1}{2}tr\left((\Psi + C)\Sigma^{-1}\right) \\ & -\frac{1}{2}(n+K)(\boldsymbol{\mu}-\boldsymbol{\nu})^T\Sigma^{-1}(\boldsymbol{\mu}-\boldsymbol{\nu}) - \frac{1}{2}tr\left(\left(K\boldsymbol{\xi}\boldsymbol{\xi}^T + n\overline{\mathbf{x}}\overline{\mathbf{x}}^T - \boldsymbol{\nu}\boldsymbol{\nu}^T\right)\Sigma^{-1}\right)\}.\end{aligned}$$

Clearly, we can separate a term that involves $\boldsymbol{\mu}$ as a function of Σ, along with a term that involves Σ only, as follows

$$\pi(\mu, \Sigma|\xi, \Psi, m, \mathbf{x}_{1:n}) \propto \left|\frac{1}{n+K}\Sigma\right|^{-1/2} \exp\left\{-\frac{1}{2}(n+K)(\mu-\nu)^T\Sigma^{-1}(\mu-\nu)\right\}$$
$$|\Sigma|^{-((n+m)+p+1)/2} \exp\left\{-\frac{1}{2}tr\left(B\Sigma^{-1}\right)\right\},$$

with $B = \Psi + C + \frac{nK}{n+K}(\overline{\mathbf{x}} - \xi)(\overline{\mathbf{x}} - \xi)^T$, since

$$K\xi\xi^T + n\overline{\mathbf{x}}\overline{\mathbf{x}}^T - \nu\nu^T = \frac{nK}{n+K}(\overline{\mathbf{x}} - \xi)(\overline{\mathbf{x}} - \xi)^T,$$

and hence $\pi(\mu, \Sigma|\xi, \Psi, m, K, \mathbf{x}_{1:n})$ is a $\mathcal{N}_p(\nu, (n+K)^{-1}\Sigma)$ density times an $W_p^{-1}(B, n+m)$ density so that the joint prior considered is indeed the conjugate prior.

Although conjugacy provides mathematical convenience, there are situations where it leads to priors that are inappropriate from an application point of view, or they cannot be used to include strong prior information. Such information is usually provided by experts in the field of the application (see remark 1.6.3). Therefore, we discuss general approaches to prior selection along with some desired prior properties next.

1.3.2 General Prior Selection Methods and Properties

Several approaches to prior choice are summarized next.

Remark 1.6 (Prior choice) The following approaches can be employed in prior selection.

1. Prior robustness When minor changes in the prior distribution do not change the posterior analysis substantially, then the prior chosen is called "robust." A simple way to assess robustness with respect to prior choice is as follows; if a prior distribution $\pi(\theta|\lambda)$, depends on a parameter λ, we repeat the analysis for different values of λ and record the changes in the features of the posterior distribution of θ, such as changes in the posterior mean, mode, variance and so forth. Minor changes indicate robustness to the prior chosen.

2. Subjective priors From a mathematical point of view, conjugacy is a desirable property for a prior, since it makes it easy to compute the posterior distribution. Another mathematical property that is of particular interest is that of exchangeability, meaning that the prior probability does not change under any permutation of the elements of the parameter vector $\boldsymbol{\theta}$. Both types of priors are called "subjective" priors since we impose conditions on the prior for mathematical convenience.

3. Informative vs non-informative priors A prior is called informative if there exists strong information about the parameter before we conduct the experiment. Such *a priori* information is typically the result of previous analyses of the problem or expert input by non-statisticians. For example, foresters know off hand the general dimensions of different species of trees and statisticians can incorporate that information in the prior distribution.

On the other hand, if there is no information available then we need noninformative priors and historically, Laplace was the first to use noninformative techniques to model a prior distribution. The most widely used method is that of Jeffreys' prior defined as $\pi(\boldsymbol{\theta}) \propto [\det(I_{\mathbf{X}}^F(\boldsymbol{\theta}))]^{1/2}$ and is possibly improper (see Appendix definition A.6 for the definition of the Fisher information matrix $I_{\mathbf{X}}^F(\boldsymbol{\theta})$). Note that whenever we use the data to help us create a prior distribution the resulting prior is called an Empirical Bayes prior and the corresponding analysis is called Empirical Bayes.

4. Objective Bayes Over the past few decades there has been an attempt by leading Bayesians to create "objective" Bayesian procedures, and in particular, regarding prior choice. An objective Bayesian procedure may be regarded as a default method which can be applied in cases where prior information is sparse or not well understood or differs among experimenters.

5. Prior selection methods Most methods of prior selection involve obtaining the extrema of a quantity that depends on the model distribution $f(x|\theta)$ and the entertained prior $\pi(\theta)$, that often belongs to a family of distributions. For example, the Maximum Likelihood approach to prior selection suggests that we obtain the marginal $m(x|\lambda) = \int_{\Theta} f(x|\theta)\pi(\theta|\lambda)d\theta$, for $\pi \in \Gamma_\lambda = \{\pi : \pi(\theta|\lambda), \lambda \in \Lambda\}$ and then maximize $m(x|\lambda)$ with respect to λ. The resulting prior $\pi(\theta|\widehat{\lambda})$ provides the member of the family Γ_λ that maximizes the probability of observing the datum x, where $\widehat{\lambda} = \arg\max m(x|\lambda)$.

6. Classes of priors There are many useful classes of priors that can be considered when conducting a Bayesian analysis. The ε–contamination class is the collection of priors $\Gamma_\varepsilon = \{\pi : \pi(\theta) = (1-\varepsilon)\pi_0(\theta) + \varepsilon\pi_c(\theta), \pi_c \in C\}$, where $0 < \varepsilon < 1$ and C is a class of possible contaminations. Other important classes include $\Gamma_S = \{\pi : \pi$ is symmetric about $\theta_o\}$, $\Gamma_{US} = \{\pi : \pi$ is unimodal and symmetric about $\theta_o\}$ and $\Gamma_U^+ = \{\pi : \pi$ is unimodal at θ_o with positive support$\}$.

1.3.3 Hierarchical Bayesian Models

The general structure of Hierarchical Bayesian Models (HBMs) can be described as follows:

Stage 1. Data Model: At the highest level we observe the data from some distribution [data|process, data parameters]; that is, we assume that the observed data is a result of some known process at known data model parameters.

Stage 2. Process Model: The process that governs the observed data is itself dependent on some parameters; that is, it is modeled using some distribution [process|process parameters].

Stage 3. Parameter Model: The parameters of the data and process models are modeled according to some prior distribution [data and process parameters|hyperparameters].

Stages 2 and 3 can be thought of as the prior stage in the absence of a process. A natural way to address the effect of *a priori* information to the posterior distribution is by considering hierarchical priors, an essential component of HBMs, so that the effect of the prior distribution on the posterior distribution diminishes the more stages we include. Indeed, we can think of any problem in statistics in terms of stages as the following hierarchical model:

$$\begin{aligned} X|\theta &\sim f(x|\theta),\ x \in \mathcal{X} \text{ (Stage 1, model or observation stage)} \\ \theta|\lambda_1 &\sim \pi_1(\theta|\lambda_1),\ \theta \in \Theta \text{ (Stage 2, prior stage)} \\ \lambda_1|\lambda_2 &\sim \pi_{12}(\lambda_1|\lambda_2),\ \lambda_1 \in \Lambda_1 \text{ (Stage 3, hyperprior stage)} \\ \lambda_2|\lambda_3 &\sim \pi_{23}(\lambda_2|\lambda_3),\ \lambda_2 \in \Lambda_2 \text{ (Stage 4, hyperprior stage)} \\ &\ldots \end{aligned}$$

Classical statistics recognizes stage 1 as the only valid model upon which we should base all our inference, while stages 2, 3, 4, and so forth represent the idea behind hierarchical priors. There are many posterior distributions that we can obtain in this case, but it should be noted that by construction, the variables at the kth stage affect and are affected only by variables in the $k-1$, k and $k+1$ stages and are conditionally independent of variables in any other stages.

Let us restrict to 3 stages for the sake of presentation, assuming that λ_2 is fixed. The full posterior distribution of all parameters given the data is obtained by applying Bayes' theorem and utilizing the distributions of each stage, as follows

$$\pi(\theta, \lambda_1|x) = \frac{\pi(\theta, \lambda_1, x)}{m(x)} = \frac{f(x|\theta, \lambda_1)\pi(\theta, \lambda_1)}{m(x)} = \frac{f(x|\theta)\pi(\theta|\lambda_1)\pi_{12}(\lambda_1|\lambda_2)}{m(x)},$$

since X is independent of λ_1 given θ, where

$$m(x) = \int_{\Lambda_1}\int_{\Theta} f(x|\theta)\pi_1(\theta|\lambda_1)\pi_{12}(\lambda_1|\lambda_2)d\theta d\lambda_1,$$

the normalizing constant. The full conditional distributions in this case are defined as the distributions of the parameter θ or hyperparameter λ_1, given any other variables, namely,

$$\pi(\theta|\lambda_1, x) = \frac{f(x|\theta)\pi_1(\theta|\lambda_1)}{m_1(x|\lambda_1)},$$

and

$$\pi(\lambda_1|\theta, x) = \pi(\lambda_1|\theta) = \frac{\pi_1(\theta|\lambda_1)\pi_{12}(\lambda_1|\lambda_2)}{m_2(\theta)},$$

with $m_1(x|\lambda_1) = \int_{\Theta} f(x|\theta)\pi_1(\theta|\lambda_1)d\theta$ and $m_2(\theta) = \int_{\Lambda_1} \pi_1(\theta|\lambda_1)\pi_{12}(\lambda_1|\lambda_2)d\lambda_1$. Other posterior distributions of interest include

$$\begin{aligned} \pi(\theta|x) &= \int_{\Lambda_1} \pi(\theta|\lambda_1, x)\pi_3(\lambda_1|x)d\lambda_1, \text{ and} \\ \pi_3(\lambda_1|x) &= \frac{m_1(x|\lambda_1)\pi_{12}(\lambda_1|\lambda_2)}{m(x)}, \end{aligned}$$

with $m(x) = \int_{\Lambda_1}\int_{\Theta} f(x|\theta)\pi_1(\theta|\lambda_1)\pi_{12}(\lambda_1|\lambda_2)d\theta d\lambda_1 = \int_{\Lambda_1} m_1(x|\lambda_1)\pi_{12}(\lambda_1|\lambda_2)d\lambda_1$. Note that a necessary condition for allowing these operations above is that all these densities exist and are nonzero. A classic example of an HBM is presented next.

Example 1.7 (Multivariate normal HBM) Assume that the first (data model) stage is a multivariate normal model, $\mathbf{X}|\boldsymbol{\mu}, \Sigma \sim \mathcal{N}_p(\boldsymbol{\mu}, \Sigma)$, with $\boldsymbol{\mu} \in \mathcal{R}^p$, $\Sigma > 0$, both unknown and let $\mathbf{x}_{1:n} = \{\mathbf{x}_1, \ldots, \mathbf{x}_n\}$, a random sample. For the second stage consider the joint conjugate prior of $\boldsymbol{\mu}$ and Σ as defined in equation (1.9), which depends on (hyper) parameters $\boldsymbol{\xi}$, Ψ, m and K. We treat m and K as fixed and hence require (hyper) priors for $\boldsymbol{\xi}$ and Ψ at the third stage of the hierarchy. Consider a non-informative, improper, joint prior for $\boldsymbol{\xi}$ and Ψ given by $\pi(\boldsymbol{\xi}, \Psi) \propto 1$. Therefore, we have the following hierarchical scheme for the multivariate normal HBM:

$$\begin{aligned}
\mathbf{X}|\boldsymbol{\mu}, \Sigma &\sim \mathcal{N}_p(\boldsymbol{\mu}, \Sigma),\ \mathbf{x} \in \mathcal{R}^p \text{ (Stage 1)} \\
\boldsymbol{\mu}|\boldsymbol{\xi}, K, \Sigma &\sim \mathcal{N}_p(\boldsymbol{\xi}, \frac{1}{K}\Sigma),\ \boldsymbol{\mu} \in \mathcal{R}^p \text{ (Stage 2)} \qquad (1.10) \\
\Sigma|\Psi, m &\sim W_p^{-1}(\Psi, m),\ \Sigma > 0 \text{ (Stage 2)} \\
\boldsymbol{\xi}, \Psi &\sim \pi(\boldsymbol{\xi}, \Psi) \propto 1,\ \boldsymbol{\xi} \in \mathcal{R}^p,\ \Psi > 0 \text{ (Stage 3)}.
\end{aligned}$$

In the previous example, we obtained one of the important posterior distributions in the hierarchy, namely, $\boldsymbol{\mu}, \Sigma|\boldsymbol{\xi}, \Psi, m, K, \mathbf{x}_{1:n}$. Using the second and third stage priors we obtain the posterior distribution $\boldsymbol{\xi}, \Psi|\boldsymbol{\mu}, \Sigma, K, m$, by keeping only the terms that involve $\boldsymbol{\xi}$ and Ψ, as

$$\pi(\boldsymbol{\xi}, \Psi|\boldsymbol{\mu}, \Sigma, K, m) \propto \pi(\boldsymbol{\xi}|\boldsymbol{\mu}, \Sigma, K)\pi(\Psi|\Sigma, m)$$

where we clearly recognize a $\mathcal{N}_p(\boldsymbol{\mu}, \frac{1}{K}\Sigma)$ density for $\boldsymbol{\xi}|\boldsymbol{\mu}, \Sigma, K$ and a Wishart $W_p(\Sigma, m+p+1)$ density for $\Psi|\Sigma, m$, with density

$$\pi(\Psi|\Sigma, m+p+1) = \frac{|\Sigma|^{-(m+p+1)/2}|\Psi|^{((m+p+1)-p-1)/2}\exp\left\{-\frac{1}{2}tr(\Sigma^{-1}\Psi)\right\}}{2^{(m+p+1)p/2}\Gamma_p((m+p+1)/2)},$$

for $\Psi > 0$ and 0 otherwise, where $\Gamma_p\left((m+p+1)/2\right)$ is the multivariate gamma function. Note that for a random matrix $W \sim W_p(\Sigma, n)$ with $n \leq p$, the density does not exist and we have the so-called singular Wishart distribution, which is still defined as the distribution of W since its cf exists ($\varphi_W(\Theta) = E\left[e^{itr(W\Theta)}\right] = \left|I_p - 2i\Theta\Sigma\right|^{-n/2}$, with Θ a $p \times p$ matrix), even if the density does not. Finally, we notice that even though the priors for $\boldsymbol{\xi}$ and Ψ are improper, the posterior distributions are proper and hence useful for Bayesian analysis.

1.4 Simulation Methods

We briefly discuss some of the most commonly used simulation approaches.

1.4.1 The Inverse Transform Algorithm

If the random variable X has cdf $F_X(x)$ then it is easy to see that $U = F_X(X) \sim Unif(0, 1)$ is a uniform random variable with density $f_U(u) = I(0 \leq u \leq 1)$, and therefore we can think of U as a random variable (random probability). Since $x = F_X^{-1}(u)$, for a given $U = u$, the resulting x is nothing but a generated value from a random variable having cdf $F_X(x)$.

Algorithm 1.1 (Inverse transform) The algorithm is as follows:
Step 1: Find the inverse of the cdf $F_X(x)$.
Step 2: Generate $U \sim Unif(0, 1)$ and set $X = F_X^{-1}(U)$ as a generated value from $f_X(x)$.

Example 1.8 (Exponential distribution) Assume that $X|\theta \sim Exp(\theta)$, with density $f_X(x) = \theta e^{-\theta x}$, $x > 0$, $\theta > 0$. The inverse transform method is very efficient in this case, since $F_X(x|\theta) = 1 - e^{-\theta x}$, which leads to $F_X^{-1}(x) = -\theta^{-1}\ln(1-x)$. Hence, in order to generate $X \sim Exp(\theta)$, generate $U \sim Unif(0, 1)$ and set $X = -\theta^{-1}\ln(1-U)$. Since $-\ln(1-U) \stackrel{d}{=} -\ln U(\sim Exp(1))$, we can set $X = -\theta^{-1}\ln U$.

Example 1.9 (Gamma distribution) Consider simulating from $X \sim Gamma(a, b)$, $a, b > 0$, with density $f_X(x|a,b) = \frac{x^{a-1}e^{-x/b}}{\Gamma(a)b^a}$, $x > 0$. The cdf $F_X(x|a,b)$ cannot be inverted in this case. For a an integer, there exist iid $Y_i \sim Exp(1/b)$ such that $X \stackrel{d}{=} \sum_{i=1}^{a} Y_i$, so that from the previous example we can write $Y_i = -b\ln U_i$, where the U_i are iid $Unif(0, 1)$ and hence the Y_i are iid. Therefore, in order to generate $X \sim Gamma(a, b)$, we generate independent $U_1, \ldots, U_a$ as $Unif(0, 1)$ and then simply set $X = -b\sum_{i=1}^{a}\ln U_i$. When $a > 0$, a real number, we turn to another method, known as the rejection method.

1.4.2 The Acceptance-Rejection Algorithm

Consider the continuous case for the sake of presentation. Assume that we can generate a random variable Y with pdf g (the source) and we are interested in generating values from another random variable X with pdf f (the target distribution), where X and Y have the same support. The idea behind the Acceptance-Rejection (or simply Rejection) method is to generate a variable $Y \sim g$ and then accept (or reject) it as a simulated value from f with probability proportional to $f(Y)/g(Y)$.

Algorithm 1.2 (Rejection) Let c denote a constant such that $f(Y)/g(Y) \le c$, for all y such that $g(y) > 0$, e.g., $c = \max_y\{f(Y)/g(Y)\}$.
Step 1: Generate $Y \sim g$ and $U \sim Unif(0, 1)$.
Step 2: If $U < f(Y)/(cg(Y))$, set $X = Y$, the generated value from f, otherwise return to step 1.

Remark 1.7 (Acceptance and convergence) Let us answer some important questions regarding this algorithm.

1. At each iteration the probability that we accept the value generated is $1/c$ (common for all iterations).

2. The number of iterations until the first accepted value is a geometric random variable with probability of success $1/c$. See appendix remark A.11.3 for the definition and properties of the geometric random variable. As a result, the average number of iterations until we generate a valid value from f is c.

Example 1.10 (Gamma distribution) Assume that $X \sim Gamma(a, b)$, where $a, b \in \mathcal{R}^+$, with pdf $f(x|a, b)$ (target pdf). The case of an integer a was illustrated in the previous example. For the source pdf $g(x|\mu)$ we consider an exponential with mean equal to the mean of the Gamma, i.e., $Y \sim Exp(1/ab)$, where $\mu = E(Y) = ab$, the mean of the $Gamma(a, b)$ random variable. As a result

$$h(x) = f(x|a, b)/g(x|\mu) = \frac{ax^{a-1}e^{-x(a-1)/(ab)}}{b^{a-1}\Gamma(a)},$$

and therefore the maximum is attained at $x_0 = ab$, so that

$$c = \max_x \{f(x|a, b)/g(x|\mu)\} = f(x_0|a, b)/g(x_0|\mu) = a^a e^{-a+1}/\Gamma(a).$$

As a result the algorithm becomes:
Step 1: Generate $Y \sim Exp(1/(ab))$ and $U \sim Unif(0, 1)$.
Step 2: If $U < y^{a-1}e^{-y(a-1)/(ab)}e^{a-1}/\left(a^{a-1}b^{a-1}\right)$, set $X = Y$ as the generated value, otherwise return to step 1.

Example 1.11 (Normal distribution) Assume now that $X \sim \mathcal{N}(0, 1)$, so that the target pdf is given by

$$f(x) = (2\pi)^{-1/2}e^{-x^2/2},\ x \in \mathcal{R}.$$

First we consider $|X|$ which has pdf

$$f(x) = 2(2\pi)^{-1/2}e^{-x^2/2},\ x > 0,$$

and we will use an exponential $Y \sim Exp(1)$, as the source distribution $g(x)$ to initially generate from $|X|$. We have

$$f(x)/g(x) = \sqrt{2/\pi}e^{x-x^2/2},$$

and therefore the maximum occurs at $x = 1$, i.e.,

$$c = \max_x \{f(x)/g(x)\} = f(1)/g(1) = \sqrt{2e/\pi}.$$

Thus, in order to generate from the absolute value of the standard normal $|X|$ and since

$$f(x)/(cg(x)) = \exp\left\{-(x-1)^2/2\right\},$$

the algorithm becomes:
Step 1: Generate $Y \sim Exp(1)$ and $U \sim Unif(0, 1)$.
Step 2: If $U < \exp\left\{-(y-1)^2/2\right\}$, set $X = Y$, the generated value from $|X|$, otherwise return to step 1.

Step 3: Now in order to obtain a generated value from $X \sim \mathcal{N}(0,1)$, we simply take $X = Y$ or $X = -Y$, with equal probability, i.e., generate $U_1 \sim Unif(0,1)$ and if $U_1 \leq .5$, set $X = -Y$, otherwise set $X = Y$.
Since $-\ln U \sim Exp(1)$ and $U < \exp\left\{-(y-1)^2/2\right\}$ if and only if $-\ln U > (y-1)^2/2$, we may rewrite the steps as:
Step 1: Generate independent $Y_1, Y_2 \sim Exp(1)$.
Step 2: If $Y_2 > (Y_1-1)^2/2$, then set $Y = Y_2 - (Y_1-1)^2/2$ and proceed to step 3, otherwise go to step 1.
Step 3: Generate $U \sim Unif(0,1)$ and set $X = (2I(U \leq .5)-1)Y_1$, as the generated value from $\mathcal{N}(0,1)$.
This approach is much faster since we do not have to exponentiate in the condition in step 2. Moreover, note that the random variables X and Y generated this way are independent with $X \sim \mathcal{N}(0,1)$ and $Y \sim Exp(1)$. Finally, to generate from $X \sim \mathcal{N}(\mu,\sigma^2)$, we generate $Z \sim \mathcal{N}(0,1)$ and take $X = \mu + \sigma Z$.

1.4.3 The Composition Method for Generating Mixtures

This approach is used when the target distribution is a mixture of well-known distributions. Assume that the target distribution is written as

$$f(x|\boldsymbol{\theta}_1,\ldots,\boldsymbol{\theta}_m) = \sum_{i=1}^{m} p_i f(x|\boldsymbol{\theta}_i),$$

where p_i are the mixture probabilities with $\sum_{i=1}^{m} p_i = 1$ and $f(x|\boldsymbol{\theta}_i)$ is the ith mixture component.

Algorithm 1.3 (Composite method) The general algorithm follows.
Step 1: Generate a value $U \sim Unif(0,1)$.
Step 2: If $\sum_{j=1}^{i} p_{j-1} \leq U < \sum_{j=1}^{i} p_j$, $i = 1,2,\ldots,m$, with $p_0 = 0$, then generate $X_i \sim f(x|\boldsymbol{\theta}_i)$ and set $X = X_i$ as the realization for the target distribution.

1.4.4 Generating Multivariate Normal and Related Distributions

Consider generating random vectors from a $\mathcal{N}_p(\boldsymbol{\mu},\Sigma)$, $\boldsymbol{\mu} \in \mathcal{R}^p$, $\Sigma > 0$. When Σ is singular, i.e., it has a determinant of zero, the density does not exist but the distribution is still defined as the singular p-variate normal via its cf.

Since we can always generate p independent normal random variables, we define the normal random vector $\mathbf{Z}$ with coordinates that consist of the random sample $Z_1,\ldots,Z_p \sim \mathcal{N}(0,1)$, i.e., $\mathbf{Z} = [Z_1,\ldots,Z_p]^T \sim \mathcal{N}_p(\mathbf{0}, I_p)$. To generate from a general multivariate normal $\mathcal{N}_p(\boldsymbol{\mu},\Sigma)$, we can use the Choleski decomposition of Σ, namely, there exists an upper triangular matrix C such that $\Sigma = C^T C$. Finally, we can easily see that $\mathbf{X} = \boldsymbol{\mu} + C^T\mathbf{Z}$ is distributed according to a $\mathcal{N}_p(\boldsymbol{\mu}, C^T I_p C = \Sigma)$, the target distribution.

Assuming that $\mathbf{X} \sim \mathcal{N}_p(\boldsymbol{\mu}, I_p)$, we define $\chi^2 = \mathbf{X}^T\mathbf{X} \sim \chi^2_p(\boldsymbol{\mu}^T\boldsymbol{\mu})$, the non-central χ^2 with p degrees of freedom and non-centrality parameter $\boldsymbol{\mu}^T\boldsymbol{\mu}$. Moreover, it can be shown that the pdf of χ can be written as a $Poisson(\lambda = \boldsymbol{\mu}^T\boldsymbol{\mu})$ mixture of central χ^2 random variables with $p + 2k$ degrees of freedom, for $k = 0, 1, \ldots,$ and $E(\chi^2) = p + \lambda$. For the forms of the Poisson pmf and χ^2 pdf see Appendix A.11.7 and A.8.3.

Now assume that $\mathbf{X}_1, \ldots, \mathbf{X}_n \sim \mathcal{N}_p(\mathbf{0}, \Sigma)$ and define the random matrix $\mathbf{Z} = [\mathbf{X}_1, \ldots, \mathbf{X}_n]$ (with $vec(\mathbf{Z}) \sim \mathcal{N}_{np}(\mathbf{0}, I_n \otimes \Sigma)$, where $\otimes$ denotes the Kronecker product), $n > p$, otherwise it is singular. Then if we define the random matrix $W = \mathbf{Z}\mathbf{Z}^T = \sum_{i=1}^{n} \mathbf{X}_i\mathbf{X}^T$, we have that $W \sim W_p(\Sigma, n)$, the Wishart distribution of p dimensions, n degrees of freedom and parameter matrix Σ. Note that $E(W) = n\Sigma$. In order to simulate from an inverted Wishart random matrix $A \sim W_p^{-1}(\Psi, n)$, we simulate $W \sim W_p(\Psi^{-1}, n)$ and set $A = W^{-1}$, where $E(A) = (n - p - 1)^{-1}\Psi^{-1}$.

1.5 Summary

Our rudimentary, non-measure theoretic treatment of probability and distribution theory provides the basics upon which we can start building a general framework for modeling random objects and provide statistical inference. We will discuss these topics in the next chapter, including point and interval estimation, as well as hypothesis testing, along with a plethora of texts that have appeared on these topics (see the summary of Chapter 2).

Compared to books on theoretical topics from statistics, texts on simulation methods are much fewer and more recent. With the advancement in technology, especially in the 80s and 90s, investigators were able to utilize the existing theoretical framework and illustrate it using advanced computers. Consequently, methods such as Monte Carlo, Markov Chain Monte Carlo (MCMC), Metropolis-Hasting, Gibbs sampling, Importance Sampling, Reversible Jump MCMC and Birth-Death MCMC have helped experimenters illustrate classical and Bayesian models in real-life settings. Recent texts involving general simulation methods include Robert and Casella (2004, 2010) and Thomopoulos (2013). There are many more books on simulation methods for particular areas, ranging from point process theory and stochastic geometry to statistical physics; we will refer to these texts later on appropriately. Next we discuss some additional results on the topics presented in this chapter.

Bayesian considerations

Hierarchical Bayesian Models (HBMs) were first introduced in Berliner (1996) and further studied in Wikle et al. (1998). Detailed reviews on Bayesian robustness with respect to choice of prior can be found in Berger (1984, 1985, 1994) and Robert (2007). In terms of prior selection, the investigation of conjugate priors for exponential families (see example 1.5) can be found in Diaconis and Ylvisaker

(1979). A unified framework for the simultaneous investigation of loss and prior robustness was presented in Micheas (2006).

Objective Bayes is the contemporary name for Bayesian analysis carried out in the Laplace-Jeffreys manner (Jeffreys, 1946, 1961). There is an extensive literature on the construction of objective (or reference) priors; see Kass and Wasserman (1996), Bernardo and Ramon (1998), Press (2003) and Berger et al. (2009).

Mixture models

Mixture models have their roots in a seminal work by Pearson (1894). In this article, Pearson was one of the first to incorporate the use of mixture models as well as note some of the issues involved, in particular, estimation and identifiability. Monographs concerning the subject include Everitt and Hand (1981), Titterington et al. (1985), Lindsay (1995), McLachlan and Peel (2000), Frühwirth-Schnatter (2006) and Mengersen et al. (2011).

Mixture models were discarded as possible modeling tools since they introduced major difficulties in estimation of their parameters, due to the intractable form of the likelihood; namely, for a random sample the joint distribution of the data involves the product of a sum (e.g., example 1.4). Moreover, the problems of identifiability and label switching (Jasra et al., 2005) have caused many sceptics to question the validity of the latest simulation techniques for mixture parameters. Authors have suggested various approximations over the years, but the work by Dempster et al. (1977) on the introduction of data augmentation along with an EM (Expectation-Maximization) algorithm for estimation has provided an appealing solution to the problem. We discuss the EM algorithm in the context of point estimation in Chapter 2, remark 2.8.

In particular from a Bayesian point of view, the Gibbs samplers by Tanner and Wong (1987) and Diebolt and Robert (1994) were major steps toward building a hierarchical Bayesian framework to help with estimation of the parameters of a finite mixture with a fixed number of components. Sampling methods in the variable number of components case include Reversible Jump MCMC (RJMCMC) and Birth Death MCMC (BDMCMC, see Section 6.7.3). For the development, application, evaluation and algorithms for problems involving such methods we refer to Green (1995), Richardson and Green (1997), Stephens (2000), Cappé et al. (2003), Robert and Casella (2004), Jasra et al. (2005), Dellaportas and Papageorgiou (2006), Micheas et al. (2012) and Micheas (2014). The development in these papers is with respect to additive mixture models. For the definition and applications of multiplicative mixture models, see Heller and Ghahramani (2007), Fu and Banerjee (2008), and Dey and Micheas (2014).

Non-parametric approaches have been very successful as well, using for instance Dirichlet process (DP) models or infinite mixtures. See for example Sethuraman (1994), Escobar and West (1995), Wolpert and Ickstadt (1998), Scricciolo (2006), Kottas and Sanso (2007) and Ji et al. (2009).

Multivariate analysis

The study of random vectors and matrices (arranged collections of univariate random variables) is aptly termed multivariate analysis, with the multivariate normal distribution and its related distributions playing a prominent role. Standard texts illustrating simulation, distribution theory, theoretical foundations and applications of multivariate analysis techniques include Johnson (1987), Fang and Zhang (1990), Anderson (2003), Muirhead (2005) and Johnson and Wichern (2007).

Multivariate analysis techniques do not involve only statistical inference, but also general multivariate analysis methods such as classification or clustering of observations, tests of independence, principal component analysis, canonical correlation analysis, factor analysis and specialized distribution theory, e.g., distributions of eigenvalues and eigenvectors of random matrices. See Appendix A.5 for additional topics.

Simulation methods

In the last three decades there has been an explosion in the statistics literature regarding computation and simulation methods. As technology advanced, researchers were able to visualize what theory was suggesting about Markov chains (see Chapter 6), Markov random fields (Besag, 1974) and, in general, Bayesian computation and simulation of random vectors. For instance, calculation of the normalizing constant in closed form has always been a difficult (if not impossible) problem for Bayesians before computers came along.

After the mid 80s, when computing resources were advancing, the illustration of Simulated Annealing and Gibbs Sampling to image restoration, by Geman and Geman (1984), opened up a new area in Bayesian statistics and computational statistics in general. Other pioneering papers in advancing general statistical simulation methods include Tanner and Wong (1989), Gelfand and Smith (1990), Diebolt and Robert (1994), Geyer and Møller (1994), Green (1995) and Stephens (2000). We have collected the basic ideas and methods behind simulating the simplest of random objects, a random variable or vector. Advanced simulation methods will be investigated in later chapters once we have collected the theory involved.

Statistical paradoxes

In general, a paradox is an argument that produces an inconsistency within logic or common sense. Some paradoxes have revealed errors in definitions previously assumed to be rigorous and have caused axioms of mathematics and statistics to be re-examined. In statistics, there are paradoxes with frequentist or Bayesian reasoning that have sparked great discussions, with reasonable arguments from both sides. Discussions on statistical paradoxes with examples can be found in Robert (2007).

1.6 Exercises

Probability, distribution theory and expectation

Exercise 1.1 Prove all statements of remark 1.1.

Exercise 1.2 (Total probability) For any partition $\{B_i\}$ of the sample space Ω and event $A \in 2^{\Omega}$, show that $P(A) = \sum_i P(A|B_i)P(B_i)$.

Exercise 1.3 Prove Bayes' theorem.

Exercise 1.4 Prove theorem 1.2.

Exercise 1.5 Assume that $X_1, X_2 \overset{iid}{\sim} Exp(1/\theta)$. Show that $Y = X_1 + X_2$ and X_1/Y are independent.

Exercise 1.6 Let $U \sim Unif(0, 1)$ and $Y|U = u \sim Binom(n, p = u)$, where the pmf of the binomial is given in appendix remark A.11.1. Find the (unconditional) distribution of Y.

Exercise 1.7 Show that $F_X(x) + F_Y(y) - 1 \le F_{X,Y}(x, y) \le \sqrt{F_X(x)F_Y(y)}$, for two random variables X and Y with joint cdf $F_{X,Y}$ and marginal cdfs F_X and F_Y.

Exercise 1.8 Assume that $X_1, \dots, X_n$ is a random sample from some model with density $f(x)$ and cdf $F(x)$. The ordered values of the sample, denoted by $X_{(1)} \le \dots \le X_{(n)}$, are known as the order statistics. Show that (i) $f_{X_{(1)}}(x) = n[1 - F(x)]^{n-1} f(x)$, and (ii) $f_{X_{(n)}}(x) = n[F(x)]^{n-1} f(x)$.

Exercise 1.9 Consider two iid random variables X and Y. Find $P(X < Y)$.

Exercise 1.10 Let $X_i \overset{iid}{\sim} \mathcal{N}(0, 1)$, $i = 1, 2, 3, 4$. Show that $Y = X_1X_2 - X_3X_4$ follows the Laplace distribution with pdf $f(y) = \frac{1}{2}e^{-|y|}$, $y \in \mathcal{R}$.

Exercise 1.11 Suppose that $X_1, X_2 \overset{iid}{\sim} Exp(\theta)$. Find the distribution of $W = X_1 - X_2$.

Exercise 1.12 Consider $X_i \overset{iid}{\sim} f(x|\theta_1, \theta_2) = 1/h(\theta_1, \theta_2)$, $\theta_1 \le x \le \theta_2$, $i = 1, 2, \dots, n$, where $\theta_1 < \theta_2$ are real numbers. Find the distribution of $Q = \left(X_{(n)} - X_{(1)}\right)/h(\theta_1, \theta_2)$.

Exercise 1.13 Assume that $X_1, \dots, X_n$ is a random sample from some model with density $f(x)$ and cdf $F(x)$ and let $X_{(1)} \le \dots \le X_{(n)}$ denote the order statistics. Show that the joint distribution of $X_{(r)}$ and $X_{(s)}$ is given by

$$f_{X_{(r)},X_{(s)}}(u, v) = C^n_{r-1,s-r-1,n-s} f(u)f(v)\,[F(u)]^{r-1}\,[F(v) - F(u)]^{s-r-1}\,[1 - F(v)]^{n-s},$$

$u < v$, $1 \le r < s \le n$.

Exercise 1.14 Assume that $X_i \overset{iid}{\sim} f(x|\theta_1, \theta_2) = g(x)/h(\theta_1, \theta_2)$, $\theta_1 \le x \le \theta_2$, $i = 1, 2, \dots, n$, where $\theta_1 < \theta_2$ are real numbers. Show that $Q_{rs} = h(X_{(r)}, X_{(s)})/h(\theta_1, \theta_2) \sim Beta(s-r, n-s+r+1)$, $1 \le r < s \le n$, where the density of a Beta random variable is given in Appendix A.12.1.

Exercise 1.15 Let $X_i \overset{iid}{\sim} f(x|\theta)$, $i = 1, 2, \dots, n$. Show that $E\left(X_1/\left(X_1 + \dots + X_n\right)\right) = 1/n$.

Exercise 1.16 If X is a random variable with mgf $m_X(t)$, show that $P(X > a) \le \inf\{m_{X-a}(t), t > 0\}$, for all $a \in \mathcal{R}$.

Exercise 1.17 If $X \sim \mathcal{N}(0, 1)$ then show that $P(X > a) \le \exp\{-a^2/2\}$, for all $a > 0$.

Exercise 1.18 Assume that $\mathbf{X}$ follows a (singular) normal distribution with mean $\mathbf{0}$ and covariance matrix $\Sigma = \begin{bmatrix} 4 & 2 \\ 2 & 1 \end{bmatrix}$.
(i) Prove that Σ is of rank 1.
(ii) Find $\mathbf{a}$ so that $\mathbf{X} = \mathbf{a}Y$ and Y has a nonsingular univariate normal distribution and give the density of Y.

Exercise 1.19 Consider $\mathbf{X} \sim \mathcal{N}_p(\mathbf{b} + \mathbf{\Gamma z}, \sigma^2 I_p)$, where $\mathbf{b} = (\beta, \beta, \ldots, \beta)^T$, $\mathbf{z} = (z_1, z_2, \ldots, z_p)^T$, $\mathbf{\Gamma} = diag(\gamma)$, with β, γ and $\mathbf{z}$ constants with $\sum_{i=1}^{p} z_i = 0$. Find the joint distribution of $\overline{X} = \frac{1}{p}\sum_{i=1}^{p} X_i$ and $U = \sum_{i=1}^{p} z_i X_i \Big/ \left(\sum_{j=1}^{p} z_j^2\right)$, where $\sum_{j=1}^{p} z_j^2 > 0$.

Exercise 1.20 (Chebyshev inequality) Let X be a random variable with $E(X) = \mu$. Then for any $\varepsilon > 0$, we have $P(|X - \mu| \ge \varepsilon) \le Var(X)/\varepsilon^2$.

Exercise 1.21 (Markov inequality) If $P(X \ge 0) = 1$ and $P(X = 0) < 1$ then for any $r > 0$, $P(X \ge r) \le E(X)/r$, with equality if and only if $P(X = r) = p = 1 - P(X = 0)$, $0 < p \le 1$.

Exercise 1.22 (Jensen inequality) For any random variable X, if $g(x)$ is a convex function, then $Eg(X) \ge g(EX)$, with equality if and only if for every line $a + bx$ that is tangent to $g(x)$ at $x = E(X)$, we have $P(g(X) = a + bX) = 1$.

Simulation and computation: use your favorite language to code the functions

Exercise 1.23 Simulate the toss of a 12-sided perfect die 10000 times and compute the frequency of each side. What values do you expect to see for the relative frequencies?

Exercise 1.24 Write a function that, given an integer number N, would generate values from the discrete uniform pmf that takes values N-1, N and N+1. Generate 100 values from this distribution and compute their average.

Exercise 1.25 Simulate 1000 values from a discrete random variable with: $P(X = 1) = .3$, $P(X = 2) = .2$, $P(X = 3) = .35$, $P(X = 4) = .15$. Produce a histogram of the generated sample.

Exercise 1.26 Consider a Poisson random variable with pmf $p(x|\lambda)$, truncated at a value $k > 0$ (in the function you build, make sure we can specify both λ and k as arguments). Call the truncated Poisson random variable X. How would you simulate 1000 values from this X? Note that the pmf of X is nothing but $p(x|\lambda, k) = p(x|\lambda)/P(X \le k)$, $x = 0, 1, ..., k$.

Exercise 1.27 A pair of fair dice is to be continually rolled until all the possible outcomes 2,3,...,12, have occurred at least once. Write a function that would simulate values from this discrete random variable that stops once we get all outcomes to appear at least once and displays the frequencies of each outcome in a plot.

Exercise 1.28 Give a method for generating a random variable having density $f(x) = exp(2x)I(x \leq 0) + exp(-2x)I(x > 0)$. Simulate 10000 values from $f(x)$ and produce a pdf plot of these values (not a histogram).

Exercise 1.29 Simulate 1000 values from the continuous random variable with pdf $f(x) = 30(x^2 - 2x^3 + x^4), 0 \leq x \leq 1$. Also produce the pdf plot over these values (not a histogram, i.e., order the simulated values $x_1, \ldots, x_{1000}$, and plot x_i vs $f(x_i)$ using solid lines to connect the points).

Exercise 1.30 Generate 10000 values from a central t-distribution with 50 degrees of freedom and compute their mean.

Exercise 1.31 Write a function that would generate and plot a histogram of 10000 values from a Pareto distribution with pdf $f(x|\theta) = a\theta^a/x^{a+1}$, $x \geq \theta$, scale parameter $\theta > 0$ and known shape parameter $\alpha > 0$. Run your function for $\theta = 10$ and $a = 1$.

Exercise 1.32 Write a function that simulates 2000 realizations from a random variable having cdf $F(x) = (1/3)(1 - exp(-2x) + 2x)I(0 < x < 1) + (1/3)(3 - exp(-2x))I(x > 1)$ and produce their pdf plot (not a histogram). Repeat for $F(x) = 1 - exp(-ax^b)$, $a, b, x > 0$ (use $a = .1, b = 2$ when you run the function and produce the pdf plot).

Exercise 1.33 Give three methods for generating from a continuous random variable with cdf $F(x) = x^n, 0 \leq x \leq 1$, where n is a fixed positive integer. Discuss the efficiency of the algorithms.

Exercise 1.34 Prove the two statements of remark 1.7.

Exercise 1.35 Devise and implement a rejection sampler in order to generate values from a mixture of two beta distributions: $0.7 * Beta(4, 2) + 0.3 * Beta(1.5, 3)$. Use your rejection method to simulate 1000 observations from this mixture, plot the results in a histogram and add the true density to the histogram.

Exercise 1.36 Assume that X follows a mixture of univariate normal distributions, i.e.,

$$f(x|p_1, p_2, \mu_1, \mu_2, \sigma_1^2, \sigma_2^2) = p_1\phi(x|\mu_1, \sigma_1^2) + p_2\phi(x|\mu_2, \sigma_2^2),$$

where $\phi(x|\mu, \sigma^2)$ denotes the $\mathcal{N}(\mu, \sigma^2)$ density. Write a function that would take, as arguments, the mixture parameters $\boldsymbol{\theta} = [p_1, p_2, \mu_1, \mu_2, \sigma_1^2, \sigma_2^2]$ and r, an integer representing the number of realizations requested. The function should perform the following tasks:
(i) Check if $p_1 + p_2 = 1$ and report an error and exit if not satisfied.
(ii) Plot the mixture density (not a histogram).
(iii) Simulate r realizations from the mixture and return them as an $r \times 1$ vector.
(iv) Run your function with $r = 1000$ and for the following mixtures:
(a) $\boldsymbol{\theta} = [.5, .5, 0, 4, 1, 2]$, (b) $\boldsymbol{\theta} = [.2, .8, 1, 5, 1, 1]$, (c) $\boldsymbol{\theta} = [.1, .9, 3, 4, .3, .3]$

Chapter 2

Statistical Inference

2.1 Introduction

Suppose that we entertain a model $f(x|\boldsymbol{\theta})$ for a random object X taking values in $\mathcal{X}$, where $\boldsymbol{\theta} \in \Theta \subset \mathcal{R}^k$ are the parameters of the model (typically unknown) and assume that we have a random sample $x_1, \ldots, x_n$, of size n from this model, namely we have random variables $X_i \overset{iid}{\sim} f(x|\boldsymbol{\theta})$ and $X_i(\omega_i) = x_i$, $i = 1, 2, \ldots, n$, for some outcome $\omega_i \in \Omega$. Let $\mathbf{X} = (X_1, \ldots, X_n)$ be the random vector and $\mathbf{x} = (x_1, \ldots, x_n)$ the corresponding data vector.

Since any function $\mathbf{T}(\mathbf{x})$ of the data is an estimator for a parameter $\boldsymbol{\theta}$, there is an uncountable number of estimators we can use and hence we need to somehow reduce the collection of functions to those that are useful. This can be accomplished in a classical or Bayesian framework, either by considering methods that give us forms of these functions or by requiring certain properties to be satisfied by a function of the data.

In addition, we need to evaluate estimators and find the best, in some sense, among them. The area of statistics that attempts to unify methods of evaluating estimators is called decision theory and its basic elements will be collected first.

2.2 Decision Theory

When we choose an estimator $\mathbf{T}(\mathbf{x})$ to estimate the true value of a parameter $\boldsymbol{\theta}$ we are making a decision so that $\mathbf{a} = \mathbf{T}(\mathbf{x}) \in \mathcal{A}$ is called a decision rule or an action (the term action is typically used in the absence of data), where $\mathcal{A}$ is the action space, which consists of all possible actions. Decision theory provides us with the necessary tools in order to obtain decision rules, as well as evaluate their effectiveness.

Remark 2.1 (Decision theory components) Any decision theoretic problem consists of the following elements.

1. Loss function The primary component is that of the loss function, denoted by $L(\boldsymbol{\theta}, \mathbf{a})$, which quantifies the error or loss incurred by making a specific decision and is defined as a function of the action $\mathbf{a}$ and the parameter $\boldsymbol{\theta}$. Standard choices include the square error or quadratic loss (SEL) $L(\boldsymbol{\theta}, \mathbf{a}) = (\mathbf{a} - \boldsymbol{\theta})^T(\mathbf{a} - \boldsymbol{\theta})$ and the weighted square error loss (WSEL) $L_w(\boldsymbol{\theta}, \mathbf{a}) = (\mathbf{a} - \boldsymbol{\theta})^T\mathbf{D}(\mathbf{a} - \boldsymbol{\theta})$, for some matrix $\mathbf{D}$, that may depend on $\boldsymbol{\theta}$ and is typically positive definite.

Other useful loss functions, in the univariate case in particular, include the absolute value loss $L(\theta, a) = |a-\theta|$ or the linear exponential (LINEX) loss $L_b(\theta, a) = exp\{b(a - \theta)\} - b(a - \theta) - 1$, $b \neq 0$. A rich class of loss functions can be obtained by considering the φ-divergence $D_\varphi(\boldsymbol{\theta}, \mathbf{a})$ between an entertained model $f(x|\mathbf{a})$ and the true model $f(x|\boldsymbol{\theta})$, defined by

$$D_\varphi(\boldsymbol{\theta}, \mathbf{a}) = \int_X \varphi\left(f(x|\mathbf{a})/f(x|\boldsymbol{\theta})\right) f(x|\boldsymbol{\theta})dx, \tag{2.1}$$

with φ a real, continuous convex function on $[0, +\infty)$ that satisfies the conditions $0\varphi(0/0) = 0$ and $0\varphi(t/0) = t \lim_{u\to+\infty} \varphi(u)/u$, $t > 0$. For more details on these conditions see Vajda (1989), Cressie and Pardo (2000) and Stummer and Vajda (2010). Additional assumptions, such as $\varphi(1) = 0$, can be introduced in order to have $D_\varphi(\boldsymbol{\theta}, \mathbf{a})$ take its minimum value at 0, since in general we have $\varphi(1) \leq D_\varphi(\boldsymbol{\theta}, \mathbf{a}) \leq \varphi(0) + \lim_{u\to+\infty} \varphi(u)/u$, with strict inequality if $\varphi''(u) > 0$ (see for example Lemma 2.1 of Micheas and Zografos, 2006).

2. Frequentist risk The next component that incorporates the classical paradigm in a decision theoretic framework is the risk function. The frequentist risk for a decision rule $\mathbf{T}$ is defined by

$$R(\boldsymbol{\theta}, \mathbf{T}) = E_\theta\left[L(\boldsymbol{\theta}, \mathbf{T}(\mathbf{X}))\right],$$

and is the function of $\boldsymbol{\theta}$ that quantifies the penalty incurred (on the average), when using $\mathbf{T}(\mathbf{X})$ to estimate $\boldsymbol{\theta}$. Clearly, a decision rule is good if its risk is small. For example, in the univariate case note that under SEL the risk function becomes the Mean Square Error (MSE) defined by

$$MSE = R(\theta, T) = E\left[T(X) - \theta\right]^2 = Var(T(X)) + [BIAS]^2, \tag{2.2}$$

where $BIAS = E(T(X)) - \theta$.

3. Bayes risk The Bayesian paradigm can be incorporated in a decision theoretic framework as follows; the Bayes risk of a decision rule $\mathbf{T}$ with respect to a prior distribution π on Θ is defined by

$$r(\pi, \mathbf{T}) = E^\pi[R(\boldsymbol{\theta}, \mathbf{T})], \tag{2.3}$$

and provides the average (with respect to the prior π) expected loss incurred when estimating $\boldsymbol{\theta}$ using $\mathbf{T}$. The dependence of this measure on the prior π renders it quite subjective and experimenters often conduct robustness studies with respect to the choice of π.

Based on the frequentist and Bayes risks, we can built several principles to help us evaluate estimators.

Remark 2.2 (Decision theory principles) The following principles can be used in order to evaluate estimators.

1. Admissibility principle A decision rule $\mathbf{T}$ is called R-better than a rule $\mathbf{T}^*$ if $R(\boldsymbol{\theta}, \mathbf{T}) \leq R(\boldsymbol{\theta}, \mathbf{T}^*)$, $\forall \boldsymbol{\theta} \in \Theta$, with strict inequality for at least one $\boldsymbol{\theta}$. More generally, a decision rule $\mathbf{T}$ is admissible if it is R-better than any other rule $\mathbf{T}^*$. If there is a rule $\mathbf{T}^*$ that is R-better than $\mathbf{T}$, then $\mathbf{T}$ is called inadmissible. Finding the admissible estimator can be a daunting task since we need to compare functions across all $\boldsymbol{\theta}$ in the parameter space.

2. Bayes risk principle A decision rule $\mathbf{T}_1$ is preferred to a rule $\mathbf{T}_2$ if $r(\pi, \mathbf{T}_1) < r(\pi, \mathbf{T}_2)$, with the Bayes risk function $r(\pi, .)$ given by (2.3). If $\mathbf{T}^\pi$ minimizes $r(\pi, \mathbf{T})$, then $\mathbf{T}^\pi$ is called the Bayes rule. When the prior π is improper, $\mathbf{T}^\pi$ is called the generalized Bayes rule.
A decision rule T_n^* (that may depend on n) is called extended Bayes if it is a Bayes rule in the limit, meaning that there exists a sequence of priors $\{\pi_n\}$ such that $\lim_{n\to\infty} r(\pi_n, T_n^*) < \lim_{n\to\infty} r(\pi_n, T)$, for any other decision rule T.

3. Minimax principle A decision rule $\mathbf{T}_1$ is preferred to a rule $\mathbf{T}_2$ if $\sup_{\boldsymbol{\theta}\in\Theta} R(\boldsymbol{\theta}, \mathbf{T}_1) < \sup_{\boldsymbol{\theta}\in\Theta} R(\boldsymbol{\theta}, \mathbf{T}_2)$, that is, $\mathbf{T}_1$ has smaller maximum (over all $\boldsymbol{\theta}$) average loss than $\mathbf{T}_2$. A decision rule $\mathbf{T}^*$ is called minimax if $\mathbf{T}^*$ is such that

$$\sup_{\boldsymbol{\theta}\in\Theta} R(\boldsymbol{\theta}, \mathbf{T}^*) = \inf_{\mathbf{a}\in\mathcal{A}} \sup_{\boldsymbol{\theta}\in\Theta} R(\boldsymbol{\theta}, \mathbf{a}),$$

i.e., $\mathbf{T}^*$ minimizes maximum risk for all $\mathbf{a} \in \mathcal{A}$.

Instead of working with the Bayes risk to find the Bayes rule, it is often much easier to work with the posterior distribution as we see next.

Remark 2.3 (Bayesian expected loss and the Bayes rule) If $\pi^*(\boldsymbol{\theta})$ is the believed density of $\boldsymbol{\theta}$ at the time of decision making then the Bayesian Expected Loss of an action $\mathbf{a}$ is defined as

$$\rho(\pi^*, \mathbf{a}) = E^{\pi^*}[L(\boldsymbol{\theta}, \mathbf{a})].$$

If we choose π^* to be the posterior density $\pi(\boldsymbol{\theta}|\mathbf{x})$ with respect to some prior $\pi(\boldsymbol{\theta})$, once the data has been observed, then $\rho(\pi^*, \mathbf{a})$ is called the Posterior Expected Loss (PEL) of the action $\mathbf{a}$. If $\mathbf{a}^*$ minimizes $\rho(\pi^*, \mathbf{a})$, then $\mathbf{a}^*$ is called the Bayes action. As a consequence, instead of minimizing the Bayes risk with respect to $\mathbf{T}$, we make the problem easier by minimizing the PEL. Indeed, given a prior π and assuming $\mathbf{x}$ and $\boldsymbol{\theta}$ are continuous for simplicity, we write the Bayes Risk as

$$r(\pi, \mathbf{T}) = E^{m(\mathbf{x})}\left[E^{\pi(\boldsymbol{\theta}|\mathbf{x})}(L(\boldsymbol{\theta}, \mathbf{T}(\mathbf{x})))\right],$$

with

$$\pi(\boldsymbol{\theta}|\mathbf{x}) = f(\mathbf{x}|\boldsymbol{\theta})\pi(\boldsymbol{\theta})/m(\mathbf{x}), \tag{2.4}$$

the posterior distribution of $\boldsymbol{\theta}|\mathbf{x}$, where

$$m(\mathbf{x}) = \int_{\Theta} f(\mathbf{x}|\boldsymbol{\theta})\pi(\boldsymbol{\theta})d\boldsymbol{\theta}, \tag{2.5}$$

the normalizing constant (the marginal or unconditional distribution of $\mathbf{x}$). Thus, minimizing the Bayes risk is equivalent to minimizing the PEL, which is typically easier to accomplish in closed form as well as numerically.

2.3 Point Estimation

Now we consider investigating point estimators, i.e., the rule $\mathbf{T}(\mathbf{x})$ we use to estimate $\boldsymbol{\theta}$ is a singleton in $\mathcal{X}$. We present both the classical and Bayesian approaches, starting with the former.

2.3.1 Classical Methods

First we collect the classical methods of point estimation and their properties. Based on a point estimator $\widehat{\boldsymbol{\theta}} = \widehat{\boldsymbol{\theta}}(\mathbf{x})$ of a parameter $\boldsymbol{\theta}$, classical statistics suggests that the best possible model from which these observations arise is simply $f(\mathbf{x}|\widehat{\boldsymbol{\theta}})$. Consequently, in order to generate new values for this random object $\mathbf{X}$ we need to simulate from $f(\mathbf{x}|\widehat{\boldsymbol{\theta}})$. The following remark summarizes two of the classic methods in the literature.

Remark 2.4 (Estimation methods) The following methods can be used in a classical setting in order to obtain forms of estimators $\widehat{\boldsymbol{\theta}}$.

1. Method of moments The first method is dating back at least to Karl Pearson in the late 1800's and is perhaps the first method of finding point estimators; it equates the sample with the theoretical non-central means μ_j, thus creating a system of equations involving the data and the parameters. Recall that the sample k^{th} non-central moment is defined by $m_k = \frac{1}{n}\sum\limits_{i=1}^{n} x_i^k$, $k = 1, 2, \ldots$, based on the random sample $X_i(\omega_i) = x_i$, $i = 1, 2, \ldots, n$, from $L(\boldsymbol{\theta}|\mathbf{x}) = f(\mathbf{x}|\boldsymbol{\theta})$, where $\boldsymbol{\theta} = (\theta_1, \ldots, \theta_k)$ is the parameter vector. The method of moments estimator $\widetilde{\boldsymbol{\theta}} = \widetilde{\boldsymbol{\theta}}(\mathbf{x})$ of a parameter vector $\boldsymbol{\theta}$ is defined as the solution to the system of equations $m_j \simeq \mu_j$, $j = 1, 2, \ldots, k$.

2. Maximum likelihood estimator The next method provides estimators by attempting to maximize the probability of observing the acquired sample and is based on the likelihood principle, namely, if $\mathbf{x}$ and $\mathbf{y}$ are two samples with $L(\boldsymbol{\theta}|\mathbf{x}) = k(\mathbf{x}, \mathbf{y})L(\boldsymbol{\theta}|\mathbf{y})$, for all $\boldsymbol{\theta} \in \Theta$, then any conclusion about $\boldsymbol{\theta}$ drawn from $\mathbf{x}$ and $\mathbf{y}$ should be identical. As a result, we define the maximum likelihood estimator (MLE) $\widehat{\boldsymbol{\theta}} = \widehat{\boldsymbol{\theta}}(\mathbf{x})$ of a parameter vector $\boldsymbol{\theta}$ as the value $\boldsymbol{\theta} \in \Theta$ that maximizes the likelihood function, that is, $\widehat{\boldsymbol{\theta}} = \arg\max L(\boldsymbol{\theta}|\mathbf{x})$.

Next we summarize some of the most commonly sought-after properties of estimators.

Remark 2.5 (Properties of estimators) We discuss some basic properties of estimators in a classical framework.

1. Sufficiency and minimal sufficiency A statistic $T(\mathbf{X})$ is a sufficient statistic for a parameter vector $\boldsymbol{\theta}$ (or the family of distributions $\mathcal{P} = \{f(x|\boldsymbol{\theta}) : \boldsymbol{\theta} \in \Theta\}$) if the conditional distribution of the sample $\mathbf{X}$ given $T(\mathbf{X})$ does not depend on $\boldsymbol{\theta}$. Consequently, the sufficiency principle suggests that any inference about $\boldsymbol{\theta}$ should depend on $\mathbf{X}$ only through $T(\mathbf{X})$. Note that a sufficient statistic provides reduction of the data without any loss of information about $\boldsymbol{\theta}$. Finding sufficient statistics is easy using the factorization theorem (Fisher-Neyman): $T(\mathbf{X})$ is a sufficient statistic for $\boldsymbol{\theta}$ if and only if there exist functions $g(t|\boldsymbol{\theta})$ and $h(\mathbf{x})$ such that

$$f(\mathbf{x}|\boldsymbol{\theta}) = h(\mathbf{x})g(T(\mathbf{x})|\boldsymbol{\theta}).$$

A statistic $T(\mathbf{X})$ is called the minimal sufficient statistic for a parameter vector $\boldsymbol{\theta}$ if for any other sufficient statistic $T^*(\mathbf{X})$ we have $T^*(\mathbf{X}) = g(T(\mathbf{X}))$, for some function g. Sufficient and minimal sufficient statistics are not unique; e.g., any monotone function of a sufficient statistic is also sufficient and any one-to-one function of a minimal sufficient statistic is also minimal. It can be shown that $T(\mathbf{X})$ is a minimal sufficient statistic if it satisfies the following: for two samples $\mathbf{x}$ and $\mathbf{y}$ the ratio $f(\mathbf{x}|\boldsymbol{\theta})/f(\mathbf{y}|\boldsymbol{\theta})$ does not depend on $\boldsymbol{\theta}$ if and only if $T(\mathbf{x}) = T(\mathbf{y})$.

2. Ancillary estimator A statistic $T(\mathbf{X})$ is called ancillary for a parameter vector $\boldsymbol{\theta}$ if its pdf does not depend on $\boldsymbol{\theta}$. Clearly, an ancillary statistic does not contain any information about $\boldsymbol{\theta}$ so that intuitively a minimal sufficient statistic, which summarizes all information about the parameter, is independent of any ancillary statistic. However, when we combine ancillarity with other statistics with certain properties, we obtain valuable information that can be used to conduct inference about $\boldsymbol{\theta}$.

3. Unbiased estimator An estimator $T(\mathbf{X})$ of $g(\boldsymbol{\theta})$ is called unbiased if and only if

$$E(T(\mathbf{X})) = g(\boldsymbol{\theta}). \tag{2.6}$$

If such T exists, then $g(\boldsymbol{\theta})$ is called U-estimable. There are many functions $T(\mathbf{X})$ that can be unbiased, although solving equation (2.6) can be difficult. Instead, we obtain an estimator using one of the aforementioned methods and then try to transform it into an unbiased estimator.

4. Complete statistic A statistic $T(\mathbf{X})$ is called complete for $\boldsymbol{\theta}$ (or the family of distributions $\mathcal{P} = \{f(x|\boldsymbol{\theta}) : \boldsymbol{\theta} \in \Theta\}$), if for any function h we have

$$E_{\boldsymbol{\theta}}[h(T(\mathbf{X}))] = 0 \Longrightarrow P_{\boldsymbol{\theta}}(h(T(\mathbf{X})) = 0) = 1, \tag{2.7}$$

that is, the only unbiased estimator of zero that is a function of T is the function zero a.e. When h is restricted to bounded functions, then T is called boundedly complete. It can be shown that if $T(\mathbf{X})$ is boundedly complete, sufficient and finite dimensional then T is minimal sufficient.

5. Pivotal quantity A function $Q(\mathbf{X}, \boldsymbol{\theta})$ of the data $\mathbf{X}$ and the parameter $\boldsymbol{\theta}$ is called a pivotal quantity if the distribution of $Q(\mathbf{X}, \boldsymbol{\theta})$ does not depend on $\boldsymbol{\theta}$. Good pivotal quantities are those that can be inverted, in the sense that we can solve the equation $Q(\mathbf{X}, \boldsymbol{\theta}) = q$ with respect to $\boldsymbol{\theta}$, for a given value q. Pivotal quantities always exist; indeed, if $X_i \overset{iid}{\sim} F(x|\theta)$, $i = 1, 2, \ldots, n$, where F is the cdf, then it can be shown that

$$Q(\mathbf{X}, \theta) = -\sum_{i=1}^{n} \ln F(X_i|\theta) \sim Gamma(n, 1) \equiv \frac{1}{2}\chi^2_{2n},$$

independent of θ and $Q(\mathbf{X}, \theta)$ is pivotal. Unfortunately, nothing guarantees that $Q(\mathbf{X}, \theta)$ will be invertible. For example, if $X_i \overset{iid}{\sim} Exp(\theta) \equiv Gamma(a = 1, b = 1/\theta)$, $i = 1, 2, \ldots, n$, with density $f(x|\theta) = \theta e^{-\theta x}$, $x, \theta > 0$, then $Q(\mathbf{X}, \theta) = -\sum_{i=1}^{n} \ln\left(1 - e^{-\theta x_i}\right)$, which cannot be inverted.

6. UMVUE An estimator $T(\mathbf{X})$ is called the Uniformly Minimum Variance Unbiased Estimator (UMVUE) for a function of the parameter $g(\boldsymbol{\theta})$, if $T(\mathbf{X})$ has the smallest variance among all unbiased estimators of $g(\boldsymbol{\theta})$.

We collect some additional results on MLEs and their properties below.

Remark 2.6 (Obtaining MLEs) Some comments are in order about the MLE.

1. Since the likelihood function $L(\boldsymbol{\theta}|\mathbf{x})$ and the log-likelihood $l(\boldsymbol{\theta}|\mathbf{x}) = \ln L(\boldsymbol{\theta}|\mathbf{x}) = \sum_{i=1}^{n} \ln f(x_i|\boldsymbol{\theta})$ achieve their maximum at the same point $\boldsymbol{\theta}$, it is often more convenient to maximize $l(\boldsymbol{\theta}|\mathbf{x})$. The likelihood equation (also known as an estimating equation) we need to solve is either $\nabla L(\boldsymbol{\theta}|\mathbf{x}) = \mathbf{0}$ or $\nabla l(\boldsymbol{\theta}|\mathbf{x}) = \mathbf{0}$. Once we obtain $\widehat{\boldsymbol{\theta}}$ we need to show that the Hessian matrix $\nabla^2 L(\boldsymbol{\theta}|\mathbf{x})|_{\boldsymbol{\theta}=\widehat{\boldsymbol{\theta}}}$ or $\nabla^2 l(\boldsymbol{\theta}|\mathbf{x})|_{\boldsymbol{\theta}=\widehat{\boldsymbol{\theta}}}$ is negative definite, in order to ensure that this value $\widehat{\boldsymbol{\theta}}$ is where we achieve the maximum.

2. If $\widehat{\boldsymbol{\theta}}$ is the MLE of $\boldsymbol{\theta}$ then for any invertible function $g(\boldsymbol{\theta})$ the MLE of $g(\boldsymbol{\theta})$ is $g(\widehat{\boldsymbol{\theta}})$.

3. The MLE of $\boldsymbol{\theta}$ could be unique or not exist or there could be an infinite number of MLEs for $\boldsymbol{\theta}$.

4. If the MLE is unique then it is always a function of the sufficient statistic.

In many cases the MLE cannot be computed analytically (in closed form) since we cannot solve the likelihood equation. In this case, methods such as the Netwon-Raphson can provide a numerical solution. We collect the method in the following remark.

Remark 2.7 (Newton-Raphson method) The Newton-Raphson method utilizes a simple idea to help us obtain the root of a univariate (or multivariate) equation $f(x) = 0$, with respect to x. The underlying idea is to approximate the graph of the function $f(x)$ by tangent lines. More precisely, the Newton-Raphson formula consists geometrically of extending the tangent line on the graph of $f(x)$ at a current point x_k, until it crosses zero, then setting the next guess x_{k+1} to the value where the zero-crossing occurs. The method is as follows:

Suppose that we take the Taylor expansion of $f(x)$ at the (starting) point x_0. Then we can write

$$f(x) = f(x_0) + f'(x_0)(x - x_0) + \frac{1}{2} f''(x_0)(x - x_0)^2 + \ldots,$$

and keeping only the first two terms we obtain

$$f(x) \simeq f(x_0) + f'(x_0)(x - x_0).$$

Therefore, setting $f(x) = 0$ to find the next approximation x_1 of the zero of $f(x)$ we find:

$$x_1 = x_0 - \frac{f(x_0)}{f'(x_0)}.$$

Clearly, $f'(x_0)$ must not be 0 for this to be defined, i.e., we do not want to obtain extrema of $f(x)$ at this x_0. Thus, letting $x = x_0$, some starting value (that approximates a root of the equation), we use the following recursive formula

$$x_{k+1} = x_k - \frac{f(x_k)}{f'(x_k)},$$

$k = 0, 1, \ldots$, to obtain an x_k closer and closer to the value of a root of $f(x)$. Note that the tangent line to the graph of $f(x)$ at $(x_0, f(x_0))$ is given by the equation

$$f(x_0) - f(x) = f'(x_0)(x_0 - x).$$

Eventually we expect to hit a point x_k that makes $f(x_k) = 0$. The problem with this method is that we cannot allow $f'(x_k)$ to become 0. In other words, the method will not work in an area about an extrema of $f(x)$. In order to use this method to compute the MLE numerically, we typically take the function f to be $\nabla l(\theta|\mathbf{x})$.

The Newton-Raphson method is easily generalized to the multivariate case. Indeed, assume that $\mathbf{x} = [x_1, \ldots, x_p]^T \in \mathcal{R}^p$ and consider solving the system of equations $\mathbf{f}(\mathbf{x}) = \mathbf{0}$, where $\mathbf{f}(\mathbf{x}) = [f_1(\mathbf{x}), \ldots, f_p(\mathbf{x})]^T$. Define the Jacobian matrix by

$$\mathbf{J}(\mathbf{x}) = \begin{bmatrix} \frac{\partial f_1(\mathbf{x})}{\partial x_1} & \cdots & \frac{\partial f_1(\mathbf{x})}{\partial x_p} \\ \vdots & \ddots & \vdots \\ \frac{\partial f_p(\mathbf{x})}{\partial x_1} & \cdots & \frac{\partial f_p(\mathbf{x})}{\partial x_p} \end{bmatrix}$$

and assume that $\mathbf{J}^{-1}(\mathbf{x})$ exists. Then an approximation to the root of $\mathbf{f}$ is given by the recursive formula

$$\mathbf{x}_{n+1} = \mathbf{x}_n - \mathbf{J}^{-1}(\mathbf{x}_n)\mathbf{f}(\mathbf{x}_n),$$

for $n = 0, 1, \ldots$, where $\mathbf{x}_0$ is an initial guess for the root.

An alternative approach to finding MLEs is via an appeal to the EM algorithm, which was introduced by Dempster et al. (1977) in order to estimate missing data models, e.g., recall example 1.4 where the membership indicators represent information that is missing (latent, censored or unknown variables). The following remark presents the general method and how it can be used in order to obtain MLEs.

Remark 2.8 (EM algorithm) Consider an experiment where the model distribution contains missing or incomplete information. Such models can be represented in terms of what is known as demarginalization; that is, assume that the entertained model $f(x, z|\theta)$ contains a latent variable z, with the marginal distribution (likelihood) given by

$$g(x|\theta) = \int_Z f(x, z|\theta)dz. \tag{2.8}$$

Although we have a random sample from the (marginal) distribution $g(x|\theta)$ we are really interested in the data-augmented density $f(x, z|\theta)$, i.e., we know the marginal but we want to make statements about θ based on the joint distribution $f(x, z|\theta)$ (demarginalization), including finding the MLE of θ. Note that working with $f(x, z|\theta)$

is typically easier.

In particular, suppose that $X_i \overset{iid}{\sim} g(x|\theta)$, $i = 1, 2, \ldots, n$, and we want to compute the MLE of θ, based on the likelihood $L(\theta|\mathbf{x}) = \prod_{i=1}^{n} g(x_i|\theta) = g(\mathbf{x}|\theta)$, which is hard to maximize. The idea behind the EM algorithm is to overcome the difficulty of maximizing $L(\theta|\mathbf{x})$, by exploiting the form of (2.8) and solving a sequence of easier maximization problems whose limit is the MLE of $L(\theta|\mathbf{x})$. The EM can be thought of as the predecessor of the data augmentation MCMC algorithm where we replace simulation with maximization.

Now assume that we augment the data $\mathbf{x}$ with the latent variables $\mathbf{z}$, with the completed data distribution of $(\mathbf{x}, \mathbf{z})$ being $f(\mathbf{x}, \mathbf{z}|\theta)$, so that the conditional distribution of the latent variables $\mathbf{z}$ given the data $\mathbf{x}$ and the parameter θ is given by

$$\pi(\mathbf{z}|\theta, \mathbf{x}) = \frac{f(\mathbf{x}, \mathbf{z}|\theta)}{g(\mathbf{x}|\theta)}.$$

Therefore, solving for the marginal yields

$$g(\mathbf{x}|\theta) = \frac{f(\mathbf{x}, \mathbf{z}|\theta)}{\pi(\mathbf{z}|\theta, \mathbf{x})}, \tag{2.9}$$

and we can use the latter equation can be used to connect the complete-data likelihood $L(\theta|\mathbf{x}, \mathbf{z}) = f(\mathbf{x}, \mathbf{z}|\theta)$, with the observed-data likelihood $L(\theta|\mathbf{x})$. Indeed, for any θ_0, (2.9) can be written as

$$l(\theta|\mathbf{x}) = \ln L(\theta|\mathbf{x}) = E^{\pi(\mathbf{z}|\theta_0, \mathbf{x})}\left[\ln L(\theta|\mathbf{x}, \mathbf{Z})\right] - E^{\pi(\mathbf{z}|\theta_0, \mathbf{x})}\left[\ln \pi(\mathbf{Z}|\theta, \mathbf{x})\right],$$

and since the term $E^{\pi(\mathbf{z}|\theta_0, \mathbf{x})}\left[\ln \pi(\mathbf{Z}|\theta, \mathbf{x})\right]$ is maximized at $\theta = \theta_0$, in order to maximize $l(\theta|\mathbf{x})$, we only need consider the term $E^{\pi(\mathbf{z}|\theta_0, \mathbf{x})}\left[\ln L(\theta|\mathbf{x}, \mathbf{Z})\right]$. We collect the EM algorithm next. See Robert and Casella (2004) for details on convergence issues, as well as modifications and improvements of the EM algorithm.

Algorithm 2.1 (Expectation-Maximization) Consider the setup of the previous remark. Assume that at iteration k the MLE of θ is $\widehat{\theta}^{(k)}$. The following iterations are conducted until a fixed $\widehat{\theta}^{(k+1)}$ is obtained.
Step 1 (E-step): Compute the expectation $Q\left(\theta|\widehat{\theta}^{(k)}, \mathbf{x}\right) = E^{\pi(\mathbf{z}|\theta^{(k)}, \mathbf{x})}\left[\ln L(\theta|\mathbf{x}, \mathbf{z})\right]$.
Step 2 (M-step): Maximize $Q\left(\theta|\widehat{\theta}^{(k)}, \mathbf{x}\right)$ with respect to θ and obtain $\widehat{\theta}^{(k+1)} = \arg\max_{\theta} Q\left(\theta|\widehat{\theta}^{(k)}, \mathbf{x}\right)$.

Example 2.1 (Exponential family) Recall the setup of example 1.2. The log-likelihood (ignoring the constant terms) is given by

$$l(\boldsymbol{\theta}|\mathbf{x}) = \boldsymbol{\theta}^T \mathbf{T}(\mathbf{x}) - A(\boldsymbol{\theta}) = \sum_{j=1}^{q} \theta_j T_j(\mathbf{x}) - A(\boldsymbol{\theta}). \tag{2.10}$$

Then assuming that we have augmented the data $\mathbf{x}$ using some unobserved $\mathbf{z}$, so that $(\mathbf{x}, \mathbf{z})$ has log-likelihood of the form (2.10), we have

$$Q\left(\boldsymbol{\theta}|\widehat{\boldsymbol{\theta}}^{(k)}, \mathbf{x}\right) = E^{\pi(\mathbf{z}|\boldsymbol{\theta}^{(k)},\mathbf{x})}[\ln L(\boldsymbol{\theta}|\mathbf{x}, \mathbf{z})] = \sum_{j=1}^{q} \theta_j T_j^{(k)}(\mathbf{x}) - A(\boldsymbol{\theta}),$$

where $T_j^{(k)}(\mathbf{x}) = E^{\pi(\mathbf{z}|\boldsymbol{\theta}^{(k)},\mathbf{x})}\left[T_j(\mathbf{x}, \mathbf{z})\right]$. Now maximize $Q\left(\boldsymbol{\theta}|\widehat{\boldsymbol{\theta}}^{(k)}, \mathbf{x}\right)$ with respect to $\boldsymbol{\theta}$ to obtain the complete-data MLE by solving the equations

$$0 = \frac{\partial}{\partial \theta_j} Q\left(\boldsymbol{\theta}|\widehat{\boldsymbol{\theta}}^{(k)}, \mathbf{x}\right) = T_j^{(k)}(\mathbf{x}) - \frac{\partial}{\partial \theta_j} A(\boldsymbol{\theta})$$

$j = 1, 2, \ldots, q$, and therefore, the maxima $\boldsymbol{\theta}^{(k+1)}$ are given by solving the equations

$$E^{\pi(\mathbf{z}|\boldsymbol{\theta}^{(k)},\mathbf{x})}\left[T_j(\mathbf{x}, \mathbf{Z})\right] = T_j^{(k)}(\mathbf{x}),$$

$j = 1, 2, \ldots, q$, i.e., the observed equals the expected.

The following remark summarizes four main results we can use in order to obtain UMVUEs.

Remark 2.9 (Obtaining UMVUEs) Choosing the SEL as the loss function, if $T(\mathbf{X})$ is unbiased then from equation (2.2) we have $R(\theta, T) = Var(T(\mathbf{X}))$ and $T(\mathbf{X})$ is the UMVUE if it minimizes the risk function of T. Note that if the UMVUE exists then it is unique. We discuss four basic theorems for finding the UMVUE. Let $\mathbf{X} \sim f(\mathbf{x}|\theta)$, $\theta \in \Theta$.

1. Let $T(\mathbf{X})$ be an estimator such that $E(T^2(\mathbf{X})) < +\infty$. Let also $U(\mathbf{X})$ denote any unbiased estimator of zero, such that $E(U^2(\mathbf{X})) < +\infty$. Then $T(\mathbf{X})$ is the UMVUE for its mean $E(T(\mathbf{X})) = g(\theta)$ if and only if $E_\theta(T(\mathbf{X})U(\mathbf{X})) = 0$, for all estimators U and $\theta \in \Theta$.

2. Rao-Blackwell Let $T(\mathbf{X})$ be sufficient for θ. Let $\delta(\mathbf{X})$ be an unbiased estimator of $g(\theta)$ and assume that the loss function $L(\theta, a)$ is a strictly convex function of a. If δ has finite expectation and risk $R(\theta,\delta) < \infty$, then the estimator $U(T) = E[\delta(\mathbf{X})|T]$, that depends on the data only through T, has smaller risk than δ, i.e., $R(\theta, U) < R(\theta, \delta)$ unless $\delta(X) = U(T)$ w.p. 1.

The Rao-Blackwell theorem provides R-better estimators U by conditioning an unbiased estimator δ on the sufficient statistic T. Assume further that T is complete. Then, under SEL conditioning an unbiased estimator δ of $g(\theta)$ on T, we obtain a UMVUE estimator U of $g(\theta)$, since $E(U) = E(E[\delta(\mathbf{X})|T]) = E(\delta(\mathbf{X})) = g(\theta)$.

3. Lehmann-Scheffé Let $T(\mathbf{X})$ be complete-sufficient for θ. Then every U-estimable function $g(\theta)$ has one and only one unbiased estimator that is a function of T.

4. Cramer-Rao lower bound Assume that $X_i \overset{iid}{\sim} f(x|\boldsymbol{\theta})$, $\boldsymbol{\theta} \in \Theta \subset \mathcal{R}^k$ and that the regularity conditions (see Appendix A.4) hold. Let $T(\mathbf{X})$ be an estimator such that $E_{\boldsymbol{\theta}}(T) = g(\boldsymbol{\theta})$ and $E(T^2) < +\infty$. Then $E_{\boldsymbol{\theta}}\left(\frac{\partial \ln f(\mathbf{x}|\boldsymbol{\theta})}{\partial \theta_i}\right) = 0$ and

$$Var(T) \geq \mathbf{a}^T [I_{\mathbf{x}}^F(\boldsymbol{\theta})]^{-1} \mathbf{a}, \qquad (2.11)$$

where $\mathbf{a}^T = [a_1, \ldots, a_k]$, $a_i = \frac{\partial g(\boldsymbol{\theta})}{\partial \theta_i}$, $i = 1, 2, \ldots, k$ and $I_{\mathbf{x}}^F(\boldsymbol{\theta})$ is the Fisher information. This lower bound is known as the Cramér-Rao lower bound (CR-LB). The CR-LB does not guarantee the existence or tell us how to compute the UMVUE.

Before discussing specific examples, we obtain some final remarks about point estimators and their properties.

Remark 2.10 (Point estimators and their properties) The following are straightforward to prove.

1. Exponential families Let $\mathbf{X}$ be distributed according to a full rank q-parameter exponential family (example 1.2). Then the statistic $\mathbf{T}$ is minimal sufficient and complete for $\boldsymbol{\theta}$.

2. Basu If T is complete-sufficient for the family of distributions $\mathcal{P} = \{f(x|\boldsymbol{\theta}) : \boldsymbol{\theta} \in \Theta\}$, then any ancillary statistic V is independent of T. A direct consequence of Basu's theorem is that an ancillary statistic cannot be complete.

3. Generalizing the CR-LB When estimating $\mathbf{g}(\boldsymbol{\theta})^T = [g_1(\boldsymbol{\theta}), \dots, g_k(\boldsymbol{\theta})]$, $\boldsymbol{\theta} \in \Theta \subset \mathcal{R}^p$, we can generalize the CR-LB using

$$\text{CR-LB} = D(\boldsymbol{\theta})[I_{\mathbf{x}}^F(\boldsymbol{\theta})]^{-1} D(\boldsymbol{\theta})^T,$$

where $D(\boldsymbol{\theta}) = \left[\left(\frac{\partial g_j(\boldsymbol{\theta})}{\partial \theta_i}\right)\right]$. Consequently, we can look for an unbiased estimator $\mathbf{T} = [T_1, \dots, T_k]^T$ of $\mathbf{g}(\boldsymbol{\theta})$, such that $Var(\mathbf{T})$=CR-LB and then $\mathbf{T}$ becomes the UMVUE.

4. Attaining the CR-LB Consider the univariate case for simplicity. The CR-LB is achieved if and only if

$$\frac{\partial}{\partial \theta} \ln \prod_{i=1}^{n} f(x_i|\theta) = k(\theta, n)\left[U(\mathbf{x}) - g(\theta)\right],$$

and in this case $U(\mathbf{X})$ is the UMVUE of $g(\theta)$.

Next we present several examples illustrating the aforementioned methods of point estimation.

Example 2.2 (Uniform distribution) Suppose that $X_i \overset{iid}{\sim} Unif(0, \theta)$, $i = 1, 2, \dots, n$, with pdf $f_x(x) = \frac{1}{\theta} I_{[0,\theta]}(x)$ and cdf $F_x(x) = \frac{x}{\theta} I_{[0,\theta]}(x)$, $\theta > 0$, so that the likelihood is given by

$$L(\theta|\mathbf{x}) = f(\mathbf{x}|\theta) = \theta^{-n} \prod_{i=1}^{n} I_{[0,\theta]}(x_i) = \theta^{-n} I_{[T(\mathbf{x}), +\infty)}(\theta),$$

where $T(\mathbf{X}) = X_{(n)} = \max_i X_i$, the largest order statistic. The distribution of T is given by

$$f_{X_{(n)}}(t) = n t^{n-1} \theta^{-n} I_{[0,\theta]}(t). \tag{2.12}$$

Clearly, since $L(\theta|\mathbf{x})$ is a decreasing function of θ, the maximum is attained at $\widehat{\theta} = T(\mathbf{x})$. In contrast, using the method of moments we equate the sample with the theoretical mean to obtain the estimator $\widetilde{\theta} = \frac{2}{n}\sum_{i=1}^{n} X_i$. Now we write the joint as

$$f(\mathbf{x}|\theta) = \theta^{-n} I_{[x_{(n)},+\infty)}(\theta) = h(\mathbf{x})g(T(\mathbf{x})|\theta),$$

with $h(\mathbf{x}) = 1$ and $g(T(\mathbf{x})|\theta) = \theta^{-n} I_{[x_{(n)},+\infty)}(\theta)$ and using the factorization theorem the sufficient statistic is the same as the MLE, namely, $T(\mathbf{X}) = X_{(n)}$. In order to find the minimal sufficient statistic, we consider the ratio

$$r = \frac{f(\mathbf{x}|\theta)}{f(\mathbf{y}|\theta)} = \frac{I_{[x_{(n)},+\infty)}(\theta)}{I_{[y_{(n)},+\infty)}(\theta)},$$

which is constant in θ if and only if $x_{(n)} = y_{(n)}$, so that $X_{(n)}$ is also minimal sufficient. Using equation (2.12), we apply the pdf transformation approach to obtain the distribution of $Q = Q(\mathbf{X}, \theta) = X_{(n)}/\theta = T/\theta$. The Jacobian of the transformation $T \longmapsto Q$ is given by $\frac{dT}{dQ} = \theta$ and hence

$$f_Q(q|\theta) = \theta f_T(\theta q) = \theta n(q\theta)^{n-1}\theta^{-n} I_{[0,\theta]}(q\theta) = nq^{n-1} I_{[0,1]}(q),$$

which is a *Beta*$(n, 1)$ distribution that does not depend on θ and hence Q is a pivotal quantity. Clearly, $E\left(X_{(n)}/\theta\right) = E(Q) = n/(n+1)$, which leads to an unbiased estimator $(n+1)X_{(n)}/n$ for θ. Now consider the LHS of equation (2.7) for an arbitrary function h, namely, $E_\theta[h(T(\mathbf{X}))] = 0$, which implies that

$$\int_0^\theta h(t)t^{n-1}dt = 0,$$

and upon differentiating with respect to θ both sides of the latter equation we obtain $h(\theta)\theta^{n-1} = 0$, $\theta > 0$, which is satisfied only if $h(t) = 0$, for all $t > 0$. Thus $T = X_{(n)}$ is also complete. Now consider all unbiased estimators of θ based on the complete-sufficient statistic T, that is, $E_\theta[h(T)] = \theta$, so that

$$\int_0^\theta h(t)t^{n-1}dt = \theta^{n+1}/n.$$

Differentiating with respect to θ on both sides we obtain

$$h(\theta)\theta^{n-1} = (n+1)\theta^n/n,$$

and hence all unbiased estimators based on T must satisfy $h(T) = (n+1)T/n$, so that $U = (n+1)T/n$ is thc unique unbiased estimator that is also complete-sufficient, i.e., U is the UMVUE.

Example 2.3 (A statistic that is not complete) Consider a random sample $X_i \sim Unif(\theta - 1/2, \theta + 1/2)$, $i = 1, 2, \dots, n$, $\theta \in \mathcal{R}$. Here $T = (X_{(1)}, X_{(n)})$ is minimal sufficient but T is not complete since $X_{(n)} - X_{(1)}$ is ancillary with $E_\theta\left(X_{(n)} - X_{(1)} - (n-1)/(n+1)\right) = 0$, $\forall\theta \in \mathcal{R}$ and hence we have a function of T that is an unbiased estimator of zero, but it is not the function zero.

Example 2.4 (Poisson distribution) Let $X_i \overset{iid}{\sim} Poisson(\theta)$, $i = 1, 2, \dots, n$, $\theta > 0$ and $T(\mathbf{X}) = \sum_{i=1}^{n} X_i$, so that

$$f(\mathbf{x}|\theta) = \theta^{T(\mathbf{x})}e^{-n\theta}/\prod_{i=1}^{n} x_i! = \exp\{T(\mathbf{x})\ln\theta - n\theta\}/\prod_{i=1}^{n} x_i!,$$

which can be written in canonical form (letting $\lambda = \ln\theta$), as

$$f(\mathbf{x}|\lambda) = \exp\{\lambda T(\mathbf{x}) - ne^{\lambda}\}/\prod_{i=1}^{n} x_i!,$$

and hence we have a full rank one-parameter exponential family (example 1.2) with $A(\lambda) = ne^{\lambda}$. Consequently, T is complete-sufficient for λ (and hence for θ, since the definitions of sufficiency and completeness are unaffected by one-to-one transformations on the parameter), with $E(T) = \frac{dA(\lambda)}{d\lambda} = ne^{\lambda} = n\theta$, which makes T/n the UMVUE for θ. Now note that $P(X_1 = 0) = e^{-\theta}$ and letting $\delta(\mathbf{X}) = 1$, if $X_1 = 0$ and 0, if $X_1 \neq 0$, then $E(\delta) = 1P(X_1 = 0) + 0P(X_1 \neq 0) = e^{-\theta}$, so that δ is an unbiased estimator of $e^{-\theta}$, but it is not the UMVUE since it is not a function of the complete-sufficient statistic. Using Rao-Blackwell we know that $U(t) = E(\delta|T = t)$ becomes the UMVUE. Since $\sum_{i=1}^{n} X_i \sim Poisson(n\theta)$ and $\sum_{i=2}^{n} X_i \sim Poisson((n-1)\theta)$, we obtain after some algebra

$$U(t) = E(\delta|T = t) = ((n-1)/n)^t,$$

and hence $U(T) = ((n-1)/n)^T$ is the UMVUE for $e^{-\theta}$.

Example 2.5 (Exponential family) Consider a full rank one-parameter exponential family (recall example 1.2). We can write

$$\ln f(x|\theta) = \ln h(x) + \theta T(x) - A(\theta),$$

and for a random sample x_i, $i = 1, 2, \ldots, n$, from $f(x|\theta)$, we obtain

$$\frac{\partial}{\partial\theta}\ln\prod_{i=1}^{n} f(x_i|\theta) = n\left(\frac{1}{n}\sum_{i=1}^{n} T(x_i) - \frac{dA(\theta)}{d\theta}\right).$$

From remark 2.10, with $k(\theta, n) = n$, we have that $U(\mathbf{X}) = \frac{1}{n}\sum_{i=1}^{n} T(X_i)$ is the UMVUE of $g(\theta) = \frac{dA(\theta)}{d\theta}$.

Example 2.6 (Multivariate normal) Let $\mathbf{x}_{1:n} = \{\mathbf{x}_1, \ldots, \mathbf{x}_n\}$ be a random sample from a $\mathcal{N}_p(\boldsymbol{\mu}, \boldsymbol{\Sigma})$, $\boldsymbol{\mu} \in \mathcal{R}^p$, $\boldsymbol{\Sigma} > 0$, $n > p$, with joint density

$$f(\mathbf{x}_{1:n}|\boldsymbol{\mu}, \boldsymbol{\Sigma}) = |2\pi\boldsymbol{\Sigma}|^{-n/2}\exp\left\{-\frac{1}{2}\sum_{i=1}^{n}(\mathbf{x}_i - \boldsymbol{\mu})^T\boldsymbol{\Sigma}^{-1}(\mathbf{x}_i - \boldsymbol{\mu})\right\}, \quad (2.13)$$

and considering the exponent we can write

$$\begin{aligned} -\frac{1}{2}\sum_{i=1}^{n}(\mathbf{x}_i - \boldsymbol{\mu})^T\boldsymbol{\Sigma}^{-1}(\mathbf{x}_i - \boldsymbol{\mu}) &= -\frac{1}{2}tr\left(\boldsymbol{\Sigma}^{-1}\left[\sum_{i=1}^{n}\mathbf{x}_i\mathbf{x}_i^T\right]\right) \\ &+ (\boldsymbol{\Sigma}^{-1}\boldsymbol{\mu})^T\sum_{i=1}^{n}\mathbf{x}_i + \frac{n}{4}(\boldsymbol{\Sigma}^{-1}\boldsymbol{\mu})^T\left(-\frac{1}{2}\boldsymbol{\Sigma}^{-1}\right)^{-1}\boldsymbol{\Sigma}^{-1}\boldsymbol{\mu}. \end{aligned}$$

Let $\boldsymbol{\theta} = (\boldsymbol{\theta}_1, \boldsymbol{\theta}_2) = \left(\boldsymbol{\Sigma}^{-1}\boldsymbol{\mu}, -\frac{1}{2}\boldsymbol{\Sigma}^{-1}\right)$ and write (2.13) as

$$\begin{aligned} f(\mathbf{x}_{1:n}|\boldsymbol{\mu}, \boldsymbol{\Sigma}) &= \exp\{\boldsymbol{\theta}_1^T\mathbf{T}_1 + tr(\boldsymbol{\theta}_2\mathbf{T}_2) + (n/4)\boldsymbol{\theta}_1^T\boldsymbol{\theta}_2\boldsymbol{\theta}_1 \\ &\quad -(np/2)\ln(2\pi) + (n/2)\ln(|-2\boldsymbol{\theta}_2|)\}, \end{aligned}$$

which is a member of the exponential family with $h(\mathbf{x}) = 1$, $A(\boldsymbol{\theta}) = (np/2)\ln(2\pi)$

$-(n/2)\ln(|-2\theta_2|) - \theta_1^T\theta_2^{-1}\theta_1/4$ and $\mathbf{T} = [\mathbf{T}_1, \mathbf{T}_2] = \left[\sum_{i=1}^{n} \mathbf{x}_i, \sum_{i=1}^{n} \mathbf{x}_i\mathbf{x}_i^T\right]$ is the complete-sufficient statistic. Now take the exponent once more and add and subtract $\overline{\mathbf{x}} = \frac{1}{n}\sum_{i=1}^{n} \mathbf{x}_i$, to obtain

$$\sum_{i=1}^{n}(\mathbf{x}_i - \boldsymbol{\mu})^T\boldsymbol{\Sigma}^{-1}(\mathbf{x}_i - \boldsymbol{\mu}) = \sum_{i=1}^{n}(\mathbf{x}_i - \overline{\mathbf{x}})^T\boldsymbol{\Sigma}^{-1}(\mathbf{x}_i - \overline{\mathbf{x}}) + n(\overline{\mathbf{x}} - \boldsymbol{\mu})^T\boldsymbol{\Sigma}^{-1}(\overline{\mathbf{x}} - \boldsymbol{\mu}), \quad (2.14)$$

since $\sum_{i=1}^{n}(\mathbf{x}_i - \overline{\mathbf{x}}) = \mathbf{0}$. Based on equation (2.14), we can clearly see that the likelihood $L(\boldsymbol{\mu}, \boldsymbol{\Sigma}|\mathbf{x}_{1:n}) = f(\mathbf{x}_{1:n}|\boldsymbol{\mu}, \boldsymbol{\Sigma})$ is maximized for $\widehat{\boldsymbol{\mu}} = \overline{\mathbf{X}}$. Now using the mgf approach we can easily show that $\overline{\mathbf{x}} \sim \mathcal{N}_p(\boldsymbol{\mu}, \frac{1}{n}\boldsymbol{\Sigma})$, so that $E(\overline{\mathbf{X}}) = \boldsymbol{\mu}$ and hence $\overline{\mathbf{X}}$ is the UMVUE for $\boldsymbol{\mu}$. Moreover, letting $\mathbf{y} = \sqrt{n}\boldsymbol{\Sigma}^{-1/2}(\overline{\mathbf{x}} - \boldsymbol{\mu}) \sim \mathcal{N}_p(\mathbf{0}, I_p)$, we have that $n(\overline{\mathbf{x}} - \boldsymbol{\mu})^T\boldsymbol{\Sigma}^{-1}(\overline{\mathbf{x}} - \boldsymbol{\mu}) = \mathbf{y}^T\mathbf{y} \sim \chi^2_p$, a χ^2 random variable with p degrees of freedom, where $\boldsymbol{\Sigma}^{-1} = \boldsymbol{\Sigma}^{-1/2}\boldsymbol{\Sigma}^{-1/2}$ and I_p is the $p \times p$ identity matrix. Turning to estimation of $\boldsymbol{\Sigma}$, in order to maximize the log-likelihood function with respect to $\boldsymbol{\Sigma}$ we write

$$\ln L(\widehat{\boldsymbol{\mu}}, \boldsymbol{\Sigma}|\mathbf{x}_{1:n}) = -(np/2)\ln(2\pi) - (n/2)\ln|\boldsymbol{\Sigma}| - (1/2)tr(\boldsymbol{\Sigma}^{-1}\mathbf{A}),$$

where $\mathbf{A} = \sum_{i=1}^{n}(\mathbf{X}_i - \overline{\mathbf{X}})(\mathbf{X}_i - \overline{\mathbf{X}})^T = \sum_{i=1}^{n}\mathbf{X}_i\mathbf{X}^T - n\overline{\mathbf{X}}\overline{\mathbf{X}}^T$, a random matrix that is a function of the complete-sufficient statistic and considering the maximum of the function $f(\boldsymbol{\Sigma}) = -nlog|\boldsymbol{\Sigma}| - tr(\boldsymbol{\Sigma}^{-1}\mathbf{A})$, $\boldsymbol{\Sigma} > 0$, we can show that it is achieved at $\widehat{\boldsymbol{\Sigma}} = \frac{1}{n}\mathbf{A}$, the MLE of $\boldsymbol{\Sigma}$. Noting that $Var(\mathbf{X}) = E(\mathbf{X}\mathbf{X}^T) - E(\mathbf{X})E(\mathbf{X})^T$, we can write

$$E(\mathbf{A}) = \sum_{i=1}^{n} E\mathbf{X}_i\mathbf{X}^T - nE\overline{\mathbf{X}}\overline{\mathbf{X}}^T = (n-1)\boldsymbol{\Sigma},$$

and hence $\mathbf{S} = \frac{1}{n-1}\mathbf{A}$ is the UMVUE for $\boldsymbol{\Sigma}$, which is different than the MLE. In addition, it can be shown that $\mathbf{A} \sim W_p(\boldsymbol{\Sigma}, n-1)$, with density

$$\pi(\mathbf{A}|\boldsymbol{\Sigma}, m) = \frac{|\boldsymbol{\Sigma}|^{-(n-1)/2}|\mathbf{A}|^{((n-1)-p-1)/2}\exp\{-\frac{1}{2}tr(\boldsymbol{\Sigma}^{-1}\mathbf{A})\}}{2^{(n-1)p/2}\Gamma_p((n-1)/2)},$$

for $\mathbf{A} > 0$ and 0 otherwise, where $\Gamma_p((n-1)/2)$ is the multivariate gamma function.

2.3.2 Bayesian Approach

Now we discuss Bayesian methods of point estimation.

Remark 2.11 (Bayesian estimation considerations) All methods created in a Bayesian framework are based on the posterior distribution $\pi(\boldsymbol{\theta}|\mathbf{x})$.

1. Sufficient statistic When calculating the posterior distribution it is convenient to condition on the sufficient statistic T instead of the whole sample, that is, using the factorization theorem, we can write

$$\pi(\boldsymbol{\theta}|T(\mathbf{x})) = \pi(\boldsymbol{\theta}|\mathbf{x}) \propto g(T(\mathbf{x})|\boldsymbol{\theta})h(\mathbf{x})\pi(\boldsymbol{\theta}) \propto g(T(\mathbf{x})|\boldsymbol{\theta})\pi(\boldsymbol{\theta}). \quad (2.15)$$

2. Posterior propriety Posterior propriety is important since if the posterior $\pi(\boldsymbol{\theta}|\mathbf{x})$ is not a proper distribution (it does not integrate to one), we cannot apply a Bayesian analysis. When the prior $\pi(\boldsymbol{\theta})$ is proper then the posterior will always be a proper distribution since the marginal $m(x)$ exists and is finite.

3. Generalized MLE The generalized MLE of $\boldsymbol{\theta}$ is the largest mode of the posterior distribution, also known as the Maximum *a Posteriori* (MAP) estimator. Note that when we do not have any information about the parameter before the experiment, then we could, in theory, consider a flat, non-informative prior $\pi(\boldsymbol{\theta}) \propto 1$, which is possibly improper, i.e., *a priori* we place the same weight on all possible values of the parameter. Then, since $\pi(\boldsymbol{\theta}|\mathbf{x}) \propto f(\mathbf{x}|\boldsymbol{\theta})$, the generalized MLE becomes the classical MLE. Note that any central tendency measure of the posterior distribution can be thought of as a point estimator of $\boldsymbol{\theta}$, including the mean and median of $\pi(\boldsymbol{\theta}|\mathbf{x})$.

Example 2.7 (Bayes rule) Suppose that $X|\theta \sim f(x|\theta)$, $\theta \in \Theta$ and we entertain *a prior distribution* $\pi(\theta)$, so that the posterior distribution is $\pi(\theta|x) \propto f(x|\theta)\pi(\theta)$. Depending on the choice of loss function $L(\theta, a)$ the Bayes rule becomes a feature of the posterior distribution. Indeed, under SEL $L(\theta, a) = (\theta - a)^2$, the PEL becomes

$$\text{PEL} = E^{\pi(\theta|x)}\left(L(\theta, a)\right) = E^{\pi(\theta|x)}(\theta^2|x) - 2aE^{\pi(\theta|x)}(\theta|x) + a^2,$$

with $\frac{d\text{PEL}}{da} = -2E^{\pi(\theta|x)}(\theta|x) + 2a$ and $\frac{d^2\text{PEL}}{da^2} = 2 > 0$, for all actions a, so that the PEL is minimized at the Bayes rule $T(X) = E^{\pi(\theta|x)}(\theta|X)$, the posterior mean. Similarly, under absolute loss $L(\theta, a) = |a - \theta|$, we can show that the Bayes rule is the median of the posterior distribution.

2.3.3 Evaluating Point Estimators Using Decision Theory

We turn to the evaluation of the estimators we have seen thus far and how they relate to each other. An important class of estimators are those with constant risk.

Definition 2.1 Equalizer rule

A rule T^* is called an equalizer if it has constant frequentist risk, i.e., $R(\theta, T^*) = c$, $\forall \theta \in \Theta$, for some constant c.

The following remark summarizes some of the classic results that can help us evaluate estimators.

Remark 2.12 (Evaluating estimators) We summarize the important results that relate admissible, Bayes and minimax rules. The proofs of many of these theorems are simplified under an exponential family model. We can show the following.

1. Every admissible rule will be either Bayes or Extended Bayes.

2. An admissible rule with constant risk is minimax.

3. A unique minimax rule is admissible.

4. Any unique Bayes rule is admissible.

5. Unbiased estimators $T(X)$ of $g(\theta)$ cannot be Bayes rules under SEL, unless $T(X) = g(\theta)$ w.p. 1.

6. Minimax rules are not necessarily unique or have constant risk or need not be admissible or exist.

7. An equalizer rule which is Bayes or Extended Bayes is minimax.

Example 2.8 (Decision theoretic estimators of a normal mean) Suppose that $X|\theta \sim \mathcal{N}(\theta, 1)$, $\theta \in \mathcal{R}$, choose the SEL $L(\theta, a) = (\theta - a)^2$ as the loss function of the problem and consider decision rules of the form $T_c(X) = cX$. Then the frequentist risk is obtained as

$$R(\theta, T_c) = E^{x|\theta}(L(\theta, a)) = E^{x|\theta}((\theta - cX)^2) = c^2 + (1 - c)^2\theta^2.$$

Since for $c > 1$, $R(\theta, T_1) = 1 < c^2 + (1 - c)^2\theta^2 = R(\theta, T_c)$, the rules T_c are inadmissible for $c > 1$. Now suppose that $\theta \sim \mathcal{N}(0, \tau^2)$, a prior distribution on θ. Then the Bayes risk with respect to this prior is given by

$$r(\pi, T_c) = E^{\pi}[R(\theta, T_c)] = E^{\pi}[c^2 + (1 - c)^2\theta^2] = c^2 + (1 - c)^2\tau^2,$$

which is minimized as a function of c at $c_0 = \frac{\tau^2}{1+\tau^2}$. Thus $T_{c_0}(X) = \frac{\tau^2}{1+\tau^2}X$, is the unique Bayes rule with respect to this prior and hence (by remark 2.12.4) admissible. In fact, one can prove that T_c is admissible for $0 \leq c \leq 1$. By remark 2.12.2, the rule $T_1(X) = X$, is admissible with constant risk and hence minimax.

We present an important result below that can be used to connect admissibility and Bayes rules, namely Blyth's lemma. First, define the complete class of decision rules.

Definition 2.2 Complete class of decision rules

A class C of rules is said to be complete if for every rule that is not a member of C, there exists a rule from C, that is R-better.

Lemma 2.1 (Blyth) Consider a nonempty open set $\Theta \subset \mathcal{R}^p$. Assume that the decision rules with continuous risk form a complete class and that for a continuous risk rule T_0, there exists a sequence of $\{\pi_n\}$ of (possibly improper) prior distributions such that

(i) $r(\pi_n, T_0)$ is finite for all n;

(ii) for every nonempty open set $C \subset \Theta$, there exists $K > 0$ and N such that for every $n \geq N$, $\pi_n(C) \geq K$; and

(iii) $\lim\limits_{n\to+\infty} [r(\pi_n, T_0) - r(\pi_n, T^*)] = 0$, where T^* is the Bayes rule with respect to π_n.

Then the decision rule T_0 is admissible.

There are two foundational elements from a decision theoretic point of view, that involve the concepts of utility and randomized decision rules. The utility function is defined as the opposite of the loss function and as such, it enjoys a different interpretation, in the sense that it allows us to talk about rewards in choosing a certain action, instead of just the loss incurred. Moreover, a randomized decision rule $T^*(x, .)$ is, for each x, a probability distribution on the action space $\mathcal{A}$, with the interpretation that if x is observed, $T^*(x, A)$ is the probability that an action in $A \subset \mathcal{A}$, will be chosen. In no-data problems, a randomized decision rule, also called a randomized action, is again a probability distribution on $\mathcal{A}$. An immediate application of this concept can be seen in randomized test procedures, for hypothesis testing problems in a decision theoretic framework (see Section 2.5.3).

Another important topic of decision theory involves the so-called shrinkage estimators. James and Stein (1961) defined an estimator for the multivariate normal mean problem, that is R-better than $T_0(\mathbf{x}) = \mathbf{x}$ under quadratic loss for $p \geq 3$, by

$$T^{JS}(\mathbf{x}) = \left(1 - (p-2)/\mathbf{x}^T\mathbf{x}\right)\mathbf{x}. \tag{2.16}$$

This estimator has a strange behavior as $\mathbf{x}$ gets near $\mathbf{0}$. The term $\left(1 - (p-2)/\mathbf{x}^T\mathbf{x}\right)$ becomes negative and even goes to $-\infty$ as $\mathbf{x}^T\mathbf{x}$ goes to zero, but $T^{JS}(\mathbf{x})$ remains R-better than $T_0(\mathbf{x})$. Further improvements (in terms of risk) based on modifications of T^{JS} have appeared in the literature (Robert, 2007, p. 98), including the general form of the admissible estimator for $p \geq 3$. All such estimators are called shrinkage estimators since they tend to shrink $\mathbf{x}$ toward $\mathbf{0}$.

2.3.4 Convergence Concepts and Asymptotic Behavior

Point estimators are obtained based on a random sample $X_1, \dots, X_n$, of size n, from some distribution $f(x|\theta)$. As n increases, the sample approaches the whole population and the estimator should approach the true value of the parameter. In theory, we could increase the sample size to infinity, but in reality we deal with finite samples. Investigating the case $n \to +\infty$ provides, at the very least, useful approximations for estimators obtained based on finite samples. We discuss the basic definitions of convergence for a sequence of random variables $X_1, \dots, X_n, \dots$, along with the Laws of Large Numbers below. We revisit these concepts under a more rigorous framework in the later chapters.

Definition 2.3 Convergence types

Let $\{X_n\}_{n=1}^{+\infty}$ be a sequence of random variables.

1. Convergence in probability We say that X_n converges in probability to a random variable X, if for every $\varepsilon > 0$,

$$\lim_{n\to\infty} P(|X_n - X| < \varepsilon) = 1.$$

We write $X_n \stackrel{p}{\to} X$.

2. Almost sure convergence We say that X_n converges almost surely (a.s.) to a random variable X, if

$$P(\lim_{n\to\infty} X_n = X) = 1.$$

We write $X_n \stackrel{a.s.}{\to} X$ or say that X_n converges to X w.p. 1 or a.s.

3. Weak convergence If X_n has cdf $F_n(x)$, then we say that X_n converges in distribution (or weakly or in law) to a random variable X, with cdf $F(x)$, if

$$\lim_{n\to\infty} F_n(x) = F(x),$$

for all continuity points x of $F(x)$. We write $X_n \overset{w}{\to} X$ (or $X_n \overset{d}{\to} X$ or $X_n \overset{\mathcal{L}}{\to} X$). It is sometimes convenient to denote weak convergence in terms of the cdfs involved by writing $F_n \overset{w}{\to} F$.

Note that it can be shown that $X_n \overset{a.s.}{\to} X \Rightarrow X_n \overset{p}{\to} X \Rightarrow X_n \overset{w}{\to} X$ (see exercise 5.4). Now we introduce an additional desirable property for a point estimator from a classical perspective, that of consistency.

Definition 2.4 Estimator consistency

An estimator $T_n(\mathbf{X})$ of a parameter $\boldsymbol{\theta}$, based on a random sample $\mathbf{X} = (X_1, \dots, X_n)$, is called weakly consistent for $\boldsymbol{\theta}$, when $T_n(\mathbf{X}) \overset{p}{\to} \theta$ and strongly consistent when $T_n(\mathbf{X}) \overset{a.s.}{\to} \boldsymbol{\theta}$.

Consistency is a desired property for MLE estimators and as we will see in Chapter 5, it is not difficult to establish. Next we collect a rudimentary theorem that can be used to show consistency of an estimator, including the MLE.

Theorem 2.1 (Consistency) Consider an estimator $T_n(\mathbf{X})$ of a parameter $\boldsymbol{\theta}$, based on a random sample $\mathbf{X} = (X_1, \dots, X_n)$, and assume that the following conditions hold:
(i) $E[T_n(\mathbf{X})] \to \theta$, as $n \to \infty$, and
(ii) $Var(T_n(\mathbf{X})) \to 0$, as $n \to \infty$.
Then $T_n(\mathbf{X})$ is consistent for $\boldsymbol{\theta}$.

The following theorems allow us to study the asymptotic behavior of averages of random variables, as well as prove under which conditions sample averages are consistent estimators of the corresponding population averages. A straightforward application of Chebyshev's inequality gives the following.

Theorem 2.2 (Weak law of large numbers (WLLN)) Let $X_1, \dots, X_n$ be a random sample from some model with $E(X_i) = \mu$ and $Var(X_i) = \sigma^2 < \infty$. Define $\overline{X}_n = \frac{1}{n}\sum_{i=1}^{n} X_i$. Then $\overline{X}_n \overset{p}{\to} \mu$.

Note that there are more general versions of the WLLN where the assumption $\sigma^2 < \infty$ is not required. In particular, requiring integrability is enough (see definition 3.16), however, the rudimentary version presented above is applicable in most practical situations.

A stronger analog of the WLLN is given next (under the same assumptions as in the WLLN theorem) and the proofs of both laws are requested as exercises. These laws illustrate the asymptotic behavior of the sample average as the sample size increases and provide additional justification for using the sample average in estimating the population mean.

Theorem 2.3 (Strong law of large numbers (SLLN)) Let $X_1, \ldots, X_n$, be a random sample with $E(X_i) = \mu$ and $Var(X_i) = \sigma^2 < \infty$. Define $\overline{X}_n = \frac{1}{n}\sum_{i=1}^{n} X_i$. Then $\overline{X}_n \overset{a.s.}{\to} \mu$.

A great consequence of the SLLN is the Monte Carlo approximation.

Remark 2.13 (Monte Carlo) Assume that $X_1, \ldots, X_n$ are iid $f(x|\theta)$, $x \in \mathcal{X}$, with $E(g(X_i)) = \lambda$, for all $i = 1, 2, \ldots, n$. Then the SLLN tells us that w.p. 1 we have

$$\frac{1}{n}\sum_{i=1}^{n} g(X_i) \to \lambda, \text{ as } n \to +\infty.$$

The SLLN states that if we generate a large number of values from $f(x|\theta)$, then in order to approximate λ all we have to do is take the average of the generated values evaluated through $g(x)$. Note that the values $g(x_1), \ldots, g(x_n)$ can be thought of as realizations from the random variable $g(X)$. The Monte Carlo integration method consists of using this idea to evaluate the integral $E(g(X))$, i.e.,

$$\lambda = E(g(X)) = \int_{\mathcal{X}} g(x) f(x|\theta) dx \simeq \frac{1}{n}\sum_{i=1}^{n} g(x_i),$$

where $X_1, \ldots, X_n$ is a random sample from $f(x|\theta)$. Clearly, the larger the n the better the approximation. Choosing $g(X) = I(X \in A)$ and noting that $E(g(X)) = P(X \in A)$, allows us to approximate probabilities, that is,

$$P(X \in A) \simeq \frac{1}{n}\sum_{i=1}^{n} I(x_i \in A).$$

2.4 Interval Estimation

Point estimators $T(\mathbf{x})$ provide a single value in an attempt to guess the parameter θ. We turn now to a natural extension of point estimation by considering a range of values to help us estimate the parameter. From a classical point of view, an interval estimator of a parameter must be a set with bounds that are affected by the observed data. This leads to the following definition.

Definition 2.5 Interval estimator

Let $X_i|\theta \overset{iid}{\sim} f(x|\theta)$, $x \in \mathcal{X}$, $i = 1, 2, \ldots, n$, $\theta \in \Theta \subseteq \mathcal{R}$. An interval estimator or confidence interval (CI) of the parameter θ is defined as the random interval $C(\mathbf{X}) = [L(\mathbf{X}), U(\mathbf{X})]$, for some real-valued statistics L and U, such that $L(\mathbf{x}) \leq U(\mathbf{x})$, for all $\mathbf{x}$. Moreover, if

$$P^{\mathbf{x}|\theta}(\theta \in [L(\mathbf{X}), U(\mathbf{X})]) = 1 - \alpha, \tag{2.17}$$

then $100(1 - \alpha)\%$ is called the confidence or coverage probability of the interval. The confidence coefficient of $C(\mathbf{X})$ is defined as the smallest coverage probability $\inf_{\theta \in \Theta} P^{\mathbf{x}|\theta}(\theta \in C(\mathbf{X}))$.

The following remark provides some insight and interpretation of confidence intervals.

Remark 2.14 (Confidence interval interpretation) Equation (2.17) involves a probability that is computed based on the model distribution $f(\mathbf{x}|\theta)$ and assumes an interpretation in terms of a relative frequency, that is, if we were to conduct the experiment 100 times, then we would expect to find the true value of the parameter $100(1 - \alpha)\%$ of the times within the CI. Classical methods are often called frequentist methods due to interpretations such as the latter.

Further note that this probability statement is computed based on the distribution of $\mathbf{X}|\theta$ and requires all $\mathbf{X} \in \mathcal{X}^n$, in order to compute it, namely, unobserved data. Thus, requiring the statement $L(\mathbf{X}) \leq \theta \leq U(\mathbf{X})$ to be satisfied w.p. $1 - \alpha$, for fixed and unknown θ, is similar to requiring from the statistician to guess realizations (data) $\mathbf{x}$ of the random vector $\mathbf{X}$ that satisfy $L(\mathbf{x}) \leq \theta \leq U(\mathbf{x})$, for the given θ. This logic is counterintuitive since the purpose of any estimation procedure in statistics is to use observed data $\mathbf{x}$ to estimate the unknown parameter θ and not the opposite.

Bayesians, on the other hand, utilize the observed data to update the prior and calculate the posterior distribution in order to obtain an interval estimator for the parameter.

Definition 2.6 Credible set

Let $X_i|\theta \overset{iid}{\sim} f(x|\boldsymbol{\theta})$, $x \in \mathcal{X}$, $i = 1, 2, \ldots, n$, $\boldsymbol{\theta} \in \Theta \subseteq \mathcal{R}^p$ and let $\pi(\boldsymbol{\theta})$ be a prior distribution on $\boldsymbol{\theta}$. Then if the posterior $\pi(\boldsymbol{\theta}|\mathbf{x})$ exists, a $100(1 - \alpha)\%$ credible set (CS) for $\theta \in \Theta$, is the subset A of Θ with

$$P(\boldsymbol{\theta} \in A|\mathbf{x}) = 1 - \alpha. \tag{2.18}$$

Clearly, there are many sets that could satisfy equations (2.17) and (2.18) and hence we need to evaluate these interval estimators in terms of their properties.

2.4.1 Confidence Intervals

We investigate first the construction of confidence intervals from a classical point of view by utilizing pivotal quantities. Let $Q = Q(\mathbf{X}, \theta) \in \mathcal{R}$ be pivotal for a real parameter $\theta \in \Theta \subseteq \mathcal{R}$ and consider the probability

$$P(q_1 \leq Q \leq q_2) = 1 - \alpha, \quad (2.19)$$

for some constants q_1 and q_2, $q_1 < q_2$, which can be written in terms of the cdf of Q, $F_Q(q) = P(Q \leq q)$, as

$$F_Q(q_2) - F_Q(q_1-) = 1 - \alpha,$$

where q_1- means we approach q_1 from the left ($q_1- = \lim_{\varepsilon \to 0}(q_1 - \varepsilon)$). Note that if $q_1 \leq Q(\mathbf{X}, \theta) \leq q_2$ if and only if $L(\mathbf{X}, q_1, q_2) \leq g(\theta) \leq U(\mathbf{X}, q_1, q_2)$, then $[L(\mathbf{X}, q_1, q_2), U(\mathbf{X}, q_1, q_2)]$ is a $100(1 - \alpha)\%$ CI for $g(\theta)$. We summarize some of the classic properties of confidence intervals next.

Remark 2.15 (Confidence interval properties) The following are the desirable properties for classical confidence intervals and are obtained for different choices of q_1 and q_2.

1. Equal tail If q_1 and q_2 are such that $F_Q(q_1) = 1 - F_Q(q_2) = \frac{\alpha}{2}$ then the resulting CI is called equal tail.

2. Minimum width Let q_1^* and q_2^* be such that they satisfy equation (2.19) and minimize the length of the interval $[L(\mathbf{X}, q_1, q_2), U(\mathbf{X}, q_1, q_2)]$. Then the interval $[L(\mathbf{X}, q_1, q_2), U(\mathbf{X}, q_1, q_2)]$ is called the Minimum Width CI among all confidence intervals of confidence $100(1 - \alpha)\%$.

3. Intervals of the form $[L(\mathbf{X}, q_1, q_2), U(\mathbf{X}, q_1, q_2)]$ are called two-sided, while intervals of the form $[L(\mathbf{X}, q_1), +\infty)$ or $(-\infty, U(\mathbf{X}, q_2)]$, are one-sided and can be obtained in a similar fashion.

A second method is based on the sufficient statistic. Let T be the sufficient statistic with distribution $f_{T|\theta}(t|\theta)$ and consider a confidence interval $[V_1, V_2]$, such that $h_1(V_1) = T$ and $h_2(V_2) = T$, $h_1(\theta) < h_2(\theta)$, with $p_1 = P^{T|\theta}(T < h_1(\theta))$ and $p_2 = P^{T|\theta}(T > h_2(\theta))$, with $0 < p_1, p_2 < 1$ and $0 < p_1 + p_2 < 1$. Then $[V_1, V_2]$ has coverage $100(1 - p_1 - p_2)\%$, since $P(h_1(\theta) \leq T \leq h_2(\theta)) = 1 - p_1 - p_2$.

2.4.2 Highest Posterior Density Credible Sets

A special class of credible sets is defined below.

Definition 2.7 HPD credible set

Let $X_i|\boldsymbol{\theta} \overset{iid}{\sim} f(x|\boldsymbol{\theta})$, $x \in \mathcal{X}$, $i = 1, 2, \ldots, n$, $\boldsymbol{\theta} \in \Theta \subseteq \mathcal{R}^p$ and let $\pi(\boldsymbol{\theta})$ be a prior distribution on $\boldsymbol{\theta}$. Then if the posterior $\pi(\boldsymbol{\theta}|\mathbf{x})$ exists, the $100(1 - \alpha)\%$ Highest Posterior Density (HPD) credible set for $\boldsymbol{\theta} \in \Theta$, is the subset $C_k(\mathbf{x})$ of Θ of the form

$$C_k(\mathbf{x}) = \{\boldsymbol{\theta} \in \Theta : \pi(\boldsymbol{\theta}|\mathbf{x}) \geq k(\alpha)\}, \quad (2.20)$$

where $k(\alpha)$ is the largest constant such that

$$P(\boldsymbol{\theta} \in C_k(\mathbf{x})|\mathbf{x}) \geq 1 - \alpha, \quad (2.21)$$

with equality when $\boldsymbol{\theta}$ is continuous. It should be noted that for unimodal $\pi(\boldsymbol{\theta}|\mathbf{x})$, it can proven that the $100(1-\alpha)\%$ HPD set is also the minimum width set for the given α.

Example 2.9 (HPD credible set for normal mean) Let $X_i|\theta \overset{iid}{\sim} \mathcal{N}(\theta, \sigma^2)$, with σ^2 known and consider the prior $\theta \sim \mathcal{N}(\mu, \tau^2)$, with μ, τ^2 known. It is easy to see that the posterior of $\theta|\mathbf{x}$ is $\mathcal{N}\left(\xi(\mathbf{x}), \eta^2(\mathbf{x})\right)$, with

$$\xi(\mathbf{x}) = \frac{n\tau^2}{n\tau^2+\sigma^2}\overline{x} + \frac{\sigma^2}{n\tau^2+\sigma^2}\mu,$$

and $\eta^2(\mathbf{x}) = \tau^2\sigma^2/(n\tau^2+\sigma^2)$. We find the $100(1-\alpha)\%$ HPD credible set for θ. We need to find a set $C_k(\mathbf{x}) = \{\theta \in \Theta : \pi(\theta|\mathbf{x}) \geq k(\alpha)\}$, with $P(\theta \in C_k(\mathbf{x})|\mathbf{x}) = 1-\alpha$. We can show that $\pi(\theta|\mathbf{x}) \geq k(\alpha)$ if and only if $|\theta - \xi(\mathbf{x})| \leq k_1$, where

$$1-\alpha = P(\theta \in C_k(\mathbf{x})|\mathbf{x}) = P\left(|Z| \leq k_1/\eta(\mathbf{x})|Z = (\theta - \xi(\mathbf{x}))/\eta(\mathbf{x}) \sim \mathcal{N}(0,1)\right),$$

and hence

$$P\left(Z > k_1/\eta(\mathbf{x})\right) = \alpha/2,$$

so that $k_1 = \eta(\mathbf{x})z_{\alpha/2}$, where $z_{\alpha/2}$ is the $\alpha/2$ percentile of a $\mathcal{N}(0,1)$ distribution. Hence the $100(1-\alpha)\%$ HPD is of the form $[\xi(\mathbf{x}) - \eta(\mathbf{x})z_{\alpha/2}, \xi(\mathbf{x}) + \eta(\mathbf{x})z_{\alpha/2}]$.

Example 2.10 (HPD using Monte Carlo) Let $X|\theta \sim Exp(\theta)$, $\theta > 0$ and consider the prior $\theta \sim Exp(\lambda)$, with λ known. We can easily show that $\theta|x \sim Gamma(2, 1/(\lambda+x))$, so that $\pi(\theta|x) = k$ if and only if $\theta e^{-(\lambda+x)\theta}(\lambda+x)^2 = k$, which leads to the equation

$$\ln\theta - (\lambda+x)\theta = \ln\left(k/(\lambda+x)^2\right), \tag{2.22}$$

which cannot be solved analytically with respect to θ. Hence, in order to find the largest k that satisfies $P(\pi(\theta|x) \geq k|x) = 1-\alpha$, we turn to Monte Carlo simulation. Following remark 2.13, we approximate the posterior probability of the event $C_k(x) = \{\theta > 0 : \pi(\theta|x) \geq k\}$, for a given k, using

$$P(\pi(\theta|x) \geq k|x) = E^{\theta|X}[I(\theta \in C_k(x))] \simeq \frac{1}{L}\sum_{i=1}^{L} I(\theta_i \in C_k(x)),$$

where L is a large integer and $\theta_1, \ldots, \theta_L$ are realizations from a $Gamma(2, 1/(\lambda+x))$. We then select the k^* that gives the largest probability below $1-\alpha$ (for continuous θ there is only one such k^*). Finally, the HPD set is defined based on values $\theta > 0$, that are the roots of equation (2.22) for $k = k^*$. In our case, since the posterior is unimodal, there will be two roots θ_1^* and θ_2^* and the $100(1-\alpha)\%$ HPD will be $[\theta_1^*, \theta_2^*]$.

Example 2.11 (Credible ellipsoid) Let $\mathbf{X}|\boldsymbol{\theta} \sim \mathcal{N}_p(\boldsymbol{\theta}, \Sigma)$, with $\boldsymbol{\theta} \sim \mathcal{N}_p(\boldsymbol{\mu}, R)$, with $\Sigma > 0$, $\boldsymbol{\mu}$, and $R > 0$, are fixed. It is straightforward to see that $\boldsymbol{\theta}|\mathbf{X} \sim \mathcal{N}_p(\mathbf{v}, \Psi)$, with $\mathbf{v} = n\Psi\Sigma^{-1}\overline{\mathbf{x}} + \Psi R^{-1}\boldsymbol{\mu}$ and $\Psi = (n\Sigma^{-1} + R^{-1})^{-1}$. Now $\pi(\theta|x) \geq k$ if and only if $q = (\boldsymbol{\theta} - \mathbf{v})^T\Psi^{-1}(\boldsymbol{\theta} - \mathbf{v}) \leq k_1$, where $q|\mathbf{x} \sim \chi^2_p$, so that $P(q \leq k_1|\mathbf{x}) = P(\pi(\boldsymbol{\theta}|\mathbf{x}) \geq k|\mathbf{x}) = 1-\alpha$, which leads to $P(q \geq k_1|\mathbf{x}) = \alpha$ and hence $k_1 = \chi^2_{p;\alpha}$. Thus a $100(1-\alpha)\%$ HPD credible set for $\boldsymbol{\theta}$ is $\{\boldsymbol{\theta} \in \mathcal{R}^p : (\boldsymbol{\theta} - \mathbf{v})^T\Psi^{-1}(\boldsymbol{\theta} - \mathbf{v}) \leq \chi^2_{p;\alpha}\}$, which is an ellipsoid in $\mathcal{R}^p$.

2.4.3 Decision Theoretic

Classical or Bayesian methods can be easily incorporated in a decision theoretic framework by choosing an appropriate loss function and treating possible actions as intervals. Let $I(\theta, C)$ denote the indicator function taking value 1, if $\theta \in C$ and 0, if $\theta \notin C$, where $C = C(x) \subseteq \mathcal{R}$ and define the loss function

$$L(\theta, C) = b\,\|C\| - I(\theta, C),$$

with $b > 0$, some constant and $\|C\| = length(C)$. The frequentist risk for an interval $C(x)$ is obtained as

$$R(\theta, C) = bE_{\theta}^{x|\theta}\left[\|C(x)\|\right] - P_{\theta}^{x|\theta}(\theta \in C(x)),$$

which depends on the length of C, $\|C(x)\|$, as well as the coverage probability $P_{\theta}^{x|\theta}(\theta \in C(x))$. Obtaining rules $C(x)$, with different properties, such as minimum width or equal tail of confidence $100(1-\alpha)\%$, reduces to minimizing the frequentist risk under different conditions.

Example 2.12 (Decision theoretic interval estimation) Let $X_i|\theta \overset{iid}{\sim} \mathcal{N}(\theta, \sigma^2)$, with σ^2 known and consider interval rules of the form $C_r = C_r(\mathbf{x}) = [\overline{x}-r\sigma, \overline{x}+r\sigma]$, $r \geq 0$, with length $l_r = \|C(\mathbf{x})\| = 2r\sigma$ and coverage probability

$$P_{\theta}^{x|\theta}(\theta \in C_r(\mathbf{x})) = P_{\theta}^{x|\theta}\left(-\sqrt{n}r \leq \sqrt{n}(\overline{X} - \theta)/\sigma \leq \sqrt{n}r\right) = 2P(Z \leq \sqrt{n}r) - 1,$$

where $Z \sim \mathcal{N}(0, 1)$. Then the frequentist risk becomes

$$R(\theta, C_r) = b2r\sigma - 2P(Z \leq \sqrt{n}r) + 1,$$

which is independent of θ, so that C_r, $r \geq 0$, are equalizer rules. Choosing $r_0 = z_{\alpha/2}/\sqrt{n}$, where $P(Z > z_{\alpha/2}) = \alpha/2$, the coverage probability becomes

$$P_{\theta}^{x|\theta}(\theta \in C_{r_0}(\mathbf{x})) = 2P(Z \leq z_{\alpha/2}) - 1 = 1 - \alpha,$$

and the corresponding risk assumes the form

$$R(\theta, C_{r_0}) = 2b\sigma z_{\alpha/2}/\sqrt{n} + 1 - \alpha.$$

In general we want to minimize risk with respect to the decision C_r in order to obtain the admissible rule within the class of rules C_r, i.e., we need to find

$$r = \arg\min_{r \geq 0} R(\theta, C_r).$$

Let Φ denote the cdf of Z. The derivative of the risk function with respect to $r \geq 0$, is given by

$$\frac{dR(\theta, C_r)}{dr} = 2b\sigma - 2\sqrt{n}\Phi'(\sqrt{n}r) = 2b\sigma - 2\sqrt{n}(2\pi)^{-1/2}e^{-nr^2/2},$$

and setting it to zero, we obtain

$$-nr^2/2 = \ln\left(b\sigma\sqrt{2\pi/n}\right). \tag{2.23}$$

Now if $b > \sqrt{n/(2\pi)}/\sigma$, then $\ln\left(b\sigma\sqrt{2\pi/n}\right)$ is defined and is positive and consequently $\frac{dR(\theta,C_r)}{dr} \geq 0$. Thus $\arg\min_{r\geq 0} R(\theta, C_r) = 0$ and $C_0(\mathbf{X}) = \{\overline{X}\}$, a degenerate random set (singleton). Now if $b < \sqrt{n/(2\pi)}/\sigma$, then $\ln\left(b\sigma\sqrt{2\pi/n}\right) < 0$ and hence equation (2.23) has a solution at $r = \pm\sqrt{-(2/n)\ln\left(b\sigma\sqrt{2\pi/n}\right)}$. Since $r \geq 0$, the only valid solution is

$$r_0 = \arg\min_{r\geq 0} R(\theta, C_r) = \sqrt{-(2/n)\ln\left(b\sigma\sqrt{2\pi/n}\right)}, \quad (2.24)$$

for $b < \sqrt{n/(2\pi)}/\sigma$. Note that when $b = \frac{1}{\sigma}\sqrt{n}(2\pi)^{-1/2}e^{-z^2_{\alpha/2}/2} < \sqrt{n/(2\pi)}/\sigma$, the solution to equation (2.23) is $r_0 = z_{\alpha/2}/\sqrt{n}$ and yields the usual $100(1-\alpha)\%$ minimum width CI. Now consider the prior distribution $\theta \sim \pi(\theta) \propto 1$ and since $\overline{X}|\theta \sim \mathcal{N}(\theta, \sigma^2/n)$, where $\overline{X}$ is the sufficient statistic, the posterior distribution of $\theta|\overline{X}$ is $\mathcal{N}(\overline{X}, \sigma^2/n)$, or $\sqrt{n}\left(\theta - \overline{X}\right)/\sigma|\overline{X} \sim \mathcal{N}(0,1)$. Then the PEL is given by

$$\text{PEL} = E^{\theta|\overline{X}}[L(\theta, C_r)] = 2rb\sigma - P(\overline{x} - r\sigma \leq \theta \leq \overline{x} + r\sigma) = 2rb\sigma + 1 - 2\Phi(r\sqrt{n}),$$

and hence the PEL is exactly the same as the frequentist risk, which is minimized at r_0 given by equation (2.24). Thus, C_{r_0} is the unique Bayes rule and from remark 2.12.4 is admissible and using remark 2.12.7 it is also minimax.

2.5 Hypothesis Testing

The last statistical inference method we will discuss is hypothesis testing.

2.5.1 Classic Methods

The classical approach involves using a random sample **x** from a model $f(\mathbf{x}|\boldsymbol{\theta})$, in order to create a statistical test. The major components of a statistical test are summarized below.

Remark 2.16 (Components of a statistical test) We collect all the components of a statistical test below.

1. Hypotheses A null hypothesis $H_0 : \boldsymbol{\theta} \in \Theta_0 \subseteq \mathcal{R}^p$ and an alternative hypothesis $H_a : \boldsymbol{\theta} \in \Theta_a$. Note that when $\Theta_a = \mathcal{R}^p \setminus \Theta_0$, the alternative hypothesis is complementary.

2. Test statistic A statistic $T(\mathbf{X})$ is called the test statistic for this hypothesis. The distribution of $T(\mathbf{X})$ under H_0 and H_a is required in order to define other important components of the test.

3. Rejection region The test we create is based on the idea that we divide the support $\mathcal{S}$ of $T(\mathbf{X})$ in two areas C and $\mathcal{S} \setminus C$ and we reject the null hypothesis if $T(\mathbf{X}) \in C$ or do not reject if $T(\mathbf{X}) \in \mathcal{S} \setminus C = C^c$. The region C is aptly named the rejection (or critical) region (RR) for this test.

4. Type I and II errors In order to evaluate a test procedure and select the best, in some sense, we need the help of the so-called Type I and II errors. The schematic of Table 2.1 provides the definition of these errors and when they occur for any hypothesis testing problem. Therefore, the probability of the type I error is given by

$$\alpha = P(\text{Type I Error}) = P(\text{Reject } H_0|H_0 \text{ is true}),$$

		Statistician's Decision	
		Accept H_0	Reject H_0
State of	H_0 describes the true state	No Error	Type I Error
Nature	H_a describes the true state	Type II Error	No Error

Table 2.1: Schematic for any hypothesis testing problem along with the occurrence of the Type I and II errors.

and similarly, we quantify the probability of a type II error using

$$\beta = P(\text{Type II Error}) = P(\text{Accept } H_0 | H_0 \text{ is false}).$$

If H_0 or H_a are simple hypotheses, e.g., $\Theta_0 = \{\boldsymbol{\theta}_0\}$ or $\Theta_a = \{\boldsymbol{\theta}_a\}$ (a single value for the parameter), then α and β take only one value. In this case, α is called the level of significance of the test. If H_0 is a composite hypothesis, i.e., Θ_0 consists of a range of values not just a singleton, then the level of significance of the test is defined as

$$\alpha = \sup_{f \in H_0} P(\text{Reject } H_0 | f),$$

where $f \in H_0$ means that f is the distribution of $T(\mathbf{X})$ under the null hypothesis. Clearly, a testing procedure will prove useful if both α and β are small. If we enlarge the RR C to a new region C^* ($C \subset C^*$), then

$$\alpha^* = P(T(\mathbf{X}) \in C^* | H_0 \text{ true}) \geq P(T(X) \in C | H_0 \text{ true}) = \alpha,$$

but on the contrary

$$\beta^* = P(T(\mathbf{X}) \notin C^* | H_a \text{ true}) \leq P(T(X) \notin C | H_a \text{ true}) = \beta,$$

and thus as α increases, β gets smaller and when α decreases, β gets larger. This means that we cannot minimize α and β at the same time. The standard approach is to fix α and then find the RR C that minimizes β.

5. Test power The power of a test is defined by

$$\rho = 1 - \beta = 1 - P(\text{Accept } H_0 | H_0 \text{ is false}) = P(\text{Reject } H_0 | H_0 \text{ is false}),$$

and hence we can write the power of a test as

$$\rho = P(\text{Reject } H_0 | H_a \text{ is true}).$$

Clearly, the more powerful a test is the better it is.

We turn now to the mathematical foundations that incorporate these ideas in a classical framework.

Definition 2.8 Randomized test function

A Randomized Test Function φ for testing $H_0 : \boldsymbol{\theta} \in \Theta \subseteq \mathcal{R}^p$ vs $H_a : \boldsymbol{\theta} \in \Theta_a$, is a function $\varphi : \mathcal{R}^n \to [0, 1]$, with

$$\varphi(\mathbf{x}) = P(\text{Reject } H_0 | \mathbf{X} = \mathbf{x}),$$

where $\mathbf{x}$ is a random sample of size n, i.e., given that we observe $\mathbf{X} = \mathbf{x}$, we flip a coin and we reject H_0 w.p. $\varphi(\mathbf{x})$.

The function $\varphi(\mathbf{x})$ allows us to describe all the necessary components of a statistical test. Clearly, $\varphi(\mathbf{X})$ is a random variable, leading to a randomized test function. If φ is such that $\varphi(\mathbf{x}) = 1$, if $\mathbf{x} \in C$, or 0, if $\mathbf{x} \notin C$, then φ is called non-randomized.

In order to obtain the probabilities of the type I and II errors, we define $\beta_\varphi(\boldsymbol{\theta}) = P_{\boldsymbol{\theta}}^{\mathbf{X}|\boldsymbol{\theta}}(\text{Reject } H_0)$, so that

$$\beta_\varphi(\boldsymbol{\theta}) = E_{\boldsymbol{\theta}}^{\mathbf{X}|\boldsymbol{\theta}}[\varphi(\mathbf{X})].$$

Now if $\boldsymbol{\theta} \in \Theta_0$ then $\beta_\varphi(\boldsymbol{\theta})$ gives the probability of the type I error, while if $\boldsymbol{\theta} \in \Theta_a$ then $\beta_\varphi(\boldsymbol{\theta})$ is the power of the test function φ at $\boldsymbol{\theta} \in \Theta_a$.

The magnitude of the test function φ is defined as $\alpha = \sup\limits_{\boldsymbol{\theta}\in\Theta_0} \beta_\varphi(\boldsymbol{\theta})$, which is nothing but the level of significance. Once we fix α, we look for $\varphi(\mathbf{x})$ such that $\beta_\varphi(\boldsymbol{\theta}) \geq \beta_{\varphi^*}(\boldsymbol{\theta})$, for any other test function φ^*, such that $\sup\limits_{\boldsymbol{\theta}\in\Theta_0} \beta_{\varphi^*}(\boldsymbol{\theta}) \leq \alpha$.

Note that when we consider a hypothesis $H : \theta = \theta_0$ we are assuming a specific model $f_0 = f(x|\theta_0)$ for the data. Hence it is useful in some cases to write hypotheses in terms of the underlying models implied, namely, $H : f = f_0$. We also write $f \in H$ meaning that f denotes a model under the hypothesis H. Test functions can be classified in two main groups as we see next.

Remark 2.17 (Properties of test functions) There are two main properties for test functions.

1. Most powerful A test function φ such that

$$\sup_{\boldsymbol{\theta}\in\Theta_0} \beta_\varphi(\boldsymbol{\theta}) = \alpha_\varphi, \tag{2.25}$$

and

$$\beta_\varphi(\boldsymbol{\theta}) \geq \beta_{\varphi^*}(\boldsymbol{\theta}), \tag{2.26}$$

$\boldsymbol{\theta} \in \Theta_a$, for any other test function φ, with $\alpha_{\varphi^*} \leq \alpha_\varphi$, is called Uniformly Most Powerful (UMP) amongst all test functions of level α_φ.

If H_0 is simple, then a test function φ that satisfies equations (2.25) and (2.26), is called simply the Most Powerful (MP) test function. Note that the word uniformly above is understood as "with respect to all $\boldsymbol{\theta} \in \Theta_0$." The nonexistence of UMP tests in some cases has led to the definition of unbiased tests.

2. Unbiased Assume that $X_i|\boldsymbol{\theta} \overset{iid}{\sim} f(x|\boldsymbol{\theta})$, $\boldsymbol{\theta} \in \Theta \subseteq \mathcal{R}^p$, $p \geq 1$. Then for testing $H_0 : \boldsymbol{\theta} \in \Theta_0$ vs $H_a : \boldsymbol{\theta} \in \Theta_a = \Theta_0^c$, at level α, a test function $\varphi(\mathbf{x})$ is called unbiased if $E_\theta[\varphi(\mathbf{X})] \leq \alpha$, $\boldsymbol{\theta} \in \Theta_0$ and $E_\theta[\varphi(\mathbf{X})] \geq \alpha$, $\boldsymbol{\theta} \in \Theta_0^c$.

Although not the most desirable property, unbiasedness guarantees two things; first, the probability of the type I error is at most α and second, the power of the test is at least α. A test function is UMP unbiased if it is UMP within the collection of all unbiased test functions. A UMP test function of level α is always

UMP unbiased since letting $\varphi^*(\mathbf{x}) = \alpha$, we have $\beta_\varphi(\theta) \geq \beta_{\varphi^*}(\theta) = \alpha$, $\theta \in \Theta_0^c$, since φ is UMP.

The following remark summarizes the classic methods of finding test functions.

Remark 2.18 (Finding test functions) There are many methods of finding test functions that lead to statistical tests with different properties and we collect the most important below.

1. Neyman-Pearson fundamental lemma For testing $H_0 : \boldsymbol{\theta} = \boldsymbol{\theta}_0$ vs $H_a : \boldsymbol{\theta} = \boldsymbol{\theta}_1$ at a level α $(0 < \alpha < 1)$ and based on $\mathbf{X}$, consider the test function

$$\varphi_{NP}(\mathbf{x}) = \begin{cases} 1 & L(\boldsymbol{\theta}_1|\mathbf{x}) > kL(\boldsymbol{\theta}_0|\mathbf{x}) \\ \gamma & L(\boldsymbol{\theta}_1|\mathbf{x}) = kL(\boldsymbol{\theta}_0|\mathbf{x}) \\ 0 & L(\boldsymbol{\theta}_1|\mathbf{x}) < kL(\boldsymbol{\theta}_0|\mathbf{x}) \end{cases}, \tag{2.27}$$

where $L(\boldsymbol{\theta}|\mathbf{x})$, the likelihood function and γ and k are unique constants such that $E_{\boldsymbol{\theta}_0}[\varphi_{NP}(\mathbf{X})] = \alpha$, or

$$P^{\mathbf{X}|\boldsymbol{\theta}_0}(L(\boldsymbol{\theta}_1|\mathbf{x}) > kL(\boldsymbol{\theta}_0|\mathbf{x})) + \gamma P^{\mathbf{X}|\boldsymbol{\theta}_0}(L(\boldsymbol{\theta}_1|\mathbf{x}) = kL(\boldsymbol{\theta}_0|\mathbf{x})) = \alpha. \tag{2.28}$$

Then $\varphi_{NP}(\mathbf{x})$ is Most Powerful (MP) amongst all other test functions of level at most α.

Neyman-Peason states that in order to test a simple H_0 against a simple H_a we should use a rejection region of level α and the most powerful among them is

$$C(\mathbf{x}) = \{\mathbf{x} : L(\boldsymbol{\theta}_0|\mathbf{x})/L(\boldsymbol{\theta}_1|\mathbf{x}) \leq k\},$$

with k given by equation (2.28). Note that $T(\mathbf{X}) = L(\boldsymbol{\theta}_0|\mathbf{x})/L(\boldsymbol{\theta}_1|\mathbf{x})$, is a random variable and hence $C(\mathbf{X}) = \{\mathbf{X} : T(\mathbf{X}) \leq k\}$ is a random set.

2. GLRT When we cannot apply the Neyman-Peason lemma and we need a test function, we use the method of the Generalized Likelihood Ratio Test (GLRT). Let $L(\boldsymbol{\theta}|\mathbf{x}) = \prod_{i=1}^{n} f(x_i|\boldsymbol{\theta})$, the likelihood function based on a sample $X_1, \ldots, X_n$ from $f(x|\boldsymbol{\theta})$, $\boldsymbol{\theta} \in \Theta \subseteq \mathcal{R}^p$. In order to test $H_0 : \boldsymbol{\theta} \in \Theta_0$ vs $H_a : \boldsymbol{\theta} \in \Theta_a$, we define the likelihood ratio

$$\lambda(\mathbf{x}) = \frac{\max\limits_{\boldsymbol{\theta} \in \Theta_0} L(\boldsymbol{\theta}|\mathbf{x})}{\max\limits_{\boldsymbol{\theta} \in \Theta} L(\boldsymbol{\theta}|\mathbf{x})},$$

with $0 \leq \lambda \leq 1$, since $\Theta_0 \subseteq \Theta$. Now if H_0 is true, then $\boldsymbol{\theta}_0 \in \Theta_0$, which means that $\max\limits_{\boldsymbol{\theta} \in \Theta_0} L(\boldsymbol{\theta}|\mathbf{x})$ would be very close to $\max\limits_{\boldsymbol{\theta} \in \Theta} L(\boldsymbol{\theta}|\mathbf{x})$ and hence we should reject H_0 for small values of λ, that is, we have the rejection region

$$C = \{x : \lambda(\mathbf{x}) \leq k\},$$

where $P(\lambda(\mathbf{X}) \leq k|\theta_0) = \alpha$, if $\Theta_0 = \{\theta_0\}$ (simple H_0) or $P(\lambda(\mathbf{X}) \leq k|\theta_0) \leq \alpha$, $\forall \theta \in \Theta_0$, with equality for at least one $\theta \in \Theta_0$, when H_0 is composite. Clearly, the test depends on knowing the distribution of the test statistic $\lambda(\mathbf{x})$ under H_0.

If $\Theta_0 \subseteq \mathcal{R}^r \subseteq \Theta \subseteq \mathcal{R}^p$, then it can be shown that for large values of n, $-2\ln\lambda(\mathbf{X}) \sim \chi^2_{p-r}$, under H_0. Then $C = \{\mathbf{x} : -2\ln\lambda(\mathbf{x}) \geq -2\ln k\}$ and a large sample GLRT is based on

$$\varphi_{GLRT}(\mathbf{x}) = \begin{cases} 1 & -2\ln\lambda(\mathbf{X}) \geq \chi^2_{p-r;\alpha} \\ 0 & -2\ln\lambda(\mathbf{X}) < \chi^2_{p-r;\alpha} \end{cases},$$

where $\chi^2_{p-r;\alpha}$ is the α percentile of the chi-square distribution with $p-r$ degrees of freedom.

3. MLR Consider the model $f(\mathbf{x}|\theta)$, $\mathbf{x} \in \mathcal{R}^p$, $k \geq 1$, $\theta \in \Theta \subseteq \mathcal{R}$ and let $\mathcal{G} = \{f(\mathbf{x}|\theta) : \theta \in \Theta\}$. The family of densities $\mathcal{G}$ is said to have the property of Monotone Likelihood Ratio (MLR), with respect to a function $V : \mathcal{R}^p \to \mathcal{R}$, if (i) the support $\{\mathbf{x} \in \mathcal{R}^p : f(\mathbf{x}|\theta) > 0\}$ is independent of θ, (ii) for any two $\theta, \theta_1 \in \Theta$, $\theta \neq \theta_1$, $f(\mathbf{x}|\theta) \neq f(\mathbf{x}|\theta_1)$ a.e. and (iii) for all $\theta, \theta_1 \in \Theta$, $\theta < \theta_1$, the likelihood ratio $f(\mathbf{x}|\theta_1)/f(\mathbf{x}|\theta) = g_{\theta,\theta_1}(V(\mathbf{x}))$, where g is a strictly monotone function of V.

Now assume that the model $f(\mathbf{x}|\theta)$, $x \in \mathcal{X}$, has the MLR property with respect to $V(\mathbf{x})$. Then for $g_{\theta,\theta_1}(V(\mathbf{x}))$ increasing in $V(\mathbf{x})$, where $\theta < \theta_1$, the UMP test for $H_0 : \theta \leq \theta_0$ vs $H_a : \theta > \theta_0$, is given by

$$\varphi_{MLR}(\mathbf{x}) = \begin{cases} 1 & V(\mathbf{x}) > k \\ \gamma & V(\mathbf{x}) = k \\ 0 & V(\mathbf{x}) < k \end{cases}, \tag{2.29}$$

where γ and k are constants such that $E_{\theta_0}[\varphi_{MLR}(\mathbf{X})] = \alpha$, or

$$P^{\mathbf{X}|\theta_0}(V(\mathbf{x}) > k) + \gamma P^{\mathbf{X}|\theta_0}(V(\mathbf{x}) = k) = \alpha.$$

When $g_{\theta,\theta_1}(V(\mathbf{x}))$ is decreasing in $V(\mathbf{x})$, where $\theta < \theta_1$, the inequalities in equation (2.29) are reversed. To test $H_0 : \theta \geq \theta_0$ vs $H_a : \theta < \theta_0$, the UMP test function $\varphi_{MLR}(\mathbf{x})$ for increasing MLR in $V(\mathbf{x})$ is given by equation (2.29), with inequalities reversed for decreasing MLR in $V(\mathbf{x})$.

Example 2.13 (MLR for exponential family) Consider the exponential family

$$f(x|\theta) = c(\theta)e^{T(x)Q(\theta)}h(x),$$

where $\theta \in \Theta \subseteq \mathcal{R}$, $x \in \mathcal{X}$, $\mathcal{X}$ does not depend on θ and Q is a strictly monotone function. For a random sample $X_1, \ldots, X_n \sim f(x|\theta)$, the joint distribution is given by

$$f(\mathbf{x}|\theta) = c_0(\theta)e^{Q(\theta)V(\mathbf{x})}h^*(\mathbf{x}),$$

where $c_0(\theta) = c^n(\theta)$, $h^*(\mathbf{x}) = \prod_{i=1}^{n} h(x_i)$ and $V(\mathbf{x}) = \sum_{i=1}^{n} T(x_i)$. Then $\mathcal{G} = \{f(\mathbf{x}|\theta) : \theta \in \Theta\}$ has the MLR property with respect to V and

$$g_{\theta,\theta_1}(V(\mathbf{x})) = \frac{c_0(\theta_1)}{c_0(\theta)} \exp\{[Q(\theta_1) - Q(\theta)]V(\mathbf{x})\},\ \theta < \theta_1.$$

The UMP test function for testing $H_0 : \theta \leq \theta_1$ or $\theta \geq \theta_2$ vs $H_a : \theta_1 < \theta < \theta_2$, is given by

$$\varphi(\mathbf{x}) = \begin{cases} 1 & c_1 < V(\mathbf{x}) < c_2 \\ \gamma_i & V(\mathbf{x}) = c_i,\ i = 1, 2 \\ 0 & V(\mathbf{x}) < c_1 \text{ or } V(\mathbf{x}) > c_2 \end{cases},$$

where γ_i and c_i, $i = 1, 2$, are constants such that $E_{\theta_1}[\varphi(\mathbf{X})] = E_{\theta_2}[\varphi(\mathbf{X})] = \alpha$. Moreover, the UMP test exists for $H_0 : \theta \leq \theta_0$ vs $H_a : \theta > \theta_0$ and $H_0 : \theta \geq \theta_0$ vs $H_a : \theta < \theta_0$, but it does not exist for $H_0 : \theta_1 \leq \theta \leq \theta_2$ vs $H_a : \theta < \theta_1$ or $\theta > \theta_2$ and $H_0 : \theta = \theta_0$ vs $H_a : \theta \neq \theta_0$. Using equation (2.29) we can obtain the forms of the test functions by replacing the wording about the monotonicity of $g_{\theta,\theta_1}(V(\mathbf{x}))$ with the corresponding wording about monotonicity of $Q(\theta)$.

The following results tell us how to use test functions created for simple vs simple hypotheses, in order to assess simple vs composite or composite vs composite hypotheses.

Remark 2.19 (MP and UMP tests) The following results relate MP and UMP test functions for simple and composite hypotheses.

1. Let H_0 be simple, H_a a composite alternative, and $\varphi(\mathbf{x})$ the MP test function of level α for H_0 against a simple alternative $H_1 : f = f_a$, $f_a \in H_a$. Assume that $\varphi(\mathbf{x})$ is the same for all $f_a \in H_a$. Then $\varphi(\mathbf{x})$ is the UMP test of level α for H_0 vs the composite H_a.

2. Let H_0 and H_a be composite hypotheses. Assume that there exists $f_0 \in H_0$ and $\varphi(\mathbf{x})$, such that $\varphi(\mathbf{x})$ is the UMP test function of level α for H_0 against H_a. If φ is of level α for all $f_0 \in H_0$, then φ is also UMP for the composite H_0 vs H_a.

Example 2.14 (UMP and GLRT tests for a uniform distribution) Let $X_i|\theta \overset{iid}{\sim} Unif(0, \theta)$, $\theta > 0$, $i = 1, 2, \ldots, n$, with $0 \leq x_{(1)} \leq x_i \leq x_{(n)} \leq \theta$, so that the joint distribution can be written as

$$f(\mathbf{x}|\theta) = \theta^{-n} I_{[x_{(n)},+\infty)}(\theta).$$

We are interested first in the UMP test of level α for $H_0 : \theta = \theta_0$ vs $H_a : \theta \neq \theta_0$. Since $f(\mathbf{x}|\theta)$ has a domain that depends on θ, we have to use the Neyman-Pearson lemma to obtain the UMP. We have

$$\begin{aligned} L(\theta_0|\mathbf{x}) &= \theta_0^{-n} I_{[x_{(n)},+\infty)}(\theta_0), \text{ and} \\ L(\theta_a|\mathbf{x}) &= \theta_a^{-n} I_{[x_{(n)},+\infty)}(\theta_a),\ \theta_a \neq \theta_0, \end{aligned}$$

and the UMP rejection region is based on

$$L(\theta_0|\mathbf{x}) \leq k L(\theta_a|\mathbf{x}),$$

which leads to

$$\theta_a^n I_{[x_{(n)},+\infty)}(\theta_0) \leq k \theta_0^n I_{[x_{(n)},+\infty)}(\theta_a). \tag{2.30}$$

If $X_{(n)} > \theta_0$ then $L(\theta_0|\mathbf{x}) = 0 \leq k L(\theta_a|\mathbf{x})$, for all k and $\theta_a \neq \theta_0$ and in this case we should reject H_0 since equation (2.30) is satisfied always. If $X_{(n)} < C < \theta_0$, then equation (2.30) becomes $\theta_a^n \leq k\theta_0^n$ and hence we reject H_0, for $k \geq \theta_a^n/\theta_0^n$ and any $\theta_a \neq \theta_0$. Notice that when $k < \theta_a^n/\theta_0^n$ we cannot reject H_0. We need to find the constant C that would yield the level α rejection region. We have

$$\begin{aligned} \alpha &= P_{\theta_0}(\text{Reject } H_0) = P_{\theta_0}(X_{(n)} < C \text{ or } X_{(n)} > \theta_0) \\ &= P_{\theta_0}(X_{(n)} < C \text{ or } X_{(n)} > \theta_0) = F_{X_{(n)}}(C) + 1 - F_{X_{(n)}}(\theta_0), \end{aligned}$$

and using equation (2.12), we have $F_{X_{(n)}}(x|\theta_0) = x^n/\theta_0^n$, $0 \leq x \leq \theta_0$, under $H_0 : \theta = \theta_0$, which leads to $\alpha = C^n/\theta_0^n$ and hence $C = \theta_0\alpha^{1/n}$, so that $L(\theta_0|\mathbf{X}) \leq kL(\theta_a|\mathbf{X})$, for some k, if and only if $\mathbf{X} \in C = \{\mathbf{X} : X_{(n)} \leq \theta_0\alpha^{1/n}$ or $X_{(n)} > \theta_0\}$ and the rejection region does not depend on θ_a. Thus, in view of remark 2.19.2, the UMP test function of level α for testing $H_0 : \theta = \theta_0$ vs $H_a : \theta \neq \theta_0$, is given by

$$\varphi(\mathbf{x}) = \begin{cases} 1 & X_{(n)} \leq \theta_0\alpha^{1/n} \text{ or } X_{(n)} > \theta_0 \\ 0 & \theta_0\alpha^{1/n} < X_{(n)} \leq \theta_0 \end{cases}.$$

Note that this test is also UMPU. Now we discuss the GLRT for this problem. We have $\theta \in \Theta = (0, +\infty)$, with $H_0 : \theta \in \Theta_0 = \{\theta_0\}$ vs $H_a : \theta \in \Theta_a = \Theta \setminus \Theta_0$, with $\widehat{\theta} = X_{(n)}$, the MLE of $\theta \in \Theta$. Then the likelihood ratio for $X_{(n)} \leq \theta_0$, is given by

$$\lambda(\mathbf{x}) = \frac{\max\limits_{\theta \in \Theta_0} L(\boldsymbol{\theta}|\mathbf{x})}{\max\limits_{\theta \in \Theta} L(\boldsymbol{\theta}|\mathbf{x})} = \frac{L(\theta_0|\mathbf{x})}{L(\widehat{\theta}|\mathbf{x})} = (x_{(n)}/\theta_0)^n \, I_{[x_{(n)},+\infty)}(\theta_0),$$

and considering $\lambda(\mathbf{X}) \leq k$, for some k, leads to $X_{(n)} \leq k_1 = \theta_0 k^{1/n}$, while for $X_{(n)} > \theta_0$, $\lambda(\mathbf{x}) = 0 \leq k$, for all $k > 0$. Then the rejection region is of the form

$$C = \{\mathbf{x} : x_{(n)} \leq c \text{ or } x_{(n)} > \theta_0\},$$

and hence the GLRT is the same as the UMP and UMPU test.

2.5.2 Bayesian Testing Procedures

When we test $H_0 : \boldsymbol{\theta} \in \Theta_0$ vs $H_1 : \boldsymbol{\theta} \in \Theta_1$, we are entertaining two different models for θ, namely, $\mathcal{M}_i = \{f(\mathbf{x}|\boldsymbol{\theta}) : \boldsymbol{\theta} \in \Theta_i\}$, $i = 1, 2$. Bayesian testing procedures are based on posterior probabilities or the Bayes factor, which involves computing the relative odds between the two models, before and after we conduct the experiment.

Definition 2.9 Bayes factor

Let $a_0 = P(\Theta_0|\mathbf{x})$ and $a_1 = P(\Theta_1|\mathbf{x})$, denote the posterior (*a posteriori*) probabilities for model Θ_0 and Θ_1, respectively, and let $\pi_0 = P(\Theta_0)$, $\pi_1 = P(\Theta_1)$, the prior (*a priori*) probabilities of the two models. Then a_0/a_1 is called the posterior odds of H_0 to H_1 and π_0/π_1 the prior odds. Moreover, the quantity

$$B = \frac{\text{posterior odds}}{\text{prior odds}} = \frac{a_0/a_1}{\pi_0/\pi_1} = \frac{a_0\pi_1}{a_1\pi_0}, \tag{2.31}$$

is called the Bayes factor in favor of Θ_0.

Some comments are in order about the usage and basic interpretations of Bayes factors.

Remark 2.20 (Using the Bayes factor) If $a_0/a_1 = 2$ then H_0 is two times as likely to be true as H_1 is, after the experiment or when $\pi_0/\pi_1 = 2$, we believe *a priori* H_0 to be two times as likely to be true as H_1, before the experiment. Consequently, the Bayes factor assumes the interpretation as the odds for H_0 to H_1 that are given by the data and if $B > 1$, H_0 is more likely, if $B < 1$, H_1 is more likely, while if $B = 1$, we cannot favor either H_0 nor H_1.

1. Suppose that $\Theta_0 = \{\boldsymbol{\theta}_0\}$ and $\Theta_1 = \{\boldsymbol{\theta}_1\}$. Then we can easily show that

$$\begin{aligned} a_0 &= \frac{\pi_0 f(\mathbf{x}|\boldsymbol{\theta}_0)}{\pi_0 f(\mathbf{x}|\boldsymbol{\theta}_0) + \pi_1 f(\mathbf{x}|\boldsymbol{\theta}_1)}, \text{ and} \\ a_1 &= \frac{\pi_1 f(\mathbf{x}|\boldsymbol{\theta}_0)}{\pi_0 f(\mathbf{x}|\boldsymbol{\theta}_0) + \pi_1 f(\mathbf{x}|\boldsymbol{\theta}_1)}, \end{aligned}$$

so that

$$\frac{a_0}{a_1} = \frac{\pi_0 f(\mathbf{x}|\boldsymbol{\theta}_0)}{\pi_1 f(\mathbf{x}|\boldsymbol{\theta}_1)},$$

and the Bayes factor becomes

$$B = \frac{a_0 \pi_1}{a_1 \pi_0} = \frac{f(\mathbf{x}|\boldsymbol{\theta}_0)}{f(\mathbf{x}|\boldsymbol{\theta}_1)},$$

the usual likelihood ratio.

2. Consider testing $H_0 : \boldsymbol{\theta} \in \Theta_0$ vs $H_1 : \boldsymbol{\theta} \in \Theta_1$ and assume a priori

$$\pi(\boldsymbol{\theta}) = \begin{cases} \pi_0 g_0(\boldsymbol{\theta}), & \boldsymbol{\theta} \in \Theta_0 \\ \pi_1 g_1(\boldsymbol{\theta}), & \boldsymbol{\theta} \in \Theta_1 \end{cases}, \tag{2.32}$$

with $\pi_0 + \pi_1 = 1, 0 \leq \pi_0, \pi_1 \leq 1$ and g_0, g_1, proper densities describing how the prior mass is spread out over the two hypotheses. As a result, the *a priori* probabilities of the two models are given by

$$\begin{aligned} \int_{\Theta_0} \pi(\boldsymbol{\theta}) d\boldsymbol{\theta} &= \pi_0 \int_{\Theta_0} g_0(\boldsymbol{\theta}) d\boldsymbol{\theta} = \pi_0, \text{ and} \\ \int_{\Theta_1} \pi(\boldsymbol{\theta}) d\boldsymbol{\theta} &= \pi_1 \int_{\Theta_1} g_1(\boldsymbol{\theta}) d\boldsymbol{\theta} = \pi_1. \end{aligned}$$

which leads to the Bayes factor

$$B = \frac{\int_{\Theta_0} f(\mathbf{x}|\boldsymbol{\theta}) g_0(\boldsymbol{\theta}) d\boldsymbol{\theta}}{\int_{\Theta_1} f(\mathbf{x}|\boldsymbol{\theta}) g_1(\boldsymbol{\theta}) d\boldsymbol{\theta}}, \tag{2.33}$$

which is a weighted ratio of the likelihoods of Θ_0 and Θ_1.

Example 2.15 (Bayes factor for point null hypothesis testing) Consider testing $H_0 : \theta = \theta_0$ vs $H_1 : \theta \neq \theta_0$ and take the prior of equation (2.32), with $g_0(\theta) = 1$. Then the marginal distribution is given by

$$m(x) = \int_{\Theta} f(x|\theta)\pi(\theta)d\theta = f(x|\theta_0)\pi_0 + (1 - \pi_0)m_1(x),$$

with the marginal for $\theta \neq \theta_0$ given by

$$m_1(x) = \int_{\{\theta \neq \theta_0\}} f(x|\theta)g_1(\theta)d\theta,$$

and hence

$$\begin{aligned} a_0 &= P(\theta = \theta_0|x) = \frac{f(x|\theta_0)\pi_0}{m(x)} = \frac{f(x|\theta_0)\pi_0}{f(x|\theta_0)\pi_0 + (1-\pi_0)m_1(x)} \\ &= \left[1 + \frac{1-\pi_0}{\pi_0}\frac{m_1(x)}{f(x|\theta_0)}\right]^{-1}, \end{aligned}$$

and $a_1 = 1 - a_0$. Thus the posterior odds is given by

$$\frac{a_0}{a_1} = \frac{\pi(\theta_0|x)}{1-\pi(\theta_0|x)} = \frac{\pi_0}{1-\pi_0}\frac{f(x|\theta_0)}{m_1(x)},$$

and the Bayes factor is given by

$$B = f(x|\theta_0)/m_1(x).$$

An important application of the Bayes factor is in Bayesian model selection.

Remark 2.21 (Bayesian model selection) The Bayes factor can be used for model selection. Indeed, consider M possible models for $x|\theta$, denoted by $\mathcal{M}_i = \{f(x|\theta) : \theta \in \Theta_i\}$, $i = 1, 2, \ldots, M$, where $\{\Theta_i\}_{i=1}^M$ is a partition of the parameter space Θ. Now consider a mixture prior distribution with M components, given by

$$\pi(\theta|\lambda) = \sum_{i=1}^M \pi_i g_i(\theta|\lambda_i) I(\theta \in \Theta_i),$$

with $\sum_{i=1}^M \pi_i = 1$, $0 \le \pi_i \le 1$, $I(\theta \in \Theta_i)$ the indicator of the set Θ_i, g_i a proper prior density over Θ_i and $\lambda = [\lambda_1, \ldots, \lambda_M]'$ hyperparameters, so that the *a priori* probability of model $\mathcal{M}_i$ is given by

$$P(\theta \in \Theta_j|\lambda) = \int_{\Theta_j} \pi(\theta|\lambda)d\theta = \pi_j \int_{\Theta_j} g_j(\theta|\lambda_j)d\theta = \pi_j,$$

$j = 1, 2, \ldots, M$, since $\Theta_i \cap \Theta_j = \emptyset$, $i \neq j$. The *a posteriori* probability of model $\mathcal{M}_i$ is computed as

$$\begin{aligned} a_i &= P(\theta \in \Theta_i|x, \lambda) = \int_{\Theta_i} \frac{f(x|\theta)\pi(\theta|\lambda)}{m(x)} d\theta \qquad (2.34) \\ &= \frac{\pi_i \int_{\Theta_i} f(x|\theta) g_i(\theta|\lambda_i) d\theta}{m(x)} = \frac{m_i(x|\lambda_i)}{m(x|\lambda)}, \end{aligned}$$

where $m_i(x|\lambda_i) = \pi_i \int_{\Theta_i} f(x|\theta) g_i(\theta|\lambda_i) d\theta$, the marginal of X with respect to model $\mathcal{M}_i$, $i = 1, 2, \ldots, M$ and $m(x|\lambda) = \int_{\Theta} f(x|\theta)\pi(\theta|\lambda)d\theta$, the marginal distribution of X. In order to compare models $\mathcal{M}_i$ and $\mathcal{M}_j$ we compute the Bayes factor

$$B_{i,j} = \frac{a_i/a_j}{\pi_i/\pi_j} = \frac{\int_{\Theta_i} f(x|\theta) g_i(\theta|\lambda_i) d\theta}{\int_{\Theta_j} f(x|\theta) g_j(\theta|\lambda_j) d\theta} = \frac{m_i(x|\lambda_i)}{m_j(x|\lambda_j)}, \qquad (2.35)$$

for $i, j = 1, 2, \ldots, M$, $i \neq j$ and select model $\mathcal{M}_k$ using the following steps:

1. Compute all Bayes factors B_{ij} based on equation (2.35), $i, j = 1, 2, \ldots, M, i \neq j$

and create the matrix $B = [(B_{ij})]$, with $B_{ii} = 0$, $i = 1, 2, \ldots, M$ and $B_{ij} = \frac{1}{B_{ji}}$. Each row of this matrix compares $\mathcal{M}_i$ with every other model.

2. Start with the first row, $i = 1$ and find $j' = \arg \max\limits_{j=1,2,\ldots,M} B_{i,j}$.
3. If $B_{i,j'} > 1$, we choose model $\mathcal{M}_i$ as the most likely among all models and we are done. Otherwise, consider the next row $i = 2, 3, \ldots,$ and repeat the process of step 2.

Since the marginal distribution is typically intractable, we often turn to Monte Carlo in order to compute the Bayes factor, as we see next.

Remark 2.22 (Monte Carlo Bayes factor) The Bayes factor of equation (2.33) can be very hard, if not impossible, to compute in closed form. We turn to Monte Carlo simulation instead and approximate the marginal distributions $m_i(x)$, using

$$m_i(x) = \int\limits_{\Theta_i} f(x|\boldsymbol{\theta}) g_i(\boldsymbol{\theta}) d\theta \simeq \frac{1}{L_i} \sum_{j=1}^{L_i} f(x|\boldsymbol{\theta}_j^{(i)}),$$

for large L_i, where $\boldsymbol{\theta}_1^{(i)}, \ldots, \boldsymbol{\theta}_{L_i}^{(i)}$, realizations from $g_i(\boldsymbol{\theta})$, $i = 0, 1$, and the Bayes factor is approximated using

$$B_{ik} \simeq \frac{\frac{1}{L_i} \sum\limits_{j=1}^{L_i} f(x|\boldsymbol{\theta}_j^{(i)})}{\frac{1}{L_k} \sum\limits_{j=1}^{L_k} f(x|\boldsymbol{\theta}_j^{(k)})}, \tag{2.36}$$

$i, k = 0, 1$, which is easy to compute and requires realizations of the corresponding prior distributions $g_0(\boldsymbol{\theta})$ and $g_1(\boldsymbol{\theta})$. See remark 6.22 (p. 266) for a general model choice example.

We end this section with the idea behind predictive inference.

Remark 2.23 (Posterior predictive inference) Suppose that we want to predict a random variable Z with density $g(z|\theta)$ (θ unknown) based on data $X = x$, with $X \sim f(x|\theta)$, e.g., x is the data in a regression problem and Z the future response variable we need to predict in the regression setup. In order to accomplish this, we compute the predictive density of $Z|X = x$, when the prior distribution for θ is $\pi(\theta)$, as follows

$$p(z|x) = \int\limits_{\Theta} g(z|\theta) \pi(\theta|x) d\theta, \tag{2.37}$$

and all predictive inference is based on this density $p(z|x)$. Since we use the posterior distribution $\pi(\theta|x)$ in equation (2.37), the distribution is known also as posterior predictive, while if we use the prior $\pi(\theta)$ it is called prior predictive.

2.5.3 Decision Theoretic

We can unify both classical and Bayesian hypothesis testing methods by considering an appropriate decision theory problem for a specific choice of loss function.

Remark 2.24 (Decision theoretic hypothesis testing) Let $X_i|\boldsymbol{\theta} \overset{iid}{\sim} f(x|\boldsymbol{\theta})$, $\boldsymbol{\theta} \in \Theta \subseteq$

$\mathcal{R}^p$, $i = 1, 2, \ldots, n$, and define the loss incurred in choosing the test function δ, to test $H_0 : \boldsymbol{\theta} \in \Theta_0$ vs $H_1 : \boldsymbol{\theta} \in \Theta_0^c$, as

$$L(\boldsymbol{\theta}, \delta) = \begin{cases} 0, & \begin{matrix} \text{if } \boldsymbol{\theta} \in \Theta_0 \text{ and } \delta = 0 \\ \text{or } \boldsymbol{\theta} \in \Theta_0^c \text{ and } \delta = 1 \end{matrix} \\ L_1, & \text{if } \boldsymbol{\theta} \in \Theta_0 \text{ and } \delta = 1 \\ L_2, & \text{if } \boldsymbol{\theta} \in \Theta_0^c \text{ and } \delta = 0 \end{cases}, \tag{2.38}$$

for decision rules (test functions) of the form

$$\delta(\mathbf{x}) = \begin{cases} 1, & \mathbf{x} \in C \\ 0, & \mathbf{x} \notin C \end{cases}, \tag{2.39}$$

with C the rejection region and L_1, L_2 constants that indicate how much we lose for wrong decisions. To further appreciate equation (2.38) recall the schematic of Table 2.1. Turning to the calculation of the risk function under this loss we write

$$R(\boldsymbol{\theta}, \delta) = E^{\mathbf{x}|\boldsymbol{\theta}}[L(\boldsymbol{\theta}, \delta)] = L(\boldsymbol{\theta}, 1)P_{\boldsymbol{\theta}}(\mathbf{X} \in C) + L(\boldsymbol{\theta}, 0)P_{\boldsymbol{\theta}}(\mathbf{X} \notin C),$$

and thus

$$R(\boldsymbol{\theta}, \delta) = \begin{cases} L_1 P_{\boldsymbol{\theta}}(\mathbf{X} \in C), & \boldsymbol{\theta} \in \Theta_0, \\ L_2 P_{\boldsymbol{\theta}}(\mathbf{X} \in C^c), & \boldsymbol{\theta} \in \Theta_0^c. \end{cases}$$

1. Admissible test function To find the admissible test we need to find δ that minimizes $R(\boldsymbol{\theta}, \delta)$ which is equivalent to minimizing $P_{\boldsymbol{\theta}}(\mathbf{X} \in C)$ with respect to $\boldsymbol{\theta} \in \Theta_0$ and $P_{\boldsymbol{\theta}}(\mathbf{X} \in C^c)$ with respect to $\boldsymbol{\theta} \in \Theta_0^c$. The latter is equivalent to maximizing $P_{\boldsymbol{\theta}}(\mathbf{X} \in C)$ with respect to $\boldsymbol{\theta} \in \Theta_0^c$, i.e., the power of the decision rule.

2. Minimax and Bayes test functions In the case of point null vs point alternative, we have $\Theta = \{\boldsymbol{\theta}_0, \boldsymbol{\theta}_1\}$, $\Theta_0 = \{\boldsymbol{\theta}_0\}$, $\Theta_0^c = \{\boldsymbol{\theta}_1\}$, with $P_{\boldsymbol{\theta}_0}(\mathbf{X} \in C) = \alpha$ and $P_{\boldsymbol{\theta}_1}(\mathbf{X} \in C) = \beta$, the probabilities of the type I and II errors and the risk function assumes the form

$$R(\boldsymbol{\theta}, \delta) = \begin{cases} L_1\alpha, & \boldsymbol{\theta} = \boldsymbol{\theta}_0, \\ L_2\beta, & \boldsymbol{\theta} = \boldsymbol{\theta}_1. \end{cases}$$

Then in order to find the minimax test function we need to find the rule δ that minimizes maximum risk, i.e., δ is such that

$$\max\{R(\boldsymbol{\theta}_0, \delta), R(\boldsymbol{\theta}_1, \delta)\} \leq \max\{R(\boldsymbol{\theta}_0, \delta^*), R(\boldsymbol{\theta}_1, \delta^*)\},$$

for any other rule δ^*. Now consider the prior distribution

$$\pi(\boldsymbol{\theta}) = \begin{cases} p_0, & \boldsymbol{\theta} = \boldsymbol{\theta}_0, \\ p_1, & \boldsymbol{\theta} = \boldsymbol{\theta}_1, \end{cases} \tag{2.40}$$

with $p_0 + p_1 = 1$, $0 \leq p_0, p_1 \leq 1$. Then in order to obtain the Bayes test function we need to minimize the Bayes risk with respect to δ, namely, minimize the function

$$r(\pi, \delta) = R(\boldsymbol{\theta}_0, \delta)p_0 + R(\boldsymbol{\theta}_1, \delta)p_1.$$

More details on decision theoretic hypothesis testing can be found in Berger (1984) and Lehmann (1986). We summarize some of the ideas of this theory in the two theorems that follow.

Theorem 2.4 (Minimax test function) Let $X_i|\theta \overset{iid}{\sim} f(x|\theta)$, $\theta \in \Theta = \{\theta_0, \theta_1\}$, $i = 1, 2, \ldots, n$ and assume we wish to test $H_0 : \theta = \theta_0$ vs $H_a : \theta = \theta_1$. Suppose that there exists a constant k such that the region

$$C = \{\mathbf{x} \in R^n : f(\mathbf{x}|\theta_1) > kf(\mathbf{x}|\theta_0)\}, \tag{2.41}$$

satisfies

$$L_1 P_{\theta_0}(\mathbf{X} \in C) = L_2 P_{\theta_1}(\mathbf{X} \in C^c).$$

Then the test function of equation (2.39) is minimax and the level of the test is $\alpha = P_{\theta_0}(\mathbf{X} \in C)$.

Theorem 2.5 (Bayes test function) Let $X_i|\theta \overset{iid}{\sim} f(x|\theta)$, $\theta \in \Theta = \{\theta_0, \theta_1\}$, $i = 1, 2, \ldots, n$, let π denote the prior given in equation (2.40) and C the critical region as defined in equation (2.41) with $k = p_0 L_1/(p_1 L_0)$. Then the test function of equation (2.39) is the Bayes test function and the level of the test is $\alpha = P_{\theta_0}(\mathbf{X} \in C)$.

It is important to note that the decision theoretic procedures of finding test functions satisfy the following result.

Remark 2.25 (Bayes, minimax and MP test functions) From both theorems and in view of the Neyman-Pearson lemma we can see that the minimax and the Bayes test functions are also the MP test functions.

Example 2.16 (Bayes test function for normal mean) Assume that $X_i|\theta \overset{iid}{\sim} \mathcal{N}(\theta, 1)$, $\theta \in \Theta = \{\theta_0, \theta_1\}$, $i = 1, 2, \ldots, n$ and write the joint distribution as

$$f(\mathbf{x}|\theta) = (2\pi)^{-n/2} \exp\left\{-\frac{1}{2}\sum_{i=1}^{n}(x_i - \overline{x})^2\right\} \exp\left\{-n(\overline{x} - \theta)^2/2\right\},$$

so that $f(\mathbf{x}|\theta_1) > cf(\mathbf{x}|\theta_0)$, leads to $\overline{x} > c_0$, with $c_0 = (\theta_1+\theta_0)/2+(\ln c) / (n(\theta_1 - \theta_0))$. Then the minimax test function is $\delta(\mathbf{X}) = I_{(c_0,+\infty)}(\overline{X})$, with c_0 such that $L_1 P_{\theta_0}(\overline{X} > c_0) = L_2 P_{\theta_1}(\overline{X} \leq c_0)$, with $\overline{X} \sim \mathcal{N}(\theta, 1/n)$. The probability of the type I error is given by

$$\alpha = P_{\theta_0}(\overline{X} > c_0) = 1 - \Phi(\sqrt{n}(c_0 - \theta_0)),$$

while the power is computed as

$$P_{\theta_1}(\overline{X} > c_0) = 1 - \Phi(\sqrt{n}(c_0 - \theta_1)).$$

Now letting θ have the prior of equation (2.40), it is easy to see that the Bayes test function is $\delta(\mathbf{X}) = I_{(c_0,+\infty)}(\overline{X})$, with

$$c_0 = (\theta_1 + \theta_0)/2 + \ln(p_0 L_1/(p_1 L_0)) / (n(\theta_1 - \theta_0)).$$

2.5.4 Classical and Bayesian *p*-values

For the sake of presentation, we consider the hypotheses $H_o : \theta \leq \theta_o$ vs $H_1 : \theta > \theta_o$, based on a test statistic X, with $X \sim f(x|\theta)$, continuous. Based on a datum

$X = x_0$, the p-value is defined as

$$p(\theta_0, x_0) = P^{f(x|\theta)}(X \geq x_0|\theta = \theta_0) = \int_{x_o}^{\infty} f(x|\theta_0)dx,$$

which is a tail area probability, with respect to the model $f(x|\theta_0)$.

Bayesians, on the other hand, argue utilizing the classical p-value $p(\theta, x_0)$ and calculate the probability of the event $A(x_0) = \{X \geq x_o\}$, based on various distributions that depend on a prior distribution $\pi(\theta)$. More precisely, the prior predictive p-value is defined as

$$\rho_1 = E^{\pi(\theta)}[p(\theta, x_0)] = P^{m(x)}(X \geq x_o),$$

so that ρ_1 is the probability of the event $A(x_0)$, with respect to the marginal distribution $m(x)$.

The posterior predictive p-value is defined as

$$\rho_2 = E^{\pi(\theta|x_o)}[p(\theta, x_0)|X = x_o] = P^{m(x|x_0)}(X \geq x_o),$$

so that ρ_2 is the probability of the event $A(x_0)$, with respect to the predictive marginal distribution $m(x|x_0)$. The two measures of Bayesian surprise presented do not come without criticism with the most crucial being that improper priors cannot be used in most cases. There are other alternatives, including the conditional predictive p-value and the partial posterior predictive p-value (see remark 2.27).

Example 2.17 (Monte Carlo p-values) Computing the p-value for a statistical test can be easily accomplished via Monte Carlo. Letting T denote the test statistic for the testing procedure and assuming the alternative hypothesis is $H_1 : \theta > \theta_o$, the classical p-value can be approximated by

$$p = P^{f(x|\theta)}(T(X) \geq t_0|\theta = \theta_0) \simeq \frac{1}{L}\sum_{j=1}^{L} I(T(x_j) \geq t_0),$$

where $t_0 = T(x)$, the value of the test statistic based on observed data $X = x_0$ and $x_1, \ldots, x_L$, are the realizations from $f(x|\theta_0)$. This procedure is called a Monte Carlo test and can be extended to more general problems.

Next we discuss the ideas behind Monte Carlo tests.

Remark 2.26 (Monte Carlo tests) Assume that we replace $g(z|\theta)$ by $f(x|\theta)$ in remark 2.23, i.e., the predictive distribution is given by

$$p(x^{(p)}|x^{(s)}) = \int_{\Theta} f(x^{(p)}|\theta)\pi(\theta|x^{(s)})d\theta, \tag{2.42}$$

where $x^{(s)}$ denotes observed data from $f(x|\theta)$ and $x^{(p)}$ future values we need to predict from the model $f(x|\theta)$. Using Monte Carlo we can approximate the predictive distribution $p(x^{(p)}|x^{(s)})$ using

$$p(x^{(p)}|x^{(s)}) \simeq \frac{1}{L}\sum_{i=1}^{L} f(x^{(p)}|\theta_i),$$

based on a sample $\theta_1, \ldots, \theta_L$, from $\pi(\theta|x^{(s)}) \propto f(x^{(s)}|\theta)\pi(\theta)$, where $\pi(\theta)$ is a prior distribution for θ and L is a large positive integer. Setting L to 1, $p(x^{(p)}|x^{(s)}) \simeq$

$f(x^{(p)}|\theta_1)$, is obviously a bad approximation to the predictive distribution. However, generating a value $x_i^{(p)}$ from the model $f(x|\theta_i)$, $i = 1, 2, \ldots, L$, can be thought of as an approximate predictive sample of size L and hence any predictive inferential method can be based on these samples.

We discuss next the creation of a Monte Carlo test (as in example 2.17) for goodness-of-fit (gof) based on this idea. In order to assess the null hypothesis H_0 : the model $f(x|\theta)$, for some θ, describes the observed sample $\mathbf{x}^{(s)} = \{x_1^{(s)}, \ldots, x_n^{(s)}\}$ of size n well, against H_a : the model $f(x|\theta)$, is not appropriate for $\mathbf{x}^{(s)}$, for any θ, we choose a test statistic $T(\mathbf{x})$ and compute the predictive p-value

$$p_{pred} = P^{p(\mathbf{x}^{(p)}|\mathbf{x}^{(s)})}(T(\mathbf{X}^{(p)}) \geq T(\mathbf{x}^{(s)})) \simeq \frac{1}{L}\sum_{i=1}^{L} I(T(\mathbf{x}_i^{(p)}) \geq T(\mathbf{x}^{(s)})), \qquad (2.43)$$

based on predictive samples $\mathbf{x}_i^{(p)} = \{x_{i1}^{(p)}, \ldots, x_{in}^{(p)}\}$, $i = 1, 2, \ldots, L$, simulated from $p(\mathbf{x}_i^{(p)}|\mathbf{x}^{(s)}) = f(\mathbf{x}|\theta_i) = \prod_{j=1}^{n} f(x_j|\theta_i)$, where θ_i, $i = 1, 2, \ldots, L$, are realizations from the posterior distribution based on the sample $\mathbf{x}^{(s)}$, given by

$$\pi(\theta|\mathbf{x}^{(s)}) \propto \left[\prod_{j=1}^{n} f(x_j^{(s)}|\theta)\right]\pi(\theta).$$

It is important to note that by construction, this test is assessing all the important assumptions including choice of model $f(x|\theta)$ and choice of prior $\pi(\theta)$, as well as the independent and identically distributed assumption of the observed data $\mathbf{x}^{(s)}$. The choice of the test statistic is crucial in this procedure and some statistics will not be able to assess gof.

We can assess the usefulness of a test statistic by simulating the ideal situation and computing p_{pred}, based on ideal samples (namely, samples from $f(x|\theta)$ that will be treated as the observed data). If a test statistic yields small values for p_{pred}, say smaller than 0.1, they should be discarded, since in the ideal situation they cannot identify that $f(x|\theta)$ is the true model. Obviously, increasing n and L yields a better approximation to the true value of the predictive p-value.

Standard test statistics can be considered as candidates, e.g., sample means, variances or order statistics. Note that for any context in statistics we can construct Monte Carlo tests in a similar way.

The discussion above leads to the creation of the following general Monte Carlo test.

Algorithm 2.2 (Monte Carlo adequacy test) The general algorithm is as follows.
Step 1: Given a sample $\mathbf{x}^{(s)} = \{x_1^{(s)}, \ldots, x_n^{(s)}\}$ from $f(\mathbf{x}|\theta)$, simulate $\theta_i|\mathbf{x}^{(s)} \sim \pi(\theta|\mathbf{x}^{(s)})$, $i = 1, 2, \ldots, L$, for large L.
Step 2: For each $i = 1, 2, \ldots, L$, simulate $\mathbf{x}_i^{(p)} \sim f(\mathbf{x}|\theta_i)$, the predictive sample.
Step 3: Choose a statistic $T(\mathbf{x})$ and calculate the p-value p_{pred} of equation (2.43) and assess gof based on p_{pred}; if $p_{pred} \simeq 0$, we have strong evidence for reject-

ing H_0 and the model is not appropriate for modeling the observed data. When $p_{pred} > 0.1$ we have a good indication that the entertained model fits the data well.

Example 2.18 (Monte Carlo adequacy test) Suppose that we entertain a $Unif(0, \theta)$, $f(x|\theta) = I_{[0,\theta]}(x)$, model for a random variable X, where $\theta > 0$, unknown and consider a subjective prior $\theta \sim Exp(\lambda)$, with $\lambda > 0$, known, which leads to the posterior distribution $\pi(\theta|\mathbf{x}^{(s)}) \propto e^{-\lambda\theta} I_{[x_{(n)}^{(s)},+\infty)}(\theta)$, where $\mathbf{x}^{(s)} = \{x_1^{(s)}, \ldots, x_n^{(s)}\}$ denotes the observed data, with $x_{(n)}^{(s)} = \max\limits_{i=1,2,..,n} x_i^{(s)}$, so that $\theta - x_{(n)}^{(s)}|\mathbf{x}^{(s)} \sim Exp(\lambda)$. In order to build the Monte Carlo model adequacy test, consider the test statistic $T(\mathbf{X})$ and simulate a predictive sample $\mathbf{x}_i^{(p)} \sim f(\mathbf{x}|\theta_i) = \prod\limits_{j=1}^{n} f(x_j|\theta_i)$, based on simulated $\theta_i \sim \pi(\theta|\mathbf{x}^{(s)})$, $i = 1, 2, \ldots, L$. The Monte Carlo predictive p-value is given by

$$p_{pred} = P^{p(\mathbf{x}^{(p)}|\mathbf{x}^{(s)})}(T(\mathbf{X}^{(p)}) \geq T(\mathbf{x}^{(s)})) \simeq \frac{1}{L}\sum_{i=1}^{L} I(T(\mathbf{x}^{(p)}) \geq T(\mathbf{x}^{(s)})). \tag{2.44}$$

Simulation results are given in Table 2.2 (see MATLAB® procedure MCModelAdequacyex1 and Appendix A.7). We use $L = 100000$ predictive samples for $n = 10, 50, 100$, observed sample points. The observed data is simulated from three models $Unif(0, 1)$, $Gamma(10, 10)$ and $\mathcal{N}(-10, 1)$. The predictive p_{pred} are computed for four candidate test statistics: $T_1(\mathbf{X}) = X_{(1)}$, $T_2(\mathbf{X}) = X_{(n)}$, $T_3(\mathbf{X}) = \overline{X}$ and $T_4(\mathbf{X}) = S^2$, the sample variance.

We would expect a good test statistic to indicate gof only for the first model, $Unif(0, 1)$, that is, large predictive p_{pred}. Based on these results, T_1 emerges as the best test statistic to assess the entertained model, since it supports the true model for the data in all cases, while it rejects the data from the alternative models. Note that the choice of $\lambda = 1$ can potentially affect the results since it directly impacts an aspect of the whole procedure, the prior distribution.

The notion of p-value in the context of hypothesis testing has been widely used but seriously criticized for a long time in the literature. We collect some of these criticisms next.

Remark 2.27 (Criticisms about p-values) The major criticism regarding the p-value is whether it provides adequate evidence against a null hypothesis or a model. Hwang et al. (1992) present a brief summary of the controversy about the classical p-value.

Criticism about the p-value has come mainly from the Bayesian viewpoint since the calculation of a p-value almost always involves averaging over sample values that have not occurred, which is a violation of the likelihood principle. There is a vast literature and discussion papers on this topic, including Cox (1977), Shafer (1982), Berger and Delampady (1987), Casella and Berger (1987), Meng (1994), Gelman et al. (1996), De La Horra and Rodrıguez-Bernal (1997, 1999, 2000, 2001), Bayarri and Berger (1999), Micheas and Dey (2003, 2007), De la Horra (2005) and

Data Model	The MC test	$n = 10$	$n = 50$	$n = 100$
	should indicate:	p_{pred} for T_i	p_{pred} for T_i	p_{pred} for T_i
$Unif(0,1)$	gof	0.4067	0.5649	0.7348
		0.9231	0.9816	0.9905
		0.8895	0.9313	0.9647
		0.8847	0.9464	0.9880
$Gamma(10,10)$	not adequate fit	0.006	0	0
		0.0514	0.2386	0.3922
		0.1848	0.0004	0.0001
		0.9845	1.0000	1.0000
$Normal(-10,1)$	not adequate fit	0.0003	0.00049	0.00038
		0.00038	0.00049	0.00038
		0.00038	0.00049	0.00038
		0.00001	0.00001	0.00001

Table 2.2: Simulations of Monte Carlo goodness-of-fit tests. We used $L = 100000$ predictive samples for $n = 10, 50, 100$ observed sample points and the data is simulated from three models $Unif(0,1)$, $Gamma(10,10)$, and $N(-10,1)$. We choose $\lambda = 1$ for the hyperparameter, and p_{pred} is provided for four statistics $T_1(\mathbf{X}) = X_{(1)}$, $T_2(\mathbf{X}) = X_{(n)}$, $T_3(\mathbf{X}) = \overline{X}$, and $T_4(\mathbf{X}) = S^2$. Based on these results T_1 emerges as the best test statistic in order to assess the entertained model.

a host of others.

In the case of point null hypothesis, the use of a p-value is highly criticized from a Bayesian point of view, whereas for a one-sided hypothesis testing scenario, the use of a p-value is argued to be sensible (see Casella and Berger, 1987). Several leading Bayesians (e.g., Box, 1980, and Rubin, 1984) have argued that the p-value approach is equivalent to the calculation of a tail-area probability of a statistic, which is useful even for checking model adequacy. The tail area probability used by Box (1980) is called the prior predictive p-value.

On the contrary, Rubin (1984) defined the posterior predictive p-value by integrating the p-value with respect to the posterior distribution. Bayarri and Berger (1999) propose two alternatives, the conditional predictive p-value and the partial posterior predictive p-value which can be argued to be more acceptable from a Bayesian (or conditional) reasoning.

2.5.5 Reconciling the Bayesian and Classical Paradigms

We have already seen some cases where the classical and Bayesian approaches in statistical inference coincide, e.g., remark 2.11.3, while in other cases they cannot be reconciled, e.g., remark 2.12.5. Obviously, a statistical procedure would be preferred if it is the common solution of a classical and a Bayesian approach, since it would enjoy interpretations from both perspectives. We discuss how one can reconcile the classical and Bayesian paradigms in the context of hypothesis testing.

Classical p-values provide a basis for rejection of a hypothesis or a model and as such they should not be discarded. In fact, we can show that for many classes of prior distributions, the infimum of the posterior probability of a composite null hypothesis, or the prior and posterior predictive p-values, are equal to the classical p-value, for very general classes of distributions. In a spirit of reconciliation, Micheas and Dey (2003, 2007) investigated Bayesian evidence, like the prior and posterior predictive p-values, against the classical p-value for the one-sided location and scale parameter hypothesis testing problem, while De la Horra (2005), examined the two-sided case. The following example (Casella and Berger, 1987), motivates this discussion.

Example 2.19 (Reconciling Bayesian and frequentist evidence) Assume that $X|\theta \sim \mathcal{N}(\theta, \sigma^2)$ and $\theta \sim \pi(\theta) \propto 1$, so that $\theta|X \sim \mathcal{N}(x, \sigma^2)$ and consider testing $H_o : \theta \leq \theta_o$ vs $H_1 : \theta > \theta_o$. The posterior probability of the null hypothesis is simply

$$a_0 = P(\theta \leq \theta_o|x) = \Phi\left((\theta_0 - x)/\sigma\right),$$

while the classical p-value against H_0 is

$$p = P(X \geq x|\theta = \theta_0) = 1 - \Phi\left((x - \theta_0)/\sigma\right) = a_0,$$

so that the classical and Bayesian approaches lead to the same result.

2.6 Summary

Our review of statistical inference in this chapter was naturally separated into three main areas, including point and interval estimation and hypothesis testing. There are a great number of texts that have been published on these core subjects in statistics, with the treatment of each subject accomplished in various depths of mathematical rigor. A theoretical treatment of the classical perspective in probability theory, distribution theory and statistical inference can be found in the classic texts by Feller (1971), Lehmann (1986) and Lehmann and Casella (1998). An excellent treatment of the classical and Bayesian approaches, that should be accessible to readers with a rudimentary mathematical background, can be found in Casella and Berger (2002). Versions of all the results presented, along with their proofs, can be found in these texts. For a general presentation of the Bayesian perspective, the reader can turn to the books by Berger (1985), Bernardo and Smith

(2000), Press (2003), Gelman et al. (2004) and Robert (2007). Standard texts on asymptotics include Ferguson (1996) and Lehmann (2004).

Decision theory

The proof of each of the parts of the important remark 2.12, can be found in Lehmann and Casella (1998). Additional theorems and lemmas that provide necessary and/or sufficient conditions for admissibility, minimaxity or the Bayes and the extended Bayes properties, can be found in Berger (1985), Anderson (2003) and Robert (2007). Robustness with respect to choice of loss function has been studied in Zellner and Geisel (1968), Brown (1975), Varian (1975), Kadane and Chuang (1978), Ramsay and Novick (1980), Zellner (1986), Makov (1994), Dey et al. (1998), Martin et al. (1998), Dey and Micheas (2000), Rios Insua and Ruggeri (2000), Micheas and Dey (2004), Micheas (2006), Jozani et al. (2012) and the references therein. A proof of Blyth's lemma can be found in Berger (1985, p. 547).

Non-parametric statistical inference

All the methods presented in this chapter involve the definition of a parametric model $f(x|\theta)$ and hence all methods produced are called parametric. Non-parametric methods do not make an assumption about the form of the underlying model for the data. Many of the procedures we have discussed have non-parametric analogs, but they involve general characteristics of the underlying distribution in the absence of parameters, such as quantiles or moments.

In general, suppose we are given a random sample $X_1, \dots, X_n$, from a distribution $F \in \mathcal{F}$, where $\mathcal{F}$ is some abstract family of distributions, e.g., distributions that have first moments or are continuous and so on. We are interested in estimating a functional $g(F)$ of F, for example, $g(F) = E(X) = \int xdF(x)$, or $g(F) = P(X \leq a) = F(a)$, for given a. Most of the definitions we have seen in the parametric setting can be considered in a non-parametric setting as well, for example, we can find the UMVUE of a functional. Non-parametric statistical inference methods can be found in the texts by Lehmann (1986), Lehmann and Casella (1998), Robert (2007) and Wasserman (2009).

2.7 Exercises

Point and interval estimation

Exercise 2.1 Prove Basu's theorem (see remark 2.10.2).

Exercise 2.2 Let $X_i \stackrel{iid}{\sim} Unif(0, \theta)$, $i = 1, 2, \dots, n$. Show that $X_{(1)}/X_{(n)}$ and $X_{(n)}$ are independent. Show that the same holds for $\sum\limits_{i=1}^{n} X_i/X_{(n)}$ and $X_{(n)}$.

Exercise 2.3 Suppose that $X_i \stackrel{iid}{\sim} \mathcal{N}(\theta_1, \theta_2)$, $\theta_1 \in \mathcal{R}$, $\theta_2 > 0$, $i = 1, 2, \dots, n$.
(i) Find the MLE $\widehat{\boldsymbol{\theta}}$ of $\boldsymbol{\theta} = (\theta_1, \theta_2)^T$.
(ii) Find the unbiased estimator of $\mathbf{g}(\boldsymbol{\theta}) = (\theta_1, \theta_1^2 + \theta_2)^T$.
(iii) Find the Fisher Information matrix $I_X^F(\boldsymbol{\theta})$ and its inverse $[I_X^F(\boldsymbol{\theta})]^{-1}$.

(iv) Find the variance of the estimator in part (ii). Is this estimator UMVUE and why?
(v) Find the UMVUE of $\boldsymbol{\theta}$ based on the CR-LB.

Exercise 2.4 Assume that (X, Y) follows a bivariate normal distribution, with $E(X) = \mu = E(Y)$, $Var(X) = \sigma_X^2$ and $Var(Y) = \sigma_Y^2$. Obtain conditions so that $I_X^F(\mu) > I_{X|Y}^F(\mu)$, where $I_{X|Y}^F(\mu)$ is computed based on the conditional distribution of X given Y.

Exercise 2.5 Assume that $(X_1, X_2, X_3) \sim Multi(n, p_1, p_2, p_3)$, a trinomial random vector, where $p_1 = \theta$, $p_2 = \theta(1-\theta)$ and $0 < \theta < 1$. Obtain the MLE of θ. Note that $p(x_1, x_2, x_3) = C^n_{x_1,x_2,x_3} p_1^{x_1} p_2^{x_2} p_3^{x_3}, 0 < p_i < 1, p_1+p_2+p_3 = 1, x_1+x_2+x_3 = n$.

Exercise 2.6 Prove the factorization theorem (see remark 2.5.1).

Exercise 2.7 Assume that $X_1, X_2, \ldots, X_n$ is a random sample from $f(x|\theta)$, $\theta \in \Theta$. Give examples of the following:
(i) The MLE $\widehat{\theta}$ of θ is unique.
(ii) The MLE $\widehat{\theta}$ of θ does not exist.
(iii) An uncountable number of MLEs $\widehat{\theta}$ of θ exist.
(iv) The MLE $\widehat{\theta}$ of θ is not a function of only the complete-sufficient statistic.
(v) A member of the exponential family of distributions where the sufficient statistic $\mathbf{T}$ lives in $\mathcal{R}^p$, the parameter $\boldsymbol{\theta}$ lives in $\mathcal{R}^q$ and $p > q$.

Exercise 2.8 Prove the statement of remark 2.10.1.

Exercise 2.9 Suppose that $X_i \overset{iid}{\sim} Exp(a, b)$, $i = 1, 2, \ldots, n$, $a \in \mathcal{R}$, $b > 0$, (location-scale exponential) with $f(x|a, b) = \frac{1}{b}e^{-(x-a)/b}$, $x \geq a$.
(i) Find the MLE of $\boldsymbol{\theta} = (a, b)^T$.
(ii) Find the MLE of $g(\boldsymbol{\theta}) = E(\overline{X})$.
(iii) Calculate $E(\widehat{a})$ and $E(\widehat{b})$, where $\widehat{\boldsymbol{\theta}} = (\widehat{a}, \widehat{b})^T$.

Exercise 2.10 Prove the statement of remark 2.10.4 as part of the proof of the CR-LB.

Exercise 2.11 Assuming that $X_i \overset{iid}{\sim} N(\mu, \sigma^2)$, $\sigma > 0$, $\mu \in \mathcal{R}$, $i = 1, 2, \ldots, n$, find the UMVUE for $g_1(\mu, \sigma^2) = \mu\sigma^2$, $g_2(\mu, \sigma^2) = \mu/\sigma^2$ and $g_3(\mu, \sigma^2) = \mu^2/\sigma^2$.

Exercise 2.12 Let $X_i \overset{iid}{\sim} Unif(\theta, \theta + 1)$, $\theta \in \mathcal{R}$, $i = 1, 2, \ldots, n$.
(i) Show that $T(\mathbf{X}) = [X_{(1)}, X_{(n)}]^T$ is not complete.
(ii) Find the MLE of θ. Is it unique? If not, find the class of MLEs for θ. Are these estimators a function of the sufficient statistic?

Exercise 2.13 Show that if T_1 and T_2 are UMVUE for $g(\theta)$ then $T_1 = T_2$.

Exercise 2.14 Assume that $X_1, X_2, \ldots, X_n$ is a random sample from $f(x|\theta) = g(x)/h(\theta)$, $a \leq x \leq \theta$, where a is known and $g(.)$ and $h(.)$ are positive, real functions.
(i) Find the MLE $\widehat{\theta}$ of θ.
(ii) Find the distribution of $Q = h(\widehat{\theta})/h(\theta)$.
(iii) Find the UMVUE for θ.

(iv) Repeat the problem for the model $f(x|\theta) = g(x)/h(\theta)$, $\theta \leq x \leq b$, where b is known.
(v) Can you generalize these results for the family of distributions $f(x|\theta) = g(x)/h(\theta)$, $a(\theta) \leq x \leq b(\theta)$, for some functions $a(.)$ and $b(.)$?

Exercise 2.15 Assume that $X_1, X_2, \ldots, X_n$ is a random sample from $f(x|\theta) = g(x)/h(\theta)$, $a(\theta) \leq x \leq b(\theta)$, for some functions $a(.)$ and $b(.)$. If $\widehat{\theta}$ is the MLE for θ, show that the $100(1-\alpha)\%$ CI of minimum length for $h(\theta)$, based on the pivotal quantity $Q = h(\widehat{\theta})/h(\theta)$, is given by $[h(\widehat{\theta}), \alpha^{-\frac{1}{n}} h(\widehat{\theta})]$.

Exercise 2.16 Assume that a pivotal quantity $Q = Q(X, \theta)$ has a decreasing distribution $f_Q(q)$, $q > 0$. Show that the q_1 and q_2 in the minimum width $100(1-\alpha)\%$ CI for θ are given by $q_1 = 0$ and $\int_0^{q_2} f_Q(q)dq = 1 - \alpha$.

Exercise 2.17 Prove the statement of remark 2.9.1.

Exercise 2.18 Let $X_1, X_2, \ldots, X_n$ be a random sample from $f(x|\theta_1, \theta_2) = g(x)/h(\theta_1, \theta_2)$, $\theta_1 \leq x \leq \theta_2$, where $g(.)$ and $h(.)$ are positive, real functions.
(i) Show that $T(\mathbf{X}) = (X_{(1)}, X_{(n)})^T$, is the complete-sufficient statistic for $\boldsymbol{\theta} = (\theta_1, \theta_2)^T$.
(ii) Find the UMVUE for $h(\theta_1, \theta_2)$.

Exercise 2.19 Prove the Rao-Blackwell theorem (see remark 2.9.2).

Exercise 2.20 Suppose that $X_1, X_2, \ldots, X_n$ is a random sample from a $Poisson(\theta)$, with density $f(x|\theta) = e^{-\theta}\theta^x/x!$, $x = 0, 1, \ldots, \theta > 0$.
(i) If $n = 2$, show that $X_1 + 2X_2$ is not sufficient for θ.
(ii) Based on the whole sample, find the UMVUE of θ^2.
(iii) Based on the whole sample, find the general form of an unbiased estimator $U(\mathbf{X})$ of $e^{-b\theta}$, where b is a known constant.

Exercise 2.21 Prove the Lehmann-Scheffé theorem (see remark 2.9.3).

Exercise 2.22 Assume that $X_1, X_2, \ldots, X_n$ is a random sample from $f(x|\theta) = c(\theta)e^{T(x)\theta}h(x)$, $x \in A$, free of θ and $\theta > 0$.
(i) Find the MLE of $g(\theta) = c'(\theta)/c(\theta)$, where $c'(.)$ denotes a derivative with respect to θ.
(ii) Find an unbiased estimator of θ based on the sample.
(iii) Show that the Fisher information is given by $I_X^F(\theta) = -\frac{d^2 \ln c(\theta)}{d\theta^2}$.

Exercise 2.23 Prove the CR-LB inequality (see remark 2.9.4).

Exercise 2.24 Consider two independent samples $X_1, X_2, \ldots, X_m$ and $Y_1, \ldots, Y_n$ according to $Exp(a, b)$ and $Exp(a_1, b_1)$, respectively.
(i) If a, b, a_1 and b_1 are all unknown, find the sufficient and complete statistic.
(ii) Find the UMVUE estimators of $a_1 - a$ and b_1/b.

Exercise 2.25 Let $X_1, X_2, \ldots, X_n$ be a random sample from $f(x|\theta) = a(\theta)b(x)I_{(0,\theta]}(x)$, $\theta > 0$, $b(x) \neq 0$ and $a(\theta) \neq 0$, $\forall\, \theta > 0$.
(i) Find the Complete-Sufficient statistic for θ.

(ii) Show that the UMVUE of $1/a(\theta)$ is $T(\mathbf{X}) = \frac{n+1}{n}B(X_{(n)})$, where $B(x) = \int_0^x b(t)dt$, $0 \leq x \leq \theta$.
(iii) Show that $Q = a(\theta)B(X_{(n)})$ is pivotal for θ and identify its distribution.
(iv) Obtain the $100(1-\alpha)\%$ minimum width CIs for $a(\theta)$. (Hint: Use the general method based on a function of the sufficient statistic.)

Exercise 2.26 Assume that $X_1, X_2, \ldots, X_n$, is a random sample from $f(x|\theta) = c(\theta)h(x)e^{x\theta}$, $x \in A \subset \mathcal{R}$ and A is free of $\theta \in \mathcal{R}$.
(i) Find the mgf of X_i.
(ii) Find the complete and sufficient statistic.
(ii) Find the MLE $\widehat{\theta}$ of θ. What assumptions do you need to impose on $c(\theta)$ and $h(x)$ in order for $\widehat{\theta}$ to exist and be unique?

Exercise 2.27 Assuming that $X_1, X_2, \ldots, X_n$ is a random sample from $\mathcal{N}(\theta, \theta)$, $\theta > 0$, show that $\overline{X}$ is not sufficient for θ and obtain the sufficient statistic for θ.

Exercise 2.28 Let $X_1, X_2, \ldots, X_n$ be a random sample from $p(x|\theta) = c(\theta)\theta^x/h(x)$, $x = 0, 1, 2, \ldots$, $\theta \in \Theta$ (Power series family of distributions), for some non-negative functions $c(.)$ and $h(.)$.
(i) Find the complete-sufficient statistic for θ.
(ii) Obtain the MLE of θ and discuss assumptions on $c(\theta)$ and $h(x)$ for its existence and uniqueness.
(iii) Find $E(X)$.

Exercise 2.29 Suppose that $X_1, X_2, \ldots, X_n$ is a random sample from $f(x|\theta, \sigma) = \sigma e^{-(x-\theta)\sigma}$, $x \geq \theta$, $\sigma > 0$ and $\theta \in \mathcal{R}$.
(i) Obtain the UMVUE for θ, with σ known.
(ii) Show that $Q = X_{(1)} - \theta$, is pivotal for θ, when σ is known.
(iii) Assuming θ is fixed, obtain the UMVUE of $1/\sigma$ using the Cramér-Rao Inequality.
(iii) Assuming σ is fixed, obtain the $100(1-\alpha)\%$ minimum width and equal tail CIs and compare them. For what value of σ will they be equal?

Exercise 2.30 Assuming that $X_i \overset{iid}{\sim} Exp(\theta)$, $\theta > 0$, $i = 1, \ldots, n$, find the UMVUE of θ^r, $r > 0$.

Exercise 2.31 Let $X_i \overset{iid}{\sim} N(\theta, a)$, $a > 0$, $\theta \in \mathcal{R}$, with a known, $i = 1, \ldots, n$.
(i) Obtain the UMVUE of θ^2.
(ii) Show that the CR-LB is not attained in this case.

Exercise 2.32 If $X_1, \ldots, X_n \overset{iid}{\sim} B(1, p)$, find the complete-sufficient statistic T for p and the UMVUE $\delta(T)$ of p^m, $m \leq n$.

Exercise 2.33 Suppose that $X_i \overset{iid}{\sim} Unif(\theta, a)$, $a \in \mathcal{R}$, $\theta \leq a$, with a known, $i = 1, \ldots, n$. Find $Q = Q(\mathbf{X}, \theta)$, a pivotal quantity for θ and obtain the $100(1-\alpha)\%$ equal tail and minimum width CIs based on Q and compare them. Solve the problem for $X_i \overset{iid}{\sim} Unif(a, \theta)$, $a \in \mathcal{R}$, $a \leq \theta$, with a known, $i = 1, \ldots, n$.

Exercise 2.34 Let $X_1, X_2, \ldots, X_n$ be a random sample from $\mathcal{N}(0, \theta^2)$, $\theta > 0$.
(i) Find a pivotal quantity for θ that is a function of $\sum_{i=1}^{n} X_i$.
(ii) Find the $100(1-\alpha)\%$ equal tail CI for θ.

Exercise 2.35 Let X be a random variable with $f(x|\theta) = \theta(1+x)^{-(1+\theta)}$, $x > 0$ and $\theta > 0$.
(i) Obtain a pivotal quantity for θ.
(ii) Obtain the $100(1-\alpha)\%$ equal tail CI for θ based on the pivotal quantity from part (i).
(iii) Obtain the $100(1-\alpha)\%$ minimum width CI for θ based on the pivotal quantity from part (i) and compare the result with the equal tail CI from part (ii).

Exercise 2.36 Let X be a random variable with $f(x|\theta) = |x-\theta|$, $\theta - 1 \leq x \leq \theta + 1$, $\theta \in \mathcal{R}$.
(i) Show that $Q = X - \theta$ is a pivotal quantity for θ.
(ii) Obtain the $100(1-\alpha)\%$ equal tail CI for θ based on the pivotal quantity from part (i).
(iii) Obtain the $100(1-\alpha)\%$ minimum width CI for θ based on the pivotal quantity from part (i) and compare the result with the equal tail CI from part (ii).

Exercise 2.37 Assume that $X_1, X_2, \ldots, X_n$ is a random sample from the distribution $f(x|\theta) = g(x)/\theta^2$, $0 \leq x \leq \theta$, with $\theta > 0$.
(i) Find a pivotal quantity for θ^2.
(ii) Obtain the $100(1-\alpha)\%$ minimum width CI for θ^2.

Exercise 2.38 Consider $X \sim f(x|\theta) = \frac{1}{2\theta}e^{-\frac{1}{\theta}|x|}$, $x \in \mathcal{R}$, $\theta > 0$, a Laplace density. First show that $Q = \frac{X}{\theta}$ is pivotal for θ and then find the equal tail and minimum width CIs if they exist. Otherwise, find a CI for θ and comment on the behavior of this distribution.

Exercise 2.39 Suppose that $X_i \overset{iid}{\sim} f(x|\theta)$, $\theta \in \Theta$, $x \in \mathcal{X} \subseteq \mathcal{R}$, $i = 1, 2, \ldots, n$, with $\theta \sim \pi(\theta)$, *a priori* and that the resulting posterior distribution $\pi(\theta|\mathbf{x})$ is unimodal. Show that the $100(1-\alpha)\%$ HPD credible set is also the minimum width credible set for the given α.

Exercise 2.40 Consider $X_i \overset{iid}{\sim} Unif(0, \theta)$, $\theta > 0$, $i = 1, 2, \ldots, n$ and assume *a priori* $\theta \sim \pi(\theta|k, x_0) = kx_0^k/\theta^{k+1}$, $\theta \geq x_0$, with x_0, k, known.
(i) Find the $100(1-\alpha)\%$ HPD credible set.
(ii) Show that $Q = X_{(n)}/\theta$ is pivotal.
(iii) Compute the $100(1-\alpha)\%$ minimum width CI based on Q and compare it with the HPD set from part (i).

Exercise 2.41 Assume that $X_1, X_2, \ldots, X_n$ is a random sample from $\mathcal{N}(\theta, 1)$ and assume we entertain a $\mathcal{N}(\mu, 1)$ prior for θ, with $\mu \in \mathcal{R}$, known.
(i) Derive the posterior distribution of $\theta|\mathbf{x}$, where $\mathbf{x} = (x_1, \ldots, x_n)^T$ is the observed sample.
(ii) Find the general form of the $100(1-\alpha)\%$ HPD credible set for θ.

Exercise 2.42 Let $f(x)$ be a unimodal pdf and assume that the interval $[L, U]$ is such that: a) $\int_L^U f(x)dx = 1 - \alpha$, b) $f(L) = f(U) > 0$, and c) $L \leq x^* \leq U$, where x^* is the mode of $f(x)$. Show that $[L, U]$ is the shortest among all intervals that satisfy a).

Hypothesis testing

Exercise 2.43 Assume that $X \sim HyperGeo(N = 6, n = 3, p)$ (see Appendix A.11.4 for the pmf of the hypergeometric random variable). Find the MP test function for testing $H_0 : p = \frac{1}{3}$ vs $H_a : p = \frac{2}{3}$, and apply it for $\alpha = 0.05$ and $X = 1$.

Exercise 2.44 Suppose that $X_i \overset{iid}{\sim} Exp(\frac{1}{\theta}), \theta > 0, i = 1, 2, \ldots, n$.
(i) Find the UMP test for $H_0 : \theta \geq \theta_0$ vs $H_a : \theta < \theta_0$ of level α.
(ii) Find the smallest value of the sample size n in order for the power of the test in (i) to be at least 95%, when $\theta_0 = 1000$ and $\alpha = 0.05$ and the alternative has value $\theta_1 = 500$.

Exercise 2.45 Prove the Neyman-Pearson fundamental lemma (remark 2.18.1).

Exercise 2.46 Suppose that (X_i, Y_i), $i = 1, 2, \ldots, n$, is a random sample from $f(x, y|\lambda, \mu) = \lambda\mu e^{-\lambda x - \mu y}$, $x, y > 0$, $\lambda, \mu > 0$.
(i) Test the hypothesis $H_0 : \lambda \geq k\mu$ vs $H_a : \lambda < k\mu$ at level α, where k is some known constant.
(ii) What are the properties of the test you found in (i)? (UMP or UMPU or GLRT?)

Exercise 2.47 Let $X_i \overset{iid}{\sim} f(x|\theta)$, $x \in \mathcal{X}$, $i = 1, 2, \ldots, n$, with $\theta \in \Theta = \{\theta_0, \theta_1\}$. Define the Hellinger divergence between $f(x|\theta_0)$ and $f(x|\theta_1)$ by

$$\rho = \rho(\theta_0, \theta_1) = \int_{\mathcal{X}} \sqrt{f(x|\theta_0) f(x|\theta_1)} dx. \tag{2.45}$$

(i) Show that $0 \leq \rho \leq 1$ with $\rho = 1$ if and only if $f(x|\theta_0) = f(x|\theta_1)$.
(ii) Let $\varphi(\mathbf{x})$ denote the test function of the GLRT for testing $H_0 : \theta = \theta_0$ vs $H_a : \theta = \theta_1$, defined as

$$\varphi(\mathbf{x}) = \begin{cases} 1 & \lambda(\mathbf{x}) = \frac{L(\theta_1|\mathbf{x})}{L(\theta_0|\mathbf{x})} > c \\ 0 & \lambda(\mathbf{x}) < c \end{cases},$$

where $\lambda(\mathbf{x}) = \frac{L(\theta_1|\mathbf{x})}{L(\theta_0|\mathbf{x})}$, $L(\theta|\mathbf{x})$ is the likelihood function and let α and β denote the probabilities of the type I and II errors. Show that

$$\sqrt{c}\alpha + \frac{1}{\sqrt{c}}\beta \leq \rho^n.$$

(iii) Find ρ under the models: $Exp(\lambda)$, $\mathcal{N}(\mu, 1)$ and $Poisson(\lambda)$ (replace the integral sign in 2.45 with the summation sign for the discrete model).

Exercise 2.48 Prove the statement of remark 2.19.1.

Exercise 2.49 Suppose that $X_i \overset{iid}{\sim} Exp(\frac{1}{\theta_1})$, $\theta_1 > 0$, $i = 1, 2, \ldots, m$, independent of $Y_j \overset{iid}{\sim} Exp(\frac{1}{\theta_2})$, $\theta_2 > 0$, $j = 1, 2, \ldots, n$.
(i) Obtain the GLRT test function for $H_0 : \theta_1 = \theta_2$ vs $H_a : \theta_1 \neq \theta_2$ of level α, and

show that it is a function $\frac{\overline{X}}{\overline{Y}}$.
(ii) Find the distribution of $\frac{\overline{X}}{\overline{Y}}$.

Exercise 2.50 Let $X_1, X_2, \ldots, X_n$ be a random sample from $p(x|\theta) = c(\theta)\theta^x/h(x)$, $x = 0, 1, 2, \ldots$, where $\theta \in \Theta$.
(i) Use the Neyman-Pearson lemma to derive the form of the randomized level α UMP test function for testing $H_0 : \theta = \theta_0$ vs $H_a : \theta > \theta_0$.
(ii) Under what conditions has this family the MLR property? Obtain the UMP test for $H_0 : \theta = \theta_0$ vs $H_a : \theta > \theta_0$ (using MLR theory) and compare with that in part (i).

Exercise 2.51 Consider $X_1, X_2, \ldots, X_n$ a random sample from $f(x|\theta) = c(\theta)e^{T(x)\theta}h(x)$, $x \in A$, free of θ and $\theta > 0$.
(i) For what assumptions does f have the MLR property?
(ii) Find the UMP test function for $H_o : \theta \leq \theta_o$ vs $H_a : \theta > \theta_o$ and a form for its power.

Exercise 2.52 Suppose that $X_1, X_2, \ldots, X_n$ is a random sample from $\mathcal{N}(\mu, \sigma^2)$.
(i) Find the UMPU test function for $H_o : \mu = \mu_o$ vs $H_a : \mu \neq \mu_o$, σ^2 unknown.
(ii) Derive the UMP test function for $H_o : \sigma^2 = \sigma_o^2$ vs $H_a : \sigma^2 < \sigma_o^2$, μ unknown.
(iii) What is the power of the test in part (ii)?

Exercise 2.53 Prove the statement of remark 2.19.2.

Exercise 2.54 Let $X_i \stackrel{iid}{\sim} Unif(\theta, a)$, $a \in \mathcal{R}$, $\theta \leq a$, with a known, $i = 1, \ldots, n$. Derive the GLRT for $H_0 : \theta = \theta_0$ vs $H_a : \theta \neq \theta_0$. Solve the problem for $X_i \stackrel{iid}{\sim} Unif(a, \theta)$, $a \in \mathcal{R}$, $a \leq \theta$, with a known, $i = 1, \ldots, n$.

Exercise 2.55 Consider the linear model $\mathbf{Y} = X\boldsymbol{\beta} + \boldsymbol{\varepsilon}$, $\boldsymbol{\varepsilon} \sim \mathcal{N}_n(\mathbf{0}, \sigma^2 I_n)$, $rank(X_{n\times p}) = p$, where $\boldsymbol{\beta} = \begin{bmatrix} \underset{q\times 1}{\boldsymbol{\beta}_1} \\ \underset{r\times 1}{\boldsymbol{\beta}_2} \end{bmatrix}$, $p = q + r$. Obtain the GLRT for the hypothesis $H_o : \underset{s\times q}{A}\boldsymbol{\beta}_1 = \underset{s\times r}{B}\boldsymbol{\beta}_2$ vs $H_a : \underset{s\times q}{A}\boldsymbol{\beta}_1 \neq \underset{s\times r}{B}\boldsymbol{\beta}_2$, for some matrices A and B.

Exercise 2.56 Suppose that $X_1, X_2, \ldots, X_n \stackrel{iid}{\sim} \mathcal{N}(\mu_x, nm)$ and $Y_1, Y_2, \ldots, Y_m \stackrel{iid}{\sim} \mathcal{N}(\mu_y, nm)$, two independent samples. Derive the form of the power, of the level α test for $H_o : \mu_x = \mu_y$ vs $H_a : \mu_x = \mu_y + 1$, that rejects H_o when $\overline{X} - \overline{Y} \geq c$, where c is some constant to be found.

Exercise 2.57 Assume that $X_1, X_2, \ldots, X_n$ is a random sample from an exponential distribution with location parameter θ, i.e., $f(x|\theta) = e^{-(x-\theta)}$, $x \geq \theta \in \mathcal{R}$. Consider testing $H_o : \theta \leq \theta_o$ vs $H_a : \theta > \theta_o$. Find the UMP test function of level α and the classical p-value.

Exercise 2.58 Suppose that $X_i \stackrel{iid}{\sim} Gamma(a, \theta)$, $i = 1, 2, \ldots, n$, where $\theta > 0$ is the scale parameter and $\alpha > 0$ is a known shape parameter. Show that this gamma has the MLR property and obtain the UMP test function of level α for testing $H_o : \theta \geq \theta_o$ vs $H_a : \theta < \theta_o$. In addition, find the classical p-value.

Exercise 2.59 Consider a random sample $X_1, X_2, \ldots, X_n$ from a Pareto distribution with pdf $f(x|\theta) = a\theta^a / x^{a+1}$, $x \geq \theta$, scale parameter $\theta > 0$ and known shape parameter $\alpha > 0$. For testing $H_o : \theta \leq \theta_o$ vs $H_a : \theta > \theta_o$, find the classical p-value, the prior and the posterior predictive p-value assuming a priori that $\theta \sim Exp(\lambda)$, with $\lambda > 0$, known. Compare the three measures of evidence against H_0.

Exercise 2.60 Let $X_i \overset{iid}{\sim} Unif(-\theta, \theta)$, $\theta > 0$, $i = 1, 2, \ldots, n$ and consider a prior for $\theta \sim Gamma(a, b)$, with $a > 2$, $b > 0$, known. Find the Bayes factor for testing $H_o : 0 < \theta \leq \theta_o$ vs $H_a : \theta > \theta_o$.

Exercise 2.61 Consider testing a point null $H_0 : \theta = \theta_0$ vs $H_0 : \theta > \theta_0$, based on a random sample $X_1, \ldots, X_n \sim f(x|\theta)$, $\theta > 0$, using a prior from the $\varepsilon-$contamination class $\Gamma_\varepsilon = \{\pi : \pi(\theta) = (1-\varepsilon)\delta_{\theta_0}(\theta) + \varepsilon\pi_{a,b}(\theta)\}$, where $\delta_{\theta_0}(\theta) = I(\theta = \theta_0)$, a point mass distribution at θ_0 (Dirac distribution) and $\pi_{a,b}$ is a $Gamma(a, b)$, density, a, $b > 0$. Find the general form of the Bayes factor, the prior and the posterior predictive p-values.

Decision theoretic

Exercise 2.62 Suppose that $X_i \overset{iid}{\sim} Exp(\theta)$, $i = 1, 2, \ldots, n$. Consider the loss function $L(\theta, \delta) = \theta^{-2}(\theta - \delta)^2$ (WSEL) and define the class of decision rules $\delta_k = \delta_k(\mathbf{X}) = 1/\overline{X}_k$, $\overline{X}_k = \frac{1}{k}\sum_{i=1}^{k} X_i$, $k = 1, 2, \ldots, n$.
(i) Obtain the (frequentist) risk in estimating θ using the above δ_k and under $L(\theta, \delta)$, for any $k = 1, 2, \ldots, n$.
(ii) Find the admissible estimator of θ in the class of decisions δ_k, $k = 1, 2, \ldots, n$.
(iii) Consider the prior distribution on θ given by $\pi(\theta) = \lambda e^{-\lambda\theta}$, $\theta > 0$ and λ a fixed positive constant. Obtain the Bayes rule for θ with respect to this prior.
(iv) Will the two estimators you found in parts (ii) and (iii) ever be equal?

Exercise 2.63 Prove theorem 2.4.

Exercise 2.64 Suppose that $X_1, X_2, \ldots, X_n$ is a random sample from $Poisson(\lambda)$, where $\lambda \sim Gamma(a, b)$, with a, $b > 0$, known. Find the Bayes rule for λ under the weighted square error loss function $L(\lambda, \delta) = \lambda^2(\lambda - \delta)^2$.

Exercise 2.65 Suppose that $T(X)$ is unbiased for $g(\theta)$, with $Var(T(X)) > 0$. Show that $T(X)$ is not Bayes under SEL for any prior π.

Exercise 2.66 Assume that $X_1, X_2, \ldots, X_n$ are independent where $X_i \sim f(x|\theta_i)$. Show that, for $i = 1, 2, \ldots, n$, if $\delta_i^{\pi_i}(X_i)$ is the Bayes rule for estimating θ_i using loss $L_i(\theta_i, a_i)$ and prior $\pi_i(\theta_i)$, then $\delta^{\pi}(\mathbf{X}) = (\delta_1^{\pi_1}(X_n), \ldots, \delta_n^{\pi_n}(X_n))$ is a Bayes rule for estimating $\boldsymbol{\theta} = (\theta_1, \ldots, \theta_n)$ using loss $\sum_{i=1}^{n} L_i(\theta_i, a_i)$ and prior $\pi(\boldsymbol{\theta}) = \prod_{i=1}^{n} \pi_i(\theta_i)$.

Exercise 2.67 Prove theorem 2.5.

Exercise 2.68 Suppose that $\mathbf{X} = (X_1, \ldots, X_p)' \in \mathcal{X} \subseteq \mathcal{R}^p$ is distributed according to the exponential family: $f(\mathbf{x}|\boldsymbol{\theta}) = c(\boldsymbol{\theta})e^{\boldsymbol{\theta}'\mathbf{x}}h(\mathbf{x})$, with $h(\mathbf{x}) > 0$, on $\mathcal{X}$. The parameter $\boldsymbol{\theta} \in \Theta = \{\boldsymbol{\theta} : \int_{X} h(\mathbf{x})e^{\boldsymbol{\theta}'\mathbf{x}}d\mathbf{x} < \infty\}$, is to be estimated under quadratic loss $L(\boldsymbol{\theta}, \delta) =$

$(\boldsymbol{\theta}-\boldsymbol{\delta})'W(\boldsymbol{\theta}-\boldsymbol{\delta})$, for some weight matrix W.
(i) Show that the (generalized) Bayes rule with respect to a (possibly improper) prior π, is given by

$$\delta^{\pi}(\mathbf{x}) = E(\boldsymbol{\theta}|\mathbf{x}) = \frac{1}{m(\mathbf{x}|\pi)} \int_{\Theta} \boldsymbol{\theta} f(\mathbf{x}|\boldsymbol{\theta})\pi(\boldsymbol{\theta})d\boldsymbol{\theta} = \nabla \ln m(\mathbf{x}|\pi) - \nabla \ln h(\mathbf{x}),$$

where $m(\mathbf{x}|\pi) = \int_{\Theta} f(\mathbf{x}|\boldsymbol{\theta})\pi(\boldsymbol{\theta})d\boldsymbol{\theta}$, the marginal with respect to π. What assumptions are needed to help you prove this?
(ii) Suppose that $\mathbf{X} \sim \mathcal{N}_p(\boldsymbol{\theta}, \Sigma)$, with $\Sigma > 0$, known. Find the form of $\delta^{\pi}(\mathbf{x})$ in part (i).

Exercise 2.69 Consider $X \sim \mathcal{N}(\theta, 1)$ and we wish to estimate θ under SEL. Use Blyth's lemma to show that $\delta(x) = x$ is admissible.

Exercise 2.70 Assume that $X_i \overset{iid}{\sim} \mathcal{N}(\theta_i, 1)$, $i = 1, 2, \ldots, p$, $p > 2$ and define an estimator $\boldsymbol{\delta}_c = (\delta_{1c}, \ldots, \delta_{pc})'$ of $\boldsymbol{\theta} = (\theta_1, \ldots, \theta_p)$, by $\delta_{ic} = \left(1 - c(p-2)/S^2\right) X_i$, where $S^2 = \sum_{i=1}^{p} X_i^2$.
(i) Show that the risk of $\delta_c(\mathbf{x})$ under the average SEL function $L(\boldsymbol{\theta}, \mathbf{d}) = \frac{1}{p}\sum_{i=1}^{p}(d_i - \theta_i)^2$ is $R(\boldsymbol{\theta}, \boldsymbol{\delta}_c) = 1 - \frac{(p-2)^2}{p} E\left(\left(2c - c^2\right)/S^2\right)$.
(ii) Show that the estimator $\delta_c(\mathbf{X})$ dominates $\delta_0(\mathbf{X}) = \mathbf{X}$ (for $c = 0$), provided that $0 < c < 2$ and $r \geq 3$.
(iii) Show that the James-Stein estimator $\delta_1(\mathbf{X}) = \delta^{JS}(\mathbf{X})$ (for $c = 1$), dominates all estimators δ_c, with $c \neq 1$.

Exercise 2.71 Let δ be the Bayes rule (UMVUE, minimax, admissible) for $g(\theta)$ under SEL. Show that $a\delta + b$ is the Bayes rule (respectively, UMVUE, minimax, admissible) for $ag(\theta) + b$, where a, b are real constants.

Exercise 2.72 Prove all parts of remark 2.12.

Exercise 2.73 Assume that $X \sim \mathcal{N}(\theta, 1)$ and we wish to estimate θ under absolute value loss, $L(\theta, a) = |\theta - a|$. Show that $\delta(x) = x$ is admissible.

Exercise 2.74 Show that a Bayes estimator is always a function of the complete-sufficient statistic.

Exercise 2.75 Assume that $\mathbf{X}_i \overset{iid}{\sim} \mathcal{N}_p(\boldsymbol{\mu}, I_p)$, $i = 1, 2, \ldots, n$ and consider estimation of $\boldsymbol{\mu}$ under quadratic loss: $L(\boldsymbol{\mu}, \boldsymbol{\delta}) = (\boldsymbol{\mu} - \boldsymbol{\delta})^T(\boldsymbol{\mu} - \boldsymbol{\delta}) = \|\boldsymbol{\mu} - \boldsymbol{\delta}\|^2$.
(i) Find the risk function for $\boldsymbol{\delta}(\mathbf{X}) = \overline{\mathbf{X}} = \frac{1}{n}\sum_{i=1}^{n} \mathbf{X}_i$.
(ii) Show that $\overline{\mathbf{X}}$ is not admissible, $p \geq 3$.
(iii) Suppose that $\boldsymbol{\mu} \sim \mathcal{N}_p(\mathbf{0}, I_p)$. Find the Bayes rule for $\boldsymbol{\mu}$ and compare its frequentist risk with that in part (i).

Exercise 2.76 Prove Blyth's lemma.

Asymptotics

Exercise 2.77 Prove theorem 2.1.

Exercise 2.78 Assuming that $X \sim Binom(n, p)$, show that $T_n = (X+1)/(n+2) \overset{p}{\rightarrow} p$.

Exercise 2.79 Suppose that $\{X_n\}$ is an iid sequence of random variables with $P(X_n = k/n) = k/n, k = 1, 2, \ldots, n$. Show that $X_n \overset{w}{\rightarrow} X \sim Unif(0, 1)$.

Exercise 2.80 Assume that $X_i \overset{iid}{\sim} Unif(0, \theta), \theta > 0, i = 1, 2, \ldots, n$. Show that the MLE of θ is a consistent estimator.

Exercise 2.81 Prove theorem 2.2.

Exercise 2.82 Prove the usual Central limit theorem (CLT), namely, let $\mathbf{X}_1, \mathbf{X}_2 \ldots,$ be iid random vectors with mean $\boldsymbol{\mu}$ and finite covariance matrix, $\boldsymbol{\Sigma}$, and prove that $\mathbf{Y}_n = \sqrt{n}(\overline{\mathbf{X}}_n - \boldsymbol{\mu}) \overset{w}{\rightarrow} \mathcal{N}_p(\mathbf{0}, \boldsymbol{\Sigma})$, where $\overline{\mathbf{X}}_n = \frac{1}{n}\sum_{i=1}^{n} \mathbf{X}_i$. Hint: Use Taylor's theorem (appendix remark A.1.8) on the cf of $\mathbf{Y}_n$.

Exercise 2.83 Prove theorem 2.3.

Simulation and computation: use your favorite language to code the functions

Exercise 2.84 Let $U \sim Unif(0, 1)$. Use simulation to approximate: a) $Cov(U, \sqrt{1 - U^2})$, b) $Cov(U^2, \sqrt{1 - U^2})$, c) $Cov(U, e^U)$.

Exercise 2.85 Use the Monte Carlo approach to approximate the integrals: a) $\int_{-\infty}^{+\infty} exp(-x^2)dx$, b) $\int_{0}^{+\infty} x(1 + x^2)^{-2}dx$, c) $\int_{-2}^{2} exp(x + x^2)dx$, d) $\int_{0}^{+\infty} e^{-x/2}dx$, e) $\int_{-\infty}^{0}\int_{-\infty}^{1} e^{-x^2-y^2}dxdy$, and f) $\int_{1}^{10} \frac{x}{\Gamma(x)}dx$.

Exercise 2.86 Write functions that would:
(i) generate a random sample from $\mathcal{N}(\mu, \sigma^2)$,
(ii) compute the MLE values of the parameters and then the values of the likelihood and the log-likelihood.
(iii) Run your code for the models: $\mathcal{N}(0, 1)$, $\mathcal{N}(10, 9)$ and comment on the results.

Exercise 2.87 Let $X_i \overset{iid}{\sim} Exp(\theta), i = 1, 2, \ldots, n$. Write a function that would:
(i) take as argument data $X_1, \ldots, X_n$ and compute and print the MLE of θ (in the console). Run your function for $n = 20$ realizations from an $Exp(10)$ distribution.
(ii) take as argument data $X_1, \ldots, X_n$ and constants $a, b > 0$, and compute and display the MAP of θ under a *Gamma(a, b)* prior. Run your function for $n = 10$ realizations from an $Exp(5)$ distribution and $a = 2, b = 3$.

Exercise 2.88 Now consider $X_i \overset{iid}{\sim} Gamma(a, b), i = 1, 2, \ldots, n, a, b > 0$ unknown. Compute the MLEs and then the values of the likelihood and the log-likelihood. Use the Newton-Raphson method if you cannot get the MLEs in closed form (recall remark 2.7). Run the code for the models $Gamma(a = 2, b = 1)$ and $Gamma(a = 10, b = 10)$, by first simulating a sample of size $n = 100$ and then pass the generated values to your function.

Exercise 2.89 Write a function that would approximate and plot the prior and posterior predictive distribution (recall remark 2.26 and use equation (2.42)) for a model $X \sim \mathcal{N}(\mu, \sigma^2)$, with σ^2 known, and prior $\mu \sim \mathcal{N}(0, 1)$. Produce the plots for $\sigma^2 = 1, 2, 3$.

Exercise 2.90 Using the setup of exercise 2.59, write a function that would compute and return the classical p-value, the prior and the posterior predictive p-value, as well as return the power of the test at some θ_1. Run the function for $n = 50$ realizations from a Pareto model with $a = 5$, $\theta = 10$, and assume that $\theta_o = 9$, $\theta_1 = 9.5$, with the prior having hyperparameter $\lambda = 0.1$. Comment on the results.

Exercise 2.91 Write a function that computes the HPD credible set of example 2.10. Generate a sample of size $n = 100$ from an $Exp(\theta = 5)$, let $\lambda = 1$, and plot the posterior along with two vertical lines at the values of θ that give the $100(1-\alpha)\%$ HPD and a horizontal line at the k^* level. Use $\alpha = 0.1, 0.05$, and 0.01 when you run the function (make α an argument of the function).

Exercise 2.92 Using the setup of exercise 2.60, write a function that computes the classical p-value, the prior and posterior predictive p-values and the Bayes factor. Generate and pass to your function $n = 100$ realizations from a $Unif(-10, 10)$, $\theta = 10$, $\pi_0 = \pi_1 = \frac{1}{2}$, and consider a prior with $a = 10$, $b = 1$. Use $\theta_0 = 8$ and find the power of the test at $\theta_1 = 9$. Comment on your findings and draw conclusions with respect to the hypothesis at hand.

Exercise 2.93 Let $X_i \overset{iid}{\sim} Binom(m, p)$, $i = 1, 2, \dots, n$, and consider a prior $p \sim Beta(a, b)$. Write a function that would take as arguments the data $X_1, \dots, X_n$ the hyper-parameters $a, b > 0$ and would perform the following tasks:
(i) Plot the posterior distribution in solid line and the prior distribution in a dashed line on the same plot.
(ii) Obtain the $100(1-\alpha)\%$ CS for p given some α.
(iii) Obtain the $100(1-\alpha)\%$ HPD CS for p given some α.
(iv) Obtain the Bayes factor for testing $H_0 : p = \frac{1}{2}$ vs $H_1 : p \neq \frac{1}{2}$ (assume $p \sim Beta(a, b)$ for $p \neq \frac{1}{2}$).
(v) Run your function for the following cases and report on your findings: (a) $n = 20$ generated data from $Binom(5, .45)$, $a = 2$, $b = 2$, (b) $n = 10$ generated data from $Binom(10, .95)$, $a = 1$, $b = 1$, and (c) $n = 100$ generated data from $Binom(20, .5)$, $a = 2$, $b = 1$.

Exercise 2.94 Consider an iid sample $X_1, \dots, X_n$ from a $\mathcal{N}(0, 3)$, and write code that would help you conduct a Monte Carlo goodness-of-fit test based on two test statistics, the sample mean and variance (see remark 2.26 and example 2.18). Run your code for data generated from four models: $\mathcal{N}(0, 1)$, $\mathcal{N}(0, 3)$, $\mathcal{N}(0, 4)$ and $\mathcal{N}(0, 9)$ and report on your findings (use $n = 30$).

Chapter 3

Measure and Integration Theory

3.1 Introduction

Measure theory provides tools that allow us to efficiently quantify or measure sets of points, where by point here we could mean anything, from real vectors, to matrices and functions, to points that are collections of points themselves. A measure on a set provides a systematic way to assign a number to each suitable subset of that set, intuitively interpreted as its size, and consequently, we can think of a measure as a generalization of the concepts of length, area and volume.

The exposition of this chapter is mathematically demanding, requiring knowledge of topology, real analysis, advanced calculus and integration theory in order to fully comprehend the underlying structures of the spaces, σ-fields, functions and integrals involved. Consequently, as we study material from this chapter, it would help us to have some classic texts from mathematics nearby (such as Rubin, 1984, Royden, 1989, Dudley, 2002 and Vestrup, 2003).

3.2 Deterministic Set Theory

We begin with some results on deterministic sets. We collect basics of deterministic mappings in Appendix A.1.

Remark 3.1 (Deterministic set theory basics) Let $\mathcal{X}$, $\mathcal{Y}$ be some spaces and consider subsets $A \subseteq \mathcal{X}$ and $B \subseteq \mathcal{X}$. Denote by $\emptyset$ the empty set (set with no points from $\mathcal{X}$) and $A^c = \{x \in \mathcal{X} : x \notin A\}$, the complement of A.

1. The Cartesian product of A and B is defined as $A \times B = \{(x, y) : x \in A, y \in B\}$. We easily generalize to the n-fold product by defining $\mathcal{X}^n = \mathcal{X} \times \cdots \times \mathcal{X} = \{(x_1, \ldots, x_n) : x_i \in \mathcal{X}, i = 1, 2, \ldots n\}$.

2. The symmetric difference between the sets A and B is defined by $A \triangle B = (A \setminus B) \cup (B \setminus A)$, where $A \setminus B = \{x \in \mathcal{X} : x \in A \text{ and } x \notin B\} = A \cap B^c$.

3. Let C be a collection of sets from $\mathcal{X}$. Then we define

$$\underset{A \in C}{\cap} A = \cap\{A : A \in C\} = \{x \in \mathcal{X} : if A \in C \Longrightarrow x \in A\},$$

and

$$\underset{A\in C}{\cup} A = \cup\{A : A \in C\} = \{x \in \mathcal{X} : \exists A \text{ such that } A \in C \text{ and } x \in A\},$$

and consequently, we can show that

$$\left(\underset{A\in C}{\cup} A\right)^c = \underset{A\in C}{\cap} A^c \text{ and } \left(\underset{A\in C}{\cap} A\right)^c = \underset{A\in C}{\cup} A^c.$$

4. If $A_i \subseteq \mathcal{X}$, $i = 1, 2, \ldots$, then we define

$$\bigcup_{i=1}^{+\infty} A_i = \{x \in \mathcal{X} : \exists i \text{ such that } x \in A_i\} \text{ and } \bigcap_{i=1}^{+\infty} A_i = \{x \in \mathcal{X} : x \in A_i, \forall i\}.$$

5. A nonempty collection $\mathcal{P}$ of subsets of a nonempty space $\mathcal{X}$ is called a π–system over $\mathcal{X}$ if and only if $A, B \in \mathcal{P} \Longrightarrow A \cap B \in \mathcal{P}$, that is, $\mathcal{P}$ is closed under intersections. Note that this implies that $\mathcal{P}$ is closed under finite intersections.

6. A nonempty collection $\mathcal{L}$ of subsets of a nonempty space $\mathcal{X}$ is called a λ–system over $\mathcal{X}$ if and only if: (i) $\mathcal{X} \in \mathcal{L}$, (ii) if $A \in \mathcal{L} \Longrightarrow A^c \in \mathcal{L}$, (iii) for every disjoint sequence $\{A_n\}_{n=1}^{+\infty}$, of $\mathcal{L}$-sets, we have $\bigcup\limits_{i=1}^{+\infty} A_i \in \mathcal{L}$, that is, $\mathcal{L}$ is closed under countable disjoint unions.

7. We denote by $2^{\mathcal{X}}$ the collection of all subsets of the space $\mathcal{X}$.

8. Note that the empty set $\emptyset$ need not be a member of a collection of sets, even if by definition $\emptyset \subseteq \mathcal{X}$.

9. Assume that $A_1, A_2, \ldots,$ is a sequence of sets from $\mathcal{X}$ and define the limit inferior and superior by

$$\overline{\lim} A_n = \lim\sup_n A_n = \bigcap_{n=1}^{+\infty} \bigcup_{k=n}^{+\infty} A_k$$

and

$$\underline{\lim} A_n = \lim\inf_n A_n = \bigcup_{n=1}^{+\infty} \bigcap_{n=k}^{+\infty} A_k.$$

Then $\lim\limits_{n\to\infty} A_n$ exists when

$$\lim_{n\to\infty} A_n = \underline{\lim} A_n = \overline{\lim} A_n.$$

We attach the additional interpretation to $\overline{\lim} A_n$ by saying A_n infinitely often (i.o.) and to $\underline{\lim} A_n$ by saying A_n eventually (ev.).

10. Set-theoretic operations on events are easily represented using indicator functions. In particular, we have the following: $I_{A^c} = 1 - I_A$, $I_{A \smallsetminus B} = I_A(1 - I_B)$, $I_{A\cap B} = I_A \wedge I_B = I_A I_B$, $I_{A\cup B} = I_A \vee I_B = I_A + I_B - I_A I_B$, $I_{\cap_n A_n} = \inf\limits_n I_{A_n} = \prod\limits_n I_{A_n}$, $I_{\cup_n A_n} = \sup\limits_n I_{A_n}$ and $I_{A \triangle B} = I_A + I_B - 2 I_A I_B$. In terms of indicator functions we may rewrite the limit definition of a sequence of sets as $\lim\limits_{n\to\infty} A_n = A$ if and only if $I_A = \lim\limits_{n\to\infty} I_{A_n}$.

A first useful collection of subsets from a space $\mathcal{X}$ is defined next.

Definition 3.1 Field

A collection $\mathcal{A}$ of subsets of $\mathcal{X}$ is called a field (or algebra) of sets or simply a field, if (i) $\emptyset \in \mathcal{A}$, (or $\mathcal{X} \in \mathcal{A}$), (ii) $A^c \in \mathcal{A}, \forall A \in \mathcal{A}$, and (iii) $A \cup B \in \mathcal{A}, \forall A, B \in \mathcal{A}$.

Some comments are in order regarding this important definition.

Remark 3.2 (Field requirements) Requirement (i) guarantees that $\mathcal{A}$ is nonempty. Some texts do not require it since it is implied by (ii) and (iii), when there is at least one set in the collection $\mathcal{A}$. Indeed, if $A \in \mathcal{A}, \mathcal{X} = A \cup A^c \in \mathcal{A}$, so that $\mathcal{X} \in \mathcal{A}$ and $\emptyset^c = \mathcal{X} \in \mathcal{A}$. Noting that $(A^c \cup B^c)^c = A \cap B$, for any $A, B \in \mathcal{A}$, we see that $\mathcal{A}$ is a π−system.

Next we collect an existence theorem regarding a minimal field based on a collection of subsets of a space $\mathcal{X}$.

Theorem 3.1 (Generated field) Given a collection C of subsets of a space $\mathcal{X}$, there exists a smallest field $\mathcal{A}$ such that $C \subseteq \mathcal{A}$, i.e., if there exists a field $\mathcal{B}$ such that $C \subseteq \mathcal{B}$, then $\mathcal{A} \subseteq \mathcal{B}$.

Proof. Take $\mathcal{F}$ as the family of all fields that contain C and define $\mathcal{A} = \cap\{\mathcal{B} : \mathcal{B} \in \mathcal{F}\}$. ■

3.3 Topological Spaces and σ-fields

The collections of subsets from a space $\mathcal{X}$ that will allow us to efficiently assign numbers to their elements are known as σ-fields.

Definition 3.2 σ-field

A collection $\mathcal{A}$ of subsets of a space $\mathcal{X}$ is called a σ-field if
(i) $\emptyset \in \mathcal{A}$ (or $\mathcal{X} \in \mathcal{A}$),
(ii) $A^c \in \mathcal{A}, \forall A \in \mathcal{A}$, and
(iii) every countable collection of sets from $\mathcal{A}$, is in $\mathcal{A}$, i.e., $\forall A_i \in \mathcal{A}, i = 1, 2, \ldots,$ we have $\bigcup\limits_{i=1}^{+\infty} A_i \in \mathcal{A}$.
A σ-field $\mathcal{A}_0$ that satisfies $\mathcal{A}_0 \subseteq \mathcal{A}$, where $\mathcal{A}$ is a σ-field, is called a sub-σ-field.

As we will see later in this chapter, the σ-fields we end up using in our definitions are typically generated σ-fields based on a smaller collection of easy-to-use sets. The following is analogous to the concept of a generated field and is given as a definition since there always exists a σ-field containing any collection of sets C, the σ-field $2^{\mathcal{X}}$.

Definition 3.3 Generated σ-field

The smallest σ-field $\mathcal{A}$ containing a collection of sets C is called the generated σ-field from C and is denoted by $\mathcal{A} = \sigma(C)$. We may also write $\mathcal{A} = \bigcap_{C \subseteq \mathcal{B}} \{\mathcal{B} : \mathcal{B} \text{ a } \sigma - field\}$.

Let us collect some comments and important results on σ-fields.

Remark 3.3 (Properties of σ-fields) We note the following.

1. The generated σ-field for a collection of sets C exists always, since there exists a σ-field that contains all sets of $\mathcal{X}$, namely, $2^{\mathcal{X}}$, the largest σ-field of subsets of $\mathcal{X}$.

2. Sub-σ-fields play a very important role in probability theory since we can attach to them the interpretation of partial information. See for example Section 6.3.1 and the idea behind filtrations.

3. We can easily show that $C \subseteq \sigma(C)$, for any collection C of subsets of $\mathcal{X}$. By definition, $C = \sigma(C)$ when C is a σ-field.

4. If C_1 and C_2 are collections of subsets of X then $C_1 \subseteq C_2 \Longrightarrow \sigma(C_1) \subseteq \sigma(C_2)$. To see this, note that

$$\sigma(C_1) = \bigcap_{C_1 \subseteq \mathcal{B}} \{\mathcal{B} : \mathcal{B} \text{ a } \sigma\text{-field}\} \subseteq \bigcap_{C_2 \subseteq \mathcal{B}} \{\mathcal{B} : \mathcal{B} \text{ a } \sigma\text{-field}\} = \sigma(C_2),$$

since in the second intersection we intersect over a smaller number of σ-fields (since $C_1 \subseteq C_2$) and thus we obtain larger collections of sets.

5. If $C_1 \subseteq \sigma(C_2)$ and $C_2 \subseteq \sigma(C_1)$ then $\sigma(C_1) = \sigma(C_2)$. This provides us with an elegant way of showing that two σ-fields are equal. To prove this take $C_1 \subseteq \sigma(C_2)$ and use part 4 of the remark to obtain $\sigma(C_1) \subseteq \sigma(\sigma(C_2)) = \sigma(C_2)$, with the latter equality the result of part 3 of the remark. Similarly, the assumption $C_2 \subseteq \sigma(C_1)$ leads to $\sigma(C_2) \subseteq \sigma(C_1)$, which establishes the other direction, so that $\sigma(C_1) = \sigma(C_2)$.

6. Let $C = \{\mathcal{G}_k : k \in K\}$ be a collection of sub-σ-fields of a σ-field $\mathcal{G}$. The intersection of all elements of C is easily shown to be a σ-field with $\bigcap_{k \in K} \mathcal{G}_k = \{G : G \in \mathcal{G}_i, \text{ for all } i \in K\}$, whereas the union, $\bigcup_{k \in K} \mathcal{G}_k = \{G : G \in \mathcal{G}_i, \text{ for some } i \in K\}$, is not necessarily a σ-field. We write $\sigma(\mathcal{G}_1, \mathcal{G}_2, \dots)$ for $\sigma\left(\bigcup_{k \in K} \mathcal{G}_k\right)$, the generated σ-field based on all members of the collection C.

7. Dynkin's $\pi - \lambda$ theorem. Many uniqueness arguments can be proven based on Dynkin's $\pi - \lambda$ theorem: if $\mathcal{P}$ is a π-system and $\mathcal{L}$ is a λ-system, then $\mathcal{P} \subseteq \mathcal{L}$ implies $\sigma(\mathcal{P}) \subseteq \mathcal{L}$.

8. Let $f : \mathcal{X} \to \mathcal{Y}$, some map and C a collection of subsets of $\mathcal{Y}$ and define the inverse map of the collection C by $f^{-1}(C) = \{E \subset \mathcal{X} : E = f^{-1}(C)$ for some $C \in C\}$. Then $f^{-1}(\sigma(C))$ is a σ-field on $\mathcal{X}$ and $\sigma(f^{-1}(C)) = f^{-1}(\sigma(C))$.

Example 3.1 (Union of σ-fields) Consider the space $\mathcal{X} = \{a, b, c\}$ and let $C = \{\{a\}, \{b\}, \{c\}\}$, the collection of all singletons of $\mathcal{X}$. Then $2^{\mathcal{X}} = \sigma(C) = \{\emptyset, \mathcal{X}, \{a\}, \{b\}, \{c\}, \{a, b\}, \{a, c\}, \{b, c\}\}$, is the largest σ-field of subsets of $\mathcal{X}$. Now let $\mathcal{A}_1 = \sigma(\{a\}) = \{\emptyset, \mathcal{X}, \{a\}, \{b, c\}\}$ and $\mathcal{A}_2 = \sigma(\{b\}) = \{\emptyset, \mathcal{X}, \{b\}, \{a, c\}\}$, two sub-$\sigma$-fields of $2^{\mathcal{X}}$. Then $\mathcal{A}_1 \cup \mathcal{A}_2 = \{\emptyset, \mathcal{X}, \{a\}, \{b\}, \{b, c\}, \{a, c\}\}$, with $\{a\} \in \mathcal{A}_1 \cup \mathcal{A}_2$, $\{b\} \in \mathcal{A}_1 \cup \mathcal{A}_2$, but $\{a\} \cup \{b\} = \{a, b\} \notin \mathcal{A}_1 \cup \mathcal{A}_2$, so that $\mathcal{A}_1 \cup \mathcal{A}_2$ is not a σ-field. Note that $\mathcal{A}_1 \cap \mathcal{A}_2 = \{\emptyset, \mathcal{X}\}$, which is a trivial σ-field.

A topological space is another important collection of subsets of a space $\mathcal{X}$ and it helps us define what we mean by open sets. In particular, the celebrated Borel σ-field $\mathcal{B}(\mathcal{X})$ of the space $\mathcal{X}$ is the σ-field generated by the open sets of $\mathcal{X}$, i.e., $\mathcal{B}(\mathcal{X}) = \sigma(O(\mathcal{X}))$. Appendix A.2 presents details on topological spaces and the Borel σ-field.

Definition 3.4 Borel σ-field

Let $O = O(\mathcal{X})$, denote the collection of open sets of a space $\mathcal{X}$, so that $(\mathcal{X}, O)$ is a topological space. The Borel σ-field $\mathcal{B}$ of a space $\mathcal{X}$ is defined by $\mathcal{B} = \sigma(O)$, that is, $\mathcal{B}$ is the generated σ-field from the collection of open sets. We write $\mathcal{B} = \mathcal{B}(\mathcal{X})$ and members of $\mathcal{B}$ are called Borel sets.

See appendix remark A.3 for additional results on Borel σ-fields. Next we collect one of the most important generated Borel σ-fields, that over $\mathcal{R}$.

Theorem 3.2 (Generating the Borel σ-field in $\mathcal{R}$) The Borel σ-field in $\mathcal{R}$, $\mathcal{B}(\mathcal{R})$, can be generated by the collection of all open intervals of $\mathcal{R}$, i.e., $\mathcal{B}(\mathcal{R}) = \sigma(\{(a, b) : \forall a < b \in \mathcal{R}\})$.

Proof. Recall that $\mathcal{B} = \mathcal{B}(\mathcal{R}) = \sigma(O)$, where O denotes the collection of all open sets of $\mathcal{R}$. Note that the collection of open intervals satisfies $\mathcal{I} = \{(a, b) : \forall a < b \in \mathcal{R}\} \subseteq O$, so that remark 3.3.4 yields $\sigma(\mathcal{I}) \subseteq \sigma(O) = \mathcal{B}$. For the other direction it is enough to point out that each $O \in O$, can be written using remark A.3.4 as $O = \bigcup_{n=1}^{+\infty} (a_n, b_n) \in \sigma(\mathcal{I})$, so that $\sigma(O) \subseteq \sigma(\mathcal{I})$ and hence $\mathcal{B} = \sigma(O) = \sigma(\mathcal{I})$. ■

Some consequences of this important theorem are given next.

Remark 3.4 (Generating the Borel σ-field) This important result can be extended to many other collections of intervals of $\mathcal{R}$. In particular, we can show that

$$\begin{aligned} \mathcal{B}(\mathcal{R}) &= \sigma(\{[a, b]\}) = \sigma(\{(a, b]\}) = \sigma(\{[a, b)\}) = \sigma(\{(a, b)\}) \\ &= \sigma(\{(-\infty, b)\}) = \sigma(\{[a, +\infty)\}) = \sigma(\{(a, +\infty)\}) \\ &= \sigma(\{\text{all closed subsets of } \mathcal{R}\}) = \sigma\left(\{(j2^{-n}, (j+1)2^{-n}),\ j, n,\ \text{integers}\}\right), \end{aligned}$$

and more importantly, $\mathcal{B}(\mathcal{R}) = \sigma(\{(-\infty, b]\})$, that is, the Borel σ-field is generated by sets we would use to define the cdf of a random variable X, with $F_X(x) = P(X \leq x)$. Important results like theorem 3.2 and their extensions, allow us to define probability measures by assigning probabilities to these sets only (a much smaller class of sets compared to $\mathcal{B}(\mathcal{R})$) and based on them, obtain probabilities for any Borel set.

Example 3.2 (Cylinder sets) Assume that we equip $\mathcal{R}^p$ with the Euclidean metric

$$\rho(\mathbf{x}, \mathbf{y}) = \sqrt{\sum_{i=1}^{p} (x_i - y_i)^2},$$

for $\mathbf{x} = (x_1, \ldots, x_p)^T$, $\mathbf{y} = (y_1, \ldots, y_p)^T$, vectors in $\mathcal{R}^p$. This leads to the definition of open balls (see Appendix remark A.5), which leads to the description of $\mathcal{O}(\mathcal{R}^p)$, the open sets in $\mathcal{R}^p$, which in turn induces a topology in $\mathcal{R}^p$, so that we can define the Borel sets $\mathcal{B}(\mathcal{R}^p) = \sigma(\mathcal{O})$. Hence, we have an idea about how to create the open sets in $\mathcal{R}^p$. But how do we define the open sets in $\mathcal{R}^\infty$? Let us create a topology without the use of a metric. Let $i_1 < i_2 < \cdots < i_n$, be arbitrary integers, $n \geq 1$ and consider open sets $O_{i_1}, \ldots, O_{i_n}$, in $\mathcal{R}$. We construct special sets of elements from $\mathcal{R}^\infty$ given by

$$\begin{aligned} C_n &= \mathop{\times}_{i=1}^{i_1-1} \mathcal{R} \times O_{i_1} \times \mathcal{R} \times \cdots \times \mathcal{R} \times O_{i_2} \times \cdots \times O_{i_N} \times \mathcal{R} \times \mathcal{R} \times \ldots \\ &= \{(x_1, x_2, \ldots) : x_{i_1} \in O_{i_1}, x_{i_2} \in O_{i_2}, \ldots, x_{i_n} \in O_{i_n}\}, \end{aligned}$$

known as cylinder open sets. Then any set $O \in \mathcal{O}(\mathcal{R}^\infty)$, can be represented as a union of cylinder open sets and the generated σ-field yields the Borel σ-field $\mathcal{B}(\mathcal{R}^\infty) = \sigma(\mathcal{O}(\mathcal{R}^\infty))$. Cylinder open sets provide the necessary ingredient of a product topology and we discuss the general framework next.

3.4 Product Spaces

The space $\mathcal{X}$ considered in the last section has elements of any type, including real numbers, vectors, matrices, functions and so forth and can be used to describe the values of a random variable of an experiment. For example, when we collect a random sample $X_1, \ldots, X_p \sim \mathcal{N}(0, 1)$, in order to describe the probability distribution of the random vector $\mathbf{X} = (X_1, \ldots, X_p)^T \sim \mathcal{N}_p(\mathbf{0}, I_p)$ and investigate its properties, we need to consider the p-fold Cartesian product of $\mathcal{R}$, leading to the product space $\mathcal{R}^p = \mathcal{R} \times \cdots \times \mathcal{R}$.

Since we have studied the case of $\mathcal{R}$ (one dimension), we would like to extend results from $\mathcal{R}$ to the space $\mathcal{R}^p$. Example 3.2 shows that this generalization is not always obvious, especially when it comes to σ-fields like the Borel σ-field. If we consider a random sample $\mathbf{X}_1, \ldots, \mathbf{X}_n$, from $\mathcal{N}_p(\mathbf{0}, I_p)$, the joint distribution of the random vector $\mathbf{X}_n = (\mathbf{X}_1^T, \ldots, \mathbf{X}_n^T)^T \sim \mathcal{N}_{np}(\mathbf{0}, I_{np})$, must be studied in the space $\mathcal{R}^{np} = \mathcal{R}^p \times \cdots \times \mathcal{R}^p = (\mathcal{R} \times \cdots \times \mathcal{R}) \times \cdots \times (\mathcal{R} \times \cdots \times \mathcal{R})$. Sending n to $+\infty$, results in an infinite dimensional random vector $\mathbf{X}_{\infty} = (\mathbf{X}_1^T, \dots, \mathbf{X}_n^T, \dots)^T$, which lives in the space $\mathcal{R}^{\infty}$ and needs to be handled appropriately.

The general definition of the product space and topology is given next.

Definition 3.5 Product space and topology

Let $\mathcal{J}$ be an arbitrary index set and assume that $(\mathcal{X}_j, \mathcal{O}_j)$, $j \in \mathcal{J}$, is a topological space. The product space is defined as the Cartesian product

$$\mathcal{X} = \underset{j\in\mathcal{J}}{\times} \mathcal{X}_j = \{x = (x_1, x_2, \dots) \in \mathcal{X} : x_j \in \mathcal{X}_j, j \in \mathcal{J}\}.$$

The product topology $\mathcal{O} = \mathcal{O}(\mathcal{X})$ on $\mathcal{X}$ consists of the collection of all (arbitrary) unions of members from the collection of sets

$$\mathcal{N} = \{\underset{j\in\mathcal{J}}{\times} O_j : O_j \in \mathcal{O}_j \text{ and } O_j = \mathcal{X}_j, \text{ for all but finitely many } j \in \mathcal{J}\},$$

that is, $\mathcal{O} = \{O \subseteq \mathcal{X} : O = \bigcup_{i\in\mathcal{I}} N_i, N_i \in \mathcal{N}, i \in \mathcal{I}\}$. Then the pair $(\mathcal{X}, \mathcal{O})$ is a topological product space. We use the generic term "open rectangles" or "cylinder open" to describe the sets of $\mathcal{N}$.

We collect some comments and classic results on product spaces below.

Remark 3.5 (Product space and topology) The construction above is not unique. It should be noted that this definition appears in some texts as a lemma that needs to be proven.

1. Tychonoff's theorem A product of compact spaces is a compact space.

2. Since $\overline{\mathcal{R}}$ is compact, $\overline{\mathcal{R}}^{\infty} = \overset{+\infty}{\underset{j=1}{\times}} \overline{\mathcal{R}}$, is compact, that is, the set of all sequences in $\overline{\mathcal{R}}$ is compact.

3. The phrase "and $O_j = \mathcal{X}_j$, for all but finitely many $j \in \mathcal{J}$" is included in the definition to stress the fact that we do not select as an open set O_j the space X_j, for all j. See for example the cylinder sets of example 3.2.

4. For any product space $\mathcal{X} = \overset{p}{\underset{j=1}{\times}} \mathcal{X}_j$, we define the ith projection function as the map $Z_i : \mathcal{X} \to \mathcal{X}_i$, that yields the ith coordinate, namely, $Z_i(x) = X(i)$, $X = (X(1), \dots, X(p)) \in \mathcal{X}$. We can think of X as a random vector and $Z_i(X)$ as a random variable and consider that by construction X is known if and only if all the projection maps are known. This relationship allows us, for example, to immediately generalize results for random variables to random vectors by studying the projections that are obviously easier to handle.

5. Writing $\mathcal{X} = \underset{j\in\mathcal{J}}{\times} \mathcal{X}_j$, leaves no room for questioning the definition of $\mathcal{X}$ as a Cartesian product. However, writing $\mathcal{D} = \underset{j\in\mathcal{J}}{\times} \mathcal{O}_j$ has only notational usefulness and should not be misunderstood as the product topology $\mathcal{O}(\mathcal{X})$. Indeed, if $O_j \in \mathcal{O}_j$, then we can write $O = \underset{j\in\mathcal{J}}{\times} O_j \in \mathcal{D}$, if we were to follow the Cartesian product definition. However, O as defined is only one of the possible sets in $\mathcal{N}$, so that $\mathcal{D} \subsetneq \mathcal{O}(\mathcal{X})$, with $\mathcal{O}(\mathcal{X})$ being a much larger collection of sets. To see this, consider $\mathcal{X} = \mathcal{R}^2$, with $\mathcal{O}(\mathcal{R}^2) = \{O = \bigcup N_i : N_i \in \mathcal{N}\}$, where $\mathcal{N} = \{N : N = O_1 \times O_2, O_1, O_2 \in \mathcal{O}(\mathcal{R})\}$, contains the open rectangles of $\mathcal{R}^2$ and $\mathcal{O}(\mathcal{R})$ the usual Euclidean topology of $\mathcal{R}$. Let $N_1 = (0,1) \times (0,2), N_2 = (1,2) \times (0,3) \in \mathcal{O}(\mathcal{R}^2)$, so that $N_1 \cup N_2 \in \mathcal{O}(\mathcal{R}^2)$, where $N_1 \cup N_2 = \{(x,y) \in \mathcal{R}^2 : x \in (0,1) \cup (1,2), y \in (0,2) \cup (0,3)\}$. Thus we have $N_i \in \mathcal{O}(\mathcal{R}) \times \mathcal{O}(\mathcal{R}) = \mathcal{O}(\mathcal{R})^2 = \{O_1 \times O_2 : O_1, O_2 \in \mathcal{O}(\mathcal{R})\}$, $i = 1,2$, but $N_1 \cup N_2$ does not belong in $\mathcal{O}(\mathcal{R})^2$ and hence $\mathcal{O}(\mathcal{R})^2$ is not a topology. This happens because $\bigcup\limits_{i=1}^{n}(O_{i1} \times \cdots \times O_{ip}) \underset{\neq}{\subset} \bigcup\limits_{i=1}^{n} O_{i1} \times \cdots \times \bigcup\limits_{i=1}^{n} O_{ip}$, in general. To unify the notation and avoid the confusion with the Cartesian product, we will write $\left(\underset{j\in\mathcal{J}}{\times} \mathcal{X}_j, \bigotimes\limits_{j\in\mathcal{J}} \mathcal{O}_j\right)$ for the topological product space $(\mathcal{X}, \mathcal{O})$.

6. Suppose that $\mathcal{X} = \mathcal{R}^p$. The bounded rectangles are defined as

$$R = \{\mathbf{x} = (x_1, \ldots, x_p) \in \mathcal{R}^p : a_i < x_i \leq b_i, i = 1, 2, \ldots, p\},$$

and play a similar role in $\mathcal{R}^p$, as the intervals $\{(a,b)\}$ play in $\mathcal{R}$ (see theorem 3.2), that is, they generate the Borel σ-field in $\mathcal{R}^p$. When a_i, b_i are rationals, we say that R is a rational rectangle. Now if O is an open set in $\mathcal{R}^p$ and $y \in O$, then there is a rational rectangle A_y such that $y \in A_y \subset O$. But then $O = \bigcup\limits_{y\in O} A_y$ and since there are only countably many rational rectangles, the latter is a countable union. Consequently, the open sets of $\mathcal{R}^p$ are members of the generated σ-field from these rational rectangles, i.e., $\mathcal{O} \subseteq \sigma(\{R\})$ and hence $\mathcal{B}(\mathcal{R}^p) = \sigma(\mathcal{O}) \subseteq \sigma(\{R\})$. The other direction is trivial, since $R \in \mathcal{O}(\mathcal{R}^p)$, so that $\{R\} \subseteq \mathcal{O}$ and hence $\sigma(\{R\}) \subseteq \sigma(\mathcal{O})$.

7. When $\mathcal{J} = \{1,2,\ldots\}$, the infinite product space $\mathcal{X} = \underset{j=1}{\overset{+\infty}{\times}} \mathcal{X}_j$, can be thought of as the space of all infinite sequences, namely, $\mathcal{X} = \{(x_1, x_2, \ldots) : x_i \in \mathcal{X}_i$, for all $i = 1,2,\ldots\}$. An alternative definition of cylinder sets in this case can be given as follows: choose a set $B^{(n)} \subset \underset{j=1}{\overset{+\infty}{\times}} \mathcal{X}_j$ and define the cylinder set $B_n \subset \mathcal{X}$, with base $B^{(n)}$ to be $B_n = \{x \in \mathcal{X} : (x_1, x_2, \ldots, x_n) \in B^{(n)}\}$. If $B^{(n)} = A_1 \times A_2 \times \cdots \times A_n$, $A_i \subset \mathcal{X}_i$, then the base is called a rectangle and need not be open.

8. An alternative definition of the infinite product σ-field $\bigotimes\limits_{n=1}^{+\infty} \mathcal{A}_n$ (see definition 3.6), can be given in terms of the generated σ-field from the collection of all measurable cylinders $\{B_n\}$, where a cylinder is called measurable if its base $B^{(n)} \in \mathcal{A}_n$, is measurable. The resulting product σ-fields are the same, no matter what collection of sets we choose to generate it, be it measurable rectangles or measurable cylinders.

9. When working with product spaces it is useful to define the cross section of a set. If A is a subset of $\mathcal{X} = \mathcal{X}_1 \times \mathcal{X}_2$ and x a point of $\mathcal{X}_1$, we define the x cross section A_x by $A_x = \{y \in \mathcal{X}_2 : (x, y) \in A\}$ and similarly for the y cross section of A, we set $A_y = \{x \in \mathcal{X}_1 : (x, y) \in A\}$. The following are straightforward to prove: (i) $I_{A_x}(y) = I_A(x, y)$, (ii) $(A^c)_x = (A_x)^c$, and (iii) $\left(\bigcup_i A_i\right)_x = \bigcup_i (A_i)_x$, for any collection $\{A_i\}$.

Products of larger classes of sets than topologies, i.e., σ-fields, are considered next.

Definition 3.6 Product σ-field

Suppose that $\mathcal{A}_j$, $j \in \mathcal{J}$, are σ-fields with $\mathcal{J}$ some index set. A measurable rectangle is a member of $\mathcal{R}^* = \underset{j\in\mathcal{J}}{\times} \mathcal{A}_j = \{A : A = \underset{j\in\mathcal{J}}{\times} A_j, A_j \in \mathcal{A}_j\}$. The product σ-field $\mathcal{A}$ is defined as the generated σ-field from the collection of measurable rectangles, namely, $\mathcal{A} = \sigma\left(\underset{j\in\mathcal{J}}{\times} \mathcal{A}_j\right)$ and we write $\mathcal{A} = \bigotimes_{j\in\mathcal{J}} \mathcal{A}_j$.

Next we collect two important results that utilize measurable rectangles.

Remark 3.6 (Rectangles and the Borel product σ-field) Some comments are in order.

1. Let $\mathcal{U}$ denote the collection of all finite disjoint unions of measurable rectangles, that is, if $U \in \mathcal{U}$ then $U = \bigcup_{i=1}^{n} R_i$, where $R_i \in \mathcal{R}^*$. Then it can be shown that $\mathcal{U}$ is a field on $\mathcal{X}_1 \times \cdots \times \mathcal{X}_n$ that generates $\mathcal{A}$, that is, $\mathcal{A} = \sigma(\mathcal{U})$.

2. The Borel product σ-field of $\mathcal{R}^p$ is given by $\mathcal{B}(\mathcal{R}^p) = \bigotimes_{j=1}^{p} \mathcal{B}(\mathcal{R}) = \sigma\left(\underset{j=1}{\overset{p}{\times}} \mathcal{B}(\mathcal{R})\right)$ and it will be denoted by $\mathcal{B}_p = \mathcal{B}(\mathcal{R}^p)$. Note that the Borel product space in $\mathcal{R}^2$ is not the Cartesian product of the Borel σ-field on $\mathcal{R}$ with itself, that is, $\mathcal{B}(\mathcal{R}) \times \mathcal{B}(\mathcal{R}) \underset{\neq}{\subset} \mathcal{B}(\mathcal{R}^2)$.

Example 3.3 (Borel product σ-field) We show that $\mathcal{B}_p = \bigotimes_{j=1}^{p} \mathcal{B}_1 = \sigma\left(\underset{j=1}{\overset{p}{\times}} \mathcal{B}_1\right)$, without the use of definition 3.6. Let $\mathcal{E}_1 = \{A \subseteq \mathcal{R}^p : A \text{ is a bounded rectangle}\}$ and let $\mathcal{E}_2 = \{A = \underset{j=1}{\overset{p}{\times}} A_j : A_j \in \mathcal{B}_1\} = \underset{j=1}{\overset{p}{\times}} \mathcal{B}_1$. For the easy direction, we note that $\mathcal{E}_1 \subseteq \mathcal{E}_2$, so that $\mathcal{B}_p = \sigma(\mathcal{E}_1) \subseteq \sigma(\mathcal{E}_2) = \bigotimes_{j=1}^{p} \mathcal{B}_1$, in view of remark 3.3.5. For the other direction, take a set $B = A_1 \times \cdots \times A_p \in \mathcal{E}_2$, with $A_i \in \mathcal{B}_1$, $i = 1, 2, \ldots, p$ and write it as $B = \bigcap_{i=1}^{n} C_i$, where $C_i = \underset{j=1}{\overset{i-1}{\times}} \mathcal{R} \times A_i \times \underset{j=i+1}{\overset{p}{\times}} \mathcal{R}$, so that C_i are cylinder sets (not necessarily open). We need to show that

$$\{C_i : A_i \in \mathcal{B}_1\} \subseteq \mathcal{B}_p, \tag{3.1}$$

$i = 1, 2, \ldots, n$, which would imply that $B = \bigcap_{i=1}^{n} C_i \in \mathcal{B}_k$, so that $\mathcal{E}_2 \subseteq \mathcal{B}_p$ leads to the desired other direction $\sigma(\mathcal{E}_2) \subseteq \mathcal{B}_p$. To prove equation (3.1), consider a typical element $\mathbf{x} = (\mathbf{x}_1, x_i, \mathbf{x}_2)$ of $\mathcal{R}^p$, where $\mathbf{x}_1 \in \mathcal{R}^{i-1}$, $x_i \in \mathcal{R}$, $\mathbf{x}_2 \in \mathcal{R}^{p-i-1}$ and define the projection mapping $f : \mathcal{R}^p \to \mathcal{R}$, by $f((\mathbf{x}_1, x_i, \mathbf{x}_2)) = x_i$, which is continuous. Therefore, the inverse image of an open set in $\mathcal{R}$, through f, is an open set in $\mathcal{R}^p$, so that $f^{-1}(O) \subseteq T$, with $O \in O(\mathcal{R})$ and $T \in O(\mathcal{R}^p)$. Consequently, we can write

$$f^{-1}(\mathcal{B}_1) = f^{-1}(\sigma(O(\mathcal{R}))) = \sigma(f^{-1}(O(\mathcal{R}))) \subseteq \sigma(O(\mathcal{R}^p)) = \mathcal{B}_p, \quad (3.2)$$

using remark 3.3.8 for the second equality. But for any $A_i \in \mathcal{B}_1$, we have

$$C_i = \mathop{\times}_{j=1}^{i-1} \mathcal{R} \times A_i \times \mathop{\times}_{j=i+1}^{p} \mathcal{R} = \{\mathbf{x} \in \mathcal{R}^p : f(\mathbf{x}) \in A_i\} = f^{-1}(A_1) \in f^{-1}(\mathcal{B}_1),$$

for each $i = 1, 2, \ldots, n$ and using equation (3.2), we establish equation (3.1).

3.5 Measurable Spaces and Mappings

The whole purpose of building topologies, fields, σ-fields and generated σ-fields is to create collections of sets that are easy to describe, derive their properties and use them in building a mathematical theory. The next definition is an important component of the measure theory mosaic.

Definition 3.7 Measurable space

A measurable space is a pair $(\Omega, \mathcal{A})$, where Ω is a nonempty space and $\mathcal{A}$ is a σ-field of subsets of Ω. Sets in $\mathcal{A}$ are called measurable sets or $\mathcal{A}$-measurable or $\mathcal{A}$-sets.

Example 3.4 (Coin flip spaces) Let us revisit example 1.1.

1. Recall that the sample space for a single flip of the coin is the set $\Omega_0 = \{\omega_0, \omega_1\}$, where the simple events $\omega_0 = \{Heads\}$, $\omega_1 = \{Tails\}$, are recoded as $\omega_0 = 0$ and $\omega_1 = 1$, so that $\Omega_0 = \{0, 1\}$. Note that $2^{\Omega_0} = \{\emptyset, \Omega_0, \{0\}, \{1\}\}$ is the largest σ-field of the coin flip space and consequently, $(\Omega_0, 2^{\Omega_0})$ can be thought of as the largest measurable space. Further note that $2^{\Omega_0} = \sigma(\{0\}) = \sigma(\{1\})$, so that we can generate the largest σ-field in the space based on the measurable set $\{0\}$ or the set $\{1\}$. What is the topology of Ω_0? We need to describe the collection of open subsets of Ω_0. From appendix definition A.1, we can easily see that both collections $\mathcal{T}_1^* = \{\emptyset, \Omega_0, \{0\}\}$ and $\mathcal{T}_2^* = \{\emptyset, \Omega_0, \{1\}\}$, induce a topology on Ω_0, so that the simple events $\{0\}$ and $\{1\}$ are both open and closed sets in Ω_0. Hence, the Borel σ-field of Ω_0 in this case is $\mathcal{B}(\Omega_0) = 2^{\Omega_0}$.

2. Now consider the experiment of n successive flips of a coin, with sample space $\Omega_n = \{\omega = (\omega_1, \ldots, \omega_n) \in \Omega_0^n : \omega_i \in \{0, 1\}, i = 1, 2, \ldots, n\} = \{\omega_1, \ldots, \omega_{2^n}\}$, consisting of 2^n simple events that are n-dimensional sequences of $0s$ and $1s$, i.e., $\omega_1 = (0, \ldots, 0, 0)$, $\omega_2 = (0, \ldots, 0, 1)$, $\omega_3 = (0, \ldots, 1, 0), \ldots, \omega_{2^n} = (1, \ldots, 1, 1)$.

Clearly, we can think of Ω_n as a product space $\Omega_n = \overset{n}{\underset{j=1}{\times}} \Omega_0$ and the largest σ-field 2^{Ω_n} consists of all subsets of Ω_n. The topology $\mathcal{T}_n$ on Ω_n is the one induced by the product topology $\bigotimes_{j=1}^{n} \mathcal{T}_1^*$ or $\bigotimes_{j=1}^{n} \mathcal{T}_2^*$ and the corresponding Borel σ-field $\mathcal{B}(\Omega_n)$ of Ω_n is obtained by generating the σ-field over the collection of open sets $\mathcal{T}_n$ or the Borel product space, namely,

$$\mathcal{B}(\Omega_n) = \sigma(\mathcal{T}_n) = \sigma\left(\overset{n}{\underset{j=1}{\times}} \mathcal{B}(\Omega_0)\right) = 2^{\Omega_n}.$$

The first equality of the latter holds by the definition of a Borel σ-field, while the second equality holds from definition 3.6. To show that $\mathcal{B}(\Omega_n) = 2^{\Omega_n}$, we observe first that $\mathcal{T}_n \subseteq 2^{\Omega_n}$. For the other direction, consider a set $A \in 2^{\Omega_2}$, where A is either $\emptyset(= \overset{n}{\underset{j=1}{\times}} \emptyset)$, or Ω_n, or $A = \{\omega_{i_1}, \ldots, \omega_{i_p}\}$, if $card(A) = p \leq 2^n = card(\Omega_n)$. In the first two trivial cases, $A \in \mathcal{B}(\Omega_n)$. For the last case, take a typical point $\omega \in A$, where $\omega = (\omega_1, \ldots, \omega_n)$, with $\omega_i = 0$ or 1, so that $\omega_i \in \mathcal{B}(\Omega_0)$, for all i, which implies $\omega \in \overset{n}{\underset{j=1}{\times}} \mathcal{B}(\Omega_0)$, for all $\omega \in A$ and therefore $A \in \mathcal{B}(\Omega_n)$, which establishes the other direction.

3. We turn now to the experiment of infinite coin flips, with sample space $\Omega_\infty = \{\omega = (\omega_1, \omega_2, \ldots) \in \Omega_0^\infty : \omega_i \in \{0, 1\}, i = 1, 2, \ldots\}$, consisting of ∞-dimensional sequences of 0*s* and 1*s*. Following the product space definition, $\Omega_\infty = \overset{+\infty}{\underset{j=1}{\times}} \Omega_0$ and the largest σ-field 2^{Ω_∞} consists of all subsets of Ω_∞, while the topology $\mathcal{T}_\infty$ on Ω_∞ is the one induced by the product topology $\bigotimes_{j=1}^{\infty} \mathcal{T}_1^*$ or $\bigotimes_{j=1}^{\infty} \mathcal{T}_2^*$, with the corresponding Borel σ-field being $\mathcal{B}(\Omega_\infty) = \sigma(\mathcal{T}_\infty) = \sigma\left(\overset{\infty}{\underset{j=1}{\times}} \mathcal{B}(\Omega_0)\right)$, by definitions 3.4 and 3.6. Using similar arguments as in part 2 of this example we can show that $\mathcal{B}(\Omega_\infty) = 2^{\Omega_\infty}$.

Example 3.5 (Exponential spaces) Consider a bounded subset $\mathcal{X} \subset \mathcal{R}^p$ and define $\mathcal{X}^n = \overset{n}{\underset{i=1}{\times}} \mathcal{X}$, $n = 1, 2, \ldots$, with $\mathcal{X}^0 = \emptyset$. If $n \neq k$, then $\mathcal{X}^k$ and $\mathcal{X}^n$ are disjoint, where the extra dimensions are filled in with copies of $\mathcal{X}$. We define the exponential space of $\mathcal{X}$ by $\mathcal{X}_e = \bigcup_{n=0}^{\infty} \mathcal{X}^n$. Every set $B \in \mathcal{X}_e$ can be expressed uniquely as the union of disjoint sets, that is, $B = \bigcup_{n=0}^{\infty} B^{(n)}$, where $B^{(n)} = B \cap \mathcal{X}^n$. Let $\mathcal{A}^{(n)}$ be the generated σ-field in $\mathcal{X}^n$ from all Borel measurable rectangles $B_1 \times B_2 \times \cdots \times B_n$, with $B_i \in \mathcal{B}(\mathcal{X})$, $i = 1, 2, \ldots, n$ and define $\mathcal{A}_e$ as the class of all sets $\bigcup_{n=0}^{\infty} B^{(n)} \in \mathcal{X}_e$, such that $B^{(n)} \in \mathcal{A}^{(n)}$. Then $\mathcal{A}_e$ is the smallest σ-field of sets in $\mathcal{X}_e$ generated by sets $B^{(n)} \in \mathcal{A}^{(n)}, n = 0, 1, \ldots$. The measurable space $(\mathcal{X}_e, \mathcal{A}_e)$ is called an exponential space (Carter and Prenter, 1972) and can be used in defining an important random object, namely, a point process.

Example 3.6 (Hit-or-miss topology) Let $\mathbb{F}$ denote the collection of all closed subsets of $\mathcal{R}^p$. In order to describe random objects that are random closed sets (RACS), we need to equip $\mathbb{F}$ with a σ-field, preferably $\mathcal{B}(\mathbb{F})$, so that we can build a measurable space $(\mathbb{F}, \mathcal{B}(\mathbb{F}))$. This can be accomplished in several ways, depending on the topology we introduce in $\mathbb{F}$. In particular, just as open intervals can be used to build a topology in $\mathcal{R}$ and generate $\mathcal{B}(\mathcal{R})$, we can use a similar approach for the space $\mathbb{F}$. First define $\mathbb{O}$ and $\mathcal{K}$ to be the collection of open and compact subsets of $\mathcal{R}^p$, respectively. Then for any $A \subseteq \mathcal{R}^p$, $O_1, \ldots, O_n \in \mathbb{O}$ and $K \in \mathcal{K}$, define

$$\mathbb{F}_A = \{F \in \mathbb{F} : F \cap A \neq \emptyset\},$$

the subsets of $\mathbb{F}$ that hit A,

$$\mathbb{F}^A = \{F \in \mathbb{F} : F \cap A = \emptyset\},$$

the subsets of $\mathbb{F}$ that miss A,

$$\mathbb{F}^K_{O_1,\ldots,O_n} = \mathbb{F}^K \cap \mathbb{F}_{O_1} \cap \cdots \cap \mathbb{F}_{O_n},$$

the subsets of $\mathbb{F}$ that miss K but hit all the open sets $O_1, \ldots, O_n$ and set

$$\Upsilon = \left\{\mathbb{F}^K_{O_1,\ldots,O_n} : K \in \mathcal{K}, O_1, \ldots, O_n \in \mathbb{O}, n \geq 1\right\}.$$

Note that by definition $\mathbb{F}^\emptyset = \mathbb{F}$, $\mathbb{F}_\emptyset = \emptyset$ and $\mathbb{F}_O = \mathbb{F}^\emptyset_O$, $\forall O \in \mathbb{O}$. The collection of sets Υ can be shown to be the base of a topology $\mathcal{T}$ (see Appendix A.2) on $\mathbb{F}$, known as the hit-or-miss topology. It can be shown that the corresponding topological space $(\mathbb{F}, \mathcal{T})$ is compact, Hausdorff and separable and therefore $\mathbb{F}$ is metrizable. Moreover, the Borel σ-field $\mathcal{B}(\mathbb{F}) = \sigma(\mathcal{T})$ on $\mathbb{F}$ can be generated by the collections $\left\{\{\mathbb{F}_O, \forall O \in \mathbb{O}\}, \{\mathbb{F}^K, \forall K \in \mathcal{K}\right\}\}$ and the collections $\{\mathbb{F}_K, \forall K \in \mathcal{K}\}$, which form a sub-base of the topology. In particular, $\mathcal{B}(\mathbb{F}) = \sigma(\{\mathbb{F}_K, \forall K \in \mathcal{K}\})$ is known as the Effros σ-field, while $\mathcal{B}(\mathbb{F}) = \sigma(\{\mathbb{F}_O, \forall O \in \mathbb{O}\}, \{\mathbb{F}^K, \forall K \in \mathcal{K}\})$ is known as the Fell topology. The two constructions may lead to different topologies once we depart from a well-behaved space like $\mathcal{R}^p$. More details on such constructions can be found in Molchanov (2005) and Nguyen (2006).

We collect another important component of measure theory, that of a measurable mapping. Any random object is essentially described by a mapping that is measurable.

Definition 3.8 Measurable map

Let $(\Omega, \mathcal{A})$ and $(\mathcal{X}, \mathcal{G})$ be two measurable spaces and define a mapping $X : \Omega \to \mathcal{X}$. Then X is called measurable $\mathcal{A}|\mathcal{G}$ if $X^{-1}(G) \in \mathcal{A}$, for all $G \in \mathcal{G}$, where $X^{-1}(G) = \{\omega \in \Omega : X(\omega) \in G\}$. We may write $X : (\Omega, \mathcal{A}) \to (\mathcal{X}, \mathcal{G})$ or simply say that X is $\mathcal{A}$-measurable if the other symbols are clear from the context.

Let us collect some comments and consequences of this important definition.

Remark 3.7 (Measurability) Based on definition 3.8 we can show the following.

1. In probability theory Ω plays the role of the sample space and $\mathcal{A}$ contains all the events of the experiment. When $\mathcal{X} = \mathcal{R}$, we typically take as the target σ-field the Borel σ-field $\mathcal{G} = \mathcal{B}(\mathcal{R})$ and X is called a random variable. For $\mathcal{X} = \mathcal{R}^p$-valued maps, we take $\mathcal{G} = \mathcal{B}(\mathcal{R}^p)$ and X is called a random vector and if $\mathcal{X} = \mathcal{R}^\infty$, we take $\mathcal{G} = \mathcal{B}(\mathcal{R}^\infty)$ and X is called a random sequence of real elements.

2. If $X : \Omega \to \mathcal{X}$ is measurable $\mathcal{A}|\mathcal{G}$ and $Y : \mathcal{X} \to \Upsilon$ is measurable $\mathcal{G}|\mathcal{H}$, then $Y \circ X : \Omega \to \Upsilon$ is measurable $\mathcal{A}|\mathcal{H}$.

3. Assume that $X : \Omega \to \mathcal{X}$. If $X^{-1}(H) \in \mathcal{A}$, for each $H \in \mathcal{H}$ where $\mathcal{G} = \sigma(\mathcal{H})$, then X is measurable $\mathcal{A}|\mathcal{G}$.

4. If $X : (\Omega, \mathcal{A}) \to (\mathcal{X}, \mathcal{B}(\mathcal{X}))$ then X is called a Borel mapping.

5. If $f : (\mathcal{R}^k, \mathcal{B}_k) \to (\mathcal{R}^p, \mathcal{B}_p)$, with $f = (f_1, \dots, f_p)$, then f is called a Borel function or simply measurable (when a measurable space is not specified). It can be shown that a vector function $f : \mathcal{R}^k \to \mathcal{R}^p$ is measurable if and only if each f_i is measurable.

6. If the Borel function $f : \mathcal{R}^k \to \mathcal{R}^p$ is continuous then it is measurable.

7. If f and g are Borel functions then so are the functions $f \pm g$, $f \pm c$, cf, $|f|$ and fg, where c is some constant. If $f : \mathcal{R} \to \mathcal{R}$ is increasing then it is measurable.

8. If $f_1, f_2, \dots,$ are $\mathcal{A}$-measurable functions then so are the functions $\inf\limits_n f_n$, $\sup\limits_n f_n$,

$$\limsup_n f_n = \overline{\lim} f_n = \inf_n \sup_{k \geq n} f_k,$$

$$\liminf_n f_n = \underline{\lim} f_n = \sup_n \inf_{k \geq n} f_k,$$

$\sup\limits_\omega \{f_1(\omega), \dots, f_n(\omega)\}$ and $\inf\limits_\omega \{f_1(\omega), \dots, f_n(\omega)\}$. If $f = \lim\limits_n f_n$ exists for all ω, i.e., $f = \overline{\lim} f_n = \underline{\lim} f_n$, then it is $\mathcal{A}$-measurable. The set $\{\omega : \overline{\lim} f_n(\omega) = \underline{\lim} f_n(\omega)\}$ is $\mathcal{A}$-measurable. If f is $\mathcal{A}$-measurable then $\{\omega : \lim\limits_n f_n(\omega) = f(\omega)\}$ is $\mathcal{A}$-measurable.

9. The indicator function $I_A(\omega) = I(\omega \in A) = \begin{cases} 1 & \text{if } \omega \in A \\ 0 & \text{if } \omega \notin A \end{cases}$, is $\mathcal{A}$-measurable if the set A is $\mathcal{A}$-measurable.

10. A map $X : \Omega \to A \subseteq \mathcal{R}$, is called a set function, which should not be confused with a set-valued map $X : \Omega \to C$, where C is a collection of subsets of a space $\mathcal{X}$, so that $X(\omega) = c$, with $\omega \in \Omega$ and $c \in C$, is a subset of $\mathcal{X}$ for the given ω. For example, let $X : \mathcal{R}^+ \to C$, where $C = \{b(\mathbf{0}, r) \subset \mathcal{R}^p : r > 0\}$, the collection of all open balls in $\mathcal{R}^p$ centered at the origin of radius r. Then X is a set-valued map with $X(r) = c = b(\mathbf{0}, r)$, that takes a positive real number and returns an open ball with the specified real number as a radius. In order to introduce measurability of such maps one has to turn to the hit-or-miss topology and measurable space (example 3.6). Measurable set-valued maps are explored in the *TMSO-PPRS* text.

Many of the results of the previous remark can be shown using the following theorem (Royden, 1989, p. 66) and definition.

Theorem 3.3 (Measurable $\overline{\mathcal{R}}$-valued functions) Let $f : E \to \overline{\mathcal{R}}$, where E is an $\mathcal{A}$-measurable set. The following statements are equivalent;
(i) $\forall a \in \mathcal{R}$ the set $\{x : f(x) > a\}$ is $\mathcal{A}$-measurable.
(ii) $\forall a \in \mathcal{R}$ the set $\{x : f(x) \geq a\}$ is $\mathcal{A}$-measurable.
(iii) $\forall a \in \mathcal{R}$ the set $\{x : f(x) < a\}$ is $\mathcal{A}$-measurable.
(iv) $\forall a \in \mathcal{R}$ the set $\{x : f(x) \leq a\}$ is $\mathcal{A}$-measurable.
These statements imply:
(v) If $a \in \overline{\mathcal{R}}$ the set $\{x : f(x) = a\}$ is $\mathcal{A}$-measurable.

Definition 3.9 Measurable function

An extended real-valued function f is said to be $\mathcal{G}$-measurable if its domain is $\mathcal{G}$-measurable and it satisfies one of (i)-(iv) from theorem 3.3.

A very important class of measurable functions are the so-called simple functions.

Definition 3.10 Simple function

Let $\{A_i\}_{i=1}^n$ be a finite partition of Ω and $\{y_i\}_{i=1}^n$, real, distinct nonzero numbers. A step function $f : \Omega \to \mathcal{R}$ is defined by

$$f(x) = \sum_{i=1}^{n} y_i I_{A_i}(x), \ x \in \Omega,$$

and it is called a (canonical) simple function or $\mathcal{A}$–measurable simple function, if every set A_i is $\mathcal{A}$-measurable, where $A_i = \{x \in \Omega : f(x) = y_i\}$, $i = 1, 2, \ldots, n$. When the $\{y_i\}$ are not distinct or $\{A_i\}$ are not a partition of Ω, then $f(x)$ is still called a simple function but it is in a non-canonical form.

The usefulness of sequences of simple functions is illustrated in the following theorem.

Theorem 3.4 (Monotone limit of simple functions) If f is $\overline{\mathcal{R}}$-valued and $\mathcal{A}$-measurable, there exists a sequence $\{f_n\}$ of simple $\mathcal{A}$-measurable functions such that

$$0 \leq f_n(\omega) \uparrow f(\omega), \text{ if } f(\omega) \geq 0,$$

and

$$0 \geq f_n(\omega) \downarrow f(\omega), \text{ if } f(\omega) \leq 0,$$

as $n \to \infty$, where $f_n \uparrow f$ means that f_n increases toward f from below and $f_n \downarrow f$ means that f_n decreases toward f from above.

Proof. Define the sequence of simple functions with the desired properties by

$$f_n(\omega) = \begin{cases} -n & -\infty \leq f(\omega) \leq -n, \\ -(k-1)2^{-n} & -k2^{-n} < f(\omega) \leq -(k-1)2^{-n}, \\ (k-1)2^{-n} & (k-1)2^{-n} \leq f(\omega) < k2^{-n}, \\ n & n \leq f(\omega) \leq \infty, \end{cases} \tag{3.3}$$

where $1 \leq k \leq n2^n$, $n = 1, 2, \ldots$ We note that f_n as defined covers the possibilities $f(\omega) = \infty$ and $f(\omega) = -\infty$. ■

Example 3.7 (Function measurability) Using remark 3.7 we can show measurability for a wide variety of functions.

1. Let $\Omega = [0, 1]^2$, $\mathcal{B}(\Omega)$ be the Borel sets of Ω and consider the function $X(\omega_1, \omega_2) = (\omega_1 \wedge \omega_2, \omega_1 \vee \omega_2)$, where $\omega_1 \wedge \omega_2 = \min\{\omega_1, \omega_2\}$ and $\omega_1 \vee \omega_2 = \max\{\omega_1, \omega_2\}$. Both coordinates of X are continuous functions so that X is $\mathcal{B}(\Omega)|\mathcal{B}(\Omega)$ measurable using remark 3.7.

2. Recall example 3.4 and define the mapping $f : \Omega_0^\infty \to \Omega_0^n$ from the measurable space $(\Omega_0^\infty, \mathcal{B}(\Omega_0^\infty))$ into the measurable space $(\Omega_0^n, \mathcal{B}(\Omega_0^n))$ by $f((\omega_1, \omega_2, \ldots)) = (\omega_1, \omega_2, \ldots, \omega_n)$. Then using remark 3.7 f is $\mathcal{B}(\Omega_0^\infty)|\mathcal{B}(\Omega_0^n)$ measurable.

3.6 Measure Theory and Measure Spaces

Now that we have a good idea about measurable sets, spaces and functions, we are ready to define a general set function that will allow us to build general measure theory.

Definition 3.11 Measure

A set function μ on a field $\mathcal{A}$ of subsets of a space Ω is called a measure if it satisfies: (i) $\mu(\emptyset) = 0$, (ii) $\mu(A) \in [0, +\infty]$, $\forall A \in \mathcal{A}$, (iii) if $A_1, A_2, \ldots,$ is a disjoint sequence of $\mathcal{A}$-sets and if $\bigcup\limits_{n=1}^{\infty} A_n \in \mathcal{A}$, then

$$\mu\left(\bigcup_{n=1}^{\infty} A_n\right) = \sum_{n=1}^{\infty} \mu(A_n).$$

If $\mathcal{A}$ is a σ-field, then the pair $(\Omega, \mathcal{A})$ is a measurable space and the triple $(\Omega, \mathcal{A}, \mu)$ is called a measure space.

The following remark summarizes some of the most important definitions and properties in general measure theory.

Remark 3.8 (Properties of measures) We collect some consequences of this definition and some standard properties of measures.

1. The measure μ is called finite if $\mu(\Omega) < +\infty$, infinite when $\mu(\Omega) = +\infty$ and a probability measure if $\mu(\Omega) = 1$. Definition 1.1 is essentially definition 3.11 when $\mu(\Omega) = 1$ and remark 1.1 contains immediate consequences of the definition.

2. σ-finite measure If $\Omega = \cup A_n$ for some finite or countable sequence $\{A_n\}$ of $\mathcal{A}$-sets satisfying $\mu(A_n) < +\infty$ then μ is called σ-finite. The $\{A_n\}$ do not have to be disjoint but we can turn them into a disjoint sequence $\{B_n\}$ by setting $B_1 = A_1$, $B_n = A_n \setminus \left(\bigcup\limits_{i=1}^{n-1} A_i\right)$, $n \geq 2$ and thus we can assume wlog that the $\{A_n\}$ are a partition of Ω. A finite measure is by definition σ-finite, although a σ-finite measure may be finite or infinite. If Ω is not a finite or countable union of $\mathcal{A}$-sets, then no measure can be σ-finite on $\mathcal{A}$. It can be shown that if μ is σ-finite on a field $\mathcal{A}$ then $\mathcal{A}$ cannot contain an uncountable disjoint collection of sets of positive μ-measure.

3. Complete measure space A measure space is said to be complete if $\mathcal{A}$ contains all subsets of sets of measure zero, that is, if $B \in \mathcal{A}$ with $\mu(B) = 0$ and $A \subset B$ imply $A \in \mathcal{A}$.

4. Finite additivity Condition (iii) is known as countable additivity and it implies finite additivity: $\mu\left(\bigcup\limits_{i=1}^{n} A_i\right) = \sum\limits_{i=1}^{n} \mu(A_i)$, for $\bigcup\limits_{i=1}^{n} A_i \in \mathcal{A}$, $\{A_i\}$ disjoint sets from the field $\mathcal{A}$.

5. Almost everywhere A property is said to hold almost everywhere with respect to μ and we write a.e. $[\mu]$, if the set of points where it fails to hold is a set of measure zero, i.e., if the property is expressed in terms of a collection of $\omega \in \Omega$ forming the set A, then A a.e. $[\mu]$ if and only if $\mu(A^c) = 0$. If the measure μ is understood by the context, we simply say that the property holds a.e. If $\mu(A^c) = 0$ for some $A \in \mathcal{A}$, then A is a support of μ and μ is concentrated on A. For a finite measure μ, A is a support if and only if $\mu(A) = \mu(\Omega)$.

6. Singularity Measures μ and v on $(\Omega, \mathcal{A})$ are called mutually singular (denoted by $\mu \perp v$), if there are disjoint sets $A, B \in \mathcal{A}$ such that $\Omega = A \cup B$ and $\mu(A) = v(B) = 0$.

7. Absolute continuity A measure v is said to be absolutely continuous with respect to the measure μ if $v(A) = 0$ for each set A for which $\mu(A) = 0$. We write $v \ll \mu$. If $v \ll \mu$ and a property holds a.e. $[\mu]$, then it holds a.e. $[v]$.

8. Semifinite measure A measure is called semifinite if each measurable set of infinite measure contains measurable sets of arbitrarily large finite measure. This is a wider class of measures since every σ-finite measure is semifinite, while the measure that assigns 0 to countable subsets of an uncountable set A and ∞ to the uncountable sets is not semifinite.

9. Transformations Let $(\Omega_1, \mathcal{A}_1)$ and $(\Omega_2, \mathcal{A}_2)$ be measurable spaces and suppose that $T : \Omega_1 \to \Omega_2$ is measurable. If μ is a measure on $\mathcal{A}_1$ then define a set function

μT^{-1} on $\mathcal{A}_2$ by $\mu T^{-1}(A_2) = \mu(T^{-1}(A_2))$, $\forall A_2 \in \mathcal{A}_2$. We can show the following: (i) μT^{-1} is a measure, (ii) if μ is finite then so is μT^{-1}, and (iii) if μ is a probability measure then so is μT^{-1}.

10. Convergence almost everywhere We say that a sequence $\{f_n\}$ of measurable functions converges almost everywhere on a set E to a function f if $\mu(\{\omega \in E \subset \Omega : f_n(\omega) \to f(\omega), \text{ as } n \to \infty\}) = 1$.

11. Atom If $(\Omega, \mathcal{A}, \mu)$ is a measure space, then a set $A \in \mathcal{A}$ is called an atom of μ if and only if $0 < \mu(A) < +\infty$ and for every subset $B \subset A$ with $B \in \mathcal{A}$, either $\mu(B) = 0$ or $\mu(B) = \mu(A)$. A measure without atoms is called nonatomic.

12. Radon measure A measure μ on a Borel σ-field $\mathcal{B}$ is called a Radon measure if $\mu(C) < \infty$ for every compact set $C \in \mathcal{B}$.

13. Locally finite measure The measure μ on $(\Omega, \mathcal{A})$ is called locally finite if $\mu(A) < \infty$, for all bounded sets $A \in \mathcal{A}$.

14. Lebesgue decomposition Let $(\Omega, \mathcal{A}, \mu)$ be a σ-finite measure space and ν a σ-finite measure defined on $\mathcal{A}$. Then we can find a measure ν_0 singular with respect to μ and a measure ν_1 absolutely continuous with respect to μ such that $\nu = \nu_0 + \nu_1$. The measures ν_0 and ν_1 are unique. This theorem connects singularity and absolute continuity. The proof mirrors that of the Radon-Nikodym theorem (theorem 3.20) and it will be discussed there.

Example 3.8 (Counterexample for non-additive measure) There are set functions that are not additive on 2^{Ω}. Indeed, let $\omega_1 \neq \omega_2$ in Ω and set $\mu(A) = 1$, if A contains both ω_1 and ω_2 and $\mu(A) = 0$, otherwise. Then μ is not additive on 2^{Ω}.

Example 3.9 (Discrete measure) A measure μ on $(\Omega, \mathcal{A})$ is discrete if there exist finitely or countably many points $\{\omega_i\}$ in Ω and masses $m_i \geq 0$, such that $\mu(A) = \sum\limits_{\omega_i \in A} m_i$, for all $A \in \mathcal{A}$. The measure μ is infinite, finite or a probability measure if $\sum\limits_i m_i = +\infty, < +\infty$, or converges to 1, respectively.

Example 3.10 (Counting measure) Let $(\Omega, \mathcal{A}, \mu)$ be a measure space where for $A \in \mathcal{A}$ finite μ is defined by

$$\mu(A) = card(A) = |A| = \#\text{ of points in } A,$$

and if A is not finite assume $\mu(A) = +\infty$. The measure μ is called the counting measure and is (σ-)finite if and only if Ω is (countable) finite. Let us verify definition 3.11. Clearly, $\mu(\emptyset) = 0$ and $\mu(A) \in [0, +\infty]$, for all $A \in \mathcal{A}$. Letting $\{A_n\}$ disjoint $\mathcal{A}$-sets we have

$$\mu\left(\bigcup_{i=1}^{n} A_i\right) = \left|\bigcup_{i=1}^{n} A_i\right| = \sum_{i=1}^{n} |A_i| = \sum_{i=1}^{n} \mu(A_i).$$

Note that the counting measure on an uncountable set, e.g., $[0, 1]$ or $\mathcal{R}$, is an example of a measure that is not σ-finite.

We discuss and prove five important properties of measures below.

Theorem 3.5 (Properties of measures) Let μ be a measure on a field $\mathcal{A}$.
(i) Monotonicity: μ is monotone, that is, $A \subset B \Longrightarrow \mu(A) \leq \mu(B)$.
(ii) Continuity from below: If A_n, $n = 1, 2, \ldots$ and A are $\mathcal{A}$-sets with $A_n \uparrow A$ then $\mu(A_n) \uparrow \mu(A)$.
(iii) Continuity from above: If A_n, $n = 1, 2, \ldots$ and A are $\mathcal{A}$-sets with $\mu(A_1) < \infty$ and $A_n \downarrow A$ then $\mu(A_n) \downarrow \mu(A)$.
(iv) Countable subadditivity (Boole's Inequality): For a sequence of $\mathcal{A}$-sets $\{A_n\}$ with $\bigcup\limits_{n=1}^{\infty} A_n \in \mathcal{A}$ we have

$$\mu\left(\bigcup_{n=1}^{\infty} A_n\right) \leq \sum_{n=1}^{\infty} \mu(A_n).$$

(v) Inclusion-exclusion Formula: For $\{A_i\}_{i=1}^{n}$ any $\mathcal{A}$-sets with $\mu(A_i) < \infty$ we have

$$\begin{aligned}\mu\left(\bigcup_{i=1}^{n} A_i\right) &= \sum_{k=1}^{n} \sum \left\{(-1)^{k-1}\mu\left(\bigcap_{j=1}^{k} A_{i_j}\right) : 1 \leq i_1 < \cdots < i_k \leq n\right\} \qquad (3.4)\\ &= \sum_i \mu(A_i) - \sum_{i<j} \mu(A_i \cap A_j) + \cdots + (-1)^{n+1}\mu(A_1 \cap \cdots \cap A_n).\end{aligned}$$

Proof. (i) For sets $A, B \in \mathcal{A}$, with $A \subset B$ and $B \smallsetminus A \in \mathcal{A}$, using finite additivity we can write: $\mu(B) = \mu(A \cup (B \smallsetminus A)) = \mu(A) + \mu(B \smallsetminus A)$, so that $0 \leq \mu(B \smallsetminus A) = \mu(B) - \mu(A)$.
(ii) Consider A_n, $A \in \mathcal{A}$, with $A_n \uparrow A$. Since $A_n \subseteq A_{n+1}$ we see that $\mu(A_n) \leq \mu(A_{n+1})$. Moreover, we see that $\bigcup\limits_{k=1}^{+\infty} A_k \subset \bigcup\limits_{k=2}^{+\infty} A_k \subset \bigcup\limits_{k=3}^{+\infty} A_k \subset \ldots$ and remark 3.1.9 yields $\bigcap\limits_{n=1}^{+\infty}\bigcup\limits_{k=n}^{+\infty} A_k = \bigcup\limits_{n=1}^{+\infty} A_n$, so that $A = \underline{\lim}_{n\to\infty} A_n = \overline{\lim} A_n = \bigcup\limits_{n=1}^{+\infty} A_n$. Let $B_n = A_{n+1} \smallsetminus A_n$ and note that $B_i \cap B_j = \emptyset$, $i \neq j$ and $A = \bigcup\limits_{n=1}^{+\infty} A_n = A_1 \cup \bigcup\limits_{n=1}^{+\infty} B_n$. Applying countable additivity to the sequence $\{B_n\}_{n=0}^{+\infty}$, with $B_0 = A_1$, we obtain

$$\begin{aligned}\mu\left(\bigcup_{n=0}^{+\infty} B_n\right) &= \mu(A_1) + \sum_{n=1}^{+\infty} \mu(B_n) = \mu(A_1) + \lim_{n\to\infty} \sum_{k=1}^{n} [\mu(A_{k+1}) - \mu(A_k)]\\ &= \mu(A_1) + \lim_{n\to\infty} [(\mu(A_2) - \mu(A_1)) + \ldots + (\mu(A_{n+1}) - \mu(A_n))]\\ &= \mu(A_1) + \lim_{n\to\infty} \mu(A_{n+1}) - \mu(A_1),\end{aligned}$$

which leads to the desired $\lim\limits_{n\to\infty} \mu(A_n) = \mu(A)$.
(iii) Now let A_n, $A \in \mathcal{A}$, with $A_n \supseteq A_{n+1}$, such that $A_n \downarrow A$, where $A = \lim\limits_{n\to\infty} A_n = \underline{\lim} A_n = \bigcup\limits_{n=1}^{+\infty}\bigcap\limits_{n=k}^{+\infty} A_k = \bigcap\limits_{n=1}^{+\infty} A_n$. Setting $B_n = A_1 \smallsetminus A_n$, we note that $B_n \subseteq B_{n+1}$, with $B_n \uparrow B = A_1 \smallsetminus A = \bigcup\limits_{n=1}^{+\infty}(A_1 \smallsetminus A_n)$ and from part (ii) we obtain $\mu(B_n) \uparrow \mu(B)$, which

leads to $\lim_{n\to\infty}\mu(A_1 \smallsetminus A_n) = \mu(A_1 \smallsetminus A)$, so that $\mu(A_1) - \lim_{n\to\infty}\mu(A_n) = \mu(A_1) - \mu(A)$ and the desired result is established, since $\mu(A_1) < \infty$ allows us to write the measure of the set difference as the difference of the measures.

(iv) Suppose that $\{A_n\}_{n=1}^{+\infty}$ are $\mathcal{A}$-sets with $\bigcup_{n=1}^{\infty} A_n \in \mathcal{A}$ and let $B_1 = A_1$,

$$B_n = A_n \smallsetminus \left[\bigcup_{i=1}^{n-1} A_i\right] = A_n \cap \left[\bigcap_{i=1}^{n-1} A_i^c\right] = A_n \cap A_1^c \cap A_2^c \cap \ldots \cap A_{n-1}^c,$$

for $n > 1$, a possibly improper set difference since we might be subtracting a larger set, in which case $B_n = \emptyset$. Clearly, $B_n \subset A_n$ so that by monotonicity $\mu(B_n) \leq \mu(A_n)$, for all n. In addition, note that the $\{B_n\}$ are disjoint with

$$\begin{aligned}\bigcup_{n=1}^{\infty} B_n &= A_1 \cup \bigcup_{n=2}^{\infty}\left(A_n \cap \left[\bigcap_{i=1}^{n-1} A_i^c\right]\right) = A_1 \cup \left(\bigcup_{n=2}^{\infty} A_n \cap \bigcup_{n=2}^{\infty}\left[\bigcap_{i=1}^{n-1} A_i^c\right]\right)\\ &= \bigcup_{n=1}^{\infty} A_k \cap \left(A_1 \cup \bigcup_{n=2}^{\infty}\left[\bigcap_{i=1}^{n-1} A_i^c\right]\right),\end{aligned}$$

since

$$\begin{aligned}A_1 \cup \bigcup_{n=2}^{\infty}\left[\bigcap_{i=1}^{n-1} A_i^c\right] &= A_1 \cup \bigcup_{n=2}^{\infty}\left[\left(\bigcup_{i=1}^{n-1} A_i\right)^c\right] = A_1 \cup \left[\bigcap_{n=2}^{\infty}\left(\bigcup_{i=1}^{n-1} A_i\right)\right]^c\\ &= A_1 \cup [A_1 \cap (A_1 \cup A_2) \cap (A_1 \cup A_2 \cup A_3) \cap \ldots]^c\\ &= A_1 \cup A_1^c = \Omega,\end{aligned}$$

and appealing to countable additivity we obtain

$$\mu\left(\bigcup_{n=1}^{\infty} A_n\right) = \mu\left(\bigcup_{n=1}^{\infty} B_n\right) = \sum_{n=1}^{\infty} \mu(B_n) \leq \sum_{n=1}^{\infty} \mu(A_n),$$

as claimed.

(v) We use induction on n. For $n = 1$, equation (3.4) becomes $\mu(A_1) = \mu(A_1)$ and it holds trivially. For $n = 2$ we first write

$$A_1 \cup A_2 = [A_1 \smallsetminus (A_1 \cap A_2)] \cup [A_2 \smallsetminus (A_1 \cap A_2)] \cup (A_1 \cap A_2),$$

a disjoint union and then apply finite additivity to obtain

$$\begin{aligned}\mu(A_1 \cup A_2) &= \mu(A_1 \smallsetminus (A_1 \cap A_2)) + \mu(A_2 \smallsetminus (A_1 \cap A_2)) + \mu(A_1 \cap A_2)\\ &= \mu(A_1) - \mu(A_1 \cap A_2) + \mu(A_2) - \mu(A_1 \cap A_2) + \mu(A_1 \cap A_2)\\ &= \mu(A_1) + \mu(A_2) - \mu(A_1 \cap A_2),\end{aligned}$$

thus establishing the formula for $n = 2$. Now assume equation (3.4) holds for n. We show that it holds for $n + 1$. Using the result for $n = 2$ we can write

$$\begin{aligned}\mu\left(\bigcup_{i=1}^{n+1} A_i\right) &= \mu\left(\bigcup_{i=1}^{n} A_i \cup A_{n+1}\right)\\ &= \mu\left(\bigcup_{i=1}^{n} A_i\right) + \mu(A_{n+1}) - \mu\left(\bigcup_{i=1}^{n} A_i \cap A_{n+1}\right)\\ &= \mu\left(\bigcup_{i=1}^{n} A_i\right) + \mu(A_{n+1}) - \mu\left(\bigcup_{i=1}^{n} (A_i \cap A_{n+1})\right),\end{aligned}$$

and using the formula for n we have

$$\begin{aligned}\mu\left(\bigcup_{i=1}^{n+1} A_i\right) &= \sum_{k=1}^{n}\sum\left\{(-1)^{k-1}\mu\left(\bigcap_{j=1}^{k} A_{i_j}\right) : 1 \le i_1 < ... < i_k \le n\right\} + \mu(A_{n+1})\\ &\quad - \sum_{k=1}^{n}\sum\left\{(-1)^{k-1}\mu\left(\bigcap_{j=1}^{k} A_{i_j} \cap A_{n+1}\right) : 1 \le i_1 < ... < i_k \le n\right\}\\ &= \sum_{k=1}^{n+1}\sum\left\{(-1)^{k-1}\mu\left(\bigcap_{j=1}^{k} A_{i_j}\right) : 1 \le i_1 < ... < i_k \le n+1\right\},\end{aligned}$$

with the last equality established after some tedious (not hard) algebra once we expand the terms in the first equality and simplify. ■

The uniqueness-of-measure theorem that follows, is the result that allows us to uniquely determine measures on σ-fields when the measures are defined and agree in smaller collections of sets, i.e., a π-system. This result is particularly useful in probability theory, since it helps us to uniquely identify probability distributions of random variables.

Theorem 3.6 (Uniqueness of measure) Let μ_1 and μ_2 denote measures defined on $\sigma(\mathcal{P})$, where $\mathcal{P}$ is a π-system. If μ_1 is σ-finite with respect to $\mathcal{P}$ and if μ_1 and μ_2 agree on $\mathcal{P}$ then μ_2 is σ-finite and μ_1 and μ_2 agree on $\sigma(\mathcal{P})$.

Proof. Since μ_1 is σ-finite on $\mathcal{P}$ we may write $\Omega = \bigcup_{n=1}^{\infty} B_n$, for some sequence $\{B_n\}_{n=1}^{\infty}$ of $\mathcal{P}$-sets, with $\mu_1(B_n) < +\infty$, for all $n = 1, 2, ...$ Since $\mu_2 = \mu_1$ on $\mathcal{P}$ we have $\mu_2(B_n) < +\infty$, for all n and hence μ_2 is σ-finite on $\mathcal{P}$.

Now choose an arbitrary $A \in \sigma(\mathcal{P})$. We need to show that $\mu_1(A) = \mu_2(A)$. First note that using continuity from below we can write

$$\begin{aligned}\mu_i(A) &= \mu_i\left(\bigcup_{k=1}^{\infty}(B_k \cap A)\right) = \mu_i\left(\lim_{n\to\infty}\bigcup_{k=1}^{n}(B_k \cap A)\right)\\ &= \lim_{n\to\infty}\mu_i\left(\bigcup_{k=1}^{n}(B_k \cap A)\right),\end{aligned}$$

for $i = 1, 2$ and therefore we need to show that

$$\lim_{n\to\infty}\mu_1\left(\bigcup_{k=1}^{n}(B_k \cap A)\right) = \lim_{n\to\infty}\mu_2\left(\bigcup_{k=1}^{n}(B_k \cap A)\right),$$

which in turn will hold if we show that

$$\mu_1\left(\bigcup_{k=1}^{n}(B_k \cap A)\right) = \mu_2\left(\bigcup_{k=1}^{n}(B_k \cap A)\right).$$

Use the inclusion-exclusion formula to write

$$\mu_i\left(\bigcup_{k=1}^{n}(B_k \cap A)\right) = \sum_{k=1}^{n}\sum\left\{(-1)^{k-1}\mu_i\left(\bigcap_{j=1}^{k} B_{i_j} \cap A\right) : 1 \le i_1 < ... < i_k \le n\right\},$$

for all n, so that we need to show that

$$\mu_1\left(\bigcap_{j=1}^{k} B_{i_j} \cap A\right) = \mu_2\left(\bigcap_{j=1}^{k} B_{i_j} \cap A\right), \tag{3.5}$$

for $k = 1, 2, ..., n$, $1 \leq i_1 < ... < i_k \leq n$ and all $n = 1, 2, ...$ For fixed k, n and i_j, we have that $\bigcap_{j=1}^{k} B_{i_j} \in \mathcal{P}$, since $\mathcal{P}$ is a π-system and $B_{i_j} \in \mathcal{P}$, $j = 1, 2, .., k$. By monotonicity and the fact that μ_1 and μ_2 agree on $\mathcal{P}$, we have that $\mu_1\left(\bigcap_{j=1}^{k} B_{i_j}\right) = \mu_2\left(\bigcap_{j=1}^{k} B_{i_j}\right) < +\infty$. Define the class of sets in $\sigma(\mathcal{P})$

$$\mathcal{L}_B = \{A \in \sigma(\mathcal{P}) : \mu_1(B \cap A) = \mu_2(B \cap A)\},$$

for some $B \in \mathcal{P}$. Then $\mathcal{L}_B$ is a λ-system on Ω and since $\mathcal{P} \subseteq \mathcal{L}_B$, Dynkin's $\pi - \lambda$ theorem implies $\sigma(\mathcal{P}) \subseteq \mathcal{L}_B$. Since $A \in \sigma(\mathcal{P})$ and $\sigma(\mathcal{P}) \subseteq \mathcal{L}_{B_{i_1} \cap ... \cap B_{i_k}}$, we have $A \in \mathcal{L}_{B_{i_1} \cap ... \cap B_{i_k}}$, so that equation (3.5) holds and the desired result is established. ■

Example 3.11 (Counterexample for uniqueness of measure) Letting $\mathcal{P} = \emptyset$ we have that $\mathcal{P}$ is a π-system and $\sigma(\mathcal{P}) = \{\emptyset, \Omega\}$. Then any finite measures have to agree on $\mathcal{P}$ but need not agree on $\sigma(\mathcal{P})$.

Example 3.12 (Probability measure via the cdf) Let $(\Omega, \mathcal{A}) = (\mathcal{R}, \mathcal{B}_1)$ and consider the collection of sets $\mathcal{P} = \{(-\infty, x], x \in \mathcal{R}\}$. Then $\mathcal{P}$ is a π-system and two finite measures that agree on $\mathcal{P}$ also agree on $\mathcal{B}_1$. The consequence of uniqueness of measure in this case illustrates that the cdf uniquely determines the probability distribution of a random variable, since defining the cdf $F(x) = P((-\infty, x])$ based on a probability measure P uniquely defines P on all Borel sets in $\mathcal{R}$. This is also the result of the Lebesgue-Stieltjes theorem: there exists a bijection between cdfs F on $\mathcal{R}$ and probability measures P on $\mathcal{B}(\mathcal{R})$ via $F(x) = P((-\infty, x])$.

3.6.1 Signed Measures and Decomposition Theorems

Decomposition theorems in measure theory provide results that connect different types of measures and have great applications in all manner of proofs. We study below three of those measure decomposition theorems.

The first two theorems involve the concept of a signed measure. If we have two measures μ_1 and μ_2 on the same measurable space $(\Omega, \mathcal{A})$ then owing to condition (ii) of definition 3.11, μ_1 and μ_2 are called nonnegative signed measures. We can easily see that $\mu_3 = \mu_1 + \mu_2$ is a measure defined on $(\Omega, \mathcal{A})$. But what happens if set $\mu_3 = \mu_1 - \mu_2$? Clearly, μ_3 will not be nonnegative always and it is possible that it will be undefined if for some $A \in \mathcal{A}$ we have $\mu_1(A) = \mu_2(A) = \infty$. To avoid this we could ask for at least one of the measures to be finite. The next definition takes into account this discussion by replacing condition (ii) of definition 3.11 and taking care of the consequences in condition (iii).

Definition 3.12 Signed measure

A signed measure μ on a measurable space $(\Omega, \mathcal{A})$, where $\mathcal{A}$ is a σ-field, is an extended real-valued set function defined for sets of $\mathcal{A}$ that has the following properties: (i) $\mu(\emptyset) = 0$, (ii) μ assumes at most one of the values $+\infty$ or $-\infty$, and (iii) if $A_1, A_2, \ldots$ is a disjoint sequence of $\mathcal{A}$-sets then

$$\mu\left(\bigcup_{i=1}^{\infty} A_i\right) = \sum_{n=1}^{\infty} \mu(A_n),$$

with equality understood to mean that the series on the right converges absolutely if $\mu\left(\bigcup_{i=1}^{\infty} A_i\right)$ is finite and that it properly diverges otherwise.

If we replace condition (ii) of definition 3.11 with (ii)′ $\mu(A) \in [-\infty, 0]$, $\forall A \in \mathcal{A}$, then μ is called a signed nonpositive measure. Clearly, a measure is a special case of a signed measure, however a signed measure is not in general a measure. We say that a set A is a positive set with respect to a signed measure μ if A is measurable and for every measurable subset of E of A we have $\mu(E) \geq 0$. Every measurable subset of a positive set is again positive and the restriction of μ to a positive set yields a measure.

Similarly, a set B is called a negative set if it is measurable and every measurable subset of B has nonpositive μ measure. A set that is both positive and negative with respect to μ has zero measure. We need to be careful in this case. A null set with respect to a signed measure is not only a null set in the sense of remark 3.12, i.e., has measure zero, but with the above definitions, it contains all its subsets by construction.

While every null set must have measure zero in the case of signed measures, a set of measure zero may be a union of two sets whose measures are not zero but cancel each other. Clearly, for the nonnegatively signed measure in remark 3.12 we do not have this situation and measure space completion is required.

We collect first some useful lemmas and request their proofs as exercises.

Lemma 3.1 Every measurable subset of a positive set is itself positive. The union of a countable collection of positive sets is positive.

Lemma 3.2 Let E be a measurable set such that $0 < \mu(E) < \infty$. Then there is a positive set A contained in E with $\mu(A) > 0$.

The following decomposition theorem allows us to work with a specific partition of the space.

Theorem 3.7 (Hahn decomposition) Let μ be a signed measure on the measurable space $(\Omega, \mathcal{A})$. Then there is a positive set A^+ and a negative set A^- such that $\Omega = A^+ \cup A^-$ and $A^+ \cap A^- = \emptyset$. The collection $\{A^+, A^-\}$ is called a Hahn decomposition of Ω with respect to μ and is not unique.

Proof. Without loss of generality, assume that μ does not assign the value $+\infty$

to any $\mathcal{A}$-set. Let a be the supremum of $\mu(A)$ over all sets A that are positive with respect to μ. Since the empty set is a positive set, $a \geq 0$. Let $\{A_n\}$ be a sequence of positive sets such that $a = \lim\limits_{n\to\infty}\mu(A_n)$ and set $A^+ = \bigcup\limits_{n=1}^{\infty} A_n$. By lemma 3.1, the set A^+ is itself a positive set and hence $a \geq \mu(A^+)$. But $A^+ \smallsetminus A_n \subset A^+$ yields $\mu(A^+ \smallsetminus A_n) \geq 0$, so that

$$\mu(A^+) = \mu(A_n) + \mu(A^+ \smallsetminus A_n) \geq \mu(A_n),$$

which implies that $\mu(A^+) \geq a$ and hence $\mu(A^+) = a$, with $a < \infty$.

Now let $A^- = (A^+)^c$ and suppose that E is a positive subset of A^-. Then E and A^+ are disjoint and $A^+ \cup E$ is a positive set. Therefore

$$a \geq \mu(A^+ \cup E) = \mu(A^+) + \mu(E) = a + \mu(E),$$

so that $\mu(E) = 0$, since $0 \leq a < \infty$. Thus A^- does not contain any positive subsets of positive measure and therefore, no subset of positive measure by lemma 3.2. Therefore, A^- is a negative set. ■

The second decomposition theorem presented next, allows decomposition of a measure into two mutually singular signed measures.

Theorem 3.8 (Jordan decomposition) Let μ be a signed measure on the measurable space $(\Omega, \mathcal{A})$. Then there are two mutually singular measures μ^+ and μ^- on $(\Omega, \mathcal{A})$ such that $\mu = \mu^+ - \mu^-$. Moreover, μ^+ and μ^- are unique.

Proof. Let $\{A^+, A^-\}$ be a Hahn decomposition of Ω with respect to μ and define two finite measures μ^+ and μ^- for any $A \in \mathcal{A}$ by

$$\mu^+(A) = \mu(A \cap A^+), \text{ and}$$
$$\mu^-(A) = -\mu(A \cap A^-),$$

so that $\mu = \mu^+ - \mu^-$ and μ^+ and μ^- are mutually singular, since they have disjoint supports. To show uniqueness, note that if $E \subset A$ then $\mu(E) \leq \mu^+(E) \leq \mu^+(A)$, with equality if $E = A \cap A^+$. Therefore, $\mu^+(A) = \sup\limits_{E\subset A}\mu(E)$ and following similar arguments $\mu^-(A) = -\inf\limits_{E\subset A}\mu(E)$. The unique measures μ^+ and μ^- thus created are called the upper and lower variations of μ and the measure $|\mu|$ defined by $|\mu| = \mu^+ + \mu^-$ is called the total variation. ■

The following lemma is useful in proving subsequent results.

Lemma 3.3 If μ and v are finite measures (i.e., signed nonnegative) that are not mutually singular then there exists a set A and $\varepsilon > 0$ such that $\mu(A) > 0$ and $\varepsilon\mu(E) \leq v(E)$, for all $E \subset A$.

Proof. Let $\{A_n^+, A_n^-\}$ be a Hahn decomposition of Ω with respect to the set function $u = v - \frac{1}{n}\mu$ and set $B = \bigcup\limits_{n=1}^{\infty} A_n^+$, so that $B^c = \bigcap\limits_{n=1}^{\infty} A_n^-$. Since $B^c \subset A_n^-$ we must have $u(B^c) \leq 0$ or $v(B^c) \leq \frac{1}{n}\mu(B^c)$, for arbitrary n and hence $v(B^c) = 0$. This means that B^c supports v and since $\mu \perp v$, B^c cannot support μ, i.e., $\mu(B^c) = 0$. Consequently $\mu(B) > 0$ and therefore $\mu(A_n^+) > 0$ for some n. Set $A = A_n^+$ and $\varepsilon = \frac{1}{n}$ to obtain the claim. ■

The results of this section can be used to prove the Radon-Nikodym and the Lebesgue Decomposition theorems, as well as the all-important Carathéodory extension theorem as we discuss next.

3.6.2 Carathéodory Measurability and Extension Theorem

It is much easier to define a measure on a π-system or a field and then extend it somehow to assign values to the generated σ-field, since the former is a much smaller and easier to describe collection of sets. This extension can be accomplished in a unique way using outer measures and the general construction is due to the Greek mathematician Konstantinos Carathéodory (1873-1950).

Definition 3.13 Outer measure

An outer measure is a set function μ^* defined for all subsets of a space Ω that satisfies: (i) $\mu(A) \in [0, +\infty]$, $\forall A \subset \Omega$, (ii) $\mu^*(\emptyset) = 0$, (iii) μ^* is monotone, that is, $A \subset B \Longrightarrow \mu^*(A) \leq \mu^*(B)$, and (iv) μ^* is countably subadditive, that is

$$\mu^*\left(\bigcup_{i=1}^{\infty} A_i\right) \leq \sum_{n=1}^{\infty} \mu^*(A_n).$$

Note that condition (i) is forcing the outer measure to be non-negative (signed) and is not required in general. While the outer measure, as defined, has the advantage that it is defined for all subsets of the space Ω, it is not countably additive. It becomes countably additive once we suitably reduce the family of sets on which it is defined. Carathéodory's general definition of a measurable set with respect to μ^* provides us with a collection of such well-behaved sets.

Definition 3.14 Carathéodory measurability

A set $E \subset \Omega$ is measurable with respect to μ^* if for every set $A \subset \Omega$ we have

$$\mu^*(A) = \mu^*(A \cap E) + \mu^*(A \cap E^c).$$

We denote the collection of all Carathéodory measurable sets by $\mathcal{M}$ or $\mathcal{M}(\mu^*)$.

We note the following.

Remark 3.9 (Showing measurability) Since μ^* is subadditive, to show measurability we only need to show the other direction

$$\mu^*(A) \geq \mu^*(A \cap E) + \mu^*(A \cap E^c),$$

where $A = (A \cap E) \cup (A \cap E^c)$. The inequality is trivially true when $\mu^*(A) = \infty$ and so we need only establish it for sets A with $\mu^*(A) < \infty$.

The following theorem establishes some important properties of the class $\mathcal{M}$ of μ^*-measurable sets, as well as the measure μ^*.

Theorem 3.9 (Properties of the class $\mathcal{M}$) The class $\mathcal{M}$ of μ^*-measurable sets is a σ-field. If $\overline{\mu}$ is the restriction of μ^* to $\mathcal{M}$, then $\overline{\mu}$ is a measure on $\mathcal{M}$.

Proof. Clearly, $\emptyset \in \mathcal{M}$, and the symmetry of definition 3.14, shows that $E^c \in \mathcal{M}$, whenever $E \in \mathcal{M}$. Now consider two $\mathcal{M}$-sets, E_1 and E_2. From the measurability of E_2, we have

$$\mu^*(A) = \mu^*(A \cap E_2) + \mu^*(A \cap E_2^c),$$

and applying the measurability condition for E_1, we obtain

$$\mu^*(A) = \mu^*(A \cap E_2) + \mu^*(A \cap E_2^c \cap E_1) + \mu^*(A \cap E_2^c \cap E_1^c),$$

for any set $A \subset \Omega$. Since

$$A \cap (E_1 \cup E_2) = (A \cap E_2) \cup (A \cap E_1 \cap E_2^c),$$

using subadditivity we can write

$$\mu^*(A \cap (E_1 \cup E_2)) \leq \mu^*(A \cap E_2) + \mu^*(A \cap E_1 \cap E_2^c),$$

which leads to

$$\mu^*(A) \geq \mu^*(A \cap (E_1 \cup E_2)) + \mu^*(A \cap E_1^c \cap E_2^c),$$

and hence $E_1 \cup E_2$ is measurable, since $(E_1 \cup E_2)^c = E_1^c \cap E_2^c$. Now using induction, the union of any finite number of $\mathcal{M}$-sets is measurable, and thus $\mathcal{M}$ is a field. To show that $\mathcal{M}$ is a σ-field, let $E = \bigcup\limits_{i=1}^{\infty} E_i$, where $\{E_i\}$ is a disjoint sequence of $\mathcal{M}$-sets. We need to show that E is an $\mathcal{M}$-set. Clearly, if $G_n = \bigcup\limits_{i=1}^{n} E_i$, then $G_n \in \mathcal{M}$, and since $G_n \subset E \Rightarrow E^c \subset G_n^c$, using monotonicity we can write

$$\mu^*(A) = \mu^*(A \cap G_n) + \mu^*(A \cap G_n^c) \geq \mu^*(A \cap G_n) + \mu^*(A \cap E^c).$$

Now $G_n \cap E_n = E_n$ and $G_n \cap E_n^c = G_{n-1}$, and by the measurability of E_n, we have

$$\mu^*(A \cap G_n) = \mu^*(A \cap E_n) + \mu^*(A \cap G_{n-1}).$$

By induction

$$\mu^*(A \cap G_n) = \sum_{i=1}^{n} \mu^*(A \cap E_i),$$

so that

$$\mu^*(A) \geq \sum_{i=1}^{n} \mu^*(A \cap E_i) + \mu^*(A \cap E^c).$$

Letting $n \to \infty$, noting that $A \cap E \subset \bigcup\limits_{i=1}^{\infty}(A \cap E_i)$, and using subadditivity, we can write

$$\begin{aligned}
\mu^*(A) &\geq \sum_{i=1}^{\infty} \mu^*(A \cap E_i) + \mu^*(A \cap E^c) \\
&\geq \mu^*\left(\bigcup_{i=1}^{\infty}(A \cap E_i)\right) + \mu^*(A \cap E^c),
\end{aligned}$$

and hence

$$\mu^*(A) \geq \mu^*(A \cap E) + \mu^*(A \cap E^c).$$

The latter shows that E is measurable, so that $\mathcal{M}$ is a σ-field.

We next demonstrate the finite additivity of $\overline{\mu}$. Let E_1 and E_2 be disjoint measurable sets. Then the measurability of E_2 implies that

$$\begin{aligned} \overline{\mu}(E_1 \cup E_2) &= \mu^*(E_1 \cup E_2) \\ &= \mu^*([E_1 \cup E_2] \cap E_2) + \mu^*([E_1 \cup E_2] \cap E_2^c) \\ &= \mu^*(E_2) + \mu^*(E_1), \end{aligned}$$

and finite additivity follows by induction. If E is the disjoint union of the measurable sets $\{E_n\}$, then

$$\overline{\mu}(E) \geq \overline{\mu}\left(\bigcup_{i=1}^{n} E_i\right) = \sum_{i=1}^{n} \overline{\mu}(E_i),$$

so that

$$\overline{\mu}(E) \geq \sum_{i=1}^{\infty} \overline{\mu}(E_i).$$

By the subadditivity of μ^* we have

$$\overline{\mu}(E) \leq \sum_{i=1}^{\infty} \overline{\mu}(E_i),$$

and thus $\overline{\mu}$ is countably additive. Since $\overline{\mu}(E) \geq 0$ for all $E \in \mathcal{M}$, and $\overline{\mu}(\emptyset) = \mu^*(\emptyset) = 0$, we conclude that $\overline{\mu}$ is a measure on $\mathcal{M}$. ■

Next we discuss the idea behind the extension of measure problem.

Remark 3.10 (Extension of measure) Let us formally state the extension problem: given a measure μ on a field $\mathcal{A}$ of subsets of a space Ω, we wish to extend it to a measure on a σ-field $\mathcal{M}$ containing $\mathcal{A}$. To accomplish this, we construct an outer measure μ^* and show that the measure $\overline{\mu}$ (the restriction of μ^* to $\mathcal{M}$), induced by μ^*, is the desired extension defined on $\sigma(\mathcal{A})$ that agrees with μ on sets from the field $\mathcal{A}$. This extension can be accomplished in a unique way. In particular, for any set $A \subset \Omega$, we define the set function

$$\mu^*(A) = \inf_{\{A_i\}} \sum_{i=1}^{\infty} \mu(A_i), \tag{3.6}$$

where $\{A_i\}$ ranges over all sequences of $\mathcal{A}$-sets such that $A \subset \bigcup_{i=1}^{\infty} A_i$. We call μ^* the outer measure induced by μ. Carathéodory (1918) was the first to define outer measures. The proofs of the following results concerning μ^* can be found in Royden (1989, pp. 292-295).

1. If $A \in \mathcal{A}$ and $\{A_i\}$ is any sequence of $\mathcal{A}$-sets such that $A \subset \bigcup_{i=1}^{\infty} A_i$, then $\mu(A) \leq \sum_{i=1}^{\infty} \mu(A_i)$.

2. If $A \in \mathcal{A}$ then $\mu^*(A) = \mu(A)$.

3. The set function μ^* is an outer measure.

4. If $A \in \mathcal{A}$ then A is measurable with respect to μ^*, i.e., $\mathcal{A} \subseteq \mathcal{M}$.

5. Let μ be a measure on a field $\mathcal{A}$ and μ^* the outer measure induced by μ and E any set. Then for $\varepsilon > 0$ there is a set $A \in \mathcal{A}$ with $E \subset A$ and $\mu^*(A) \leq \mu^*(E) + \varepsilon$. There is also a set $B \in \mathcal{A}_{\sigma\delta}$ with $E \subset B$ and $\mu^*(E) = \mu^*(B)$, where

$$\mathcal{A}_{\sigma\delta} = \left\{ \bigcap_{i=1}^{\infty} A_i : A_i \in \mathcal{A}_\sigma \right\},$$

and

$$\mathcal{A}_\sigma = \left\{ \bigcup_{i=1}^{\infty} A_i : A_i \in \mathcal{A} \right\}.$$

6. Let μ be a σ-finite measure on a field $\mathcal{A}$ and μ^* be the outer measure generated by μ. A set E is μ^*-measurable if and only if E is the proper difference $A \setminus B$ of a set $A \in \mathcal{A}_{\sigma\delta}$ and a set B with $\mu^*(B) = 0$. Each set B with $\mu^*(B) = 0$ is contained in a set $C \in \mathcal{A}_{\sigma\delta}$, with $\mu^*(C) = 0$.

7. The following relationship is true in general: $\mathcal{A} \subseteq \sigma(\mathcal{A}) \subseteq \mathcal{M}(\mu^*) \subseteq 2^\Omega$.

The results of the previous remark are summarized in Carathéodory's theorem (Royden, 1989, p. 295).

Theorem 3.10 (Carathéodory) Let μ be a measure on a field $\mathcal{A}$ and μ^* the outer measure induced by μ. Then the restriction $\overline{\mu}$ of μ^* to the μ^*-measurable sets is an extension of μ to a σ-field containing $\mathcal{A}$. If μ is finite or σ-finite then so is $\overline{\mu}$. If μ is σ-finite, then $\overline{\mu}$ is the only measure on the smallest σ-field containing $\mathcal{A}$, i.e., on $\sigma(\mathcal{A})$, which is an extension of μ.

3.6.3 Construction of the Lebesgue Measure

The construction of the Lebesgue measure that follows is a special case of the extension of measure by Carathéodory. We illustrate the extension approach by defining the Lebesgue measure over $\Omega = \mathcal{R}$. A natural way of measuring an open interval $I = (a, b)$ in $\mathcal{R}$ is its length, that is, $l(I) = |I| = b - a$ and as we have seen, the Borel σ-field $\mathcal{B}_1$ on $\mathcal{R}$ is generated by the collection $\mathcal{I} = \{(a, b) : a < b\}$. We would like to extend the idea of length for open intervals to help us describe the "length" of any Borel set. First, define the collection of all finite disjoint unions of open intervals of $\mathcal{R}$ by

$$\mathcal{A} = \{A : A = \bigcup_{i=1}^{n} I_i,\ I_i = (a_i, b_i),\ a_i < b_i,\ i = 1, 2, \ldots, n\},$$

with $\mathcal{I} \subseteq \mathcal{A}$. We view the empty set $\emptyset$ as an element of $\mathcal{I}$ of length 0, while $\mathcal{R} = (-\infty, +\infty)$ is a member of $\mathcal{I}$ with length ∞. Using these conventions, it is easy to see that $\mathcal{A}$ is a field.

Now we can define the Lebesgue measure on the field $\mathcal{A}$ by

$$\lambda(A) = \sum_{i=1}^{n} l(I_i) = \sum_{i=1}^{n}(b_i - a_i).$$

Note that even if A assumes two different representations $A = \bigcup_{i=1}^{n_1} I_{1i} = \bigcup_{i=1}^{n_2} I_{2i}$, $\lambda(A)$ remains unchanged.

The next step is to extend the definition of length to all Borel sets, namely, $\mathcal{B}_1 = \sigma(\mathcal{I})$. For any set $A \subseteq \mathcal{R}$, define $\mathcal{E}_A$ to be the collection of collections of open intervals of $\mathcal{R}$ that cover A, that is

$$\mathcal{E}_A = \{\{I_i\} : A \subset \bigcup_{j \in J} I_j,\ I_i = (a_j, b_j),\ a_j < b_j, \text{ for all } j\}.$$

Finally, we define the outer measure μ induced by l on the class $\mathcal{E}_A$, $\forall A \subseteq \mathcal{R}$ by

$$\mu(A) = \inf_{A \subset \bigcup_{j \in J} I_j} \sum_{j \in J} l(I_j). \tag{3.7}$$

This outer measure μ is called the Lebesgue outer measure, or simply the Lebesgue measure in $\mathcal{R}$. It is defined for all subsets of $\mathcal{R}$ and it agrees with l on $\mathcal{A}$.

Example 3.13 (Lebesgue measure in $(0, 1]$) Consider $\Omega = (0, 1]$ and let $\mathcal{B}_0$ denote the collection of finite disjoint unions of intervals $(a, b] \subseteq \Omega$, $0 < a < b \leq 1$, augmented by the empty set. Then $\mathcal{B}_0$ is a field but not a σ-field, since it does not contain, for instance, the singletons $\{x\} \left(= \bigcap_n (x - 1/n, x]\right)$. The Lebesgue measure μ as defined (restricted on $\mathcal{B}_0$) is a measure on $\mathcal{B}_0$, with $l(\Omega) = 1$ and it is the Lebesgue measure on $\mathcal{B}(\Omega) = \sigma(\mathcal{B}_0)$, the Borel σ-field in $\Omega = (0, 1]$. Since intervals in $(0, 1]$ form a π-system generating $\mathcal{B}(\Omega)$, λ is the only probability measure on $\mathcal{B}(\Omega)$ that assigns to each interval its length as its measure.

Example 3.14 (p-dimensional Lebesgue measure) Let $\mathbf{x} = (x_1, \ldots, x_p) \in \mathcal{R}^p$ and define the collection of bounded rectangles $R = \{R(\mathbf{a}_n, \mathbf{b}_n),\ \mathbf{a}_n, \mathbf{b}_n \in \mathcal{R}^p,\ a_{ni} < b_{ni}, i = 1, 2, \ldots, p\}$, augmented by the empty set, with $R(\mathbf{a}_n, \mathbf{b}_n) = \{\mathbf{x} : a_{ni} < x_i \leq b_{ni}, i = 1, 2, \ldots, p\}$. Then the σ-field $\mathcal{B}_p$ of Borel sets of $\mathcal{R}^p$ is generated by $R \cup \{\emptyset\}$ and the k-dimensional Lebesgue measure μ_p is defined on $\mathcal{B}_p$ as the extension of the ordinary volume λ_p in $\mathcal{R}^p$, where

$$\lambda_p(R(\mathbf{a}_n, \mathbf{b}_n)) = \prod_{i=1}^{p}(b_{ni} - a_{ni}),$$

and

$$\mu_p(B) = \inf_{B \subset \bigcup_n R(\mathbf{a}_n, \mathbf{b}_n)} \sum_n \lambda_p(R(\mathbf{a}_n, \mathbf{b}_n)),$$

for any $B \in \mathcal{B}_p \subseteq \mathcal{M}(\mu_p) \subseteq 2^{\mathcal{R}^p}$. The measure μ_p defined on $\mathcal{M}(\mu_p)$ will be denoted by $\overline{\mu}_p$ and is also called the p-dimensional Lebesgue measure. The Carathéodory sets $\mathcal{M}(\mu_p)$ of $\mathcal{R}^p$ are called (p-dimensional) Lebesgue measurable sets or simply Lebesgue sets. Any function is called Lebesgue measurable if and only if sets of the form of theorem 3.3 are Lebesgue measurable sets.

Some properties of the Lebesgue measure are collected below.

Remark 3.11 (Lebesgue measure) Clearly, the p-dimensional ordinary volume λ_p is translation invariant, i.e., for any bounded rectangle $R(\mathbf{a}, \mathbf{b})$ and point $\mathbf{x} \in \mathcal{R}^p$, $R(\mathbf{a}, \mathbf{b}) \oplus \mathbf{x} = R(\mathbf{a} + \mathbf{x}, \mathbf{b} + \mathbf{x})$, so that

$$\lambda_p(R(\mathbf{a}, \mathbf{b})) = \prod_{i=1}^{p}(b_i - a_i) = \prod_{i=1}^{p}(b_i + x_i - (a_i + x_i)) = \lambda_p(R(\mathbf{a} + \mathbf{x}, \mathbf{b} + \mathbf{x})).$$

1. Translation invariance This property can be extended to the Lebesgue measure defined on $\mathcal{B}_p$ (or even $\mathcal{M}(\mu_p)$), that is, for every $B \in \mathcal{B}_p$ and $\mathbf{x} \in \mathcal{R}^p$ we have $\mu_p(B \oplus \mathbf{x}) = \mu_p(B)$. In fact, it can be proven that every translation-invariant Radon measure v on $(\mathcal{R}^p, \mathcal{B}_p)$ must be a multiple of the Lebesgue measure, i.e., $v(B) = c\mu_p(B)$, for some constant c.

2. Lebesgue in lower dimensional space If $A \in \mathcal{B}_{p-1}$ then A has measure 0 with respect to the Lebesgue measure μ_p defined on $\mathcal{B}_p$.

One of the consequences of the Carathéodory extension theorem is discussed next.

Remark 3.12 (Measure space completion) Recall that a (signed nonnegative) measure space $(\Omega, \mathcal{A}, \mu)$, $\mu(A) \geq 0$, $A \in \mathcal{A}$, is said to be complete if $B \in \mathcal{A}$, with $\mu(B) = 0$ and $A \subset B$, imply $A \in \mathcal{A}$. Denote by $Null(\mu) = \{A \in \mathcal{A} : \mu(A) = 0\}$, all the measurable sets of measure zero with respect to μ, called null sets, and $Null(\mu)$ is called the null space.

Measure space completion refers to the process by which a measure space becomes complete, namely, it contains all subsets of null sets. Hence the original σ-field $\mathcal{A}$ needs to be augmented with the subsets of the null sets, otherwise we have the logical paradox of a set having zero measure, while its subset does not.

It can be shown (see for example Vestrup, 2003, pp. 88-96) that the Carathéodory extension and uniqueness theorem for a measure μ on a field $\mathcal{A}$, makes the measure space $(\Omega, \mathcal{M}(\mu^*), \mu^*_{\mathcal{M}(\mu^*)})$ the completion of the measure space $(\Omega, \sigma(\mathcal{A}), \mu^*_{\sigma(\mathcal{A})})$, where $\mu^*_{\sigma(\mathcal{A})}$ is the restriction of μ^* on $\sigma(\mathcal{A})$ and $\mu^*_{\mathcal{M}(\mu^*)}$ is the restriction of μ^* on $\mathcal{M}(\mu^*)$.

To satisfy our curiosity that a measure space is not complete by definition, we can show that the Lebesgue measure μ_1 restricted to the Borel σ-field induces a measure space $(\mathcal{R}, \mathcal{B}(\mathcal{R}), \mu_1)$ that is not complete, whereas, $(\mathcal{R}, \mathcal{M}(\mu_1), \mu_1)$ is a complete measure space.

Example 3.15 (Nonmeasurable set) A set will be nonmeasurable in $[0, 1)$ (with respect to the Lebesgue measure on $\mathcal{B}([0, 1))$) if it lives in the collection of sets between $\mathcal{B}([0, 1))$ and $2^{[0,1)}$, i.e., outside the Borel sets of $[0, 1)$. Let us construct such a Lebesgue nonmeasurable set. First, if $x, y \in [0, 1)$, define the sum modulo 1 of x and y, denoted by $x \dotplus y$, by $x \dotplus y = x + y$, if $x + y < 1$ and $x \dotplus y = x + y - 1$, if $x + y \geq 1$.

It can be shown that μ_1 is translation invariant with respect to modulo 1, that is, if $E \subset [0, 1)$ is a measurable set, then for all $y \in [0, 1)$ the set $E \dotplus y$ is measurable and $\mu_1(E \dotplus y) = \mu_1(E)$. Now if $x - y$ is a rational number, we say that x and y are equivalent and write $x \sim y$. This is an equivalence relation and hence partitions $[0, 1)$ into equivalence classes, that is, classes such that any two elements of one class differ by a rational number, while any two elements of different classes differ by an irrational number.

By the Axiom of Choice, there is a set A which contains exactly one element from each equivalence class. Let $\{r_i\}_{i=0}^{\infty}$ denote all the rational numbers in $[0, 1)$ with $r_0 = 0$ and define $A_i = A \dotplus r_i$, so that $A_0 = A$. Let $x \in A_i \cap A_j$. Then $x = a_i \dotplus r_i = a_j \dotplus r_i$, with $a_i, a_j \in A$. But $a_i - a_j = r_j - r_i$ is a rational number and hence $a_i \sim a_j$. Since A has only one element from each equivalence class, we must have $i = j$, which implies that if $i \neq j$, $A_i \cap A_j = \emptyset$, that is, $\{A_i\}$ is a pairwise disjoint collection of sets.

On the other hand, each real number $x \in [0, 1)$ is in some equivalence class and so is equivalent to an element in A. But if x differs from an element in A by the rational number r_i, then $x \in A_i$. Thus $\bigcup_i A_i = [0, 1)$. Since each A_i is a translation modulo 1 of A, each A_i will be measurable if A is and will have the same measure. This implies that

$$\mu_1([0, 1)) = \sum_{i=1}^{\infty} \mu_1(A_i) = \sum_{i=1}^{\infty} \mu_1(A),$$

and the right side is either zero or infinite, depending on whether $\mu_1(A)$ is zero or positive. But since $\mu_1([0, 1)) = 1$, this leads to a contradiction, and consequently, A cannot be measurable. This construction is due to Vitali and shows that there are sets in $[0, 1)$ that are not Borel sets.

3.7 Defining Integrals with Respect to Measures

The last major component of a measure theoretic framework is the concept of the integral with respect to a measure. The modern notion of the integral is due to Professeur Henri Lebesgue (1902). We define and study the integral

$$\int f d\mu = \int_{\Omega} f(\omega) d\mu(\omega) = \int_{\Omega} f(\omega)\mu(d\omega),$$

where $f : \Omega \to \mathcal{X}$ is some measurable function in a measure space $(\Omega, \mathcal{A}, \mu)$, where all the integrals above are equivalent forms and denote the same thing. We need to make some conventions before we proceed. We agree that $0 * \infty = \infty * 0 = 0$, $x * \infty = \infty * x = \infty$, $x > 0$, $x - \infty = -\infty$, $\infty - x = \infty$, $x \in \mathcal{R}$, $\infty * \infty = \infty$ and we say that $\infty - \infty$ is undefined. Moreover, we agree that $\inf \emptyset = \infty$. Although we can define the integral for any mapping, we consider measurable extended real-valued functions in this exposition.

Definition 3.15 Integral

Let $(\Omega, \mathcal{A}, \mu)$ be a measure space and $\{A_i\}_{i=1}^{n}$ a partition of Ω.

1. Simple function Let $h : \Omega \to \mathcal{R}$ be a simple function $h(\omega) = \sum_{i=1}^{n} h_i I_{A_i}(\omega)$, $\omega \in \Omega$, with $\{h_i\}_{i=1}^n \in \mathcal{R}$ and $A_i = \{\omega \in \Omega : h(\omega) = h_i\}$ is $\mathcal{A}$-measurable. The integral of h with respect to μ is defined by

$$I_h^{\mathcal{R}} = \int h d\mu = \sum_{i=1}^{n} h_i \mu(A_i).$$

2. $\overline{\mathcal{R}}_0^+$-valued function Let $f : \Omega \to \overline{\mathcal{R}}_0^+$ be a measurable, non-negative, extended real-valued function. We define the integral of f with respect to μ by

$$I_f^{\overline{\mathcal{R}}^+} = \int f d\mu = \sup_{h \le f} \int h d\mu = \sup_{\{A_i\}_{i=1}^n} \sum_{i=1}^{n} \left[\inf_{\omega \in A_i} f(\omega) \right] \mu(A_i),$$

where the first supremum is taken over all non-negative measurable simple functions h such that $0 \le h \le f$, while the second is taken over all partitions $\{A_i\}_{i=1}^n$ of Ω.

3. $\overline{\mathcal{R}}$-valued function The integral of an $\overline{\mathcal{R}}$-valued function f with respect to μ is defined in terms of f^+ and f^-, the positive and negative parts of f (see Appendix A.1) by

$$I_f^{\overline{\mathcal{R}}} = \int f d\mu = \int f^+ d\mu - \int f^- d\mu,$$

provided that the two integrals on the right are not infinity at the same time, in which case $\int f d\mu$ is undefined. If f vanishes outside a measurable set $A \subset \Omega$, we define

$$\int_A f d\mu = \int f I_A d\mu = \int f^+ I_A d\mu - \int f^- I_A d\mu.$$

Some discussion is in order about when we say that an integral exists.

Remark 3.13 (Lebesgue integral and integral existence issues) If the measure μ is the Lebesgue measure, then the integral is called the Lebesgue integral. By definition 3.11, μ is a nonnegative signed measure, i.e., $\mu(A_i) \ge 0$, with $y_i = \inf_{\omega \in A_i} f(\omega)$ and therefore the signs of the integrals $I_h^{\mathcal{R}}$ and $I_f^{\overline{\mathcal{R}}_0^+}$ depend on the values y_i. When μ is a finite or σ-finite measure then the integral is finite. If f is bounded in $\mathcal{R}$ then the integral is finite, but if f is, say, $\overline{\mathcal{R}}$-valued then it is possible that some $y_i = \infty$ or $-\infty$, in which case we say that the integral $I_f^{\overline{\mathcal{R}}_0^+}$ exists but is infinite.

In the case where $y_i\mu(A_i) = \infty$ and $y_j\mu(A_j) = -\infty$, for some $i \neq j$, we say that the integral does not exist or that it is undefined, since $\infty - \infty$ is undefined. Finally, note that the value of $I_h^{\mathcal{R}}$ is independent of the representation of h we use. Indeed, if $h(x) = \sum_j y_j I_{B_j}(x)$ is a non-canonical representation of h, we have $B_j = \{x : h(x) = y_j\} = \bigcup_{i: x_i = y_j} A_i$, so that $y_j\mu(B_j) = \sum_{i: x_i = y_j} x_i\mu(A_i)$, by the additivity of μ and hence

$$\sum_j y_j\mu(B_j) = \sum_j \sum_{i: x_i = y_j} x_i\mu(A_i) = \sum_{i=1}^{n} x_i\mu(A_i) = \int h d\mu.$$

Example 3.16 (Integrating with respect to counting measure) Let $\Omega = \{1, 2, \ldots, n\}$ and take μ to be the counting measure assigning measure 1 to each singleton, so that $(\Omega, 2^{\Omega}, \mu)$ is a measure space. Consider the function $f(\omega) = \omega$, $\forall \omega \in \Omega$ and let $\{A_i\}_{i=1}^{n}$ be the partition of Ω into its singletons, where $A_i = \{i\}$, with $\mu(A_i) = 1$, $i = 1, 2, \ldots, n$. Clearly, f is simple since it can be written as $f(\omega) = \sum\limits_{i=1}^{n} i I_{A_i}(\omega)$ and its integral is therefore given by

$$\int f d\mu = \sum_{i=1}^{n} i\mu(A_i) = \sum_{i=1}^{n} i = n(n+1)/2.$$

The following theorem illustrates how we can prove some basic results for integrals of simple functions.

Theorem 3.11 Let $h = \sum\limits_{i=1}^{n} x_i I_{A_i}$ and $g = \sum\limits_{j=1}^{m} y_j I_{B_j}$ be two real-valued simple functions in the measure space $(\Omega, \mathcal{A}, \mu)$ and a and b real constants. Then we can show that
(i) $\int (ah + bg) d\mu = a \int h d\mu + b \int g d\mu$,
(ii) if $g \leq h$ a.e. $[\mu]$ then $\int g d\mu \leq \int h d\mu$, and
(iii) if h is bounded then $\inf\limits_{h \leq \varphi} \int \varphi d\mu = \sup\limits_{\psi \leq h} \int \psi d\mu$, where φ and ψ range over all simple functions.

Proof. We prove only the first two parts here. The last part is requested as an exercise.
(i) First note that the collection of intersections $\{E_k\} = \{A_i \cap B_j\}$ form a finite disjoint collection of $N = nm$ measurable sets since $\{A_i\}$ and $\{B_j\}$ are the sets in the canonical representations of h and g, respectively, so that we may write $h = \sum\limits_{k=1}^{N} a_k I_{E_k}$ and $g = \sum\limits_{k=1}^{N} b_k I_{E_k}$. Therefore in view of remark 3.13 we can write

$$ah + bg = \sum_{k=1}^{N} (aa_k + bb_k) I_{E_k} = \int (ah + bg)\, d\mu,$$

and the claim is established.
(ii) To prove this part we note that

$$\int h d\mu - \int g d\mu = \int (h - g) d\mu \geq 0,$$

since the integral of a simple function, which is greater than or equal to zero a.e., is nonnegative by the definition of the integral. ■

Clearly, owing to the sequential form of definition 3.15, we can obtain results first for $I_h^{\mathcal{R}}$, which is easier to do, and then extend them to the $\overline{\mathcal{R}}_0^+$ and $\overline{\mathcal{R}}$ cases. When this extension is not straightforward we require additional tools (see remark 3.15). First, we summarize some of these results for $\overline{\mathcal{R}}_0^+$-valued integrands below.

Remark 3.14 (Properties of $I_f^{\overline{\mathcal{R}}_0^+}$) Assume that f, g are $\overline{\mathcal{R}}_0^+$-valued in what follows.

1. If $f = 0$ a.e. $[\mu]$ then $\int f d\mu = 0$.

2. Monotonicity If $0 \leq f(\omega) \leq g(\omega)$, $\forall \omega \in \Omega$, then $\int f d\mu \leq \int g d\mu$. Moreover, if $0 \leq f \leq g$ a.e. $[\mu]$ then $\int f d\mu \leq \int g d\mu$.

3. Linearity If $a, b \in \mathcal{R}_0^+$ then $\int (af + bg) d\mu = a \int f d\mu + b \int g d\mu$.

4. If $f = g$ a.e. $[\mu]$ then $\int f d\mu = \int g d\mu$.

5. If $\mu(\{\omega : f(\omega) > 0\}) > 0$ then $\int f d\mu > 0$.

6. If $\int f d\mu < \infty$ then $f < \infty$ a.e. $[\mu]$.

7. Assume that $\mu(A) = 0$. Then $\int f d\mu = \int\limits_A f d\mu$ by definition, since μ vanishes outside A.

8. If f is bounded on a measurable set $E \subset \mathcal{R}$ with $\mu_1(E) < \infty$, then $\sup\limits_{h \leq f} \int\limits_E h d\mu_1 = \inf\limits_{f \leq \varphi} \int\limits_E \varphi d\mu_1$, for all simple functions h and φ if and only if f is measurable.

The additional tool we need in order to generalize results from simple functions to any $\overline{\mathcal{R}}_0^+$-valued function and then to any $\overline{\mathcal{R}}$-valued function is the following.

Theorem 3.12 (Monotone convergence theorem (MCT)) Let $\{f_n\}$ be a sequence of nonnegative measurable functions such that $0 \leq f_n(\omega) \uparrow f(\omega)$, $\forall \omega \in \Omega$. Then $0 \leq \int f_n d\mu \uparrow \int f d\mu$.

Proof. Since $f_n \leq f$, using remark 3.14.2 we have that the sequence $a_n = \int f_n d\mu$ is nondecreasing and bounded above by $\int f d\mu = \sup\limits_{\{A_i\}_{i=1}^k} \sum\limits_{i=1}^k \left[\inf\limits_{\omega \in A_i} f(\omega)\right] \mu(A_i)$. Now if $a_n = \infty$ for any positive integer n then the claim is trivially true.

Therefore, assuming that $a_n < \infty$, $\forall n = 1, 2, \ldots$, to complete the proof it suffices to show that

$$\lim_n a_n = \lim_n \int f_n d\mu \geq \sum_{i=1}^k x_i \mu(A_i), \tag{3.8}$$

for any partition $\{A_i\}_{i=1}^k$ of Ω, with $x_i = \inf\limits_{\omega \in A_i} f(\omega)$. Fix an $\varepsilon > 0$, with $\varepsilon < x_i$ and set $A_{in} = \{\omega \in A_i : f_n(\omega) > x_i - \varepsilon\}$, $i = 1, 2, \ldots, k$. Since $f_n \uparrow f$ we have that $A_{in} \uparrow A_i$, as $n \to \infty$, with $0 < \mu(A_{in}) < \mu(A_i) < \infty$. Now decomposing Ω into $A_{1n}, \ldots, A_{kn}$ and $\left(\bigcup\limits_{i=1}^k A_{in}\right)^c$ we note that

$$\int f_n d\mu \geq \sum_{i=1}^k (x_i - \varepsilon)\mu(A_{in}) \to \sum_{i=1}^k (x_i - \varepsilon)\mu(A_i) = \sum_{i=1}^k x_i \mu(A_i) - \varepsilon \sum_{i=1}^k \mu(A_i),$$

for arbitrary ε. Sending $\varepsilon \to 0$ we have that equation (3.8) holds. ■

Example 3.17 (Series of real numbers and MCT) Consider a countable space $\Omega = \{1, 2, \dots, \infty\}$ and take μ to be the counting measure so that $(\Omega, 2^\Omega, \mu)$ is a measure space. Now any $\overline{\mathcal{R}}^+$ function defined on Ω can be thought of as a sequence $x_m \geq 0$, $m = 1, 2, \dots$. Consider the sequence of functions $x_m^{(n)} = x_m$, if $m \leq n$ and 0, if $m > n$ and let $\{A_i\}_{i=1}^{n+1}$ be a partition of Ω, with $A_i = \{i\}$, $i = 1, 2, \dots, n$ and $A_{n+1} = \{n+1, n+2, \dots\}$. Then we can write $x_m^{(n)} = x_m I_{[1,n]}(m) = \sum\limits_{k=1}^{n} x_k I_{A_k}(m)$, $m, n = 1, 2, \dots$, a sum of simple functions $f_m^{(k)} = I_{A_k}(m)$, so that its integral is given by

$$\int x_m^{(n)} d\mu = \sum_{k=1}^{n} x_k \int f_m^{(k)} d\mu = \sum_{k=1}^{n} x_k \mu(A_k) = \sum_{k=1}^{n} x_k.$$

Now applying the MCT on the sequence $\{x_m^{(n)}\}$, with $0 \leq x_m^{(n)} \uparrow \sum\limits_{k=1}^{\infty} x_k I_{A_k}(m)$ yields $\sum\limits_{k=1}^{n} x_k \uparrow \sum\limits_{k=1}^{\infty} x_k$, where the limit $\sum\limits_{k=1}^{\infty} x_k$ may be finite or infinite.

Definition 3.16 Integrability

Let $f : \Omega \to \overline{\mathcal{R}}$ be defined on a measure space $(\Omega, \mathcal{A}, \mu)$. We say that f is integrable with respect to μ if

$$\int |f| d\mu = \int f^+ d\mu + \int f^- d\mu < +\infty.$$

Example 3.18 (Integrability counterexample) Integrability not only implies that the integral is defined but that it is also finite. In particular, we must have $\int f^+ d\mu < +\infty$ and $\int f^- d\mu < +\infty$. Consider the setup of the previous example but now assume that x_m is $\mathcal{R}$-valued. The function $x_m = (-1)^{m+1}/m$, $m = 1, 2, \dots$, is not integrable by definition 3.16 since $\sum\limits_{m=1}^{\infty} x_m^+ = \sum\limits_{m=1}^{\infty} x_m^- = +\infty$, i.e., the integral is undefined but the alternative harmonic series $\sum\limits_{m=1}^{\infty} (-1)^{m+1}/m$, does converge since $\lim\limits_{n} \sum\limits_{m=1}^{n} (-1)^{m+1}/m = \log 2$. This illustrates why integrability of f requires both $\int f^+ d\mu$ and $\int f^- d\mu$ to be finite.

Next we discuss how we can use the MCT and integration to the limit in order to prove statements about a nonnegative measurable function.

Remark 3.15 (Using the MCT) Note that from theorem 3.4 any nonnegative function f can be written as a limit of nonnegative increasing simple functions, the requirement of the MCT. Consequently, the MCT allows us to pass the limit under the integral sign. Many claims in integration and probability theory can be proven this way. The steps are summarized as follows.

Step 1: Show that the claim holds for the simple function $\varphi = I_A$.

Step 2: By linearity of $I_h^{\mathcal{R}}$, we have the claim for any nonnegative simple function, including sequences of simple functions like those of theorem 3.4.

Step 3: The claim holds for any nonnegative measurable function since there is a sequence of simple functions $\{f_n\}$ such that $0 \leq f_n(\omega) \uparrow f(\omega)$ and the MCT allows us to prove the claim by passing the limit under the integral sign.

Example 3.19 (Using the MCT for proofs) We illustrate the approach of remark 3.15.

1. Consider the linearity property (remark 3.14.3). The claim holds trivially for nonnegative simple functions (steps 1 and 2). By theorem 3.4, there exist nonnegative sequences of simple functions $\{f_n\}$ and $\{g_n\}$ such that $0 \leq f_n \uparrow f$ and $0 \leq g_n \uparrow g$ and therefore $0 \leq af_n + bg_n \uparrow af + bg$, so that linearity of the integral for simple functions yields

$$a \int f_n d\mu + b \int g_n d\mu = \int (af_n + bg_n) d\mu,$$

and the MCT allows us to pass the limit under the integral sign and obtain the claim.

2. If $0 \leq f_n \uparrow f$ a.e. $[\mu]$ then $0 \leq \int f_n d\mu \uparrow \int f d\mu$. To see this consider $0 \leq f_n \uparrow f$ on the set $A = \{\omega \in \Omega : 0 \leq f_n(\omega) \uparrow f(\omega)\}$, with $\mu(A^c) = 0$, so that $0 \leq f_n I_A \uparrow f I_A$ holds everywhere. Then the MCT gives

$$\int f_n d\mu = \int f_n I_A d\mu \uparrow \int f I_A d\mu = \int f d\mu.$$

We are now ready to collect integral properties for extended real-valued integrands.

Remark 3.16 (Properties of $I_f^{\overline{\mathcal{R}}}$) For what follows, assume that all functions are $\overline{\mathcal{R}}$-valued and measurable in $(\Omega, \mathcal{A}, \mu)$ where $a, b, c \in \mathcal{R}$.

1. Linearity We have $\int (af + bg) d\mu = a \int f d\mu + b \int g d\mu$, provided that the expression on the right is meaningful. All integrals are defined when f and g are integrable.

2. If $f = c$ a.e. $[\mu]$ then $\int f d\mu = c\mu(\Omega)$.

3. If $f = g$ a.e. $[\mu]$ then either the integrals of f and g with respect to μ are both defined and are equal or neither is defined.

4. Monotonicity If $f \leq g$ a.e. $[\mu]$ and $\int f d\mu$ is defined and is different from $-\infty$ or $\int g d\mu$ is defined and is different from ∞ then both integrals are defined and $\int f d\mu \leq \int g d\mu$.

5. If $\int f d\mu = \int g d\mu < +\infty$ and $f \leq g$ a.e. $[\mu]$ then $f = g$ a.e. $[\mu]$.

6. If $\int f d\mu$ is defined then $\left|\int f d\mu\right| < \int |f| d\mu$.

7. If $\int f d\mu$ is undefined then $\int |f| d\mu = +\infty$.

8. $\int |f+g| d\mu \leq \int |f| d\mu + \int |g| d\mu$.

9. If $f_n \geq 0$ then $\int \sum_{n=1}^{+\infty} f_n d\mu = \sum_{n=1}^{+\infty} \int f_n d\mu$.

10. If $\sum_{n=1}^{+\infty} f_n$ converges a.e. $[\mu]$ and $\left|\sum_{k=1}^{n} f_k\right| \leq g$ a.e. $[\mu]$, where g is integrable, then $\sum_{n=1}^{+\infty} f_n$ and the f_n are integrable and $\int \sum_{n=1}^{+\infty} f_n d\mu = \sum_{n=1}^{+\infty} \int f_n d\mu$.

11. If $\sum_{n=1}^{+\infty} \int |f_n| d\mu < \infty$ then $\sum_{n=1}^{+\infty} f_n$ converges absolutely a.e. $[\mu]$ and is integrable and $\int \sum_{n=1}^{+\infty} f_n d\mu = \sum_{n=1}^{+\infty} \int f_n d\mu$.

12. If $|f| \leq |g|$ a.e. $[\mu]$ and g is integrable then f is integrable as well.

13. If $\mu(\Omega) < \infty$, then a bounded function f is integrable.

The following theorems can be proven easily using the MCT and provide three more ways of passing the limit under the integral sign.

Theorem 3.13 (Fatou's lemma) Let $\{f_n\}$ be a sequence of nonnegative measurable functions. Then

$$\int \underline{\lim} f_n d\mu \leq \underline{\lim} \int f_n d\mu.$$

Proof. Set $g_n = \inf_{n \geq k} f_k$ and note that $0 \leq g_n \uparrow g = \underline{\lim} f_n$ so that the fact that $g_n \leq f_n$ and the MCT gives

$$\int f_n d\mu \geq \int g_n d\mu \uparrow \int g d\mu,$$

as claimed. ■

Theorem 3.14 (Lebesgue's dominated convergence theorem (DCT)) Let $\{f_n\}$, f be $\overline{\mathcal{R}}$-valued and $g \geq 0$, all measurable functions in (Ω, A, μ). If $|f_n| \leq g$ a.e. $[\mu]$, where g is integrable and if $f_n \to f$ a.e. $[\mu]$, then f and f_n are integrable and $\int f_n d\mu \to \int f d\mu$.

Proof. Apply Fatou's Lemma on the sequences of functions $g - f_n$ and $g + f_n$. ■

Theorem 3.15 (Bounded convergence theorem (BCT)) Suppose that $\mu(\Omega) < \infty$ and that the f_n are uniformly bounded, i.e., there exists $M > 0$ such that

$|f_n(\omega)| \leq M < \infty$, $\forall \omega \in \Omega$ and $n = 1, 2, \ldots$. Then $f_n \to f$ a.e. $[\mu]$ implies $\int f_n d\mu \to \int f d\mu$.

Proof. This is a special case of theorem 3.14. ■

The last approach we will discuss that allows us to pass the limit under the integral sign is based on the property of Uniform Integrability. Suppose that f is integrable. Then $|f|I_{[|f|\geq a]}$ goes to 0 a.e. $[\mu]$ as $a \to \infty$ and is dominated by $|f|$ so that

$$\lim_{a\to\infty} \int_{|f|\geq a} |f| d\mu = 0. \tag{3.9}$$

Definition 3.17 Uniform integrability

A sequence $\{f_n\}$ is uniformly integrable if (3.9) holds uniformly in n, that is,

$$\lim_{a\to\infty} \sup_n \int_{|f_n|\geq a} |f_n| d\mu = 0.$$

We collect some results involving uniformly integrable sequences of functions below.

Remark 3.17 (Uniform integrability) For what follows assume that $\mu(\Omega) < \infty$ and $f_n \to f$ a.e. $[\mu]$.

1. If $\{f_n\}$ are uniformly integrable then f is integrable and

$$\int f_n d\mu \to \int f d\mu. \tag{3.10}$$

2. If f and $\{f_n\}$ are nonnegative and integrable then equation (3.10) implies that $\{f_n\}$ are uniformly integrable.

3. If f and $\{f_n\}$ are integrable then the following are equivalent (i) $\{f_n\}$ are uniformly integrable, (b) $\int |f - f_n| d\mu \to 0$, and (c) $\int |f_n| d\mu \to \int |f| d\mu$.

We end this section with a note on treating the integral as an operator.

Remark 3.18 (Integral operator) We can define an integral operator on the collection of all $\overline{\mathcal{R}}$-valued functions by $T(f) = \int f d\mu$, which is a functional that takes as arguments functions and yields their integral with respect to μ. Since $T(af + bg) = aT(f) + bT(g)$, $T(f)$ is called a linear operator.

3.7.1 Change of Variable and Integration over Sets

An important consequence of transformation of measure in integration theory is change of variable, which can be used to give tractable forms for integrals that are otherwise hard to compute. Suppose that $(\Omega_1, \mathcal{A}_1, \mu)$ is a measure space and let $(\Omega_2, \mathcal{A}_2)$ be a measurable space. Consider a mapping $T : \Omega_1 \to \Omega_2$ that is $\mathcal{A}_1|\mathcal{A}_2$-measurable and define the induced measure μT^{-1} on $\mathcal{A}_2$ (see remark 3.8.9), by

$$\mu T^{-1}(A_2) = \mu(T^{-1}(A_2)), \ \forall A_2 \in \mathcal{A}_2.$$

Theorem 3.16 (Change of variable: general measure) If $f : \Omega_2 \to \overline{\mathcal{R}^+}$ is $\mathcal{A}_2|\mathcal{B}(\overline{\mathcal{R}^+})$-measurable then $f \circ T : \Omega_1 \to \overline{\mathcal{R}^+}$ is $\mathcal{A}_1|\mathcal{B}(\overline{\mathcal{R}^+})$-measurable and

$$\int_{\Omega_1} f(T(\omega))\mu(d\omega) = \int_{\Omega_2} f(t)\mu T^{-1}(dt). \tag{3.11}$$

If $f : \Omega_2 \to \overline{\mathcal{R}}$ is $\mathcal{A}_2|\mathcal{B}(\overline{\mathcal{R}})$-measurable then $f \circ T : \Omega_1 \to \overline{\mathcal{R}}$ is $\mathcal{A}_1|\mathcal{B}(\overline{\mathcal{R}})$-measurable and it is integrable $[\mu T^{-1}]$ if and only if $f \circ T$ is integrable $[\mu]$, in which case (3.11) and

$$\int_{T^{-1}(A_2)} f(T(\omega))\mu(d\omega) = \int_{A_2} f(t)\mu T^{-1}(dt), \tag{3.12}$$

for all $A_2 \in \mathcal{A}_2$, hold, in the sense that if the left side is defined then the other is defined and they are the same. When f is $\overline{\mathcal{R}^+}$-valued (3.11) always holds.

Proof. The measurability arguments follow by remark 3.7.2. For the trivial simple function $f = I_{A_2}$, $A_2 \in \mathcal{A}_2$, we have $f \circ T = I_{T^{-1}(A_2)}$ and (3.11) reduces to equation (3.11). By linearity of the integral, (3.11) holds for all nonnegative simple functions and if f_n are simple functions such that $0 \le f_n \uparrow f$, then $0 \le f_n \circ T \uparrow f \circ T$ and (3.11) follows by the MCT.

Applying (3.11) to $|f|$ establishes the assertion about integrability and for integrable f, (3.11) follows by decomposition into positive and negative parts. Finally, replacing f by fI_{A_2} in (3.11) reduces to (3.12). ■

Example 3.20 (Change of variable) Let $(\Omega_2, \mathcal{A}_2) = (\mathcal{R}, \mathcal{B}_1)$ and let $T = \varphi$, a $\mathcal{B}_1$-measurable, real-valued function. If $f(x) = x$ then (3.11) becomes

$$\int_{\Omega_1} \varphi(\omega)\mu(d\omega) = \int_{\mathcal{R}} t\mu\varphi^{-1}(dt).$$

For a simple function $\varphi = \sum_i x_i I_{A_i}$ we have that $\mu\varphi^{-1}$ has mass $\mu(A_i)$ at x_i so that the latter integral reduces to $\sum_i x_i \mu(A_i)$.

We summarize some results on integration over sets in the next theorem (Billingsley, 2012, p. 226).

Theorem 3.17 (Integration over sets) We can show the following.

(i) If $A_1, A_2, \ldots,$ are disjoint with $B = \bigcup_{n=1}^{+\infty} A_n$ and if f is either nonnegative or integrable then

$$\int_B f d\mu = \sum_{n=1}^{+\infty} \int_{A_n} f d\mu.$$

(ii) If f and g are nonnegative and $\int_A f d\mu = \int_A g d\mu$ for all $A \in \mathcal{A}$ and if μ is σ-finite then $f = g$ a.e. $[\mu]$.

(iii) If f and g are integrable and $\int\limits_A f d\mu = \int\limits_A g d\mu$ for all $A \in \mathcal{A}$ and if μ is σ-finite then $f = g$ a.e. $[\mu]$.
(iv) If f and g are integrable and $\int\limits_A f d\mu = \int\limits_A g d\mu$ for all $A \in \mathcal{P}$, where $\mathcal{P}$ is a π-system generating $\mathcal{A}$ and Ω is a finite or countable union of $\mathcal{P}$-sets, then $f = g$ a.e. $[\mu]$.

3.7.2 Lebesgue, Riemann and Riemann-Stieltjes Integrals

Recall that for any function $f : [a, b] \to \mathcal{R}$, the Riemann integral $R - \int\limits_a^b f(x)dx$ is defined over an arbitrary finite partition $\{A_i = (\xi_{i-1}, \xi_i)\}_{i=1}^n$, with $a = \xi_0 < \xi_1 < \cdots < \xi_n = b$, as the common value of the upper and lower Riemann integrals, that is

$$R - \int\limits_a^b f(x)dx = \sup_{\{\xi_i\}_{i=0}^n} \sum_{i=1}^n \left[\inf_{\xi_{i-1}<x<\xi_i} f(x) \right] (\xi_i - \xi_{i-1}). \tag{3.13}$$

On the other hand, the Lebesgue integral of f denoted by $\mathcal{L} - \int\limits_a^b f(x)dx$ (or $\int f I_{[a,b]} d\mu_1 = \int\limits_a^b f d\mu_1$) is defined using equation (3.7) by

$$\int f I_{[a,b]} d\mu_1 = \sup_{\{A_i\}_{i=1}^n} \sum_{i=1}^n \left[\inf_{x \in A_i} f(x) \right] \mu_1(A_i) \tag{3.14}$$

where the Lebesgue measure in $\mathcal{R}$ for the open interval A_i is simply $\mu_1(A_i) = \xi_i - \xi_{i-1}$, $i = 1, 2, \ldots, n$. Obviously, equations (3.13) and (3.14) appear to arrive at the same result, although the two integral definitions are different. Does this happen always? The answer is no, as we see in the next classic example.

Example 3.21 (Riemann integral counterexample) Let $f(x) = 0$, if x irrational, or 1, if x rational, be the indicator function of the set of rationals in $(0, 1]$. Then the Lebesgue integral $\mathcal{L} - \int\limits_0^1 f(x)dx$ is 0 because $f = 0$ *a.e.* $[\mu_1]$. But for an arbitrary partition $\{A_i = (\xi_{i-1}, \xi_i)\}$ of $(0, 1]$, $M_i = \sup_{\xi_{i-1}<x<\xi_i} f(x) = 1$ and $m_i = \inf_{\xi_{i-1}<x<\xi_i} f(x) = 0$ for all i, so that the upper and lower Riemann sums are given by

$$S = \sum_{i=1}^n (\xi_i - \xi_{i-1})1 = b - a,$$

and

$$s = \sum_{i=1}^n (\xi_i - \xi_{i-1})0 = 0,$$

and hence $\overline{R} - \int_a^b f(x)dx = \inf_{\{\xi_i\}_{i=0}^n} S = b - a$, while $\underline{R} - \int_a^b f(x)dx = \sup_{\{\xi_i\}_{i=0}^n} s = 0$, so that f is not Riemann integrable.

Next we discuss when the Lebesgue and Riemann integrals coincide.

Remark 3.19 (Lebesgue and Riemann integrals) Clearly, the Lebesgue integral (when defined) can be thought of as a generalization of the Riemann integral. For what follows in the rest of the book, Lebesgue integrals will be computed using their Riemann versions, whenever possible, since Riemann integrals are easier to compute. Let $f : \Omega \to [a, b]$, $\Omega \subset \mathcal{R}^p$, be a bounded, measurable function in the space $(\Omega, \mathcal{B}_p, \mu_p)$. The following results clarify when the two integrals coincide.

1. If f is Riemann integrable on $[a, b]$ then it is measurable and $R - \int_a^b f(x)dx = \mathcal{L} - \int_a^b f(x)dx$.

2. The function f is Riemann integrable if and only if f is continuous on Ω a.e. $[\mu_p]$.

3. If f is Riemann integrable then $R - \int_\Omega f(\mathbf{x})d\mathbf{x} = \mathcal{L} - \int_\Omega f d\mu_p$.

Example 3.22 (Riemann does not imply Lebesgue integrability) In order to appreciate that Riemann integrability does not imply Lebesgue integrability always, consider integrating $f(x) = \frac{\sin x}{x}$, $x \in (0, +\infty)$. It can be shown (see example 3.23) that $R - \int_0^{+\infty} f(x)dx = \lim_{t\to\infty} \int_0^t \frac{\sin x}{x} dx = \pi/2$, but the Lebesgue integral is undefined since $f(x)$ is not Lebesgue integrable over $(0, +\infty)$ because the integrals of f^+ and f^- are both $+\infty$.

The Riemann-Stieltjes (RS) integral $RS - \int_a^b f(x)dg(x)$ or simply $\int_a^b f(x)dg(x)$ is defined using similar arguments as in the case of the Riemann integral. First recall the $\varepsilon - \delta$ definition of the Riemann integral: a bounded function $f : [a, b] \to \mathcal{R}$ is Riemann integrable with $R - \int_a^b f(x)dx = r$ if and only if $\forall \varepsilon > 0$, $\exists \delta > 0$, such that for any finite partition $\{A_i = (\xi_{i-1}, \xi_i)\}_{i=1}^n$, with $a = \xi_0 < \xi_1 < \cdots < \xi_n = b$ and $\xi_i - \xi_{i-1} < \delta$ we have $\left| r - \sum_{i=1}^n f(x_i)(\xi_i - \xi_{i-1}) \right| < \varepsilon$, for some $x_i \in A_i$.

Similarly, a bounded function $f : [a, b] \to \mathcal{R}$ is RS integrable with respect to any bounded function $g : [a, b] \to \mathcal{R}$, if $\forall \varepsilon > 0$, $\exists \delta > 0$, such that for any finite partition $\{A_i = (\xi_{i-1}, \xi_i)\}_{i=1}^n$, with $a = \xi_0 < \xi_1 < \cdots < \xi_n = b$ and $g(\xi_i) - g(\xi_{i-1}) < \delta$ we have $\left| r - \sum_{i=1}^{n} f(x_i)(g(\xi_i) - g(\xi_{i-1})) \right| < \varepsilon$, for some $x_i \in A_i$.

Note that under certain conditions, such as Riemann integrability of f and f being RS integrable with respect to g over all bounded intervals, we can extend these definitions of the Riemann and RS integrals to any subset $A \subseteq \mathcal{R}$. Moreover, properties such as linearity, monotonicity and theorems like the MCT as easily seen for the RS integral using the definition and similar arguments like the ones we used for the construction of the general integral in the previous sections.

The following theorems are straightforward to prove using the definitions of the Riemann and RS integrals (e.g., see Fristedt and Gray, 1997, pp. 705, 707).

Theorem 3.18 (Riemann-Stieltjes and Riemann integrals) Let g be a function with a continuous first derivative on an interval $[a, b]$ and f a bounded $\overline{\mathcal{R}}$-valued function on $[a, b]$. Then fg' is Riemann integrable on $[a, b]$ if and only if f is RS integrable with respect to g on $[a, b]$ in which case we have

$$RS - \int_a^b f(x)dg(x) = R - \int_a^b f(x)g'(x)dx. \tag{3.15}$$

Theorem 3.19 (Integration by parts) Suppose that a function f is RS integrable with respect to a function g on an interval $[a, b]$. Then g is RS integrable with respect to f on $[a, b]$ and

$$\int_a^b f(x)dg(x) = f(b)g(b) - f(a)g(a) - \int_a^b g(x)df(x). \tag{3.16}$$

A direct consequence of the latter theorem is the following fundamental result.

Remark 3.20 (Fundamental theorem of calculus) Take $f(x) = 1$ and $g(x) = H(x)$ with $H'(x) = h(x)$ in equation (3.16) to obtain the Fundamental Theorem of Calculus

$$\int_a^b h(x)dx = \int_a^b dH(x) = H(b) - H(a),$$

since $\int_a^b H(x)d1 = 0$ by definition and $RS - \int_a^b dH(x) = R - \int_a^b h(x)dx$, by equation (3.15).

We end this section by connecting the RS integral with the Lebesgue integral on the Borel σ-field in $\mathcal{R}$. First recall that $\mathcal{B}(\mathcal{R}) = \sigma(\{(-\infty, x]\})$ and set $F(x) = \mu_1((-\infty, x])$, where μ_1 is the Lebesgue measure defined on the measure space $(\mathcal{R}, \mathcal{B}_1, \mu_1)$. Clearly, $F(x)$ is nonnegative and nondecreasing and we notice that for $a < b$ we have

$$\begin{aligned}\int_{\mathcal{R}} I_{(a,b]}(x)dF(x) &= \int_a^b dF(x) = F(b) - F(a) = \mu_1((-\infty, b]) - \mu_1((-\infty, a]) \\ &= \mu_1((-\infty, b] \smallsetminus (-\infty, a]) = \mu_1((a, b]) = \int_{\mathcal{R}} I_{(a,b]}(x)\mu_1(dx) = \int_a^b d\mu_1,\end{aligned}$$

so that the RS and Lebesgue integrals coincide when integrating the trivial simple function $I_{(a,b]}$. Omitting all the details involved, we can show that by linearity the two integrals coincide on any nonnegative simple function $\varphi = \sum_{i=1}^{n} a_i I_{A_i}$, i.e.,

$$\int_{\mathcal{R}} \sum_{i=1}^{n} a_i I_{A_i} dF(x) = \int_{\mathcal{R}} \sum_{i=1}^{n} a_i I_{A_i} \mu_1(dx)$$

so that the MCT applied on both sides and for both types of integrals yields that for any nonnegative, measurable function f the two integrals coincide. Therefore, for any $\overline{\mathcal{R}}$-valued measurable function $f = f^+ - f^-$ we have

$$\int_{\mathcal{R}} f(x)dF(x) = \int_{\mathcal{R}} f(x)\mu_1(dx), \tag{3.17}$$

provided that all integrals involved are defined. This construction can be easily extended to $\mathcal{R}^p$ by considering all bounded rectangles and the Lebesgue measure μ_p on $\mathcal{B}_p$. It can be shown that (3.17) holds for any Radon measure, not just the Lebesgue measure.

3.7.3 Radon-Nikodym Theorem

The theorem that allows us to define and work with densities is the Radon-Nikodym theorem. First note that for any measure μ and $f \geq 0$ measurable function defined on a measure space $(\Omega, \mathcal{A}, \mu)$, we can define a new measure $v(A) = \int_A f d\mu$, $A \in \mathcal{A}$, which will be finite if and only if f is integrable. Since the integral over a set of μ−measure zero is zero, we have that $v \ll \mu$. The converse of this result plays an important role in probability theory. In particular, the next theorem shows that every absolutely continuous measure v is obtained in this fashion from μ, subject to σ-finiteness restrictions.

Theorem 3.20 (Radon-Nikodym) Let $(\Omega, \mathcal{A}, \mu)$ be a σ-finite measure space and let v be a σ-finite measure defined on $\mathcal{A}$, which is absolutely continuous with respect to μ. Then there exists a nonnegative $\mathcal{A}|\mathcal{B}(\mathcal{R}_0^+)$-measurable function f such that for each set $A \in \mathcal{A}$ we have

$$v(A) = \int_A f d\mu. \tag{3.18}$$

The function f is unique in the sense that if g is any measurable function with this property then $g = f$ a.e. $[\mu]$. The function f is called the Radon-Nikodym (R-N) derivative with respect to μ and is denoted by $f = \left[\frac{dv}{d\mu}\right]$.

Proof (Finite case). Consider first the finite case for the measures ν and μ and assume that $\nu \ll \mu$. Let $\mathcal{G}$ be the collection of nonnegative functions g such that $\int_A g d\mu \leq \nu(A)$, for all $A \in \mathcal{A}$ and let $a = \sup_{g \in \mathcal{G}} \int g d\mu \leq \nu(\Omega)$. Choose $g_n^* \in \mathcal{G}$ such that $\int g_n^* d\mu > a - \frac{1}{n}$ and set $f_n = \max\{g_1^*, ..., g_n^*\}$ with $\lim_n f_n = f$ (more precisely $f_n \uparrow f$). We claim that this f has the desired property, i.e., $f = \left[\frac{d\nu}{d\mu}\right]$, so that equality in $\int_A g d\mu \leq \nu(A)$ is achieved for $g = f$. First we show that $f \in \mathcal{G}$.

Take any $g_1, g_2 \in \mathcal{G}$ and write

$$\begin{aligned}\int_A \max\{g_1, g_2\} d\mu &= \int_{A \cap \{g_1 \geq g_2\}} \max\{g_1, g_2\} d\mu + \int_{A \cap \{g_1 < g_2\}} \max\{g_1, g_2\} d\mu \\ &\leq \nu(A \cap \{g_1 \geq g_2\}) + \nu(A \cap \{g_1 < g_2\}) = \nu(A),\end{aligned}$$

so that $\mathcal{G}$ is closed under the formation of finite maxima. Now take any $g_n \in \mathcal{G}$ such that $g_n \uparrow g$ and use the MCT to obtain

$$\int_A g d\mu = \lim_n \int_A g_n d\mu \leq \nu(A),$$

which implies that $\mathcal{G}$ is closed under nondecreasing limits. Consequently, since $f = \lim_n \max\{g_1^*, ..., g_n^*\}$, we have $f \in G$ and since

$$\int f d\mu = \lim_n \int f_n d\mu \geq \lim_n \int g_n^* d\mu > \lim_n \left(a - \frac{1}{n}\right) \geq a,$$

the integral $\int f d\mu$ is maximal, i.e., $\int f d\mu = a$.

Let $\nu_0(A) = \int_A f d\mu$ and note that $\nu_1(A) = \nu(A) - \nu_0(A) \geq 0$, for all $A \in \mathcal{A}$, since $f \in \mathcal{G}$. Furthermore, ν_0 is absolutely continuous with respect to μ by definition. Writing

$$\nu(A) = \nu_0(A) + \nu_1(A), \tag{3.19}$$

we see that ν_0 and ν_1 are finite and nonnegative measures. If we show that $\nu_1(A) = 0$, for all $A \in \mathcal{A}$, then $\nu_0(A) = \int_A f d\mu = \nu(A)$, for all $A \in \mathcal{A}$, as claimed. Suppose that ν_1 is not singular with respect to μ. From lemma 3.3, there is a set A_0 and $\varepsilon > 0$ such that $\mu(A_0) > 0$ and $\varepsilon\mu(E) \leq \nu_1(E)$, for all $E \subset A_0$. Then for all E we can write

$$\begin{aligned}\int_E (f + \varepsilon I_{A_0}) d\mu &= \int_E f d\mu + \varepsilon\mu(E \cap A_0) \leq \int_E f d\mu + \nu_1(E \cap A_0) \\ &= \int_{E \cap A_0} f d\mu + \nu_1(E \cap A_0) + \int_{E \cap A_0^c} f d\mu \\ &\leq \nu(E \cap A_0) + \int_{E \cap A_0^c} f d\mu \leq \nu(E \cap A_0) + \nu(E \cap A_0^c) = \nu(E),\end{aligned}$$

so that $f + \varepsilon I_{A_0} \in \mathcal{G}$. Since $\int (f + \varepsilon I_{A_0}) d\mu = a + \varepsilon\mu(A_0) > a$, f is not maximal. This contradiction arises since we assumed ν_1 is not singular with μ and so we must have $\nu_1 \perp \mu$, which means that there is some set $S \subset \Omega$ such that $\nu_1(S) = \mu(S^c) = 0$.

But since $\nu \ll \mu$, $\mu(S^c) = 0 \implies \nu(S^c) = 0$ and hence $\nu_1(S^c) = \nu(S^c) - \nu_0(S^c) \leq \nu(S^c) = 0$, so that $\nu_1(S) + \nu_1(S^c) = 0$, or $\nu_1(\Omega) = 0$. Hence $\nu_1(A) = 0$ for all $A \in \mathcal{A}$ and (3.19) yields the claim. Note that the absolute continuity $v \ll \mu$ was not used in the construction of v_0 and v_1, thus showing that v can be decomposed always into an absolutely continuous part and a singular part with respect to μ. Therefore, we can think of the presentation up to the last paragraph as the proof for the finite measure case of the Lebesgue decomposition theorem (remark 3.8.14). ■

Proof (σ-finite case). To prove the σ-finite case, recall remark 3.8.2 and the alternative definition of a σ-finite measure. Since μ and v are σ-finite measures, there are partitions $\{A_i\}$ and $\{B_j\}$ of Ω, such that $\mu(A_i) < \infty$ and $v(B_j) < \infty$. Note that the collection of sets $C = \{C : C = A_i \cap B_j\}$ form a partition of Ω. Consequently, we can write the σ-finite measures μ and v, as $\mu = \sum\limits_{i=1}^{+\infty} \mu_i$ and $v = \sum\limits_{i=1}^{+\infty} v_i$, where $\mu_i(A) = \mu(A \cap C_i) < \infty$ and $v_i(A) = v(A \cap C_i) < \infty$, $\forall A \in \mathcal{A}$, $i = 1, 2, \ldots$, are sequences of finite, mutually singular measures on $(\Omega, \mathcal{A})$, i.e., $\mu_i \perp \mu_j$, $i \neq j$ and $v_i \perp v_j$, $i \neq j$. Since μ_i and v_i are finite, the Lebesgue decomposition theorem for the finite case we proved above allows the decomposition of v_i as: $v_i = v_{i0} + v_{i1}$, where $v_{i0} \ll \mu_i$ and $v_{i1} \perp \mu_i$. Then for all $A \in \mathcal{A}$ we can write

$$\begin{aligned} v(A) &= \sum_{i=1}^{+\infty} v_i(A) = \sum_{i=1}^{+\infty} [v_{i0}(A) + v_{i1}(A)] = \sum_{i=1}^{+\infty} v_{i0}(A) + \sum_{i=1}^{+\infty} v_{i1}(A) \\ &= v_0(A) + v_1(A), \end{aligned}$$

where $v_0 = \sum\limits_{i=1}^{+\infty} v_{i0}$ and $v_1 = \sum\limits_{i=1}^{+\infty} v_{i1}$ and it is straightforward to show that v_0 and v_1 are σ-finite measures.

Now if $A \in \mathcal{A}$ with $\mu_i(A) = 0$, then $v_{i0}(A) = 0$, for all i and hence $v_0 \ll \mu_i$. Similarly, since $v_{i1} \perp \mu_i$, there exists $A \in \mathcal{A}$ such that $\mu_i(A^c) = v_{i1}(A) = 0$, so that $v_1 \perp \mu_i$, for all i.

Moreover, if $A \in \mathcal{A}$ with $\mu(A) = 0$, then since μ_i is finite and nonnegative we must have $\mu_i(A) = 0$, for all i and therefore $v_0(A) = 0$, which implies that $v_0 \ll \mu$. Consider now $v_1 \perp \mu_i$, for all i, so that there exists $A \in \mathcal{A}$ with $\mu_i(A^c) = v_1(A) = 0$. But $\mu(A^c) = \sum\limits_{i=1}^{+\infty} \mu_i(A^c) = 0\infty = 0$ and hence $v_1 \perp \mu$. The proof up to this point establishes the Lebesgue decomposition theorem in the general case.

Now assume that $v \ll \mu$, that is, for all $A \in \mathcal{A}$, $\mu(A) = 0 \Longrightarrow v(A) = 0$. But this implies that $v_i \ll \mu_i$, since $\mu(A) = 0 \Longrightarrow \mu_i(A) = 0$, for all i and $\mu_i(A) = \mu(A \cap C_i) = 0 \Longrightarrow v(A \cap C_i) = 0 \Longrightarrow v_i(A) = 0$. From the finite measure case of the Radon-Nikodym theorem, there exist nonnegative $\mathcal{A}|\mathcal{B}(\mathcal{R}_0^+)$-measurable functions $\{f_i\}$, such that $f_i = \left[\frac{dv_i}{d\mu_i}\right]$, $i = 1, 2, \ldots$, or $v_i(A) = \int\limits_A f_i d\mu_i$, $A \in \mathcal{A}$. Set $f = \sum\limits_{i=1}^{+\infty} f_i I_{A_i}$, so that $f = f_i$ on A_i. Clearly, f is nonnegative, finite valued and $\mathcal{A}|\mathcal{B}(\mathcal{R}_0^+)$-measurable. Now since $v = \sum\limits_{i=1}^{+\infty} v_i$ we can write

$$v(A) = \sum_{i=1}^{+\infty} v_i(A) = \sum_{i=1}^{+\infty} \int_A f_i d\mu_i = \sum_{i=1}^{+\infty} \int_A f_i I_{A_i} d\mu = \int_A \sum_{i=1}^{+\infty} f_i I_{A_i} d\mu = \int_A f d\mu,$$

where we used remark 3.16.9 to pass the sum under the integral sign and thus the claim is established. ■

The R-N derivative plays the role of the density of a probability measure and the following general properties can be useful in establishing a variety of results for a probability measure.

Remark 3.21 (R-N derivative properties) Let μ, λ and v be σ-finite measures on the same measurable space $(\Omega, \mathcal{A})$.

1. If $v \ll \mu$ and f is a nonnegative, measurable function then

$$\int f dv = \int f \left[\frac{dv}{d\mu}\right] d\mu.$$

2. If $v \ll \mu$ and $\lambda \ll \mu$ then

$$\left[\frac{d(v+\lambda)}{d\mu}\right] = \left[\frac{dv}{d\mu}\right] + \left[\frac{d\lambda}{d\mu}\right].$$

3. If $v \ll \mu \ll \lambda$ then

$$\left[\frac{dv}{d\lambda}\right] = \left[\frac{dv}{d\mu}\right]\left[\frac{d\mu}{d\lambda}\right].$$

4. If $v \ll \mu$ and $\mu \ll v$ then

$$\left[\frac{dv}{d\mu}\right] = \left[\frac{d\mu}{dv}\right]^{-1}.$$

3.7.4 Product Measure and Fubini Theorem

Extending integration theory from one to higher dimensions can be accomplished by defining product measures. The tool that we can use to compute integrals over products of σ-finite measure spaces is Professore Guido Fubini's famous theorem which allows us to calculate double integrals as iterated integrals.

Recall that if $(\Omega_1, \mathcal{A}_1)$ and $(\Omega_2, \mathcal{A}_2)$ are two measurable spaces then the product measurable space is defined by $(\Omega, \mathcal{A})$, where $\Omega = \Omega_1 \times \Omega_2$, $\mathcal{A} = \mathcal{A}_1 \bigotimes \mathcal{A}_2$ and $\mathcal{A}$ is generated by the collection of all measurable rectangles $A = A_1 \times A_2 \subset \Omega$, for $A_1 \in \mathcal{A}_1$, $A_2 \in \mathcal{A}_2$. Suppose that we equip the measurable spaces with two σ-finite measures μ_1, μ_2, so that $(\Omega_1, \mathcal{A}_1, \mu_1)$ and $(\Omega_2, \mathcal{A}_2, \mu_2)$ are σ-finite measure spaces.

To motivate the construction of the product measure, let us use a probabilistic point of view where we assume that μ_1, μ_2 are probability measures, so that $\mu_1(\Omega_1) = \mu_2(\Omega_1) = 1$. In this case, we would like the product measure we create on $(\Omega, \mathcal{A})$ to coincide with μ_1 and μ_2 for measurable rectangles of the form $A_1 \times \Omega_2$ and $\Omega_1 \times A_2$. An obvious candidate is the set function $\mu = \mu_1 \times \mu_2$ defined by

$$\mu(A_1 \times A_2) = \mu_1(A_1)\mu_2(A_2), \tag{3.20}$$

and we investigate its general properties next. Note that in $\mathcal{R}^p$ we have to work with the collection of all measurable rectangles since $\mathcal{B}_p$ is generated by these sets.

We begin our discussion with some useful results that will lead to the general definition of the product measure and corresponding product space. Recall remark 3.5.9 and the definition of the section of the set. In a similar fashion, we can define the ω_1 section of an $\mathcal{A}$-measurable function f by $f(\omega_1, .)$, for every $\omega_1 \in \Omega_1$. Based on these definitions we can prove the following.

Lemma 3.4 Let $(\Omega_1, \mathcal{A}_1, \mu_1)$ and $(\Omega_2, \mathcal{A}_2, \mu_2)$ be σ-finite measure spaces and consider the product measurable space $(\Omega, \mathcal{A})$ with $\Omega = \Omega_1 \times \Omega_2$ and $\mathcal{A} = \mathcal{A}_1 \bigotimes \mathcal{A}_2$ and let $\mu = \mu_1 \times \mu_2$ as defined by equation (3.20).
(i) If A is $\mathcal{A}$-measurable then for each fixed $\omega_1 \in \Omega_1$, the ω_1 section of A is $\mathcal{A}_2$-measurable, i.e., $A_{\omega_1} \in \mathcal{A}_2$ and similarly, for each $\omega_2 \in \Omega_2$ we have $A_{\omega_2} \in \mathcal{A}_1$.
(ii) If f is measurable $\mathcal{A}|\overline{\mathcal{B}}_1$ then for each fixed $\omega_1 \in \Omega_1$, the ω_1 section of f, $f(\omega_1, .)$, is measurable $\mathcal{A}_2|\overline{\mathcal{B}}_1$ and for each fixed $\omega_2 \in \Omega_2$ we have that the ω_2 section of f, $f(., \omega_2)$, is measurable $\mathcal{A}_1|\overline{\mathcal{B}}_1$.
(iii) Let $\{A_{i1} \times A_{i2}\}$ be a countable, disjoint collection of measurable rectangles whose union is a measurable rectangle $A_1 \times A_2 = \bigcup\limits_i (A_{i1} \times A_{i2})$. Then $\mu(A_1 \times A_2) = \sum\limits_i \mu(A_{i1} \times A_{i2})$, so that μ is a measure on the measurable rectangles $\mathcal{R}^* = \mathcal{A}_1 \times \mathcal{A}_2$.
(iv) For any $A \in \mathcal{A}$, the function $g_A(\omega_1) = \mu_2(A_{\omega_1})$, $\forall \omega_1 \in \Omega_1$, is measurable $\mathcal{A}_1|\overline{\mathcal{B}}_1$ and the function $h_A(\omega_2) = \mu_1(A_{\omega_2})$, $\forall \omega_2 \in \Omega_2$, is measurable $\mathcal{A}_2|\overline{\mathcal{B}}_1$. Moreover,

$$\int\limits_{\Omega_1} g_A d\mu_1 = \int\limits_{\Omega_2} h_A d\mu_2. \tag{3.21}$$

Proof. (i) Fix $\omega_1 \in \Omega_1$ and consider the mapping $T_{\omega_1} : \Omega_2 \to \Omega_1 \times \Omega_2$, defined by $T_{\omega_1}(\omega_2) = (\omega_1, \omega_2)$. If $A = A_1 \times A_2 \in \mathcal{A}$ is a measurable rectangle, then $T_{\omega_1}^{-1}(A) = \{\omega_2 \in \Omega_2 : (\omega_1, \omega_2) \in A\} = A_{\omega_1}$ is equal to A_2 if $\omega_1 \in A_1$ or $\emptyset$ if $\omega_1 \notin A_1$ and in either case $T_{\omega_1}^{-1}(A) \in \mathcal{A}_2$. Consequently, since the measurable rectangles generate $\mathcal{A}$, remark 3.7.3 yields that T_{ω_1} is measurable $\mathcal{A}_2|\mathcal{A}_1 \times \mathcal{A}_2$ and hence $T_{\omega_1}^{-1}(A) = A_{\omega_1} \in \mathcal{A}_2$. By symmetry we have $A_{\omega_2} \in \mathcal{A}_1$.
(ii) By remark 3.7.2 if f is measurable $\mathcal{A}|\overline{\mathcal{B}}_1$ then $f \circ T_{\omega_1}$ is measurable $\mathcal{A}_2|\overline{\mathcal{B}}_1$ so that $f(\omega_1, .) = (f \circ T_{\omega_1})(.)$ is measurable $\mathcal{A}_2$. Similarly, by the symmetry of these statements, we have that $f(., \omega_2)$ is measurable $\mathcal{A}_1$, for each fixed $\omega_2 \in \Omega_2$.
(iii) Clearly, $\mu(\emptyset) = 0$ and $\mu(A) \geq 0$, for all $A \in \mathcal{R}^*$. Fix a point $\omega_1 \in A_1$. Then for each $\omega_2 \in A_2$, the point (ω_1, ω_2) belongs to exactly one rectangle $A_{i1} \times A_{i2}$. Thus A_2 is the disjoint union of those A_{i2} such that ω_1 is in the corresponding A_{i1}. Hence

$$\sum_i \mu_2(A_{i2}) I_{A_{i1}}(\omega_1) = \mu_2(A_2) I_{A_1}(\omega_1),$$

since μ_2 is countably additive. Therefore

$$\sum_i \int \mu_2(A_{i2}) I_{A_{i1}}(\omega_1) d\mu_1 = \int \mu_2(A_2) I_{A_1}(\omega_1) d\mu_1,$$

or

$$\sum_i \mu_2(A_{i2}) \mu_1(A_{i1}) = \mu_2(A_2) \mu_1(A_1),$$

as claimed.
(iv) From part (i) for any $\mathcal{A}$-set A we have $A_{\omega_2} \in \mathcal{A}_1$ and $A_{\omega_1} \in \mathcal{A}_2$ so that the

functions $\mu_1(A_{\omega_2})$ and $\mu_2(A_{\omega_1})$, for fixed $(\omega_1, \omega_2) \in \Omega$, are well defined. Due to symmetry we only consider the measurability statements about $g_A(\omega_1)$, for fixed $\omega_1 \in \Omega_1$. Consider the collection $\mathcal{L}$ of sets $A \in \mathcal{A}$, such that the function $g_A(\omega_1) = \mu_2(A_{\omega_1})$, for fixed $\omega_1 \in \Omega_1$, is measurable $\mathcal{A}_1|\overline{\mathcal{B}}_1$ and equation (3.21) holds. Clearly, $\mathcal{L} \subseteq \mathcal{A}$. We need to show the other direction.

First recall that the collection of measurable rectangles $\mathcal{R}^* = \mathcal{A}_1 \times \mathcal{A}_2 = \{A = A_1 \times A_2 \subset \Omega : A_1 \in \mathcal{A}_1 \text{ and } A_2 \in \mathcal{A}_2\}$ form a π-system and generate $\mathcal{A}$, i.e., $\mathcal{A} = \sigma(\mathcal{A}_1 \times \mathcal{A}_2)$ by definition 3.6. We show that $\mathcal{R}^* \subseteq \mathcal{L}$. Indeed, write

$$(A_1 \times A_2)_{\omega_1} = \begin{cases} A_2, & \omega_1 \in \Omega_1 \\ \emptyset, & \omega_1 \notin \Omega_1 \end{cases},$$

so that $g_A(\omega_1) = \mu_2(A_{\omega_1}) = I_{A_1}(\omega_1)\mu_2(A_2)$. Since $A_1 \in \mathcal{A}_1$, the function $I_{A_1}(\omega_1)$ is $\mathcal{A}_1|\overline{\mathcal{B}}_1$-measurable and therefore $g_A(\omega_1)$ is $\mathcal{A}_1|\overline{\mathcal{B}}_1$-measurable (since $\mu_2(A_2) \in \overline{\mathcal{B}}_1$). In addition

$$\int_{\Omega_1} I_{A_1}(\omega_1)\mu_2(A_2)d\mu_1(\omega_1) = \mu_1(A_1)\mu_2(A_2) = \int_{\Omega_2} I_{A_2}(\omega_2)\mu_1(A_1)d\mu_2(\omega_2), \tag{3.22}$$

so that equation (3.21) is established for measurable rectangles.

Next we show that $\mathcal{L}$ is a λ-system. Clearly, $\Omega \in \mathcal{L}$ since Ω is a measurable rectangle. Suppose that $A = A_1 \times A_2 \in \mathcal{L}$ so that $g_A(\omega_1)$ is $\mathcal{A}_1|\overline{\mathcal{B}}_1$-measurable and therefore for all $F \in \overline{\mathcal{B}}_1$ we have $g_A^{-1}(F) \in \mathcal{A}_1$. We can write

$$g_{A^c}^{-1}(F) = \{\omega_1 \in \Omega_1 : \mu_2((A^c)_{\omega_1}) \in F\} = \{\omega_1 \in \Omega_1 : \mu_2((A_{\omega_1})^c) \in F\}.$$

If $\mu_2(\Omega_2) < \infty$ then

$$\begin{aligned} g_{A^c}^{-1}(F) &= \{\omega_1 \in \Omega_1 : \mu_2(\Omega_2) - \mu_2(A_{\omega_1}) \in F\} \\ &= \{\omega_1 \in \Omega_1 : \mu_2(A_{\omega_1}) \in -F + \mu_2(\Omega_2)\} = g_A^{-1}(G), \end{aligned}$$

where $G = -F + \mu_2(\Omega_2) = \{r \in \overline{\mathcal{R}} : r = -\omega_1 + \mu_2(\Omega_2), \omega_1 \in F\} \in \overline{\mathcal{B}}_1$, is well defined, so that $g_A^{-1}(G) \in \mathcal{A}_1$, i.e., $g_{A^c}(\omega_1)$ is $\mathcal{A}_1|\overline{\mathcal{B}}_1$-measurable and hence $A^c \in \mathcal{L}$. When $\mu_2(\Omega_2) = +\infty$, we have

$$g_{A^c}^{-1}(F) = \{\omega_1 \in \Omega_1 : \infty \in F\} = \begin{cases} \Omega_1, & \infty \in F \\ \emptyset, & \infty \notin F \end{cases},$$

and in both cases $g_{A^c}^{-1}(F) \in \mathcal{A}_1$, so that $g_{A^c}(\omega_1)$ is $\mathcal{A}_1|\overline{\mathcal{B}}_1$-measurable and similarly $h_{A^c}(\omega_1)$ is $\mathcal{A}_2|\overline{\mathcal{B}}_1$-measurable. To show that $A^c \in \mathcal{L}$ it remains to show equation (3.21). Since $A \in \mathcal{L}$ we have

$$\int_{\Omega_1} \mu_2(A_{\omega_1})d\mu_1(\omega_1) = \int_{\Omega_2} \mu_1(A_{\omega_2})d\mu_2(\omega_2),$$

so that

$$\begin{aligned} \int_{\Omega_1} g_{A^c}d\mu_1 &= \int_{\Omega_1} \mu_2((A^c)_{\omega_1})d\mu_1(\omega_1) = \int_{\Omega_1} \mu_2((A_{\omega_1})^c)d\mu_1(\omega_1) \\ &= \int_{\Omega_1} [\mu_2(\Omega_2) - \mu_2(A_{\omega_1})]d\mu_1(\omega_1) \\ &= \mu_2(\Omega_2)\int_{\Omega_1} d\mu_1(\omega_1) - \int_{\Omega_1} \mu_2(A_{\omega_1})d\mu_1(\omega_1) \\ &= \mu_2(\Omega_2)\mu_1(\Omega_1) - \mu(A), \end{aligned}$$

and similarly

$$\int_{\Omega_2} h_{A^c} d\mu_2 = \mu_1(\Omega_1)\mu_2(\Omega_2) - \mu(A).$$

This shows that $A^c \in \mathcal{L}$.

Now consider a disjoint sequence $\{B_i\}$ of $\mathcal{L}$-sets, with $B = \bigcup_i B_i$. Then the mappings $\mu_2((B_i)_{\omega_1})$ are $\mathcal{A}_1|\overline{\mathcal{B}}_1$-measurable and therefore

$$\sum_i \mu_2((B_i)_{\omega_1}) = \mu_2\left(\bigcup_i (B_i)_{\omega_1}\right) = \mu_2(B_{\omega_1}) = g_B(\omega_1)$$

is $\mathcal{A}_1|\overline{\mathcal{B}}_1$-measurable and using similar arguments $h_B(\omega_1)$ is $\mathcal{A}_2|\overline{\mathcal{B}}_1$-measurable. Moreover

$$\begin{aligned}
\int_{\Omega_1} g_B d\mu_1 &= \int_{\Omega_1} \mu_2(B_{\omega_1}) d\mu_1(\omega_1) = \int_{\Omega_1} \sum_i \mu_2((B_i)_{\omega_1}) d\mu_1(\omega_1) \\
&= \sum_i \int_{\Omega_1} \mu_2((B_i)_{\omega_1}) d\mu_1(\omega_1) = \sum_i \int_{\Omega_2} \mu_1((B_i)_{\omega_2}) d\mu_2(\omega_2) \\
&= \int_{\Omega_2} \sum_i \mu_1((B_i)_{\omega_2}) d\mu_2(\omega_2) = \int_{\Omega_2} h_B d\mu_2.
\end{aligned}$$

The last part shows that $\mathcal{L}$ is closed under countable, disjoint unions and hence $\mathcal{L}$ is a λ-system. Consequently, appealing to Dynkin's $\pi - \lambda$ theorem, we have that $\mathcal{L} = \mathcal{A}$ and the claims are established. ■

We now have enough motivation to introduce a well-defined product measure μ. The theorem that follows the definition demonstrates some key properties of μ.

Definition 3.18 Product measure space

Suppose that $(\Omega_1, \mathcal{A}_1, \mu_1)$ and $(\Omega_2, \mathcal{A}_2, \mu_2)$ are σ-finite measure spaces. The product measure space is defined as the triple $(\Omega, \mathcal{A}, \mu)$, where $\Omega = \Omega_1 \times \Omega_2$, $\mathcal{A} = \mathcal{A}_1 \bigotimes \mathcal{A}_2$ and μ is defined by

$$\mu(A) = (\mu_1 \times \mu_2)(A) = \int_{\Omega_1} \mu_2(A_{\omega_1}) d\mu_1(\omega_1) = \int_{\Omega_2} \mu_1(A_{\omega_2}) d\mu_2(\omega_2),$$

where $A \in \mathcal{A}$.

Theorem 3.21 (Product measure properties) Consider the setup of the previous definition. Then we can show
(i) μ is a σ-finite measure on $\mathcal{A}$.
(ii) $\mu(A) = \mu_1(A_1)\mu_2(A_2)$, for all $A_1 \in \mathcal{A}_1$ and $A_2 \in \mathcal{A}_2$.
(iii) The measure μ is the only measure on $\mathcal{A}$ such that $\mu(A_1 \times A_2) = \mu_1(A_1)\mu_2(A_2)$ on the measurable rectangles.
(iv) If μ_1 and μ_2 are probability measures then so is μ.

Proof. (i) Clearly, μ is nonnegative and $\mu(\emptyset) = 0$ by definition. Let $\{E_n\}_{n=1}^{+\infty}$ be a disjoint sequence of $\mathcal{A}$-sets so that $\{(E_n)_{\omega_1}\}_{n=1}^{+\infty}$ is a disjoint sequence of $\mathcal{A}_2$-sets.

Since μ_2 is a measure, using remark 3.14.9 we can swap the sum under the integral sign and write

$$\begin{aligned}\mu\left(\bigcup_{n=1}^{\infty} E_n\right) &= \int_{\Omega_1} \mu_2\left(\left(\bigcup_{n=1}^{\infty} E_n\right)_{\omega_1}\right) d\mu_1(\omega_1) = \int_{\Omega_1} \mu_2\left(\bigcup_{n=1}^{\infty} (E_n)_{\omega_1}\right) d\mu_1(\omega_1) \\ &= \int_{\Omega_1} \sum_{n=1}^{+\infty} \mu_2\left((E_n)_{\omega_1}\right) d\mu_1(\omega_1) = \sum_{n=1}^{+\infty} \int_{\Omega_1} \mu_2\left((E_n)_{\omega_1}\right) d\mu_1(\omega_1) \\ &= \sum_{n=1}^{+\infty} \mu(E_n),\end{aligned}$$

so that μ is a measure on $\mathcal{A}$. Next we treat σ-finiteness. Assume that $\{A_n\}$ is a disjoint partition of Ω_1, with $A_n \in \mathcal{A}_1$ and $\mu_1(A_n) < \infty$ and $\{B_n\}$ a disjoint partition of Ω_2, with $B_n \in \mathcal{A}_2$ and $\mu_2(B_n) < \infty$. As a consequence, $\Omega_1 \times \Omega_2 = \bigcup_{n=1}^{+\infty}(A_n \times B_n)$, so that

$$\begin{aligned}\mu(A_n \times B_n) &= \int_{\Omega_1} \mu_2\left((A_n \times B_n)_{\omega_1}\right) d\mu_1(\omega_1) = \int_{\Omega_1} \mu_2(B_n) I_{A_n}(\omega_1)\, d\mu_1(\omega_1) \\ &= \mu_1(A_n)\mu_2(B_n) < \infty,\end{aligned}$$

for all n, thus showing that μ is a σ-finite measure.
(ii) This was proven in equation (3.22).
(iii) Let v be another measure on $\mathcal{A}$ such that $v(A_1 \times A_2) = \mu_1(A_1)\mu_2(A_2)$ on the measurable rectangles $\mathcal{R}^*$. Since $\mathcal{R}^*$ is a π-system generating $\mathcal{A}$, any measures agreeing on $\mathcal{R}^*$ must agree on $\mathcal{A}$ (uniqueness of measure, theorem 3.6) and hence $v = \mu$ on $\mathcal{A}$.
(iv) Since $\Omega = \Omega_1 \times \Omega_2$ is a measurable rectangle, we have $\mu(\Omega) = \mu_1(\Omega_1)\mu_2(\Omega_2) = 1$. ■

Let f be an $\mathcal{A}|\overline{\mathcal{B}}^*$-measurable function and consider the integral of f with respect to the product measure $\mu = \mu_1 \times \mu_2$ denoted by

$$\int f d\mu = \int_{\Omega_1 \times \Omega_2} f d(\mu_1 \times \mu_2) = \int_{\Omega_1 \times \Omega_2} f(\omega_1, \omega_2) d(\mu_1 \times \mu_2)(\omega_1, \omega_2). \tag{3.23}$$

Extending the product space definition to higher dimensions is straightforward, but in the infinite product case we need to exercise caution as we see below.

Remark 3.22 (Infinite product measure) The infinite product measure space is defined on the infinite product measurable space $\left(\Omega = \mathop{\times}\limits_{n=1}^{+\infty} \Omega_n, \mathcal{A} = \bigotimes_{n=1}^{+\infty} \mathcal{A}_n\right)$, where the product measure is the unique measure on $(\Omega, \mathcal{A})$ such that

$$\mu(A) = \left(\mathop{\times}\limits_{n=1}^{+\infty} \mu_n\right)(A) = \prod_{n=1}^{+\infty} \mu_n(A_n),$$

for all measurable rectangles $A = \mathop{\times}\limits_{n=1}^{+\infty} A_n$, with $A_n \in \mathcal{A}_n$. The infinite product measure μ defined for all $\mathcal{A}$-sets is obtained through the extension of measure theorem by first defining μ on the measurable cylinders and uniqueness is justified since the measurable cylinders generate $\mathcal{A}$.

There are several results that allow us to calculate (3.23) using iterated integrals, depending on if: a) f is nonnegative or b) f is integrable with respect to μ. Both results (and their versions assuming complete measure spaces) can be referred to as Fubini theorems, although the theorem that treats case a) is also known as Tonelli's theorem.

Theorem 3.22 (Tonelli) Consider the product measure as in definition 3.18 and assume that $f : \Omega \to [0, \infty]$ is an $\mathcal{A}|\overline{\mathcal{B}}^+$-measurable function, with $\overline{\mathcal{B}}^+ = \mathcal{B}([0, \infty])$.
(i) For all $\omega_1 \in \Omega_1$ the mapping $f_{\omega_1}(\omega_2) = f(\omega_1, \omega_2)$ is $\mathcal{A}_2|\overline{\mathcal{B}}^+$-measurable and for all $\omega_2 \in \Omega_2$ the mapping $f_{\omega_2}(\omega_1) = f(\omega_1, \omega_2)$ is $\mathcal{A}_1|\overline{\mathcal{B}}^+$-measurable.
(ii) The mapping $\int\limits_{\Omega_2} f(\omega_1, \omega_2)d\mu_2(\omega_2)$ is $\mathcal{A}_1|\overline{\mathcal{B}}^+$-measurable and nonnegative and the function $\int\limits_{\Omega_1} f(\omega_1, \omega_2)d\mu_1(\omega_1)$ is $\mathcal{A}_2|\overline{\mathcal{B}}^+$-measurable and nonnegative.
(iii) We can write the double integral of f with respect to μ as an iterated integral, that is,

$$\int\limits_{\Omega_1}\left[\int\limits_{\Omega_2} f d\mu_2\right] d\mu_1 = \int\limits_{\Omega_1\times\Omega_2} f d(\mu_1 \times \mu_2) = \int\limits_{\Omega_2}\left[\int\limits_{\Omega_1} f d\mu_1\right] d\mu_2. \tag{3.24}$$

Proof. Due to symmetry it suffices to show the first parts of all statements (i)-(iii).
First note that the theorem holds for $f = I_A$ with $A \in \mathcal{A}$ since

$$\int\limits_{\Omega_2} f d\mu_2 = \int\limits_{\Omega_2} I_A d\mu_2 = \mu_2(A_{\omega_1}),$$

and using lemma 3.4, parts (ii) and (iv) and definition 3.18, parts (i) and (ii) hold and

$$\int f d\mu = \int I_A d\mu = \mu(A) = \int\limits_{\Omega_1} \mu_2(A_{\omega_1})d\mu_1(\omega_1) = \int\limits_{\Omega_1}\left[\int\limits_{\Omega_2} I_A d\mu_2\right] d\mu_1,$$

and (3.24) is established.

Now consider any nonnegative simple function $f = \sum\limits_{i=1}^{n} c_i I_{C_i}$, $0 \le c_1, ..., c_n < \infty$ and $\{C_i\}$ disjoint $\mathcal{A}$-sets that form a partition of Ω. Clearly, the mapping $f(\omega_1, \omega_2) = \sum\limits_{i=1}^{n} c_i I_{C_i}(\omega_1, \omega_2)$, for fixed $\omega_1 \in \Omega_1$, is a linear combination of $\mathcal{A}_2|\overline{\mathcal{B}}^+$-measurable functions $g_{\omega_1}(\omega_2) = I_{C_i}(\omega_1, \omega_2)$ and hence (i) holds. Moreover, linearity of the integral gives

$$\begin{aligned}\int\limits_{\Omega_2} f d\mu_2 &= \int\limits_{\Omega_2} \sum_{i=1}^{n} c_i I_{C_i}(\omega_1, \omega_2)d\mu_2(\omega_2) = \sum_{i=1}^{n} c_i \int\limits_{\Omega_2} I_{C_i}(\omega_1, \omega_2)d\mu_2(\omega_2) \\ &= \sum_{i=1}^{n} c_i \mu_2((C_i)_{\omega_1}),\end{aligned}$$

with $\mu_2((C_i)_{\omega_1})$ an $\mathcal{A}_1|\overline{\mathcal{B}}^+$-measurable and nonnegative function, for each $\omega_1 \in \Omega_1$, so that (ii) is established. To see equation (3.24) we write

$$\begin{aligned}\int f d\mu &= \int \sum_{i=1}^{n} c_i I_{C_i}(\omega_1, \omega_2) d\mu = \sum_{i=1}^{n} c_i \int I_{C_i}(\omega_1, \omega_2) d\mu = \sum_{i=1}^{n} c_i \mu(C_i) \\ &= \sum_{i=1}^{n} c_i \int_{\Omega_1} \mu_2((C_i)_{\omega_1}) d\mu_1(\omega_1) = \int_{\Omega_1} \sum_{i=1}^{n} c_i \mu_2((C_i)_{\omega_1}) d\mu_1(\omega_1) \\ &= \int_{\Omega_1} \sum_{i=1}^{n} c_i \mu_2((C_i)_{\omega_1}) d\mu_1(\omega_1) = \int_{\Omega_1} \left[\int_{\Omega_2} f d\mu_2 \right] d\mu_1.\end{aligned}$$

Next consider any $f \geq 0$, an $\mathcal{A}|\overline{\mathcal{B}}^+$-measurable function. From theorem 3.4 we can approximate f via nonnegative, nondecreasing simple functions $\{\varphi_n\}$, namely, $0 \leq \varphi_n(\omega_1, \omega_2) \uparrow f(\omega_1, \omega_2)$ and therefore the ω_1-sections of the functions φ_n and φ maintain this property, that is, $0 \leq \varphi_n(\omega_1, .) \uparrow f(\omega_1, .)$, for fixed $\omega_1 \in \Omega_1$. Using the MCT we can pass the limit under the integral sign whenever needed. To see (i), the mapping $f(\omega_1, \omega_2) = \lim_{n\to\infty} \varphi_n(\omega_1, \omega_2)$, for fixed $\omega_1 \in \Omega_1$, is $\mathcal{A}_2|\overline{\mathcal{B}}^+$-measurable as the limit of $\mathcal{A}_2|\overline{\mathcal{B}}^+$-measurable functions $g_{\omega_1}(\omega_2) = \varphi_n(\omega_1, \omega_2)$. In addition

$$\int_{\Omega_2} f d\mu_2 = \lim_{n\to\infty} \int_{\Omega_2} \varphi_n d\mu_2,$$

is $\mathcal{A}_1|\overline{\mathcal{B}}^+$-measurable as the limit of $\mathcal{A}_1|\overline{\mathcal{B}}^+$-measurable functions $\int_{\Omega_2} \varphi_n(\omega_1, \omega_2) d\mu_2(\omega_2)$, for each $\omega_1 \in \Omega_1$ and part (ii) is established. For the last part we apply the MCT twice to obtain

$$\begin{aligned}\int f d\mu &= \lim_{n\to\infty} \int \varphi_n d\mu = \lim_{n\to\infty} \int_{\Omega_1} \left[\int_{\Omega_2} \varphi_n d\mu_2 \right] d\mu_1 \\ &= \int_{\Omega_1} \left[\lim_{n\to\infty} \int_{\Omega_2} \varphi_n d\mu_2 \right] d\mu_1 = \int_{\Omega_1} \left[\int_{\Omega_2} f d\mu_2 \right] d\mu_1.\end{aligned}$$

■

Next we collect and proof the classic Fubini theorem.

Theorem 3.23 (Fubini) Consider the product measure as in definition 3.18 and assume that $f : \Omega \to \overline{\mathcal{R}}$ is an $\mathcal{A}|\overline{\mathcal{B}}$-measurable function that is integrable with respect to μ.
(i) For almost all ω_1, the function $f_{\omega_1}(\omega_2) = f(\omega_1, \omega_2)$ is an integrable function with respect to μ_2, and for almost all ω_2, the function $f_{\omega_2}(\omega_1) = f(\omega_1, \omega_2)$ is μ_1-integrable.
(ii) The function $\int_{\Omega_2} f(\omega_1, \omega_2) d\mu_2(\omega_2)$ defined for almost all $\omega_1 \in \Omega_1$ is μ_1-integrable and the function $\int_{\Omega_1} f(x, y) d\mu_1(x)$ defined for almost all $\omega_2 \in \Omega_2$ is

μ_2-integrable.
(iii) We can write the double integral of f with respect to μ as an iterated integral, that is,

$$\int_{\Omega_1}\left[\int_{\Omega_2} f d\mu_2\right] d\mu_1 = \int f d\mu = \int_{\Omega_2}\left[\int_{\Omega_1} f d\mu_1\right] d\mu_2.$$

Proof. The proof mirrors the Tonelli theorem proof, with the only major difference being how we can pass limit under the integral sign for *a.s.* measurable functions and so we appeal to the approach of example 3.19.2, the *a.s.* version of the MCT. To bring in Tonelli's result, note that $\int f d\mu = \int f^+ d\mu - \int f^- d\mu < \infty$, due to integrability and recall that $\int f d\mu = \int g d\mu$ if $f = g$ *a.e.* $[\mu]$. ■

We consider generalizations of Fubini and Tonelli below.

Remark 3.23 (Fubini in higher dimensional spaces) Versions of Fubini and Tonelli are easily obtained for higher dimensional product spaces, that is, we can easily iterate n-dimensional integrals in a measure space $\left(\Omega = \underset{i=1}{\overset{n}{\times}} \Omega_i, \mathcal{A} = \bigotimes_{i=1}^{n} \mathcal{A}_i, \mu = \underset{i=1}{\overset{n}{\times}} \mu_i\right)$ by

$$\int f d\mu = \int_{\Omega_1}\int_{\Omega_2} \dots \int_{\Omega_n} f d\mu_1 \dots d\mu_2 d\mu_n,$$

and we can change the order of integration as we see fit. When μ is the Lebesgue measure μ_n on $(\mathcal{R}^n, \mathcal{B}_n)$, then the integral $\int f d\mu_n$ is the n-dimensional Lebesgue integral. The extension to the infinite dimensional case is immediate and depends on how well defined the infinite product measure is.

The next two theorems are concerned with the Lebesgue integral and change of variable (Billingsley, 2012, p. 239). These are special cases of change of variable for a general measure and these theorems were illustrated in Section 3.7.1.

Theorem 3.24 (Change of variable: Lebesgue integral) Let $\varphi : I_1 \to I_2$ be a strictly increasing differentiable function, where I_1, I_2 intervals and let μ_1 be the Lebesgue measure in $(\mathcal{R}, \mathcal{B}_1)$. If $f : I_2 \to \overline{\mathcal{R}}$ is a measurable function then

$$\int_{I_2} f d\mu_1 = \int_{I_1} (f \circ \varphi)\varphi' d\mu_1.$$

We extend this result to the multivariate case below.

Theorem 3.25 (Multivariate change of variable: Lebesgue integral) Let $\varphi : O \to \mathcal{R}^p$ be an invertible, continuously differentiable function, defined on an open set $O \subset \mathcal{R}^p$. Assume that $B \in \mathcal{B}_p$ such that $B \subset O$ and let $A = \varphi^{-1}(B)$. If

$f : \mathcal{R}^p \to \overline{\mathcal{R}}$ is a measurable function then

$$\int_B f d\mu_p = \int_A (f \circ \varphi)|J| d\mu_p. \tag{3.25}$$

where $J = \left|\frac{\partial \mathbf{x}}{\partial \mathbf{y}}\right| = \left|\left[\left(\frac{\partial x_j}{\partial y_i}\right)\right]\right| = \left|\left[\left(\frac{\partial \varphi_j^{-1}(y_j)}{\partial y_i}\right)\right]\right|$ is the Jacobian determinant of the transformation $\mathbf{y} = \varphi(\mathbf{x})$, with $\varphi = (\varphi_1, \dots, \varphi_p)$ and μ_p is the p-dimensional Lebesgue measure.

Example 3.23 (Fubini theorem and change of variable) We show that $\lim_{t\to\infty} \int_0^t (\sin x)\, x^{-1} dx = \pi/2$. First note that

$$\int_0^t e^{-ux} \sin x dx = (1 + u^2)^{-1} \left[1 - e^{-ut}(u \sin t + \cos t)\right],$$

as follows from differentiation with respect to t and since

$$\int_0^t \left[\int_0^{+\infty} \left|e^{-ux} \sin x\right| du\right] dx = \int_0^t |\sin x| x^{-1} dx \le t < \infty,$$

Fubini's theorem applies to the integration of $e^{-ux} \sin x$ over $(0, t) \times \mathcal{R}^+$ so that we can write

$$\int_0^t (\sin x)\, x^{-1} dx = \int_0^{+\infty} (1 + u^2)^{-1} du - \int_0^{+\infty} e^{-ut}(1 + u^2)^{-1}(u \sin t + \cos t) du.$$

Now note that

$$\int_0^{+\infty} (1 + u^2)^{-1} du = [\tan u]_0^{+\infty} = \pi/2,$$

and making the transformation $v = ut$, we have

$$\int_0^{+\infty} e^{-ut}(1 + u^2)^{-1}(u \sin t + \cos t) du = \int_0^{+\infty} \left(1 + v^2/t^2\right)^{-1} e^{-v} \left(vt^{-1} \sin t + \cos t\right) t^{-1} dv,$$

which goes to 0 as $t \to \infty$ and the claim is established.

Example 3.24 (Fubini theorem and polar coordinate space) We compute the integral $I = \int_{-\infty}^{+\infty} e^{-x^2} d\mu_1(x)$, where μ_1 is the Lebesgue measure on $(\mathcal{R}, \mathcal{B}_1)$, using a double integral. Consider the polar coordinate space $(\rho, \theta) \in U = \{(\rho, \theta) : \rho > 0, 0 < \theta < 2\pi\}$ and define the transformation $T(\rho, \theta) = (\rho \cos \theta, \rho \sin \theta) : U \to \mathcal{R}^2$. The Jacobian of the transformation is given by $J = \left|\nabla^2 T(\rho, \theta)\right| = \rho$, with $T(U) = \mathcal{R}^2$ and using equation (3.25) we can write

$$\int_{\mathcal{R}^2} f(x, y) d\mu_2(x, y) = \int_U f(\rho \cos \theta, \rho \sin \theta) \rho d\mu_2(\rho, \theta),$$

for any measurable function f. Choose $f(x, y) = e^{-x^2 - y^2}$, a measurable function, so that

$$I^2 = \int_U \rho e^{-\rho^2} d\mu_2(\rho, \theta),$$

and using the Fubini theorem we have

$$I^2 = 2\pi \int_0^{+\infty} \rho e^{-\rho^2} d\mu_1(\rho),$$

where the function $g(\rho) = \rho e^{-\rho^2}$, $\rho > 0$, is bounded and continuous. From remark 3.19, the Lebesgue integral is the same as the Riemann integral, so that using the fundamental theorem of calculus we may write

$$I^2 = 2\pi \int_0^{+\infty} \rho e^{-\rho^2} d\rho = 2\pi \int_0^{+\infty} d\left(-e^{-\rho^2}/2\right) = 2\pi \left[-e^{-\rho^2}/2\right]_0^{+\infty} = \pi,$$

and consequently

$$\int_{\mathcal{R}} e^{-x^2} d\mu_1(x) = \sqrt{\pi}.$$

3.7.5 $\mathcal{L}^p$-spaces

The following definition extends the idea of integrability and allows us to build well-behaved spaces of functions.

Definition 3.19 $\mathcal{L}^p$-space

If $(\Omega, \mathcal{A}, \mu)$ is a complete measure space, we denote by $\mathcal{L}^p(\Omega, \mathcal{A}, \mu)$ (or simply $\mathcal{L}^p$ when the measure space is implied by the context) the space of all measurable functions f on Ω for which $\int |f|^p d\mu < +\infty$ (i.e., $|f|^p$ is integrable) and define an equivalence class on $\mathcal{L}^p$ by: $f \sim g \iff f = g$ a.e. $[\mu]$, for $f, g \in \mathcal{L}^p$.

Several results on $\mathcal{L}^p$-spaces are summarized next.

Remark 3.24 ($\mathcal{L}^p$-spaces results) In order to metrize $\mathcal{L}^p$-spaces, for different $p \geq 1$, we define the $\mathcal{L}^p$-norm by

$$\|f\|_p = \left(\int |f|^p d\mu\right)^{1/p},$$

for all $p \geq 1$ and if $p = +\infty$, we define the essential supremum norm by

$$\|f\|_\infty = ess\sup|f| = \inf_{a>0}\{a : \mu(\{\omega \in \Omega : |f(\omega)| > a\}) = 0\},$$

and $\mathcal{L}^\infty$ consists of f for which $\|f\|_\infty < +\infty$. The following are straightforward to show.

1. The space $\mathcal{L}^p$ is a linear (or vector) space, i.e., if $a, b \in \mathcal{R}$ and $f, g \in \mathcal{L}^p$, then $af + bg \in \mathcal{L}^p$.

2. Hölder's inequality The indices p and q, $1 < p, q < +\infty$, are called conjugate if $1/p + 1/q = 1$. Note that this holds trivially for $p = 1$ and $q = +\infty$. Suppose that p and q are conjugate indices and let $f \in \mathcal{L}^p$ and $g \in \mathcal{L}^q$. Then fg is integrable and

$$\left|\int fg d\mu\right| \leq \int |fg| d\mu \leq \|f\|_p \|g\|_q.$$

3. Minkowski's inequality Let $f, g \in \mathcal{L}^p$ $(1 \leq p \leq +\infty)$. Then $f + g \in \mathcal{L}^p$ and

$$\|f + g\|_p \leq \|f\|_p + \|g\|_p.$$

4. $\|f\|_p$ is a norm in $\mathcal{L}^p$ so that $\mathcal{L}^p$ is a normed vector space.

5. The functional $d(f, g) = \|f - g\|_p$, for $f, g \in \mathcal{L}^p$, $1 \leq p \leq \infty$, defines a metric on $\mathcal{L}^p$.

In order to introduce convergence in $\mathcal{L}^p$-spaces we require the following definition.

Definition 3.20 Convergence in normed vector spaces

A sequence of functions $\{f_n\}$ in a normed vector space $\mathcal{X}$ is said to converge to f if $\forall \varepsilon > 0, \exists N > 0 : \forall n > N \Longrightarrow \|f_n - f\| < \varepsilon$, where $\|.\|$ is the norm of the space $\mathcal{X}$. We write $f_n \to f$ or $\lim\limits_{n\to\infty} f_n = f$.

We are now ready to collect convergence concepts in $\mathcal{L}^p$-spaces.

Remark 3.25 (Convergence in $\mathcal{L}^p$-spaces) We collect some basic definitions and results involving convergence in $\mathcal{L}^p$-spaces below.

1. $\mathcal{L}^p$-convergence In $\mathcal{L}^p$-spaces, $f_n \stackrel{\mathcal{L}^p}{\to} f \iff \forall \varepsilon > 0, \exists N > 0 : \forall n > N \Longrightarrow \|f_n - f\|_p < \varepsilon$, i.e., $\|f_n - f\|_p \to 0$, as $n \to +\infty$. This is often referred to as convergence in the mean of order p. In probabilistic terms, a random sequence of $\mathcal{R}$-valued random variables $X = (X_1, X_2, \dots)$ defined on a probability space $(\Omega, \mathcal{A}, P)$ converges in $\mathcal{L}^p$ to a random variable X if

$$\int |X_n - X|^p dP \to 0,$$

for $0 < p < +\infty$, denoted by $X_n \stackrel{\mathcal{L}^p}{\to} X$.

2. It is important to note that the latter definition is not pointwise convergence, i.e., $\lim\limits_{n\to\infty} f_n(x) = f(x)$, for all x, since $\|f\|_p$ operates on functions f, not their arguments x (x is integrated out).

3. Cauchy sequence of functions A sequence of functions in a normed vector space of function $\mathcal{X}$ is Cauchy if and only if $\forall \varepsilon > 0, \exists N > 0 : \forall n, m > N \Longrightarrow \|f_n - f_m\| < \varepsilon$.

4. Complete normed space A normed space is complete if every Cauchy sequence of elements from the space converges to an element of the space. For functions, $\mathcal{X}$ is complete if and only if for every Cauchy sequence of functions $\{f_n\} \in \mathcal{X}$, there exists $f \in \mathcal{X} : f_n \to f$.

5. Banach space A complete normed vector space is called a Banach space.

6. Riesz-Fischer theorem The $\mathcal{L}^p$-spaces, $1 \leq p \leq +\infty$, are complete, and consequently, Banach spaces.

Additional properties of $\mathcal{L}^p$-spaces can be studied using linear functionals. We collect the general definition first.

Definition 3.21 Bounded linear functionals

A linear functional on a normed space $\mathcal{X}$ is a mapping $F : \mathcal{X} \to \mathcal{R}$ such that $F(af + bg) = aF(f) + bF(g)$, for all $a, b \in \mathcal{R}$ and $f, g \in \mathcal{X}$. The functional F is bounded if there exists $M > 0$ such that $|F(f)| \leq M\|f\|$, $\forall f \in \mathcal{X}$. We define the norm of the functional by

$$\|F\| = \sup_{f\neq 0} \frac{|F(f)|}{\|f\|}.$$

Next we present an important result about linear functionals in $\mathcal{L}^p$-spaces.

Remark 3.26 (Bounded linear functionals in $\mathcal{L}^p$-spaces) Given a complete measure space $(\Omega, \mathcal{A}, \mu)$ and any $g \in \mathcal{L}^q$ we define a functional on $\mathcal{L}^p$ by $F(f) = \int f g d\mu$, for all $f \in \mathcal{L}^p$. Then we can show that F is a bounded, linear functional with $\|F\| = \|g\|_q$. The converse of this is known as the Riesz representation theorem. Namely, let F be any bounded, linear functional on $\mathcal{L}^p$, $1 \leq p < +\infty$. Then there exists $g \in L^q$ such that $F(f) = \int f g d\mu$ and moreover $\|F\| = \|g\|_q$, where p and q are conjugate.

A special case of $\mathcal{L}^p$-spaces is for $p = 2$, and we collect some results for this case below.

Remark 3.27 ($\mathcal{L}^2$-spaces) In what follows, consider the space $\mathcal{L}^2$. First note that $p = 2$ is its own conjugate.

1. If $f, g \in \mathcal{L}^2$ then $(f, g) = \int f g d\mu$ defines an inner product on $\mathcal{L}^2$. Then using Hölder's inequality we can write

$$(f, g) = \int f g d\mu \leq \left(\int f^2 d\mu\right)^{1/2} \left(\int g^2 d\mu\right)^{1/2} = \|f\|_2 \|g\|_2,$$

i.e., the Cauchy-Schwarz inequality.

2. The $\mathcal{L}^2$-space equipped with the norm $\|f\|^2 = (f, g)$ is complete.

3. If $(f, g) = 0$ then f and g are called orthogonal, denoted by $f \perp g$. If $f_1, \ldots, f_n$ are pairwise orthogonal then

$$\left\|\sum_{i=1}^{n} f_i\right\|^2 = \left(\sum_{i=1}^{n} f_i, \sum_{j=1}^{n} f_j\right) = \sum_{i=1}^{n}\sum_{j=1}^{n}(f_i, f_j) = \sum_{i=1}^{n} \|f_i\|^2,$$

which is simply the Pythagorean theorem for functions.

3.8 Summary

In this chapter, we have presented a rigorous treatment of measure and integration theory. There is a plethora of texts on measure theory that the reader can turn to for additional results and exposition with Halmos (1950) being the standard reference for decades. Recent texts include Kallenberg (1986, 2002), Dudley (2002), Durrett (2010), Cinlar (2010), Billingsley (2012) and Klenke (2014). We briefly discuss some key ideas from this chapter and give some complementary concepts and references.

Set systems, measurability and measures

More details on set systems, the construction and cardinality of σ-fields, in particular Borel σ-fields, can be found in Chapter 1 of Vestrup (2003). Historical accounts on the development of set theory with the pioneering work of Georg Cantor and Émile Borel at the end of the 19th century can be found in Dudley (2002, p 22). Details on the Axiom of Choice and its equivalent forms along with references to their proofs can be found there as well.

We have collected an example of a set that is not Borel measurable with respect to the Lebesgue measure on $[0, 1)$, in example 3.15. To answer the question of whether or not all subsets of $\mathcal{R}$ are measurable with respect to the Lebesgue measure μ_1, Dudley (2002, Section 3.4) presents some results on the Lebesgue measurable and nonmeasurable sets along with proofs, in particular about the cardinality of Lebesgue measurable sets. References on the development of the concepts of measurability and general measures can be found in Dudley (2002, pp 111 and 112).

Integration theory

An excellent exposition of Fubini's theorem and its versions can be found in Chapter 10 of Vestrup (2003), while Dudley (2002, pp 148, 149 and 246) provides the historical account of the development of Lebesgue's integral, the MCT, the DCT, the Fubini and Tonelli theorems and Lebesgue's fundamental theorem of calculus. Most of the development was initially considered over bounded intervals. It seems natural to start by showing that an integral statement holds for simple functions, move to nonnegative functions and then finally real-valued functions, using linearity of integral and an appropriate limit theorem, like the MCT, to integrate to the limit. Much of the presentation of integration theory in this chapter was influenced by Royden (1989), Fristedt and Gray (1997) and Billingsley (2012).

3.9 Exercises

Set theory, σ-fields and measurability

Exercise 3.1 Show that for all $A, B, C \subseteq \Omega$ we have
(i) $A \triangle B = \Omega \Longleftrightarrow A = B^c$,
(ii) $A \triangle B = \emptyset \Longleftrightarrow A = B$ and
(iii) $(A \triangle B) \cap C = (A \cap C) \triangle (B \cap C)$.

Exercise 3.2 Prove Dynkin's $\pi - \lambda$ theorem.

Exercise 3.3 Show in detail that the statement of remark 3.1.3 holds.

Exercise 3.4 Let $f : \mathcal{X} \to \mathcal{Y}$ be some map and C be a collection of subsets of $\mathcal{Y}$ and define the inverse map of the collection C by $f^{-1}(C) = \{E \subset \mathcal{X} : E = f^{-1}(C)$ for some $C \in C\}$. Show that $f^{-1}(\sigma(C))$ is a σ-field on $\mathcal{X}$ with $\sigma(f^{-1}(C)) = f^{-1}(\sigma(C))$.

Exercise 3.5 Let $\mathcal{A}_1$ and $\mathcal{A}_2$ be two σ-fields of subsets of Ω. Give an example (other than example 3.1) illustrating that $\mathcal{A}_1 \cup \mathcal{A}_2$ is not a σ-field. What can be said about $\mathcal{A}_1 \cup \mathcal{A}_2$ if $\mathcal{A}_1$ and $\mathcal{A}_2$ are fields?

Exercise 3.6 Prove the statements of remark 3.3.8.

Exercise 3.7 Let $\Omega = \{\omega_1, \omega_2, \omega_3, \omega_4, \omega_5\}$. Find the σ-fields $\sigma(\{\omega_1, \omega_2\})$, $\sigma(\Omega)$ and $\sigma(\{\omega_5\})$.

Exercise 3.8 Let $f : \mathcal{X} \to \mathcal{Y}$ be a one-to-one function from $\mathcal{X}$ onto $\mathcal{Y}$. Show that for all $B \subseteq \mathcal{Y}$ we have $f[f^{-1}(B)] = B$.

Exercise 3.9 Prove statements (i)-(iii) of remark 3.5.9.

Exercise 3.10 Prove that the Borel σ-field $\mathcal{B}_1$ on the real line $\mathcal{R}$ is not generated by the following:
(i) any finite collection of subsets of $\mathcal{R}$,
(ii) the collection of real singletons, and
(iii) the collection of all finite subsets of $\mathcal{R}$.

Exercise 3.11 Prove the statement of remark 3.6.1.

Exercise 3.12 Show that the sets $\{x \in \mathcal{R}^{\infty} : \sup\limits_n x_n < a\}$ and $\{x \in \mathcal{R}^{\infty} : \sum\limits_{n=1}^{+\infty} x_n < a\}$ are $\mathcal{B}_{\infty}$-measurable, for all $a \in \mathcal{R}$.

Exercise 3.13 Prove all statements of theorem 3.3.

Exercise 3.14 Suppose that $X : \Omega \to \overline{\mathcal{R}}$ is $\mathcal{A}|\overline{\mathcal{B}}$-measurable. Show that if $\mathcal{A} = \{\emptyset, \Omega\}$ then X must be a constant.

Exercise 3.15 Prove statements 2, 3 and 5, of remark 3.7.

Exercise 3.16 Prove Tychonoff's theorem: a product of compact spaces is a compact space.

Exercise 3.17 Show that the smallest σ-field $\mathcal{A}$ on Ω such that a function $X : \Omega \to \overline{\mathcal{R}}$ is $\mathcal{A}|\overline{\mathcal{B}}$-measurable is $X^{-1}(\overline{\mathcal{B}})$.

Exercise 3.18 Show that statements 7-9 of remark 3.7 hold.

Exercise 3.19 Let $\{f_n\}_{n=1}^{+\infty}$ be a sequence of Lebesgue measurable functions with common domain E. Show that the functions $\sum\limits_{n=1}^{k} f_n(x)$ and $\prod\limits_{n=1}^{k} f_n(x)$ are also measurable for any k.

Exercise 3.20 A real-valued function f defined on $\mathcal{R}$ is upper semicontinuous at x if for each $\varepsilon > 0$ there is a $\delta > 0$ such that $|x - y| < \delta$ implies that $f(y) < f(x) + \varepsilon$. Show that if f is everywhere upper semicontinuous then is it measurable.

Exercise 3.21 Assume that $f : \mathcal{R} \to \mathcal{R}$ is differentiable. Show that f and $\frac{df(x)}{dx}$ are measurable.

Measure theory

Exercise 3.22 Let μ be a finitely additive, real-valued function on a field $\mathcal{A}$. Show that μ is countably additive if and only if μ is "continuous at $\emptyset$," that is, $\mu(A_n) \to 0$, when $A_n \downarrow \emptyset$ and $A_n \in \mathcal{A}$.

Exercise 3.23 Prove statements (i)-(iii) or remark 3.8.9.

Exercise 3.24 Show that a measure μ on $(\Omega, \mathcal{A})$ is σ-finite if and only if μ can be written as a countable sum of pairwise mutually singular finite measures on $(\Omega, \mathcal{A})$.

Exercise 3.25 Prove lemma 3.1.

Exercise 3.26 Suppose that $(\mathcal{R}^p, \mathcal{B}_p, \mu_p)$ is the Lebesgue measure space in $\mathcal{R}^p$. Show that (i) $A \oplus \mathbf{x} \in \mathcal{B}_p$ and (ii) $\mu_p(A \oplus \mathbf{x}) = \mu_p(A)$, for all Borel sets A and points $\mathbf{x} \in \mathcal{R}^p$ (i.e., the Lebesgue measure is translation invariant).

Exercise 3.27 Show that every translation-invariant Radon measure v on $(\mathcal{R}_p, \mathcal{B}_p)$ must be a multiple of the Lebesgue measure, i.e., $v(B) = c\mu_p(B)$, for some constant c.

Exercise 3.28 Prove lemma 3.2.

Exercise 3.29 (Measure regularity) Suppose that μ is a measure on $\mathcal{B}_k$ such that $\mu(A) < \infty$ if A is bounded. Show the following.
(i) For $A \in \mathcal{B}_k$ and $\varepsilon > 0$, there exists a closed set C and an open set O such that $C \subset A \subset O$ and $\mu(O \setminus C) < \varepsilon$. This property is known as measure regularity.
(ii) If $\mu(A) < \infty$, then $\mu(A) = \sup\limits_{K \subseteq A,\, K \text{ compact}} \mu(K)$.

Exercise 3.30 Let μ^* denote the Lebesgue outer measure. Show that
(i) if $\mu^*(A) = 0$ then A is μ^*-measurable,
(ii) the interval $(x, +\infty)$ is μ^*-measurable, for any $x \in \mathcal{R}$, and
(iii) every Borel set is μ^*-measurable.

Exercise 3.31 Show that if $A \in \mathcal{B}_{p-1}$ then A has measure 0 with respect to the Lebesgue measure μ_p defined on $\mathcal{B}_p$.

Exercise 3.32 Let μ^* denote the Lebesgue outer measure. Show that
(i) if A is countable then $\mu^*(A) = 0$, and
(ii) if $\mu^*(A) = 0$ then $\mu^*(A \cup B) = \mu^*(B)$, for any set $B \in \mathcal{M}(\mu^*)$.

Exercise 3.33 Prove statements 1-6 of remark 3.10.

Exercise 3.34 Let $A \subseteq \mathcal{R}$ be a (Lebesgue) μ_1-measurable set of finite measure and f_n a sequence of measurable functions defined on A. Let f be a real-valued function such that for each $x \in A$ we have $f_n(x) \to f(x)$ as $n \to \infty$. Then given $\varepsilon > 0$ and $\delta > 0$, there exists a measurable set $B \subset A$ with $\mu_1(B) < \delta$ and an integer $n_0 > 0$ such that for all $x \notin B$ and all $n > n_0$ we have $|f_n(x) - f(x)| < \varepsilon$.

Exercise 3.35 Give a detailed proof of the Carathéodory extension theorem 3.10, based on the results of remark 3.10.

Exercise 3.36 Let $(\Omega, \mathcal{A}, \mu)$ be a measure space. Show that for any sequence $\{A_n\}$ of $\mathcal{A}$-sets we have

$$\mu(\underline{\lim} A_n) \leq \underline{\lim} \mu(A_n) \leq \overline{\lim} \mu(A_n).$$

Exercise 3.37 (Convergence in measure) A sequence of functions f_n of measurable functions converges to f in the (Lebesgue) measure μ if for a given $\varepsilon > 0$ there exists $n_0 > 0$ such that for all $n > n_0$ we have

$$\mu(\{x : |f_n(x) - f(x)| \geq \varepsilon\}) < \varepsilon.$$

We write $f_n \stackrel{\mu_1}{\to} f$. Show that
(i) if $f_n \to f$ a.e. then $f_n \stackrel{\mu_1}{\to} f$, and
(ii) if $f_n \stackrel{\mu_1}{\to} f$ then there exists a subsequence f_{k_n} of f_n with $f_{k_n} \stackrel{\mu_1}{\to} f$.

Exercise 3.38 Let $(\Omega, \mathcal{A}, \mu)$ be a measure space.
(i) Show that $\mu(A \triangle B) = 0 \Longrightarrow \mu(A) = \mu(B)$, provided that $A, B \in \mathcal{A}$.
(ii) If μ is complete then show that $A \in \mathcal{A}$ and $B \subset \Omega$ with $\mu(A \triangle B) = 0$ together imply that $B \in \mathcal{A}$.

Exercise 3.39 Give an example of a measure space $(\Omega, \mathcal{A}, \mu)$ such that for $A, A_1, A_2, \dots \in \mathcal{A}$ with $A_n \downarrow A$ and $\mu(A_n) = +\infty$ we have $A = \emptyset$. What are the implications of this with regard to the continuity of measure theorem?

Exercise 3.40 Let $\Omega = \{1, 2, 3, \dots\}$ and $\mathcal{A} = 2^\Omega$. Define $\mu(A) = \sum\limits_{k \in A} 2^{-k}$, if A is a finite $\mathcal{A}$-set and let $\mu(A) = +\infty$ otherwise. Is μ finitely additive? Countably additive?

Exercise 3.41 Let $\mathcal{A} = \{A \subseteq \Omega : A \text{ is finite or } A^c \text{ is finite}\}$. Define μ on $\mathcal{A}$ by $\mu(A) = 0$ if A is finite and $\mu(A) = 1$ if A^c is finite. Show that
(i) μ fails to be well-defined when Ω is finite,
(ii) if Ω is infinite, then $\mu(A) \geq 0$, $\forall A \in \mathcal{A}$, $\mu(\emptyset) = 0$ and for all disjoint $A_1, \dots, A_n \in \mathcal{A}$, such that $\bigcup\limits_{i=1}^{n} A_i \in \mathcal{A}$, we have $\mu\left(\bigcup\limits_{i=1}^{n} A_i\right) = \sum\limits_{i=1}^{n} \mu(A_i)$ and
(iii) when Ω is uncountable, μ is a measure. Is μ a σ–finite measure?

Exercise 3.42 Let $\mathbb{Q}$ denote the rationals and μ_1 the Lebesgue measure on $\mathcal{R}$. Show that $\mu_1(\mathbb{Q}) = 0$.

Exercise 3.43 Give an example of an open and unbounded set in $\mathcal{R}$ with a finite, strictly positive Lebesgue measure.

Exercise 3.44 Assume that $f : \mathcal{R}^+ \to \mathcal{R}^+$ is increasing, g is Lebesgue integrable and let μ_1 denote the Lebesgue measure on $\mathcal{R}$. Show that $\mu_1(\{x : |g(x)| \geq c\}) \leq \frac{1}{f(c)} \int f(|g(x)|) d\mu_1(x)$, for any $c \geq 0$.

Integration theory

Exercise 3.45 Assume that $\mu_1(A) = 0$. Show that for any Lebesgue integrable function f we have $\int\limits_A f d\mu_1 = 0$.

Exercise 3.46 Prove part (iii) of theorem 3.11.

Exercise 3.47 Using the definition of the Lebesgue integral with respect to the

Lebesgue measure μ_1 in $(\mathcal{R}, \mathcal{M})$, find the integrals: (i) $\int_0^1 x\mu_1(dx)$, (ii) $\int_{-1}^1 x^2\mu_1(dx)$, (iii) $\int_1^{+\infty} x^{-3}\mu_1(dx)$, (iv) $\int_{-\infty}^{+\infty} e^{-x^2}\mu_1(dx)$, and (v) $\int_0^{+\infty} e^{-x}\mu_1(dx)$.

Exercise 3.48 Prove the Lebesgue decomposition theorem by assuming that the Radon-Nikodym theorem holds.

Exercise 3.49 (Euler's gamma function) Show that the function $f(x) = x^{a-1}e^{-x}$, $x > 0$, $a > 0$, is Lebesgue integrable in $(0, +\infty)$.

Exercise 3.50 Prove all parts of remark 3.16.

Exercise 3.51 Let $(\Omega, \mathcal{A}, \mu)$ be a measure space and g a nonnegative measurable function on Ω. Define $v(E) = \int_E g d\mu$, for all $E \subseteq \Omega$.

(i) Show that v is a measure on $(\Omega, \mathcal{A})$.

(ii) If $f \geq 0$ is a simple function on $(\Omega, \mathcal{A})$ then show that $\int f dv = \int f g d\mu$.

Exercise 3.52 Prove all the statements of remark 3.14. Assume that the corresponding results for simple functions hold, where applicable.

Exercise 3.53 Show that for a Lebesgue measurable function f, if one of the integrals $\int_F f I_E d\mu_1$ and $\int_{E\cap F} f d\mu_1$ is finite then so is the other and both integrals are equal.

Exercise 3.54 Prove all the statements of remark 3.17.

Exercise 3.55 Show that the functions $\sin\left(x^2\right)$ and $\cos\left(x^2\right)$ are Riemann integrable in $(0, +\infty)$ (and find their value) but they are not Lebesgue integrable.

Exercise 3.56 Prove all parts of theorem 3.17.

Exercise 3.57 Let μ_1 denote the Lebesgue measure on $[0, 1]$ and show that $\int_0^1 x^{-x}\mu_1(dx) = \sum_{n=1}^{+\infty} n^{-n}$.

Exercise 3.58 Prove theorem 3.24.

Exercise 3.59 Let $(\Omega, \mathcal{A}, \mu)$ be a measure space and let f be an $\overline{\mathcal{R}}^+$-valued measurable function defined on Ω. Let $B = \{x \in \Omega : f(x) > 0\}$ and define $v(A) = \mu(A \cap B)$, $\forall A \in \mathcal{A}$. Show that

(i) $\int f d\mu = \int f dv$, and

(ii) if $\int f d\mu < \infty$, then $(\Omega, \mathcal{A}, v)$ is a σ−finite measure space.

Exercise 3.60 Suppose that f_n are integrable and $\sup_n \int f_n d\mu < \infty$. Show that if $f_n \uparrow f$ then f is integrable and $\int f_n d\mu \uparrow \int f d\mu$.

Exercise 3.61 Prove all statements of remark 3.21.

Exercise 3.62 Prove the Riemann-Lebesgue Theorem: If f is integrable on $\mathcal{R}$, then

$$\lim_{n\to\infty} \int_{\mathcal{R}} f(x)\cos(nx)dx = 0.$$

Exercise 3.63 Prove theorem 3.18.

Exercise 3.64 Show all statements of remark 3.19.

Exercise 3.65 Prove the MCT for the Riemann-Stieltjes integral.

Exercise 3.66 Prove theorem 3.19.

$\mathcal{L}^p$ spaces

Exercise 3.67 Consider a function $f \in \mathcal{L}^p(\Omega, \mathcal{A}, \mu_1)$, $1 \le p < \infty$. Then there exists an $\delta > 0$ and a bounded Lebesgue measurable function f_0 with $|f_0| \le \delta$ and for any $\varepsilon > 0$ we have $\|f - f_0\|_p < \varepsilon$.

Exercise 3.68 Let $(\Omega, \mathcal{A}, \mu)$ be a measure space with $\mu(\Omega) < +\infty$ and $0 < p < q \le \infty$.
(i) Show that $\mathcal{L}^q(\Omega, \mathcal{A}, \mu_1) \subseteq \mathcal{L}^p(\Omega, \mathcal{A}, \mu_1)$.
(ii) If $f \in \mathcal{L}^q(\Omega, \mathcal{A}, \mu_1)$ then we have $\|f\|_p \le \|f\|_q \mu(\Omega)^{\frac{1}{p}-\frac{1}{q}}$.

Exercise 3.69 Prove all the statements of remark 3.24.

Exercise 3.70 Show that the $\mathcal{L}^2$-space equipped with the norm $\|f\|^2 = (f, g)$ is complete.

Exercise 3.71 Let $f, g \in \mathcal{L}^\infty = \mathcal{L}^\infty(\Omega, \mathcal{A}, \mu)$ (i.e., $\mathcal{A}|\mathcal{B}(\overline{R})$ measurable functions with $ess\sup f < \infty$). Show that
(i) $|f(\omega)| \le ess\sup f$ a.e. μ., and
(ii) $\mathcal{L}^\infty$ is a complete metric space.

Exercise 3.72 (Riesz representation theorem) Let F be any bounded linear functional on $\mathcal{L}^p$, $1 \le p < +\infty$. Then there exists $g \in L^q$ such that $F(f) = \int fgd\mu$ and moreover $\|F\| = \|g\|_q$, where p and q are conjugate.

Exercise 3.73 Prove the Riesz-Fischer theorem.

Exercise 3.74 Assume that $f, f_n \in \mathcal{L}^p(\Omega, \mathcal{A}, \mu_1)$, $1 \le p < \infty$, such that $\sum\limits_{n=1}^{+\infty} \|f_n - f\|_p < \infty$. Show that $\lim\limits_{n\to\infty} f_n(x) = f(x)$ a.e.

Exercise 3.75 Show that if $f, g \in \mathcal{L}^2$ then $(f, g) = \int fgd\mu$ defines an inner product on $\mathcal{L}^2$. Furthermore, prove that the $\mathcal{L}^2$-space equipped with the norm $\|f\|^2 = (f, g)$ is complete.

Chapter 4

Probability Theory

4.1 Introduction

In probability theory, events are sets with a non-negative length that is at most one, since we assume that the sample space Ω is the certain event and should be assigned the largest length, i.e., 1 (probability 100%). In order to develop measure theoretic probability, we use the theoretical development of the previous chapter in order to describe sets that correspond to events and then assign numbers to them in the interval [0, 1] (probabilities).

4.2 Probability Measures and Probability Spaces

A historical account with references about the development of probability theory can be found in Dudley (2002, pp. 273-278), including Andrei Nikolaevich Kolmogorov's pioneering work in 1933, which first made the definition of a probability measure and space widely known. Since probability measures are special cases of general measures, the results of the previous chapter are inherited. We will collect however definitions and results tailored to probability theory when needed.

Definition 4.1 Probability measure and space

Consider a measurable space $(\Omega, \mathcal{A})$, where $\mathcal{A}$ is a field. A probability measure on $(\Omega, \mathcal{A})$ is a set function P that satisfies: (i) $P(\emptyset) = 0$, and $P(\Omega) = 1$, (ii) $P(A) \in [0, 1]$, $\forall A \in \mathcal{A}$, and (iii) if $A_1, A_2, \ldots,$ is a disjoint sequence of $\mathcal{A}$-sets and if $\bigcup\limits_{n=1}^{\infty} A_n \in \mathcal{A}$, then

$$P\left(\bigcup_{n=1}^{\infty} A_n\right) = \sum_{n=1}^{\infty} P(A_n). \tag{4.1}$$

If $\mathcal{A}$ is a σ-field, the triple $(\Omega, \mathcal{A}, P)$ is called a probability measure space or simply a probability space.

In probability theory, the paradoxical case $P(\Omega) < 1$ occurs in various situations (in particular, when P is the limit of probability measures P_n) and in this case $(\Omega, \mathcal{A}, P)$ is called a defective probability space.

Example 4.1 (Probability for coin flips) Recall the setup of example 3.4.

1. The sample space for a single flip of the coin is the set $\Omega_0 = \{\omega_0, \omega_1\}$, $\omega_i \in \{0, 1\}$, $i = 1, 2$, so that $2^{\Omega_0} = \{\emptyset, \Omega_0, \{0\}, \{1\}\}$. We define a discrete probability measure P_1 on the discrete measurable space $(\Omega_0, 2^{\Omega_0})$ by assigning mass to the simple events $\{0\}$ and $\{1\}$. For a fair coin we must have $P_1(\{0\}) = 1/2 = P_1(\{1\})$ and consequently all simple events are equally likely. Therefore $(\Omega_0, 2^{\Omega_0}, P_1)$ is a discrete probability space.

2. Extend the experiment to n successive flips of a fair coin and let the sample space be denoted by $\Omega_n = \{\omega = (\omega_1, \ldots, \omega_n) \in \Omega_0^n : \omega_i \in \{0, 1\}, i = 1, 2, \ldots, n\} = \{\omega_1, \ldots, \omega_{2^n}\}$, consisting of 2^n simple events. We define the set function P_n on the measurable space $(\Omega_n, 2^{\Omega_n})$ by

$$P_n(A) = |A|/2^n,$$

for all $A \in 2^{\Omega_n}$, where $|A| = card(A)$. From example 3.10 $\mu(A) = |A|$ is the counting measure and since $P_n(\Omega_n) = |\Omega_n|/2^n = 1$, the triple $(\Omega_n, 2^{\Omega_n}, P_n)$ is a discrete probability space. Suppose that $k \in \{0, 1, \ldots, n\}$. In order for $\sum\limits_{i=1}^{n} \omega_i = k$ (i.e., the experiment yields k Tails), it is necessary and sufficient that exactly k of the ω_is be 1 and hence there are C_k^n simple events with this property. Clearly, the event $A =\{k$ Tails in n-flips$\}= \left\{\omega \in \Omega_n : \sum\limits_{i=1}^{n} \omega_i = k\right\}$, is an event from 2^{Ω_n} and therefore

$$P_n(A) = C_k^n 2^{-n},$$

for $k = 0, 1, \ldots, n$. Hence P_n gives us a way of describing the probability of all sets in 2^{Ω_n}.

3. Now turn to the infinite coin flip experiment with sample space $\Omega_\infty = \{\omega = (\omega_1, \omega_2, \ldots) \in \Omega_0^\infty : \omega_i \in \{0, 1\}, i = 1, 2, \ldots\}$. Since Ω_∞ has infinitely many members we cannot use $P_n(A)$ to define probabilities in this case. However, we do know what we would like a probability measure P on $(\Omega_\infty, 2^{\Omega_\infty})$ to be equal to for certain subsets of Ω_∞. In particular, following remark 3.5.8, consider the collection $\mathcal{E}$ of all k-dimensional measurable cylinders consisting of sets of the form

$$E_k = \{\omega = (\omega_1, \omega_2, \ldots) \in \Omega_0^\infty : (\omega_1, \omega_2, \ldots, \omega_k) \in A_k\},$$

where A_k is a 2^{Ω_k}-set and define a set function P on the measurable cylinders E_k by

$$P(E_k) = \sum_{(\omega_1, \omega_2, \ldots, \omega_k) \in A_k} 2^{-k}.$$

Recall that $\mathcal{E}$ generates $2^{\Omega_\infty} (= \mathcal{B}(\Omega_\infty))$, so that P has a unique extension to the Borel sets of Ω_∞ by the extension of measure theorem defined by

$$P(E) = \inf_{E \subset \cup A_k} \sum_{(\omega_1, \omega_2, \ldots, \omega_k) \in A_k} 2^{-k}.$$

4.2.1 Extension of Probability Measure

Let us restate and treat the extension problem specifically for probability measures. Suppose that we have defined a probability measure P_0 on a field $\mathcal{A}_0$ of subsets of a space Ω and set $\mathcal{A} = \sigma(\mathcal{A}_0)$. We are interested in defining a unique probability measure P so that (Ω, A, P) is a probability space and P agrees with P_0 on $\mathcal{A}_0$. The problem can be solved using the Carathéodory extension theorem (Section 3.6.2) by defining for any subset A of Ω the outer measure

$$P^*(A) = \inf_{A \subset \cup A_k} \sum_k P_0(A_k),$$

where $\{A_k\}$ are $\mathcal{A}_0$-sets so that the desired probability measure P is the restriction of P^* to the Carathéodory measurable sets $\mathcal{M}(P^*) = \{A \subset \Omega : P^*(E) = P^*(E \cap A) + P^*(E \cap A^c)$, for all $E \subset \Omega\}$. This proposed P^* is approximating the probability of A from the outside (or from above) since P^* is computed over all covers of A consisting of $\mathcal{A}_0$-sets.

In order to motivate $\mathcal{M}(P^*)$ from a probabilistic point of view, recall the complement law. The problem is that for $\{A_k\} \in \mathcal{A}_0$ we do not always have $\cup A_k \in \mathcal{A}_0$ and hence $P^*(A) = 1 - P^*(A^c)$ may be violated. Define the inner probability measure by $P_*(A) = 1 - P^*(A^c)$, $\forall A \subset \Omega$. Clearly, we could extend P_0 for sets $A \in \mathcal{A}_0$ such that $P_*(A) = P^*(A)$, in order to satisfy the complement law $P^*(A) + P^*(A^c) = 1$. This condition is weaker than that required by $\mathcal{M}(P^*)$-sets, since it is a special case for $E = \Omega$, i.e., $1 = P^*(\Omega) = P^*(\Omega \cap A) + P^*(\Omega \cap A^c) = P^*(A) + P^*(A^c)$.

Now we consider an alternative extension method based on the Sierpiński class of sets. The interested reader can find more details on this construction in Fristedt and Gray (1997, Section 7.3).

Definition 4.2 Sierpiński class

A Sierpiński class $\mathcal{S}$ is a collection of subsets of a space Ω that is closed under limits of nondecreasing sequences of sets and proper set differences, i.e., if $A_1 \subseteq A_2 \subseteq \ldots$, with $A_n \in \mathcal{S}$, then $\lim_{n\to\infty} A_n = \bigcup_n A_n \in \mathcal{S}$ and if $A, B \in \mathcal{S}$, then $A \subseteq B$ implies $B \smallsetminus A \in \mathcal{S}$. We say that $\mathcal{S}$ is an $\mathcal{S}$-class.

Next we consider when an $\mathcal{S}$-class is a σ-field.

Remark 4.1 (σ-field and $\mathcal{S}$-class) Clearly, any σ-field is an $\mathcal{S}$-class. The converse is also true; let $\mathcal{S}$ be an $\mathcal{S}$-class and suppose that (i) $\Omega \in \mathcal{S}$, and (ii) $\mathcal{S}$ is closed under pairwise intersections. Then we can easily verify that $\mathcal{S}$ is a σ-field.

The next two theorems illustrate the usefulness of the Sierpiński class in defining the probability measure extension.

Theorem 4.1 (Sierpiński class theorem) Let $\mathcal{E}$ be a collection of subsets of a space Ω and suppose that $\mathcal{E}$ is closed under pairwise intersections and $\Omega \in \mathcal{E}$. Then the smallest Sierpiński class of subsets of Ω that contains $\mathcal{E}$ equals $\sigma(\mathcal{E})$.

Proof. An obvious candidate is $\mathcal{S}^* = \bigcap_{\mathcal{E} \subseteq \mathcal{S}} \mathcal{S}$, where $\mathcal{S}$ is any $\mathcal{S}$-class containing the collection $\mathcal{E}$ and this intersection is nonempty since 2^{Ω} is clearly an $\mathcal{S}$-class. In view of remark 4.1, it suffices to show that $\mathcal{S}^*$ is an $\mathcal{S}$-class. Suppose that $A_n \in \mathcal{S}$, with $A_n \subseteq A_{n+1}$, $n = 1, 2, \ldots$, so that $A_n \in \mathcal{S}$, $\forall \mathcal{S}$, which implies that $\lim\limits_{n\to\infty} A_n = A \in \mathcal{S}$, $\forall \mathcal{S}$ and therefore $A \in \mathcal{S}^*$. Now take $A, B \in \mathcal{S}^*$ with $A \subseteq B$ so that $A, B \in \mathcal{S}$, $\forall \mathcal{S}$, which implies $B \smallsetminus A \in \mathcal{S}$, $\forall \mathcal{S}$ and thus $B \smallsetminus A \in \mathcal{S}^*$. Noting that $\mathcal{E} \subseteq \mathcal{S}^*$ and $\mathcal{S}^*$ is the smallest σ-field completes the proof. ■

The following is not hard to show and its proof is requested as an exercise.

Theorem 4.2 (Uniqueness of probability measure) Let P and Q be probability measures on the measurable space $(\Omega, \sigma(\mathcal{E}))$, where $\mathcal{E}$ is a collection of sets closed under pairwise intersections. If $P(A) = Q(A)$, $\forall A \in \mathcal{E}$, then P coincides with Q on $\sigma(\mathcal{E})$.

Using these results, we can define an extension using the following steps. Let $\mathcal{E}$ be a field of subsets of a space Ω and R a nonnegative countably additive function defined on $\mathcal{E}$ such that $R(\Omega) = 1$. Define $\mathcal{E}_1$ to be the collection of subsets from Ω that are limits of subsets of $\mathcal{E}$, which can be shown to be a field as well and set $R_1(A) = \lim\limits_{n\to\infty} R(A_n)$, for all $A = \lim\limits_{n\to\infty} A_n \in \mathcal{E}_1$, $A_n \in \mathcal{E}$. Clearly, $R_1(A) = R(A)$, if $A \in \mathcal{E}$ and therefore R_1 is an extension of R to $\mathcal{E}_1$, which can be shown to have the same properties as R. We can repeat this process and create $\mathcal{E}_2 = \{A : A = \lim\limits_{n\to\infty} A_n, A_n \in \mathcal{E}_1\}$ and $R_2(A) = \lim\limits_{n\to\infty} R_1(A_n)$ and similarly $\mathcal{E}_3, R_3$, or $\mathcal{E}_4, R_4$ and so forth.

Note that every time we extend, it is possible that we enlarge the field with more sets, i.e., $\mathcal{E}_n \subseteq \mathcal{E}_{n+1}$ and $\sigma(\mathcal{E})$ is generally strictly larger than all the fields $\mathcal{E}_n$. However, we do not need to go past the second level since it can be shown that $\mathcal{E}_2$ is such that if $B \in \sigma(\mathcal{E})$ then there exist $A, C \in \mathcal{E}_2$ such that $A \subseteq B \subseteq C$ and $R_2(A) = R_2(C)$. Thus if a set $B \in \sigma(\mathcal{E})$ has any elements outside of $\mathcal{E}$ then those elements form a set of measure zero with respect to R_2 so that nothing new appears after the extension to $\mathcal{E}_2$. The extension is defined as the probability measure $P(B) = R_2(A)$, for all $B \in \sigma(\mathcal{E})$, such that $A \subseteq B \subseteq C$, for $A, C \in \mathcal{E}_2$ with $R_2(C \smallsetminus A) = 0$. The following result summarizes the extension theorem and will be collected without proof (Fristedt and Gray, 1997, p. 94).

Theorem 4.3 (Extension of probability measure) Let $\mathcal{E}$ be a field of subsets of a space Ω and R a nonnegative countably additive function defined on $\mathcal{E}$ such that $R(\Omega) = 1$. Then there exists a unique probability measure P defined on $\sigma(\mathcal{E})$ such that $P(A) = R(A)$, for all $A \in \mathcal{E}$.

4.2.2 Defining Random Objects

We are now ready to give the general definition of a random object. The two important concepts required are measurability (of a mapping and a space) and probability spaces.

Definition 4.3 Random object

Consider two measurable spaces $(\Omega, \mathcal{A})$ and $(\mathcal{X}, \mathcal{G})$. A measurable object X from $(\Omega, \mathcal{A})$ to $(\mathcal{X}, \mathcal{G})$ is a measurable map $X : \Omega \to \mathcal{X}$, i.e., $X^{-1}(G) \in \mathcal{A}$, $\forall G \in \mathcal{G}$. If we attach a probability measure to $(\Omega, \mathcal{A})$ so that $(\Omega, \mathcal{A}, P)$ is a probability space then X is called a random object from $(\Omega, \mathcal{A}, P)$ to $(\mathcal{X}, \mathcal{G})$.

The formal definition of the distribution of a random object is given next. Note that knowing the distribution of the random object allows us to ignore the underlying structure of the probability space $(\Omega, \mathcal{A}, P)$ and work directly in the target space $(\mathcal{X}, \mathcal{G})$. As we will see in the next section, the distribution of the random object can be used to provide the cdf and the density function for a random object, which in turn makes defining and calculating important quantities about X, such as expectation, much easier.

Definition 4.4 Distribution

The probability measure P (or the random object X) induces a probability measure Q on $(\mathcal{X}, \mathcal{G})$ by

$$Q(G) = (P \circ X^{-1})(G) = P(X^{-1}(G)) = PX^{-1}(G). \tag{4.2}$$

Consequently, $(\mathcal{X}, \mathcal{G}, Q)$ is a probability space and X^{-1} is a random object from $(\mathcal{X}, \mathcal{G})$ to $(\Omega, \mathcal{A})$. The probability measure Q is called the distribution of the random object X.

Two natural questions arise at this point; which class of $\mathcal{A}$-sets should we choose that will allow us to efficiently allocate probabilities to the values of the random object X, and moreover, which class of functions $X : \Omega \to \mathcal{X}$ is acceptable as a random object? In order to answer both questions, one can turn to Baire theory. In general, Baire sets form the smallest σ-field with respect to which X is measurable by definition (see Appendix Section A.3 and definitions A.4 and A.5). Instead, we have studied Borel sets and functions in the previous chapter and we define most of our random objects to be Borel measurable functions or maps, since the concepts of Baire and Borel sets, as well as functions, are equivalent for $\mathcal{X} = \mathcal{R}^p$.

More generally, it can be shown that the Baire and Borel σ-fields coincide in any locally compact separable metric space $(\mathcal{X}, \rho)$, provided that we use the topology induced by ρ to generate the Borel σ-field (see for example Dudley, 2002, theorem 7.1.1 or Royden, 1989, Chapter 13). The following definition provides the smallest class of such well-behaved sets for a random object X.

Definition 4.5 Generated σ-field of a random object

The σ-field $\sigma(X)$ generated by a random object X is the smallest σ-field with respect to which X is measurable.

The first part of the following theorem is of extreme importance since it gives us the explicit form of the sets in $\sigma(X)$. Note that the steps in its proof (e.g., showing that inverse images form a σ-field) hold always (see for example Kallenberg, 2002, p. 3).

Theorem 4.4 (Explicit form of the generated σ-field) Let X be a random object from a probability space $(\Omega, \mathcal{A}, P)$ to a measurable space $(\mathcal{X}, \mathcal{G})$.
(i) The σ-field $\sigma(X)$ consists of the sets of the form

$$\{\omega \in \Omega : X(\omega) \in G\} = X^{-1}(G), \tag{4.3}$$

for $G \in \mathcal{G}$.
(ii) A random variable $Y : \Omega \to \mathcal{R}$ is measurable $\sigma(X)|\mathcal{B}_1$ if and only if $Y = f(X)$, for some $\mathcal{G}|\mathcal{B}_1$ measurable map $f : \mathcal{X} \to \mathcal{R}$.

Proof. (i) Let $\mathcal{S}$ denote the class of sets of the form (4.3). We prove first that $\mathcal{S}$ is a σ-field. Since X is a random object from $(\Omega, \mathcal{A}, P)$ to $(\mathcal{X}, \mathcal{G})$, we must have $X^{-1}(G) \in \mathcal{A}$, for all $G \in \mathcal{G}$. Clearly, $\Omega = X^{-1}(\mathcal{X}) \in \mathcal{S}$ and if $S = X^{-1}(G) \in \mathcal{S}$, for some $G \in \mathcal{G}$, then $S^c = \left[X^{-1}(G)\right]^c = X^{-1}(G^c) \in \mathcal{S}$, since $G^c \in \mathcal{G}$. Moreover, if $S_i = X^{-1}(G_i) \in \mathcal{S}$, for $G_i \in \mathcal{G}$, $i = 1, 2, \ldots$, then $\bigcup\limits_i S_i = \bigcup\limits_i X^{-1}(G_i) = X^{-1}(\bigcup\limits_i G_i) \in \mathcal{S}$, since $\bigcup\limits_i G_i \in \mathcal{G}$. Consequently, $\mathcal{S}$ is a σ-field. But X is $\mathcal{S}$-measurable by definition of X as a random object and hence $\sigma(X) \subset \mathcal{S}$. Moreover, since X is $\sigma(X)$-measurable by definition 4.5, $\mathcal{S} \subset \sigma(X)$ and part (i) is established.
(ii) By remark 3.7.2, $Y = f \circ X$ is $\sigma(X)|\mathcal{B}_1$ measurable.

For the other direction, assume first that $Y : \Omega \to \mathcal{R}$ is a measurable $\sigma(X)|\mathcal{B}_1$ simple random variable, i.e., $Y = \sum\limits_{i=1}^{m} y_i I_{A_i}$, where $A_i = \{\omega : Y(\omega) = y_i\}$, $i = 1, 2, \ldots, m$, are disjoint. Clearly, A_i is a $\sigma(X)$ measurable set and hence by part (i), A_i must be of the form $A_i = X^{-1}(G_i) = \{\omega : X(\omega) \in G_i\}$, for some $G_i \in \mathcal{G}$, $i = 1, 2, \ldots, m$, that need not be disjoint. Define the $\mathcal{G}|\mathcal{B}_1$ measurable simple function $f = \sum\limits_{i=1}^{m} y_i I_{G_i}$. Since the A_i are disjoint, no $X(\omega)$ can lie in more than one G_i and hence $f(X(\omega)) = Y(\omega)$.

Now consider any random variable $Y : \Omega \to \mathcal{R}$ that is measurable $\sigma(X)|\mathcal{B}_1$ and use theorem 3.4 to approximate $Y (= Y^+ - Y^-)$, using a sequence of simple random variables Y_n, that is, $Y_n(\omega) \to Y(\omega)$ for all ω. Now for each simple function Y_n there exists, as we saw above, a $\sigma(X)|\mathcal{B}_1$ measurable function f_n such that $Y_n(\omega) = f_n(X(\omega))$, for all ω. By remark 3.7.8 the set L of x in $\mathcal{X}$

for which $\{f_n(x)\}$ converges is an $\mathcal{G}$-set. Define $f(x) = \lim_{n\to\infty} f_n(x)$ for $x \in L$ and $f(x) = 0$, for $x \in L^c = \mathcal{X} \setminus L$. Since $f(x) = \lim_{n\to\infty} f_n(x)I_L(x)$ and $f_n(x)I_L(x)$ is $\mathcal{G}|\mathcal{B}_1$ measurable we have that the limit function f is also $\mathcal{G}|\mathcal{B}_1$ measurable. Now for each $\omega \in \Omega$ we have $Y(\omega) = \lim_{n\to\infty} f_n(X(\omega))$ so that $X(\omega) \in L$ and moreover $Y(\omega) = \lim_{n\to\infty} f_n(X(\omega)) = f(X(\omega))$. ■

The following remark gives the explicit forms result for random vectors.

Remark 4.2 (Borel measurable functions) Let $\mathbf{X} = (X_1, \ldots, X_p)$ be a random vector, i.e., $\mathbf{X}$ is a map from a probability space $(\Omega, \mathcal{A}, P)$ to the Borel measurable space $(\mathcal{R}^p, \mathcal{B}_p)$. The following are straightforward to show.

1. The σ-field $\sigma(\mathbf{X}) = \sigma(X_1, X_2, \ldots, X_p)$ consists exactly of the sets of the form $\mathbf{X}^{-1}(B)$, for all $B \in \mathcal{B}_p$.

2. A random variable Y is $\sigma(\mathbf{X})|\mathcal{B}_1$-measurable if and only if there exists a measurable map $f : \mathcal{R}^p \to \mathcal{R}$ such that $Y(\omega) = f(X_1(\omega), \ldots, X_p(\omega))$, for all $\omega \in \Omega$.

Example 4.2 (Random object examples) We consider some first examples of random objects. The source space $(\Omega, \mathcal{A}, P)$ over which X takes its argument can be the same in all cases. What makes the random objects different is the (target) measurable space $(\mathcal{X}, \mathcal{G})$. See Appendix Section A.7 for details on the code used to generate these objects and produce their plots.

1. Uniform random variable Assume that $\Omega = [0, 1]$, $\mathcal{A} = \mathcal{B}([0, 1])$ and $P = \mu_1$, the Lebesgue measure on $[0, 1]$, so that $(\Omega, \mathcal{A}, P)$ is a probability space and define a function $X : [0, 1] \to [0, 1]$, by $X(\omega) = \omega$, so that $(\mathcal{X}, \mathcal{G}) = ([0, 1], \mathcal{B}([0, 1]))$. Clearly, X is $\mathcal{B}([0, 1])|\mathcal{B}([0, 1])$ measurable and therefore X is a random variable that assigns length as the probability of an event, i.e., for any Borel set A we have

$$P(X \in A) = P(\{\omega : X(\omega) = \omega \in A\}) = P(A) = \mu_1(A).$$

This random object is the uniform random variable on $[0, 1]$.

2. Function-valued random variables Let $C^{\mathcal{R}}_{[0,1]}$ denote the space of continuous real-valued functions defined on the interval $[0, 1]$. We metrize this space using the metric $d(f, g) = \max_{x \in \mathcal{R}} |f(x) - g(x)|$, for all $f, g \in C^{\mathcal{R}}_{[0,1]}$ and consequently, we can talk about open sets with elements that are functions of $C^{\mathcal{R}}_{[0,1]}$, which leads to a topology on $C^{\mathcal{R}}_{[0,1]}$ and therefore allows the definition of the Borel subsets of $C^{\mathcal{R}}_{[0,1]}$, denoted by $\mathcal{B}_C = \mathcal{B}\left(C^{\mathcal{R}}_{[0,1]}\right)$.

Now a measurable functional from $(\Omega, \mathcal{A}, P)$ to $\left(C^{\mathcal{R}}_{[0,1]}, \mathcal{B}_C\right)$ can be thought of as a "random function," which is nothing but a measurable, $C^{\mathcal{R}}_{[0,1]}$-valued random variable. For a specific example, suppose that $(\Omega, \mathcal{A}, P)$ is the usual infinite coin probability space (example 4.1.3) and define for each positive integer k the random variable $X^{(k)} = (X_t^{(k)} : t \in [0, 1])$ from $(\Omega, \mathcal{A}, P)$ to $\left(C^{\mathcal{R}}_{[0,1]}, \mathcal{B}_C\right)$, by specifying the values $X_t^{(k)}(\omega)$, $\omega = (\omega_1, \omega_2, \ldots) \in \Omega$, of the continuous function $t \longmapsto X_t^{(k)}(\omega)$, at each $t \in [0, 1]$. We first specify these values for t equal to a multiple of $1/k$ and then the value of $X_t^{(k)}(\omega)$, $\forall t \in [0, 1]$ is determined by linear interpolation. More precisely, for $j = 0, 1, \ldots, k$, let

$$X_{j/k}^{(k)}(\omega) = k^{-1/2} \sum_{i=1}^{j} (2\omega_i - 1), \tag{4.4}$$

and extend via linear interpolation to obtain $X_t^{(k)}$, $\forall t \in [0, 1]$. Note that for any subset B of $C_{[0,1]}^{\mathcal{R}}$, the set $\left(X^{(k)}\right)^{-1}(B) = \{\omega : X^{(k)}(\omega) \in B\}$ consists of a finite union of cylinder sets $\{\omega : (\omega_1, \ldots, \omega_k) = (\varepsilon_1, \ldots, \varepsilon_k), \varepsilon_i \in \{0, 1\}\}$, which are certainly $\mathcal{A}$-measurable, so that $X^{(k)}$ is a random function.

Recall that $\omega_i \in \{0, 1\}$, with $\{0\} = \{Heads\}$ and $\{1\} = \{Tails\}$, so that the sum in (4.4) equals the difference between the number of tails and the number of heads after j tosses. These types of random functions will be used as the building block of a stochastic process known as Brownian motion and we will revisit them in Chapter 7.

Now let $Q^{(k)}$ denote the induced probability measure on $(C_{[0,1]}^{\mathcal{R}}, \mathcal{B}_C)$ by $X^{(k)}$, so that $(C_{[0,1]}^{\mathcal{R}}, \mathcal{B}_C, Q^{(k)})$ is a probability space, where the sample space and σ-field are the same for all k but the probability measures are different. Consider calculating the probability that we observe random functions $X^{(k)}$ with the property that $X_1^{(k)} = 0$, i.e., in k tosses, we have half heads and half tails. Using example 4.1.2, we have

$$Q^{(k)}(X_1^{(k)} = 0) = Q^{(k)}(\{g \in C_{[0,1]}^{\mathcal{R}} : g(1) = 0\}) = \begin{cases} C_{k/2}^{k} 2^{-k} & \text{if } k \text{ is even} \\ 0 & \text{otherwise} \end{cases}.$$

Figure 4.1 (a) presents the random functions of equation (4.4) for $k = 2$, 5 and 7 and a realization $\omega = (1, 1, 0, 1, 0, 0, 0, 1, 0, 0, 1, 1, 1, 1, 0, \ldots)$.

3. Random counting measure Consider a probability space $(\Omega, \mathcal{A}, P)$ and let $\mathbb{N}^f$ denote the collection of all locally finite counting measures on $\mathcal{X} \subset \mathcal{R}^p$, where if $\varphi \in \mathbb{N}^f$ and $B \in \mathcal{B}(\mathcal{X})$, then $\varphi(B)$ denotes the number of points in the set B. In order to define a random counting measure we need to create a measurable space based on a σ-field of subsets of $\mathbb{N}^f$. Therefore, consider $\mathcal{N}^f$, the generated σ-field from the collection of sets of the form $\{\varphi \in \mathbb{N}^f : \varphi(B) = n\}$, for all $B \in \mathcal{B}(\mathcal{X})$ and all $n = 0, 1, 2, \ldots$, so that $(\mathbb{N}^f, \mathcal{N}^f)$ is a measurable space. Then a mapping N from $(\Omega, \mathcal{A}, P)$ to $(\mathbb{N}^f, \mathcal{N}^f)$ can be thought of as a random object $N(\omega)(.) = \varphi(.)$, that yields counting measures φ for different $\omega \in \Omega$ and hence it is a random counting measure. The induced probability measure $\Pi_N(Y) = P(N \in Y) = P(N^{-1}(Y))$, can be used to compute probabilities over sets $Y \in \mathcal{N}$, where each element of Y is a counting measure.

In order to appreciate the usefulness of this construction, consider a countable set of spatial locations $S = \{\mathbf{x}_1, \mathbf{x}_2, \ldots\}$, $\mathbf{x}_i \in \mathcal{X}$. Knowing $\varphi(B)$ for all $B \in \mathcal{B}(\mathcal{X})$ is equivalent to knowing the spatial locations of all points in S. Indeed, $\varphi(B) = \sum_{i=1}^{+\infty} I(\mathbf{x}_i \in B)$, $\forall B \in \mathcal{B}(\mathcal{X})$, so that knowing the locations implies that we know the counting measure φ. Conversely, $\mathbf{x}$ is a point from S if $\varphi(\{\mathbf{x}\}) > 0$ and the point is distinct if $\varphi(\{\mathbf{x}\}) = 1$. Now if φ is a random counting measure, i.e., $\varphi = N(\omega)$, for some $\omega \in \Omega$, then this important relationship between φ and S defines a random collection of points (a countable random set) N, known as a point process. Alternatively, an equivalent definition for the random point process N can be given through the exponential space of example 3.5, by letting $\mathcal{X}^n = \{\varphi : \varphi(\mathcal{X}) = n\}$. Figure 4.1 (b)-(d) presents three realizations of random counting measures illustrating the types of point processes N one can encounter, namely, regular (b), random (c) and clustered (d). We study such models throughout the *TMSO-PPRS* text.

4. Random disc Consider a random variable $R : (\Omega, \mathcal{A}, P) \to (\mathcal{R}^+, \mathcal{B}(\mathcal{R}^+))$ and let C denote the collection of all open discs centered at the origin, i.e., $C = \{b(\mathbf{0}, r) \subset \mathcal{R}^2 : r > 0\}$. Clearly, there is a one-to-one relationship between $\mathcal{R}^+$ and C, that is, for each positive real number r there is a single open disc $b(\mathbf{0}, r)$ and conversely, each open disc corresponds to a single positive real number. Now if we want to define a random disc (recall example 3.6 and remark 3.7.10), we need to equip C with a σ-field $\mathcal{G}$ and then define a measurable, disc-valued map $X : (\Omega, \mathcal{A}, P) \to (C, \mathcal{G})$ by

$$X(\omega) = b(\mathbf{0}, R(\omega)), \tag{4.5}$$

i.e., for a given $\omega \in \Omega$, $X(\omega)$ is the open disc centered at the origin of radius $R(\omega)$ and $X^{-1}(G) \in \mathcal{A}, \forall G \in \mathcal{G}$. Obviously, $\mathcal{G} = 2^C$ is a valid choice but the resulting σ-field is too large and we will have a hard time defining the distribution $Q(G) = P(X^{-1}(G))$ of X, $\forall G \in \mathcal{G}$. As we see in the *TMSO-PPRS* text, in order to fully describe the distribution Q of X it suffices to use the hit-or-miss topology $\mathcal{G} = \mathcal{B}(C) = \sigma(\{\mathbb{F}_K, \forall K \in \mathcal{K}\})$ and then compute the hitting function

$$T_X(K) = P(X \cap K \neq \emptyset), \ \forall K \in \mathcal{K},$$

which is **not** a probability measure in general, but under certain conditions it can uniquely identify the distribution of X. Figure 4.1 (e) presents five realizations of a random disc centered at the origin with radii drawn from a *Gamma*(10, 1).

5. Gaussian random field Assume that $G = \{G(\mathbf{x}) \in \mathcal{R} : \mathbf{x} \in \mathcal{R}^p\}$ is a Gaussian random field (GRF), that is, G is such that the finite-dimensional distributions $P(G(\mathbf{x}_1) \in B_1, \dots, G(\mathbf{x}_n) \in B_n)$, for any Borel sets $B_1, \dots, B_n \in \mathcal{B}(\mathcal{R})$ and $n \geq 1$, are multivariate normal. Therefore, a GRF is completely characterized by its mean function

$$\mu(\mathbf{x}) = E(G(\mathbf{x})),$$

$\mathbf{x} \in \mathcal{R}^p$ and its covariance function

$$C(\mathbf{x}_1, \mathbf{x}_2) = E\left[(G(\mathbf{x}_1) - \mu(\mathbf{x}_1))(G(\mathbf{x}_2) - \mu(\mathbf{x}_2))\right],$$

$\mathbf{x}_1, \mathbf{x}_2 \in \mathcal{R}^p$. In particular, $G(\mathbf{x}) \sim N(\mu(\mathbf{x}), C(\mathbf{x}, \mathbf{x}))$. A GRF is stationary and isotropic if $\mu(\mathbf{x}) = \mu$, i.e., $\mu(\mathbf{x})$ does not depend on the location, and $C(\mathbf{x}_1, \mathbf{x}_2)$ is a function only of the distance $r = \|\mathbf{x}_1 - \mathbf{x}_2\|$ of the points $\mathbf{x}_1$ and $\mathbf{x}_2$. In this case we use the notation $C(r) = C(\|\mathbf{x}_1 - \mathbf{x}_2\|) = C(\mathbf{x}_1 - \mathbf{x}_2)$, so that $\sigma^2 = C(0)$ is the variance of the GRF G. Note that for any $\mathbf{x} \in \mathcal{R}^p$ we can write $C(r) = C(\|\mathbf{x} - \mathbf{0}\|) = C(\mathbf{x})$, where $\mathbf{0}$ is the origin and $r = \|\mathbf{x}\|$. The most commonly used class of covariance functions is the Matérn class (Matérn, 1986) given by

$$C_{\nu,\theta,\sigma^2}(r) = \sigma^2 \frac{2^{1-\nu}}{\Gamma(\nu)} \left(\sqrt{2\nu}\frac{r}{\theta}\right)^{\nu} B_{\nu}\left(\sqrt{2\nu}\frac{r}{\theta}\right), \tag{4.6}$$

for $\nu, \theta > 0$, where $B_\nu(.)$ denotes the modified Bessel function of the second kind, θ is a scale parameter and ν a smoothness parameter such that the sample functions are m times differentiable if and only if $m < \nu$. For example, for $\nu = 0.5$, we have $C_{0.5,\theta,\sigma^2}(r) = \sigma^2 e^{-r/\theta}$. In Figure 4.1 (f), we display a realization of the GRF with $\mu = 0$ and covariance function $C_{0.5,10,1}(r) = e^{-r/10}$. Random fields play a prominent role in spatial and spatiotemporal statistics (e.g., Cressie and Wikle, 2011) and will be utilized in the *TMSO-PPRS* text.

4.2.3 Distribution Functions and Densities

The distribution theory of Chapter 1 is based on rudimentary quantities such as densities and cdfs, which are related to probability measures on the measurable space $(\mathcal{R}^p, \mathcal{B}_p)$. We collect the basic definitions and properties below.

Definition 4.6 Distribution function

A real-valued function F defined on $\overline{\mathcal{R}}$ is called a (cumulative) distribution function for $\mathcal{R}$, or simply a distribution function, if it is nondecreasing and right-continuous and satisfies $F(-\infty) = \lim\limits_{x \to -\infty} F(x) = 0$ and $F(+\infty) = \lim\limits_{x \to +\infty} F(x) = 1$.

The following shows us how to obtain the cdf given a probability measure.

Theorem 4.5 (Cdf via a probability measure) Let P be a probability measure on $(\mathcal{R}, \mathcal{B}_1)$. Then the function $F(x) = P((-\infty, x])$ is a distribution function.

Proof. Since P is monotone then $x \leq y \implies (-\infty, x] \subseteq (-\infty, y] \implies P((-\infty, x]) \leq P((-\infty, y])$ and hence F is nondecreasing. To show right-continuity, suppose that $x_n \downarrow x$, with $x_n, x \in \mathcal{R}$. Since $(-\infty, x_1] \supseteq (-\infty, x_2] \supseteq \ldots$, by continuity of measure from above we have

$$\lim_{n \to +\infty} P((-\infty, x_n]) = P\left(\bigcap_{n=1}^{+\infty} (-\infty, x_n]\right) = P((-\infty, x]),$$

and hence F is right-continuous. Using similar arguments for $x_n \downarrow -\infty$, we have $F(-\infty) = 0$. Finally, letting $x_n \uparrow +\infty$ and using continuity from below we have

$$P((-\infty, x_n]) \to P\left(\bigcup_{n=1}^{+\infty} (-\infty, x_n]\right) = P(\mathcal{R}) = 1.$$

∎

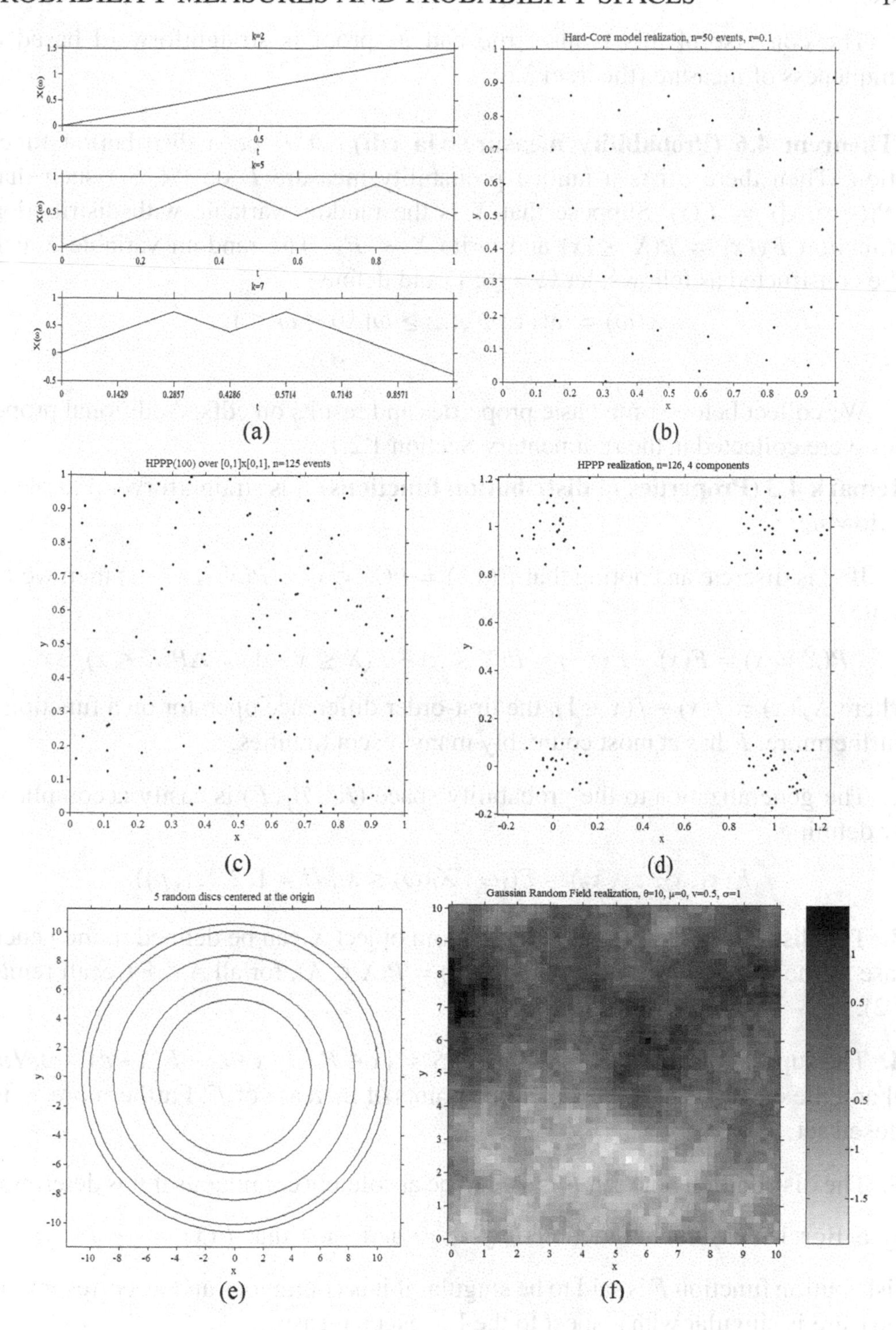

Figure 4.1: Displaying random objects: (a) random functions, (b)-(d) random counting measures showing regularity, randomness and clustering of the points, (e) random discs, and (f) Gaussian random field.

The converse of this is also true and its proof is straightforward based on uniqueness of measure (theorem 3.6).

Theorem 4.6 (Probability measure via cdf) Let F be a distribution function. Then there exists a unique probability measure P on $(\mathcal{R}, \mathcal{B}_1)$ such that $P((-\infty, x]) = F(x)$. Suppose that X is the random variable with distribution function $F_X(x) = P(X \leq x)$ and write $X \sim F_X$. The random variable X can be constructed as follows: let $\Omega = (0, 1)$ and define

$$X(\omega) = \inf\{x : F_X(x) \geq \omega\},\ 0 \leq \omega \leq 1.$$

We collect below some basic properties and results on cdfs. Additional properties were collected in the rudimentary Section 1.2.1.

Remark 4.3 (Properties of distribution functions) It is straightforward to see the following.

1. If P is discrete and noting that $F(x-) = P(X < x) = P(X \leq x - 1)$ then we can write

$$P(X = x) = F(x) - F(x-) = P(X \leq x) - P(X \leq x - 1) = \Delta P(X \leq x)/\Delta x,$$

where $\Delta f(x) = f(x) - f(x - 1)$, the first-order difference operator on a function f. Furthermore, F has at most countably many discontinuities.

2. The generalization to the probability space $(\mathcal{R}^p, \mathcal{B}_p, P)$ is easily accomplished by defining

$$F(x_1, x_2, \ldots, x_p) = P(\{\omega : X_i(\omega) \leq x_i, \forall i = 1, 2, \ldots, p\}).$$

3. The distribution function F of a random object X can be defined in the general case as the containment functional $F_X(A) = P(X \subset A)$, for all $A \in \mathbb{F}$ (recall remark 1.2).

4. The support of F is defined as the set $S = \{x \in \mathcal{R} : F(x+\varepsilon) - F(x-\varepsilon) > 0, \forall \varepsilon > 0\}$ and the elements of S are called the points of increase of F. Furthermore, S is a closed set.

5. The distribution function F is said to be absolutely continuous if it is determined by a Borel measurable function f, $f \geq 0$ and such that $F(x) = \int\limits_{-\infty}^{x} f(t)dt$. The distribution function F is said to be singular if it is continuous and its corresponding measure is singular with respect to the Lebesgue measure.

6. We can write $F = a_1F_1 + a_2F_2 + a_3F_3$, where $a_1 + a_2 + a_3 = 1$, $a_i \geq 0$ and F_1, F_2 and F_3, are discrete, absolutely continuous and singular, respectively.

7. The distribution function F is called defective if the corresponding probability measure P is defective.

Although the distribution function provides much of our rudimentary distribution theory results, we can further simplify calculations by defining any important quantity of interest in terms of the density of the random object X.

Definition 4.7 Density

Consider a random object $X : (\Omega, \mathcal{A}, P) \to (\mathcal{R}^p, \mathcal{B}_p)$ with induced probability measure $Q(B) = P(X^{-1}(B))$, $B \in \mathcal{B}_p$ and let μ be a σ-finite measure defined on $(\mathcal{R}^p, \mathcal{B}_p)$. A density for X (or for Q) with respect to μ is a nonnegative, $\mathcal{B}_p|\mathcal{B}(\mathcal{R}_0^+)$-measurable function f such that for each set $B \in \mathcal{B}_p$ we have

$$Q(B) = P(X^{-1}(B)) = P(X \in B) = \int_B f(x)d\mu(x). \tag{4.7}$$

The following remark summarizes some important results on densities.

Remark 4.4 (Densities) Some consequences of this important definition follow.

1. If $B = \mathcal{R}^p$ and $\mu = \mu_p$ then

$$1 = Q(\mathcal{R}^p) = P(X \in \mathcal{R}^p) = \int_{\mathcal{R}^p} f(\mathbf{x})d\mu_p(\mathbf{x}),$$

and consequently $f(\mathbf{x})$ represents how the probability or mass is distributed over the domain $\mathcal{R}^p$ with the total volume under the surface $y = f(\mathbf{x})$ being 1.

2. Any Lebesgue integrable, $\mathcal{B}_p|\mathcal{B}(\mathcal{R}_0^+)$-measurable function f that satisfies $\int_{\mathcal{R}^p} f(\mathbf{x})d\mu_k(\mathbf{x}) = 1$, can be thought of as the density of a random vector $\mathbf{X}$ defined on some probability space $(\Omega, \mathcal{A}, P)$ and such that $Q(B) = P(X \in B) = \int_B f(\mathbf{x})d\mu_p(\mathbf{x})$. We abbreviate this by writing $\mathbf{X} \sim f(\mathbf{x})$ (or $\mathbf{X} \sim Q$). The density does not contain any information about the measurable space $(\Omega, \mathcal{A})$ (see part 6 below).

3. If $f = g$ a.e. $[\mu]$ and f is a density for P then so is g.

4. Note that if Q has the representation (4.7) then if for some $B \in \mathcal{B}_p$ we have $\mu(B) = 0$ then $Q(B) = 0$ and hence $Q \ll \mu$. The converse is obtained using the Radon-Nikodym theorem, that is, in order for a random object X to have a density it suffices to have $Q \ll \mu$, for some σ-finite measure μ. Consequently, the density is simply the Radon-Nikodym derivative $f = \left[\frac{dQ}{d\mu}\right]$. In theory, any measure μ could be chosen but in order to simplify calculations of the integrals we typically choose the Lebesgue measure for continuous random variables and the counting measure for discrete random variables. Moreover, in view of remark 3.21.1 we can write

$$\int_B g dQ = \int_B g f d\mu,$$

for any measurable function g.

5. Recall Liebnitz's rule for differentiation of integrals (the Riemann integral version):

$$\frac{\partial}{\partial\theta}\left[\int_{a(\theta)}^{b(\theta)} f(x,\theta)dx\right] = f(b(\theta),\theta)\frac{db(\theta)}{d\theta} - f(a(\theta),\theta)\frac{da(\theta)}{d\theta} + \int_{a(\theta)}^{b(\theta)}\left[\frac{\partial}{\partial\theta}f(x,\theta)\right]dx.$$

Then assuming that X is a continuous random variable, applying the rule to $F_X(x) = \int_{-\infty}^{x} f_X(t)dt$, we obtain $f_X(x) = \frac{dF(x)}{dx}$.

6. A density f contains all the information about the probabilistic behavior of the random variable X but it contains no detailed information about the underlying probability space (Ω, A, P) over which X is defined, or about the interaction of X with other random variables defined on the space. To see this, recall example 4.1.2 and the discrete probability space $(\Omega_n, 2^{\Omega_n}, P_n)$ and consider the infinite coin flip space Ω_∞. We can define the random variable $X(\omega)$ as the number of tails in n-flips by choosing any n distinct flips from the infinite coin flip space (yielding the same sample space Ω_n for any choice) and all such variables have the same density f (the Radon-Nikodym derivative of the induced probability measure $Q_n = P_n \circ X^{-1}$, with respect to the counting measure), given by

$$f(k) = P(X(\omega) = k) = C_k^n 2^{-n},$$

where $k = 0, 1, 2, \ldots, n$. The random variable can be defined by $X(\omega) = \sum_{i=1}^{n} \omega_i$ or $X(\omega) = \sum_{i=r}^{n+r-1} \omega_i$, for any $r = 2, \ldots$ and its density $f(k)$ loses sight of this underlying structure.

The following theorem provides the rigorous definition of the marginal densities.

Theorem 4.7 (Marginal densities) Let $X = (X_1, X_2)$ be a random vector defined on $(\Omega, \mathcal{A}, P)$ that takes values in the product measure space $(\mathcal{X}, \mathcal{G}, \mu)$, where $\mathcal{X} = \mathcal{X}_1 \times \mathcal{X}_2$, $\mathcal{G} = \mathcal{G}_1 \bigotimes \mathcal{G}_2$, $\mu = \mu_1 \times \mu_2$ and μ_1, μ_2 are σ-finite measures. Suppose that the distribution of X has (joint) density f with respect to μ. Then the distributions of X_1 and X_2 have (marginal) densities

$$f_1(x_1) = \int_{\mathcal{X}_2} f(x_1, x_2)\mu_2(dx_2),$$

and

$$f_2(x_2) = \int_{\mathcal{X}_1} f(x_1, x_2)\mu_1(dx_1),$$

with respect to μ_1 and μ_2, respectively.

Proof. Let A be any $\mathcal{G}_1$-set. Then we can write

$$P(\{\omega : X_1(\omega) \in A\}) = \int_{A\times\mathcal{X}_2} f(x_1, x_2)d\mu(x_1, x_2) = \int_A \left[\int_{\mathcal{X}_2} f(x_1, x_2)d\mu_2(x_2)\right] d\mu_1(x_1),$$

by the Fubini theorem and therefore $\int\limits_{\mathcal{X}_2} f(x_1, x_2)d\mu_2(x_2) = f_1(x_1)$. Similarly for the density of X_2. ■

4.2.4 Independence

We begin exploring the concept of independence with the general definitions of independence between σ-fields and random objects below.

Definition 4.8 Independence

Consider a probability space $(\Omega, \mathcal{A}, P)$ and let $\mathcal{A}_k$, $k \in K$, be sub-σ-fields of $\mathcal{A}$.

1. If K is finite then we say that $\{\mathcal{A}_k : k \in K\}$ is (stochastically) independent if and only if

$$P\left(\bigcap_{k\in K} A_k\right) = \prod_{k\in K} P(A_k), \tag{4.8}$$

for all $A_k \in \mathcal{A}_k$.

2. If K is an infinite set, then $\{\mathcal{A}_k : k \in K\}$ is independent if and only if for every finite subset $J \subset K$, $\{\mathcal{A}_j : j \in J\}$ are independent.

3. Let $X_1, X_2, \ldots,$ be random objects from $(\Omega, \mathcal{A}, P)$ to $(\mathcal{X}, \mathcal{G})$.

(a) A finite collection $X_1, \ldots, X_n$ is said to be independent if and only if $\{\sigma(X_i), i = 1, \ldots, n\}$ are independent, where $\sigma(X_i) = \{X_i^{-1}(G) : G \in \mathcal{G}\}$, $i = 1, 2, \ldots, n$.

(b) An infinite collection $X_1, X_2, \ldots,$ is said to be independent if and only if $\{\sigma(X_i), i = 1, 2, \ldots\}$ are independent.

4. Let $A_1, A_2, \ldots,$ be $\mathcal{A}$-sets.

(a) We say that $A_1, \ldots, A_n$ are independent events if and only if $\{\sigma(A_i) : i = 1, 2, \ldots, n\}$ are independent, where $\sigma(A_i) = \{\emptyset, \Omega, A_i, A_i^c\}$, $i = 1, 2, \ldots, n$.

(b) The events of the infinite sequence $A_1, A_2, \ldots,$ are independent if and only if $\{\sigma(A_i) : i = 1, 2, \ldots\}$ are independent.

A first existence theorem for a random sequence is given next under the assumption of independence.

Theorem 4.8 (Existence of a sequence under independence) Let $(\Omega_j, \mathcal{A}_j, P_j)$ be probability spaces, $j = 1, 2, \ldots$. Then there exists a sequence of independent random objects $(X_1, X_2, \ldots)$ such that X_j has distribution P_j, for all j.

Proof. Let $\Omega_\infty = \mathop{\times}\limits_{i=1}^{\infty} \Omega_j$, $\mathcal{A}_\infty = \bigotimes\limits_{j=1}^{\infty} \mathcal{A}_j$ and $P = \mathop{\times}\limits_{i=1}^{\infty} P_j$. For $j \geq 1$, let $X_j(\omega) = \omega_j$, where $\omega = (\omega_1, \omega_2, \ldots)$, that is, define the projection map $X_j : (\Omega_\infty, \mathcal{A}_\infty, P) \to (\Omega_j, \mathcal{A}_j)$. If $B_j \in \mathcal{A}_j$ then $X_j^{-1}(B_j) = \Omega_1 \times \cdots \times \Omega_{j-1} \times B_j \times \Omega_{j+1} \times \cdots \in \mathcal{A}_\infty \Longrightarrow X_j$ is an

$\mathcal{A}_\infty|\mathcal{A}_j$ measurable function, i.e., X_j is a random object. Moreover, the distribution of X_j is given by

$$\begin{aligned} Q_j(X_j) &= P(X_j \in B_j) = P(X_j^{-1}(B_j)) \\ &= P(\Omega_1 \times \cdots \times \Omega_{j-1} \times B_j \times \Omega_{j+1} \times \ldots) = P_j(B_j), \end{aligned}$$

and hence $Q_j = P_j$.

To show independence of $X_1, \ldots, X_n, \ldots$, we have to show that $\sigma(X_1), \ldots, \sigma(X_n), \ldots$, are independent. Let $A_{j_i} \in X_{j_i}^{-1}(B_{j_i})$, where $B_{j_i} \in \mathcal{A}_{j_i}$, for some indices j_i, $i = 1, 2, \ldots, n$. Then we can write

$$\begin{aligned} P\left(\bigcap_{i=1}^{n} A_{j_i}\right) &= P\left(\bigcap_{i=1}^{n} \{\omega \in \Omega : X_{j_i} \in B_{j_i}\}\right) \\ &= P\left(\bigcap_{i=1}^{n} \left(\Omega_1 \times \cdots \times \Omega_{j-1} \times B_{j_i} \times \Omega_{j+1} \times \ldots\right)\right) \\ &= \prod_{i=1}^{n} P\left(\Omega_1 \times \cdots \times \Omega_{j-1} \times B_{j_i} \times \Omega_{j+1} \times \ldots\right) = \prod_{i=1}^{n} P_{j_i}\left(A_{j_i}\right), \end{aligned}$$

and thus $X_1, X_2, \ldots$, are independent. ■

Next we give a characterization of independence via the marginal distributions.

Theorem 4.9 (Characterization of independence) Let $X_1, \ldots, X_n$ be $\mathcal{X}$-valued, $\mathcal{A}|\mathcal{G}$-measurable random objects defined on a probability space $(\Omega, \mathcal{A}, P)$. Then $X_1, \ldots, X_n$ are independent if and only if for all $B_1, \ldots, B_n \in \mathcal{G}$ we have

$$P(X_1 \in B_1, \ldots, X_n \in B_n) = \prod_{i=1}^{n} P(X_i \in B_i). \tag{4.9}$$

Proof. By definition $X_1, \ldots, X_n$ are independent if and only if $\sigma(X_1), \ldots, \sigma(X_n)$ are independent, which holds if and only if for any $A_i \in \sigma(X_i)$, $i = 1, 2, \ldots n$, we have

$$P\left(\bigcap_{i=1}^{n} A_i\right) = \prod_{i=1}^{n} P(A_i). \tag{4.10}$$

Now $A_i \in \sigma(X_i) \iff A_i = X_i^{-1}(B_i)$, for some $B_i \in \mathcal{G}$, which implies that $P(A_i) = P(X_i \in B_i)$. Therefore equation (4.10) is equivalent to the desired equation (4.9) and the result is established. ■

The following are straightforward to prove based on the latter two theorems and definition.

Remark 4.5 (Independence) Let $\{X_n\}_{n=1}^{+\infty}$ be a sequence of independent random objects.

1. Let $\{i_1, i_2, \ldots\}$ and $\{j_1, j_2, \ldots\}$ be disjoint sets of integers. Then the σ-fields $\sigma(\{X_{i_1}, X_{i_2}, \ldots\})$ and $\sigma(\{X_{j_1}, X_{j_2}, \ldots\})$ are independent.

2. Let $J_1, J_2, \ldots,$ be disjoint sets of integers. Then the σ-fields $\mathcal{G}_k = \sigma(\{X_j : j \in J_k\})$, $k = 1, 2, \ldots,$ are independent.

3. The σ-fields $\sigma(\{X_1, X_2, \ldots, X_n\})$ and $\sigma(\{X_{n+1}, X_{n+2}, \ldots\})$ are independent.

4. If $\varphi_1, \varphi_2, \ldots,$ are Ψ-valued, $\bigotimes_{j=1}^{m} \mathcal{G}|\mathcal{Y}$–measurable functions defined on $\mathcal{X}^m$, then $Z_1 = \varphi_1(X_1, X_2, \ldots, X_m)$, $Z_2 = \varphi_2(X_{m+1}, X_{m+2}, \ldots, X_{2m}), \ldots,$ are independent.

5. If $X = (X_1, \ldots, X_n)$ has (joint) distribution Q and X_i have (marginal) distributions Q_i, $i = 1, 2, \ldots, n$, then equation (4.9) holds if and only if

$$Q(B_1 \times \cdots \times B_n) = \prod_{i=1}^{n} Q(B_i).$$

Now we illustrate the importance of the cdf in identifying independence.

Theorem 4.10 (Independence via cdfs) Let $X_1, X_2, \ldots,$ be $\mathcal{R}$–valued random variables, defined on a probability space $(\Omega, \mathcal{A}, P)$. Then the coordinates of the random vector $\mathbf{X} = (X_1, \ldots, X_n)$ are independent if and only if

$$F_{X_1,\ldots,X_n}(x_1, \ldots, x_n) = F_{X_1}(x_1) \ldots F_{X_n}(x_n), \tag{4.11}$$

$\forall n \geq 1$ and $(x_1, \ldots, x_n) \in \mathcal{R}^n$, where $F_{X_1,\ldots,X_n}, F_{X_1}, \ldots, F_{X_n}$ denote distribution functions.

Proof. ($\Longrightarrow$) If $X_1, \ldots, X_n$ are independent then equation (4.11) holds using $B_i = (-\infty, x_i]$ in theorem 4.9.

($\Longleftarrow$) For the other direction we need to show that for all $B_1, \ldots, B_n \in \mathcal{B}_1$ and for all $n = 1, 2, \ldots,$ we have

$$P(X_1 \in B_1, \ldots, X_n \in B_n) = \prod_{i=1}^{n} P(X_i \in B_i).$$

Fix $x_2, \ldots, x_n$ and define two σ-additive measures

$$Q_1(B) = P(X_1 \in B, X_2 \leq x_2, \ldots, X_n \leq x_n),$$

and

$$Q_1'(B) = P(X_1 \in B)P(X_2 \leq x_2, \ldots, X_n \leq x_n),$$

and note that on sets of the form $(-\infty, x]$, Q_1 and Q_1' agree. Therefore, by extension and uniqueness of measure, $Q_1 = Q_1'$ on $\mathcal{B}_1 = \sigma(\{(-\infty, x]\})$. Repeat the process for a fixed $B_1 \in \mathcal{B}_1$ and fixed $x_3, \ldots, x_n$ and define

$$Q_2(B) = P(X_1 \in B_1, X_2 \in B, X_3 \leq x_3, \ldots, X_n \leq x_n),$$

and

$$Q_2'(B) = P(X_1 \in B)P(X_2 \in B)P(X_3 \leq x_3, \ldots, X_n \leq x_n),$$

and since Q_2 and Q_2' agree for $B = (-\infty, x]$, they agree on $\mathcal{B}_1$. Continue this way to establish the desired equation. ■

An immediate consequence of the latter theorem is the following.

Remark 4.6 (Independence via densities) If the random variables $X_1, \ldots, X_n$ with distributions $Q_1, \ldots, Q_n$, have (marginal) densities $f_i = \left[\frac{dQ_i}{d\mu_i}\right]$ with respect to μ_i, then equation (4.11) is equivalent to

$$f_{X_1,\ldots,X_n}(x_1, \ldots, x_n) = f_{X_1}(x_1) \ldots f_{X_n}(x_n), \tag{4.12}$$

where $f_{X_1,\ldots,X_n}$ is the (joint) density of the random vector $(X_1, \ldots, X_n)$.

Example 4.3 (Random objects via densities) The following examples illustrate some random objects and their densities.

1. Uniform random point Let $X = (X_1, X_2)$ be a random point in the unit square $[0, 1]^2$ that arises by generating the coordinates X_1 and X_2 as independent and uniformly distributed random variables on $[0, 1]$. The joint density is therefore given by

$$f(x_1, x_2) = I_{[0,1]}(x_1)I_{[0,1]}(x_2).$$

For any subset $A \subseteq [0, 1]^2$ the probability that X falls in A is computed using the Radon-Nikodym theorem by

$$P(X \in A) = \int_A f(x_1, x_2)dx_1dx_2 = \int_A dx_1dx_2 = |A|,$$

the area of A and therefore probabilities in this context are measures of area. In general, let $A \subset \mathcal{R}^2$, with area $|A| > 0$. Let X be a random point uniformly distributed in A, that is,

$$f(x_1, x_2) = |A|^{-1}I_A(x_1, x_2),$$

so that for $B \subset A$, we have

$$P(X \in B) = |B|/|A|.$$

2. Uniform random lines Any straight line in $\mathcal{R}^2$ is uniquely identified by its orientation θ and signed distance ρ from the origin, namely, $L_{\theta,\rho} = \{(x, y) : x\cos\theta + y\sin\theta = \rho\}$, with $\theta \in [0, \pi)$ and $\rho \in \mathcal{R}$. Let X be a random straight line through the unit disc $b(0, 1)$, which is generated by taking θ and ρ to be independent and uniform over $[0, \pi)$ and $[-1, 1]$, respectively. Therefore, the density of the random vector (θ, ρ) is given by

$$f(\theta, \rho) = (2\pi)^{-1}I_{[0,\pi)}(\theta)I_{[-1,1]}(\rho).$$

If $C = \{(x, y) : (x - c_1)^2 + (y - c_2)^2 = r^2\}$ is a circle of radius r centered at (c_1, c_2) contained in $b(0, 1)$, then the probability that X intersects C is

$$P(X \cap C \neq \emptyset) = \int_0^{\pi}\int_{-1}^{1} I(L_{\theta,\rho} \cap C \neq \emptyset)(2\pi)^{-1}d\rho d\theta = r,$$

which is proportional to the circumference of C, regardless of the location of C within $b(0, 1)$.

The following theorem can be particularly useful in proving independence between collections of events.

Theorem 4.11 (Showing independence) If $\mathcal{G}$ and $\mathcal{L}$ are independent classes of events and $\mathcal{L}$ is closed under intersections, then $\mathcal{G}$ and $\sigma(\mathcal{L})$ are independent.

Proof. For any $B \in \mathcal{G}$ let

$$\mathcal{L}^* = \{A : A \in \sigma(\mathcal{L}) \text{ and } P(A \cap B) = P(A)P(B)\}.$$

Notice that 1) $\Omega \in \mathcal{L}$, 2) if $A_1, A_2 \in \mathcal{L}^*$ and $A_2 \subset A_1 \Longrightarrow A_1 \smallsetminus A_2 \in \mathcal{L}^*$ and 3) if $A = \lim\limits_{n\to+\infty} A_n$, with $A_n \subset A_{n+1}, n \geq 1$ and $A_n \in \mathcal{L}^* \Longrightarrow A \in \mathcal{L}^*$, since $A_n \cap B \subset A_{n+1} \cap B$ and

$$P(A \cap B) = \lim_{n\to+\infty} P(A_n \cap B) = \lim_{n\to+\infty} P(A_n)P(B) = P(A)P(B).$$

Thus $\mathcal{L}^*$ is a Sierpiński class with $\mathcal{L} \subset \mathcal{L}^*$ and by the Sierpiński class theorem $\sigma(\mathcal{L}) = \mathcal{L}^*$. As a consequence, $\mathcal{G}$ and $\sigma(\mathcal{L})$ are independent. ■

The following definition and 0-1 law helps us assess when a sequence of iid random variables converges a.s. More details on convergence of random sequences and their partial sums can be found in Chapter 5.

Definition 4.9 Tail σ-field

The tail σ-field of a sequence of random variables $\{X_n : n \geq 1\}$ on $(\Omega, \mathcal{A}, P)$ is defined by $\bigcap\limits_{n=1}^{+\infty} \sigma(\{X_j : j \geq n\})$. A subset of the tail σ-field is called a tail event.

Example 4.4 (Tail events) Tail σ-fields and events appear very often in asymptotics.

1. Let $\{X_n\}_{n=1}^{+\infty}$ be a sequence of independent random variables. The set $\{\omega \in \Omega : \sum\limits_{n=1}^{+\infty} X_n(\omega) \text{ converges}\}$ is a tail event, and, as we will see in the next theorem, this event has probability 0 or 1, which makes sense intuitively since a series either converges or it does not converge.

2. Let $\{A_n\}$ be independent. Then $X_n(\omega) = I_{A_n}(\omega)$ are independent random variables and $\bigcup\limits_{j=n}^{+\infty} A_j \in \sigma(\{X_j : j \geq m\})$, for $n \geq m \geq 1$, which implies that the event $\limsup A_n = (A_n \text{ i.o.}) = \bigcap\limits_{n=m}^{+\infty} \bigcup\limits_{j=n}^{+\infty} A_j \in \sigma(\{X_m, X_{m+1}, \dots\})$, is a tail event. Similarly, the event $\liminf A_n = (A_n \text{ ev.})$ is a tail event.

The celebrated Kolmogorov 0-1 Law is proven next.

Theorem 4.12 (Kolmogorov 0-1 Law) Tail events of a sequence $\{X_n : n \geq 1\}$ of independent random variables have probability 0 or 1.

Proof. For $n \geq 1$, $\sigma(\{X_i : 1 \leq i \leq n\})$ and $\sigma(\{X_j : j > n\})$ are independent and therefore $\sigma(\{X_i : 1 \leq i \leq n\})$ is independent of $\mathcal{D} = \bigcap_{n=0}^{+\infty} \sigma(\{X_i : j > n\})$, $\forall n \geq 1$. This implies that $\mathcal{A} = \bigcup_{n=1}^{+\infty} \sigma(\{X_i : 1 \leq i \leq n\})$ is independent of $\mathcal{D}$. We can easily see that $\mathcal{A}$ is a field and hence is closed under intersection. By theorem 4.11 $\sigma(\mathcal{A})$ is independent of $\mathcal{D}$. But $\mathcal{D} \subset \sigma(\{X_n : n \geq 1\}) = \sigma(\mathcal{A})$, which implies that the tail σ-field $\mathcal{D}$ is independent of itself and consequently $\forall B \in \mathcal{D}$, $P(B \cap B) = P^2(B) \Longrightarrow P(B) = 0$ or 1. ■

Next we discuss a few classic methods of computing the probabilities of such limiting events.

4.2.5 Calculating Probabilities for Limits of Events

Consider a probability space $(\Omega, \mathcal{A}, P)$. In order to compute the probability of an event $A \in \mathcal{A}$, it is often useful to express A in terms of other, easier to compute events (for example using remark 3.1.9). Applying Fatou's lemma (theorem 3.13) on the functions $f_n = I_{A_n}$, with $\mu = P$, so that $\int f d\mu = P(A_n)$ and noting that $(\underline{\lim} A_n)^c = \overline{\lim} A_n^c$, we have

$$P(\underline{\lim} A_n) \leq \underline{\lim} P(A_n) \leq \overline{\lim} P(A_n) \leq P(\overline{\lim} A_n).$$

Consequently, the continuity of probability measure from above and below reduces to the continuity of probability measure.

Theorem 4.13 (Continuity of probability measure) Let $\{A_n\}$ be a sequence of events in a probability space $(\Omega, \mathcal{A}, P)$. If $A = \lim_{n\to\infty} A_n$ then $P(A) = \lim_{n\to\infty} P(A_n)$.

For what follows we require an extended version of the rudimentary definition of independence between events.

Definition 4.10 Independence of events

Let A and B be two events in the probability space $(\Omega, \mathcal{A}, P)$. Then A and B are independent if and only if $P(A \cap B) = P(A)P(B)$. The events are called negatively correlated if and only if $P(A \cap B) < P(A)P(B)$ and positively correlated if and only if $P(A \cap B) > P(A)P(B)$. If the events are independent then they are called uncorrelated.

Note that for random variables, i.e., $A = \{\omega : X(\omega) = x\}$ and $B = \{\omega : Y(\omega) = y\}$, zero correlation does not necessarily imply independence (recall definition 1.12 of Pearson's correlation and see exercise 4.26). The following lemmas can be used to calculate the probability of limit superior. The proof is straightforward and is requested as an exercise.

Lemma 4.1 (Calculating probabilities in the limit) Let $\{A_n\}$ be a sequence of events in a probability space $(\Omega, \mathcal{A}, P)$ and let $A = \overline{\lim} A_n = (A_n \text{ i.o.})$.

1. Borel If $\sum\limits_{n=1}^{+\infty} P(A_n) < +\infty$, then $P(A) = 0$.

2. Kochen-Stone If $\sum\limits_{n=1}^{+\infty} P(A_n) = +\infty$, then

$$P(A) \geq \limsup_{n\to\infty} \frac{\left[\sum\limits_{k=1}^{n} P(A_k)\right]^2}{\sum\limits_{k=1}^{n}\sum\limits_{m=1}^{n} P(A_k \cap A_m)}.$$

3. Borel-Cantelli Assume that events A_i and A_j are either negatively correlated or uncorrelated. Then $\sum\limits_{n=1}^{+\infty} P(A_n) = +\infty$ implies $P(A) = 1$.

An immediate consequence of Borel's lemma is the following.

Remark 4.7 (a.s. convergence) Borel's lemma is useful in showing almost sure convergence since $X_n \overset{a.s.}{\to} X$ is equivalent to $P(|X_n - X| > \varepsilon \text{ i.o.}) = 0$, for all $\varepsilon > 0$.

Example 4.5 (Infinite coin flips) Consider the infinite coin flip example 4.1.3, and let $p_n = P(\text{Heads})$, $0 \leq p_n \leq 1$, $n \geq 1$ and $A_n = \{\text{Heads occurs at the } n^{th} \text{ flip}\}$ so that $p_n = P(A_n)$. Now if $\sum\limits_{n=1}^{+\infty} p_n < +\infty$ then $P(A_n \text{ i.o.}) = 0$, i.e., $P(\text{eventually Tails}) = 1$ and if $\sum\limits_{n=1}^{+\infty} p_n = +\infty$ then $P(A_n \text{ i.o.}) = 1$. Moreover, if $\sum\limits_{n=1}^{+\infty}(1 - p_n) < +\infty$ then $P(\text{eventually Heads}) = 1$ and if $\sum\limits_{n=1}^{+\infty}(1 - p_n) = +\infty$ then $P(A_n^c \text{ i.o.}) = 1$.

4.2.6 Expectation of a Random Object

Expectation provides a useful tool to help us study random objects and their distribution. We can think of expectation as the "average" value of the random object in the sense that we weigh the values of the random object by the corresponding probabilities we assign to those values and then aggregate these weighted values. The definition of expectation for standard random objects is given next.

Definition 4.11 Expectation

Let X be a random object defined on the probability space $(\Omega, \mathcal{A}, P)$.

1. Random variable If X takes values in the measurable space $(\mathcal{R}, \mathcal{B}_1)$, then the expected value of the random variable X is the integral of the function X with respect to P, that is

$$E(X) = \int XdP = \int_{\Omega} X(\omega)dP(\omega). \tag{4.13}$$

2. Random vector If $X = [X_1, \ldots, X_p]^T$ is a random vector in the measurable space $(\mathcal{R}^p, \mathcal{B}_p)$ then the expected value of the random vector X is the vector of expectations, that is,

$$E(X) = [EX_1, \ldots, EX_p]^T. \tag{4.14}$$

3. Random matrix If $X = [(X_{ij})]$ is a random matrix in the measurable space $(\mathcal{M}, 2^{\mathcal{M}})$, where $\mathcal{M}$ is the collection of all $m \times n$ matrices, then the expected value of the random matrix X is the matrix of expectations defined by

$$E(X) = [(EX_{ij})]. \tag{4.15}$$

We collect some consequences of this important definition next.

Remark 4.8 (Expectation) In some cases we will use the notation $E^P(X)$ in order to emphasize the probability measure with respect to which the integral is computed. Since expectation of a random variable is by definition an integral of some measurable function, it inherits all the properties we have seen from general integration theory, such as linearity and monotonicity. Note that the definition in equation (4.13) does not require X to have a density.

1. Owing to properties of integrals for measurable functions, $E(X) = E(X^+) - E(X^-)$ is defined if at least one of $E(X^+)$ or $E(X^-)$ is finite.

2. Let $X = I_A$, where A is a measurable set. Then $E(X) = P(X \in A)$.

3. Consider a random object $X : (\Omega, \mathcal{A}, P) \to (\mathcal{X}, \mathcal{G})$ and a measurable function $\varphi : (\mathcal{X}, \mathcal{G}) \to (\overline{\mathcal{R}}, \mathcal{B}(\overline{\mathcal{R}}))$, so that $\varphi \circ X$ is a random variable from $(\Omega, \mathcal{A}, P)$ into $(\overline{\mathcal{R}}, \mathcal{B}(\overline{\mathcal{R}}))$. Using change of variable (with X playing the role of T, see theorem 3.16) we have that

$$E^P(\varphi \circ X) = \int_{\Omega} \varphi(X(\omega))dP(\omega) = \int_{\overline{\mathcal{R}}} \varphi(t)dQ(t) = E^Q(\varphi), \tag{4.16}$$

where $Q = PX^{-1}$ is the distribution of X.

4. The definitions and results of Section 1.2.2 are easily obtained using definition 4.11. In particular, the definitions of moments and mgfs will be utilized in this chapter.

5. If X has distribution $Q = PX^{-1}$ and a density $f = \left[\frac{dQ}{d\mu}\right]$ with respect to a measure μ, i.e., $Q \ll \mu$, then we can write equation (4.16) as

$$E^P(\varphi \circ X) = \int_{\overline{\mathcal{R}}} \varphi(t)f(t)d\mu(t), \tag{4.17}$$

which is typically easier to compute, especially in the case where μ is a Lebesgue or the counting measure.

6. Let $X : (\Omega, \mathcal{A}, P) \to (\mathcal{R}, \mathcal{B}_1)$ be a random variable with distribution function F and let φ be an $\mathcal{R}$-valued function that is Riemann-Stieltjes integrable with respect to F on every bounded interval. Then we can show that

$$E^P(\varphi \circ X) = \int_{-\infty}^{+\infty} \varphi(x) dF(x), \tag{4.18}$$

in the sense that if one side is defined then so is the other and the two are equal.

7. If a random object X takes values in a measurable space $(\mathcal{X}, \mathcal{G})$ then we should define expectation in such a way as to have EX live in $\mathcal{X}$. This is easily accomplished for the random objects in definition 4.11. For a random collection of points (a point process) X in $\mathcal{R}^p$, the expected value must be a collection of points. Similarly, for a random set X in $\mathcal{R}^p$, the expected value must be a subset of $\mathcal{R}^p$. We discuss these two cases in more detail in the *TMSO-PPRS* text.

The following is a version of the Fubini theorem in a probabilistic setting. In particular, this result shows that under independence, two random variables are uncorrelated. The converse is not true (see exercise 4.26).

Theorem 4.14 (Independent implies uncorrelated) Let X and Y be independent, $\mathcal{R}$-valued random variables with finite expectation. Then

$$E(XY) = E(X)E(Y),$$

or equivalently $Cov(X, Y) = E(XY) - E(X)E(Y) = 0$.

Proof. The independence of X and Y is equivalent to the distribution of (X, Y) being the product measure $Q_1 \times Q_2$, where Q_1 and Q_2 are the distributions of X and Y. Appealing to the Fubini theorem we have

$$E|XY| = \int_{\mathcal{X}_1}\int_{\mathcal{X}_2} |x||y| Q_2(dy) Q_1(dx) = \int_{\mathcal{X}_1} |x| E(|Y|) Q_2(dy) = E(|X|)E(|Y|) < +\infty,$$

so that XY is integrable and hence we can use the Fubini theorem to obtain

$$E(XY) = \int_{\mathcal{X}_1}\int_{\mathcal{X}_2} xy Q_2(dy) Q_1(dx) = \int_{\mathcal{X}_1} x E(Y) Q_2(dy) = E(X)E(Y).$$

■

The following result is particularly useful and allows us to swap the expectation and product operators.

Theorem 4.15 (Expectation and independence) Let $X_1, \dots, X_n$ be independent random variables taking values in the measurable spaces $(\mathcal{X}_i, \mathcal{G}_i)$ and let f_i be $\mathcal{G}_i|\mathcal{B}_1$-measurable with $E(|f_i(X_i)|) < +\infty$, $i = 1, 2, \dots, n$. Then

$$E\left(\prod_{i=1}^{n} f_i(X_i)\right) = \prod_{i=1}^{n} E\left(f_i(X_i)\right). \tag{4.19}$$

Proof. For all $A \in \mathcal{G}_1$ let $f_1(x) = I_A(x)$ and consider the class $\mathcal{L}$ of $\mathcal{R}$-valued functions $f_2(y)$, for which equation (4.19) holds (for $n = 2$), i.e., $E(I_A(X_1)f_2(X_2)) = P(X_1 \in A)E(f_2(X_2))$. Clearly, $\mathcal{L}$ contains indicator functions $f_2(X_2) = I_B(X_2)$, for all $B \in \mathcal{G}_2$ and by linearity of expectation $\mathcal{L}$ contains $f_2(x_2) = \sum_{i=1}^{m} b_i I_{B_i}(x_2)$, any simple function. Using the MCT, $\mathcal{L}$ contains any measurable $f_2 \geq 0$ and for any $\mathcal{R}$-valued f_2, since $E(|f_2(X_2)|) < +\infty$, f_2 is integrable and hence

$$\begin{aligned} E(I_A(X_1)f_2(X_2)) &= E(I_A(X_1)f_2^+(X_2)) - E(I_A(X_1)f_2^-(X_2)) \\ &= P(X_1 \in A)E(f_2^+(X_2)) - P(X_1 \in A)E(f_2^-(X_2)) \\ &= P(X_1 \in A)E(f_2^+(X_2) - f_2^-(X_2)) \\ &= P(X_1 \in A)E(f_2(X_2)), \end{aligned}$$

so that $\mathcal{L}$ contains any $\mathcal{R}$-valued function. Now fix an $\mathcal{R}$-valued f_2 and define $\mathcal{L}'$ to be the class of $\mathcal{R}$-valued measurable functions f_1 for which $E(f_1(X_1)f_2(X_2)) = E(f_1(X_1))E(f_2(X_2))$. Then $\mathcal{L}'$ contains $f_1(x) = I_A(x)$, $A \in \mathcal{G}_1 \implies f_1(x_1) = \sum_{i=1}^{m} a_i I_{A_i}(x_1) \in \mathcal{L}'$ for any simple function and using the MCT we extend to any $f_1 \geq 0$. Following similar arguments as in the last display we can show that $\mathcal{L}'$ contains all $\mathcal{R}$-valued functions, thus establishing the result for $n = 2$. The generalization for any n is straightforward using induction. ■

4.2.7 Characteristic Functions

Let $X : (\Omega, \mathcal{A}, P) \to (\mathcal{R}, \mathcal{B}_1)$, be a random variable and recall definition 1.13 of the cf

$$\varphi_X(t) = Ee^{itX} = E\left[\cos(tX)\right] + iE\left[\sin(tX)\right],$$

for all $t \in \mathcal{R}$. Here we used DeMoivre's formula for $n = 1$, that is,

$$e^{iny} = (\cos y + i \sin y)^n = \cos(ny) + i \sin(ny),$$

$n = 0, 1, \ldots$ and $y \in \mathcal{R}$. It is sometimes convenient to say that $\varphi_X(t)$ is the cf corresponding to the probability measure P or the distribution $Q = PX^{-1}$ instead of the random variable X. Some basic properties are presented below.

Remark 4.9 (Properties of characteristic functions) If X has distribution function F_X and density f_X, then we can write

$$\varphi_X(t) = \int_{\mathcal{R}} e^{itx} P(dx) = \int_{\mathcal{R}} e^{itx} dF_X(x) = \int_{\mathcal{R}} e^{itx} f_X(x) dx,$$

where the second integral is a RS-integral and the third a standard Riemann integral. The cf $\varphi_X(t)$ is also called the Fourier transform of $f_X(x)$.

1. Clearly, $\varphi_X(0) = 1$ and $\varphi_{aX+b}(t) = e^{ibt}\varphi_X(at)$ by definition. Moreover, if X is integer valued then $\varphi_X(t + 2\pi) = \varphi_X(t)$.

2. $|\varphi_X(t)| = [\varphi_X(t)]^*\varphi_X(t) \leq 1$, where the absolute value for a complex number $z = x + iy$ is defined as $|z| = z^*z = x^2 + y^2$, the length of z, with $z^* = x - iy$ the conjugate of z. If z is a vector of complex numbers then z^* is the transpose of the conjugates of the coordinates of z, called the conjugate transpose of z.

3. φ_X is uniformly continuous on $\mathcal{R}$ with respect to the absolute value norm, that is, $\forall\varepsilon > 0$, there exists a $\delta > 0$ (that depends only on ε) such that for all $x, y \in \mathcal{R}$, with $|x - y| < \delta \Longrightarrow |\varphi_X(x) - \varphi_X(y)| < \varepsilon$.

4. The cf can generate moments since $\varphi_X^{(k)}(0) = \left[\frac{d^k\varphi_X(t)}{dt^k}\right]_{t=0} = i^k E(X^k)$, $k = 0, 1, 2, \ldots$. If EX^k exist for any k, it is easy to see that $\varphi_X(t)$ has a Taylor expansion about $t = 0$ given by

$$\varphi_X(t) = \sum_{k=0}^{+\infty} (it)^k E(X^k)/k!,$$

and hence $\varphi_X(t)$ is uniquely determined if and only if the non-central moments $E(X^k)$ are finite for all $k = 0, 1, 2, \ldots$. If

$$\sum_{k=0}^{+\infty} |t|^k E(|X|^k)/k! = E(e^{|tX|}) < \infty,$$

then the cf has a Taylor expansion if the mgf $m_X(t) = E\left(e^{tX}\right)$ of X is defined for all $t \in \mathcal{R}$.

5. If X has a density then $\varphi_X(t) \to 0$, as $|t| \to \infty$.

6. For all $n = 1, 2, \ldots$ and all complex n-tuples $(z_1, \ldots, z_n)$ and real n-tuples $(v_1, \ldots, v_n)$ we have

$$\sum_{k=1}^{n} \sum_{j=1}^{n} \varphi_X(v_k - v_j) z_j z_k^* \geq 0,$$

that is, the cf is a positive definite $\mathbb{C}$-valued function on $\mathcal{R}$, where $\mathbb{C}$ is the complex plane.

7. Bochner's theorem A continuous function φ is the cf of a random variable if and only if it is positive definite and $\varphi(0) = 1$.

Example 4.6 (Normal characteristic function) Assume that $X \sim \mathcal{N}(0, 1)$ and consider the cf for all $t \in \mathcal{R}$ given by

$$\varphi_X(t) = (2\pi)^{-1/2} \int_{\infty}^{+\infty} e^{itx} e^{-x^2/2} dx.$$

Using Liebnitz's rule we differentiate the latter with respect to t and use integration by parts to obtain $\frac{d\varphi_X(t)}{dt} = -t\varphi_X(t)$ and thus $\varphi_X(t)$ is the unique solution of a first-order, homogeneous differential equation under the condition $\varphi_X(0) = 1$. It is easy to see that the solution is given by $\varphi_X(t) = e^{-\frac{t^2}{2}}, t \in \mathcal{R}$. Since $Y = \sigma X + \mu \sim \mathcal{N}(\mu, \sigma^2)$, remark 4.9.1 yields

$$\varphi_Y(t) = e^{i\mu t}\varphi_X(\sigma t) = e^{i\mu t - \sigma^2 t^2/2}, \ t \in \mathcal{R}.$$

Uniqueness arguments for cfs can be established using the following.

Lemma 4.2 (Parseval relation) Let P and Q be two probability measures on $\mathcal{R}$ and denote their cfs by φ_P and φ_Q, respectively. Then

$$\int \varphi_Q(x - v)P(dx) = \int e^{-ivy}\varphi_P(y)Q(dy), \tag{4.20}$$

Table 4.1: Characteristic functions of commonly used continuous distributions.

Distribution	Density	Support	Parameter Space	Characteristic Function
Normal	$\frac{1}{\sqrt{2\pi}\sigma}e^{-\frac{1}{2\sigma^2}(x-\mu)^2}$	$\mathcal{R}$	$(\mu,\sigma)\in\mathcal{R}\times\mathcal{R}^+$	$e^{i\mu t-\frac{\sigma^2 t^2}{2}}$
Gamma	$\frac{x^{a-1}e^{-\frac{x}{b}}}{\Gamma(a)b^a}$	$\mathcal{R}^+$	$(a,b)\in\mathcal{R}^+\times\mathcal{R}^+$	$(1-ibt)^{-a}$
Uniform	$\frac{1}{b-a}$	$[a,b]$	$(a,b)\in\mathcal{R}^2, a<b$	$\frac{e^{ibt}-e^{iat}}{(b-a)it}=i\frac{e^{iat}-e^{ibt}}{(b-a)t}$
Double Exponential	$\frac{1}{2}e^{-\lvert x\rvert}$	$\mathcal{R}$	−	$\frac{1}{1+t^2}$
Cauchy	$\frac{1}{\pi(1+x^2)}$	$\mathcal{R}$	−	$e^{-\lvert t\rvert}$
Triangular	$1-\lvert x\rvert$	$(-1,1)$	−	$2\frac{1-\cos t}{t^2}$

for all $v\in\mathcal{R}$.

Proof. The function $f(x,y)=e^{i(x-v)y}$ is bounded and continuous and hence measurable so that the integral $I=\int f(x,y)d(P\times Q)(x,y)$ exists. Using the Fubini theorem we can write

$$I=\int e^{i(x-v)y}d(P\times Q)(x,y)=\int\left[\int e^{i(x-v)y}dP(x)\right]dQ(y)=\int e^{-ivy}\varphi_P(y)dQ(y),$$

and on the other hand

$$I=\int\left[\int e^{i(x-v)y}dQ(y)\right]dP(x)=\int\varphi_Q(x-v)dP(x),$$

and the result is established. ∎

Based on the Parseval relation we collect the following.

Remark 4.10 (Special Parseval relation) Consider the special case of the Parseval relation for Q corresponding to a $\mathcal{N}(0,\sigma^2)$ random variable. In this case, equation (4.20) reduces to

$$\int_{-\infty}^{+\infty} e^{-\sigma^2(x-v)^2/2}P(dx)=\left(2\pi\sigma^2\right)^{-1/2}\int_{-\infty}^{+\infty} e^{-ivy}e^{-y^2/(2\sigma^2)}\varphi_P(y)dy. \tag{4.21}$$

If P has distribution function F_P, then we can write

$$\int_{-\infty}^{+\infty} e^{-\sigma^2(x-v)^2/2}F_p(dx)=\left(2\pi\sigma^2\right)^{-1/2}\int_{-\infty}^{+\infty} e^{-ivy}e^{-y^2/(2\sigma^2)}\varphi_P(y)dy.$$

We will refer to these equations as the special Parseval relation applied to $\{P,\varphi_P\}$ or $\{F_P,\varphi_P\}$. Equation (4.21) uniquely determines P for a given φ_P, as shown in the next theorem.

Now we are ready to discuss the uniqueness of probability measures via uniqueness in characteristic functions.

Theorem 4.16 (Uniqueness via the characteristic function) If P and Q are probability measures on $\mathcal{R}$ with the same cf φ then $P = Q$.

Proof. Since P and Q have the same characteristic function φ, the latter remark implies

$$\int_{-\infty}^{+\infty} e^{-\frac{\sigma^2(x-v)^2}{2}} P(dx) = \int_{-\infty}^{+\infty} e^{-\frac{\sigma^2(x-v)^2}{2}} Q(dx),$$

for $v \in \mathcal{R}$ and $\sigma > 0$. Multiply both sides with $\frac{\sigma}{\sqrt{2\pi}}$, integrate over $(-\infty, a)$ with respect to the Lebesgue measure and apply the Fubini theorem to obtain

$$\int_{-\infty}^{+\infty} \left[\int_{-\infty}^{a} \frac{\sigma}{\sqrt{2\pi}} e^{-\frac{\sigma^2(x-v)^2}{2}} dv \right] P(dx) = \int_{-\infty}^{+\infty} \left[\int_{-\infty}^{a} \frac{\sigma}{\sqrt{2\pi}} e^{-\frac{\sigma^2(x-v)^2}{2}} dv \right] Q(dx),$$

so that we integrate a normal density with mean x and variance $\frac{1}{\sigma^2}$, that is,

$$\int_{-\infty}^{a} \frac{\sigma}{\sqrt{2\pi}} e^{-\frac{\sigma^2(x-v)^2}{2}} dv = F_v(a),$$

where F_v is the distribution function of a $\mathcal{N}\left(x, \frac{1}{\sigma^2}\right)$ random variable. Letting $\sigma \to \infty$, we have $F_v(a) \to 1$, if $x < a$, 0 if $x > a$ and $F_v(a) \to \frac{1}{2}$, if $x = a$. Then the BCT yields

$$P((-\infty, a)) + \frac{1}{2}P(\{a\}) = Q((-\infty, a)) + \frac{1}{2}Q(\{a\}),$$

$\forall a \in \mathcal{R}$. Since there can be at most countably many values such that either $P(\{a\})$ or $Q(\{a\})$ is nonzero, we conclude that there is a dense set of real numbers a such that $P((-\infty, a)) = Q((-\infty, a))$ and consequently an appeal to the uniqueness of measure theorem establishes the result. ∎

4.3 Conditional Probability

The conditional probability and expectation given the (fixed) value of a random variable $X(\omega) = x$ is not a random variable since the ω is fixed. We would like to see what happens if we replace the event $G = \{\omega : X(\omega) = x\}$, with a larger class of sets, for example a σ-field $\mathcal{G}$. Therefore, when we write $P(A|X)$ or $E(Y|X)$ (without setting the value $X(\omega) = x$ for a specific ω) we are referring to a random probability $P(A|\sigma(X))$ or a random expectation $E(Y|\sigma(X))$, where A is some event and Y is some random variable. Consequently, we need to define probability and expectation conditionally with respect to a σ-field. First we introduce conditioning given the value of a random variable via rigorous arguments.

4.3.1 Conditioning on the Value of a Random Variable

For what follows, let $(\Omega, \mathcal{A}, P)$ denote a probability space and define $X : (\Omega, \mathcal{A}, P) \to (\mathcal{X}, \mathcal{G})$ a random variable with distribution Q_X. We use the Radon-Nikodym theorem to give a general definition of conditional probability and motivate the definition using the following theorem. The material in this section will be

Table 4.2: Characteristic functions of commonly used discrete distributions.

Distribution	Density	Support	Parameter Space	Cf
Binomial	$C^n_x p^x q^{n-x}$	$\{0, 1, \ldots, n\}$	$n \in Z^+$ $p \in [0, 1]$ $q = 1 - p$	$(pe^{it} + q)^n$
Poisson	$\frac{\lambda^x e^{-\lambda}}{x!}$	$\{0, 1, \ldots\}$	$\lambda \in \mathcal{R}^+$	$e^{-\lambda(1-e^{it})}$
Geometric	qp^x	Z^+	$p \in [0, 1]$ $q = 1 - p$	$\frac{1-p}{1-pe^{it}}$
Negative Binomial	$C^{x-1}_{r-1} p^r q^{x-r}$	$\{r, r + 1, \ldots\}$	$r \in Z^+$ $p \in [0, 1]$ $q = 1 - p$	$\left(\frac{q}{1-pe^{it}}\right)^r$
Dirac (Delta)	1	$\{0\}$	–	1

better appreciated if we have in the back of our minds equations (4.13), (4.16) and (4.17), that illustrate how one can write expectations either as an integral $[P]$ or the distribution $Q_X = PX^{-1}$ or the density f_X (when the density exists).

Theorem 4.17 (Conditional probability via the Radon-Nikodym theorem) Let $A \in \mathcal{A}$ be a fixed set. There exists a unique (in the a.e. $[Q_X]$ sense), real-valued, Borel measurable function g defined on $(\mathcal{X}, \mathcal{G})$ such that

$$P((X \in B) \cap A) = \int_B g(x) dQ_X(x), \tag{4.22}$$

for all $B \in \mathcal{G}$.

Proof. Let $\lambda(B) = P((X \in B) \cap A)$, $B \in \mathcal{G}$ and note the λ is a σ-finite measure on $\mathcal{G}$ with $\lambda \ll Q_X$. Therefore, it follows from the Radon-Nikodym theorem that there exists a function g satisfying the statements of the theorem. ■

The latter theorem motivates the following.

Definition 4.12 Conditional probability given $X = x$

The conditional probability of a set $A \in \mathcal{A}$ given that $X = x$ (denoted by $A|X = x$) is defined as the function

$$g(x) = P(A|X = x),$$

for all $A \in \mathcal{A}$ where g is the Radon-Nikodym derivative of the latter theorem, that is, for all $B \in \mathcal{G}$ we have

$$P((X \in B) \cap A) = \int_B P(A|X = x)dQ_X(x). \tag{4.23}$$

We write $P(A|X)$ to denote the random variable $g(X)$ so that $P(A|X)$ is thought of as a random probability. If $A = (Y \in G)$ for some random variable $Y : (\Omega, \mathcal{A}, P) \to (\mathcal{Y}, \mathcal{G})$, $G \in \mathcal{G}$, then $g(x) = P(Y \in G|X = x)$ is called the conditional probability distribution of Y given $X = x$ and is denoted by $Q_Y(.|X = x)$.

Example 4.7 (Conditional probability) We should think of this definition as if we are working backward, in the sense that we want to find or guess some $g(x)$ that satisfies equation (4.22). To appreciate this further take $(\Omega, \mathcal{A}, P)$ to be a discrete probability space. In particular, assume that X takes values in $\mathcal{X} = \{x_1, x_2, \dots\}$ and set $\mathcal{G} = 2^{\mathcal{X}}$, with $P(X = x_i) = p_i > 0$, $i = 1, 2, \dots$ and $\sum_{i=1}^{+\infty} p_i = 1$. In this case Q_X has a density $f_X(x) = P(X = x)$, $x \in \mathcal{X}$, with respect to the counting measure.

1. We claim that for any $A \in \mathcal{A}$ we have that

$$g(x_i) = P(A|X = x_i) = \frac{P(A \cap \{\omega : X(\omega) = x_i\})}{P(X = x_i)}, \tag{4.24}$$

is the form of the conditional probability. We take a $B \in \mathcal{G}$ and show that this g is the desired one by showing that equation (4.23) is satisfied. Indeed, we can write

$$\begin{aligned}
\int_B g(x)dQ_X(x) &= \int_{\mathcal{X}} g(x)I_B(x)dQ_X(x) = \sum_{i=1}^{+\infty} g(x_i)I_B(x_i)P(X = x_i) \\
&= \sum_{x_i \in B} \frac{P(A \cap \{\omega : X(\omega) = x_i\})}{P(X = x_i)} P(X = x_i) \\
&= \sum_{x_i \in B} P(A \cap \{\omega : X(\omega) = x_i\}) = P(A \cap (X \in B)),
\end{aligned}$$

and therefore the function $g(x_i)$ as defined above is the conditional probability $P(A|X = x_i)$.

2. Note that the definition we have given agrees with the rudimentary $P(A|B) = P(A \cap B)/P(B)$, for $P(B) > 0$. To see this, set $X(\omega) = I_B(\omega)$ in equation (4.24).

3. Now take $A = \{\omega : Y(\omega) = y\}$ for some discrete random variable Y, let $B = \mathcal{X}$ and write (4.23) as

$$P((X \in \mathcal{X}) \cap (Y = y)) = \sum_{x \in \mathcal{X}} P(Y = y|X = x)f_X(x),$$

which reduces to the marginal density of the random vector (X, Y) with respect to X, i.e.,

$$P(Y = y) = \sum_{x \in \mathcal{X}} P(Y = y, X = x),$$

where $g(x) = P(Y = y|X = x) = P(X = x, Y = y)/P(X = x)$.

Example 4.8 (Regular conditional probability) Let $(\Omega_i, \mathcal{A}_i)$, $i = 1, 2$, be measurable spaces and define $\Omega = \Omega_1 \times \Omega_2$ and $\mathcal{A} = \mathcal{A}_1 \otimes \mathcal{A}_2$. Let $X(\omega_1, \omega_2) = \omega_1$ and $Y(\omega_1, \omega_2) = \omega_2$ be the projection functions. Assume that we are given Q_X a probability measure on $(\Omega_1, \mathcal{A}_1)$ and $P(x, B)$, $x \in \Omega_1$, $B \in \mathcal{A}_2$, a probability measure in $(\Omega_2, \mathcal{A}_2)$, for any $x \in \Omega_1$. Further assume that $P(., B)$ is a measurable function on $(\Omega_1, \mathcal{A}_1)$. Applying the definition of the product measure space (definition 3.18) along with the Fubini theorem, there exists a unique measure P on $(\Omega, \mathcal{A})$ such that

$$P(X \in A, Y \in B) = \int_A P(x, B) dQ_X(x),$$

so that $P(x, B) = P(B|X = x)$. The function $P(x, B)$ is called a version of a regular conditional probability and any two versions are equal except on a set of probability 0. This follows from our definition by setting $\Omega = \Omega_1 \times \Omega_2$, $\mathcal{X} = \Omega$, $\mathcal{A} = \mathcal{A}_1 \otimes \mathcal{A}_2$ and $\mathcal{G} = \mathcal{A}$. Note that in order to develop conditional probability for random objects that are vectors we typically take $\mathcal{X} = \mathcal{R}^p$ and $\mathcal{G} = \mathcal{B}_p$.

We motivate the definition of conditional expectation using the following theorem.

Theorem 4.18 (Conditional expectation via the Radon-Nikodym theorem) Let Y be a random variable on $(\Omega, \mathcal{A}, P)$ and let $X : (\Omega, \mathcal{A}, P) \to (\mathcal{X}, \mathcal{G})$ be a random object with distribution $Q_X = PX^{-1}$. If $E(Y)$ exists then there exists a unique Borel measurable function $g : (\mathcal{X}, \mathcal{G}) \to (\mathcal{R}, \mathcal{B}_1)$, such that for all $B \in \mathcal{G}$ we have

$$\int_{\{X \in B\}} Y dP = \int_B g(x) dQ_X(x). \tag{4.25}$$

Proof. The proof is the same as that of theorem 4.12 once we set $\lambda(B) = \int_{\{X \in B\}} Y dP$, $B \in \mathcal{G}$. ■

The definition of conditional expectation is now naturally introduced.

Definition 4.13 Conditional expectation given $X = x$

The conditional expectation of a random variable Y given $X = x$ (or $\{\omega : X(\omega) = x\}$), denoted by $E(Y|X = x)$, is the function of x defined as the Radon-Nikodym derivative $g(x)$ given in the latter theorem, namely, it is the unique Borel measurable function defined on $(\mathcal{X}, \mathcal{G})$ such that for all $B \in \mathcal{G}$ we have

$$\int_{\{X \in B\}} Y dP = \int_B E(Y|X = x) dQ_X(x). \tag{4.26}$$

We write $E(Y|X)$ to denote the random variable $g(X)$ so that $E(Y|X)$ is thought of as a random expectation.

An immediate consequence of the latter is the following.

Corollary 4.1 In the notation of theorem 4.18 we can show that

(i) $E^P(Y) = E^{Q_X}[E(Y|X)]$, and
(ii) $E(I_A|X = x) = P(A|X = x)$ a.e. $[Q_X]$, for any $A \in \mathcal{A}$.

Proof. (i) Let $B = \mathcal{X}$ so that $\{X \in \mathcal{X}\} = \{\omega : X(\omega) \in \mathcal{X}\} = \Omega$, in theorem 4.18.
(ii) Let $Y = I_A$ in the definition of $E(Y|X = x)$ to obtain

$$P((X \in B) \cap A) = \int_B E(I_A|X = x)dQ_X(x),$$

for all $B \in \mathcal{G}$, which is exactly the definition of the conditional probability $P(A|X = x)$. ■

Example 4.9 (Finite mixtures) The mixtures of Section 1.2.3 are special cases of the theory in this section. In particular, if $Q_1, \dots, Q_n$ are candidate distributions for the random variable $Y : (\Omega, \mathcal{A}, P) \to (\mathcal{Y}, \mathcal{G})$ and if $p_i = P(X = x_i)$, $0 \le p_i \le 1$, $i = 1, 2, \dots, n$, with $\sum_{i=1}^{n} p_i = 1$, is the density of the discrete random variable $X \in \mathcal{X} = \{x_1, \dots, x_n\}$ with distribution Q_X, then

$$Q(G) = P(Y \in G) = \sum_{i=1}^{n} p_i Q_i(G),$$

for $G \in \mathcal{G}$ defines a discrete mixture of the distributions $Q_1, \dots, Q_n$, where each Q_i can be thought of as a conditional distribution. Indeed, equation (4.22) can be employed with $A = (Y \in G)$, $B = \mathcal{X}$, $Q_X(\{x_i\}) = p_i$ and $Q_i(G) = P(Y \in G|X = x_i)$ to help us write

$$P(Y \in G) = \int_{\mathcal{X}} P(Y \in G|X = x)dQ_X(x) = \sum_{i=1}^{n} Q_i(G)p_i = Q(G).$$

Next we collect some results on conditional densities.

Remark 4.11 (Conditional expectation) From definition 4.12 the conditional probability distribution $Q(.|X = x)$ of Y given $X = x$ can be used to define the conditional expectation of $Y|X = x$ as a function of x by

$$E(Y|X = x) = \int_{\mathcal{Y}} yQ(dy|X = x),$$

where $\mathcal{Y}$ is the domain of Y. To see that this is equivalent to (4.26) integrate both sides above with respect to the (marginal) distribution Q_X of X and over a set B to obtain

$$\int_B E(Y|X = x)dQ_X(x) = \int_B \left[\int_{\mathcal{Y}} yQ(dy|X = x) \right] dQ_X(x). \tag{4.27}$$

Since by definition 4.13 the left side of (4.27) is

$$\int_B E(Y|X = x)dQ_X(x) = \int_{\{X \in B\}} YdP,$$

the right side of (4.27) becomes

$$\int_{\{X \in B\}} YdP = \int_B \left[\int_{\mathcal{Y}} yQ(dy|X = x) \right] dQ_X(x),$$

and thus $g(x) = \int_{\mathcal{Y}} yQ(dy|X = x)$ satisfies (4.26) as desired.

4.3.2 Conditional Probability and Expectation Given a σ-field

Now we collect the definition of conditional probability given a σ-field.

Definition 4.14 Conditional probability given a σ-field

Let A be an event in a probability space $(\Omega, \mathcal{A}, P)$ and let $\mathcal{H}$ be a sub-σ-field of $\mathcal{A}$. The conditional probability of A given $\mathcal{H}$, denoted by $P(A|\mathcal{H})$, is a random variable that satisfies:
(i) $P(A|\mathcal{H})$ is measurable in $\mathcal{H}$ and integrable, and
(ii) for all $H \in \mathcal{H}$ we have

$$\int_H P(A|\mathcal{H})dP = P(A \cap H). \tag{4.28}$$

There are many random variables $P(A|\mathcal{H})$ that satisfy the requirements of the latter definition, but any two of them are equal a.e. $[P]$. This follows from the Radon-Nikodym theorem since $P(H) = 0$ implies $v(H) = \int_H P(A|\mathcal{H})dP = 0$, that is, $v \ll P$ and $P(A|\mathcal{H})$ is simply the Radon-Nikodym derivative. An alternative, yet equivalent definition that does not require the existence of a density can be given in terms of expectations.

Remark 4.12 (Alternative definition) If $Y = P(A|\mathcal{H})$ *a.s.*, then condition (ii) above can be written as

$$E^P(YI_H) = E^P(I_A I_H), \tag{4.29}$$

for all $H \in \mathcal{H}$, where the expectations are taken with respect to P, so that the integrals are computed over the space Ω. The expectation on the left hand side is finite when Y is integrable $[P]$. Hence we may rewrite the conditions of the definition as:
(i) $Y = P(A|\mathcal{H})$ is measurable in $\mathcal{H}$, and
(ii) $E^P(YI_H) = E^P(I_A I_H)$, $\forall H \in \mathcal{H}$,
with condition (ii) now requiring the integrability condition of Y in order for the two sides to be finite (since $E^P(I_A I_H) = P(A \cap H) \leq 1$) and equal. We will adapt these versions of the conditions for what follows.

Example 4.10 (Special cases for $\mathcal{H}$) Consider the setup of the latter definition.

1. Suppose that $A \in \mathcal{H}$ (this is always the case when $\mathcal{H} = \mathcal{A}$). Then the random variable $X(\omega) = I_A(\omega)$, satisfies conditions (i) and (ii), so that $P(A|\mathcal{H}) = I_A$ *a.s.*, that is, $A \in \mathcal{H}$ is equivalent to knowing in advance whether or not A has occurred.

2. If $\mathcal{H} = \{\emptyset, \Omega\}$ then every function that is $\mathcal{H}$-measurable must be a constant since

$$\int_\emptyset P(A|\mathcal{H})dP = P(\emptyset \cap A) = P(\emptyset) = 0,$$

and

$$\int_{\Omega} P(A|\mathcal{H})dP = P(\Omega \cap A) = P(A),$$

that is, $P(A|\mathcal{H})(\omega) = P(A)$, for all $\omega \in \Omega$ and hence we do not learn anything from $\mathcal{H}$ about the probability of the event A.

3. An event A is independent of the σ-field $\mathcal{H}$ if and only if $P(A \cap H) = P(A)P(H), \forall H \in \mathcal{H}$, or equivalently

$$P(A \cap H) = \int_{H} P(A)dP,$$

and therefore A is independent of $\mathcal{H}$ if and only if $P(A|\mathcal{H})(\omega) = P(A)$ $a.s.$

4. Suppose that $B \in \mathcal{A}$ and take $\mathcal{H} = \sigma(B) = \{\emptyset, \Omega, B, B^c\}$, with $0 < P(B) < 1$. We claim that

$$Y(\omega) = P(A|\mathcal{H})(\omega) = \frac{P(A \cap B)}{P(B)} I_B(\omega) + \frac{P(A \cap B^c)}{P(B^c)} I_{B^c}(\omega) \ a.s.$$

First notice that $Y(\omega)$ as defined is $\sigma(B)$-measurable. To check the second condition we only have to consider the four members of $\sigma(B)$. Clearly, $E(YI_{\emptyset}) = 0 = P(A \cap \emptyset)$ and

$$E(YI_{\Omega}) = E(YI_B) + E(YI_{B^c}) = P(A \cap B) + P(A \cap B^c) = P(A \cap \Omega).$$

Moreover

$$E(YI_B) = \frac{P(A \cap B)}{P(B)} E(I_B I_B) = P(A \cap B),$$

and similarly

$$E(YI_{B^c}) = \frac{P(A \cap B^c)}{P(B^c)} E(I_{B^c} I_{B^c}) = P(A \cap B^c).$$

This example shows how $P(A|\mathcal{H})$ in definition 4.14 becomes a generalization of definition 4.12 (recall example 4.7.2).

5. If A and B are events in $(\Omega, \mathcal{A}, P)$ with $A \in \mathcal{H}$ then we claim that $P(A \cap B|\mathcal{H}) = P(B|\mathcal{H})I_A$ $a.s.$ To see this, let us show that $Y = P(B|\mathcal{H})I_A$ satisfies conditions (i) and (ii). Since $P(B|\mathcal{H})$ is $\mathcal{H}$-measurable by definition and $A \in \mathcal{H}$ we have that Y is $\mathcal{H}$-measurable. We need to show that $E(YI_H) = P((A \cap B) \cap H)$, for all $H \in \mathcal{H}$, or equivalently that

$$E(P(B|\mathcal{H})I_{A \cap H}) = P(B \cap (A \cap H)).$$

The latter holds by the definition of $P(B|\mathcal{H})$ and since $A \cap H \in \mathcal{H}$.

The conditional probability of an event A given a random variable X is defined as the random variable $P(A|\sigma(X))(\omega)$, where $\sigma(X) = \{X^{-1}(B) : B \in \mathcal{B}_1\}$, but will be denoted by $P(A|X)$ for brevity. If the σ-field is generated by many random variables $X_1, X_2, \ldots, X_n$ then we write $P(A|X_1, \ldots, X_n)$ instead of $P(A|\sigma(X_1, \ldots, X_n))$. If we have a set of random variables $\{X_t : t \in T\}$, for some index set T, then we will write $P(A|X_t : t \in T)$ instead of $P(A|\sigma(X_t : t \in T))$, where $\sigma(X_t : t \in T) = \sigma(\{X_t^{-1}(B) : B \in \mathcal{B}_1, t \in T\})$.

Properties and results for unconditional probabilities carry over to the conditional case as we see next. Since we are given the forms of the conditional probabilities, all we need to do to prove these results is verify conditions (i) and (ii). This is straightforward in all cases and will be left as an exercise.

Theorem 4.19 (Properties of conditional probability given a σ-field) Let $(\Omega, \mathcal{A}, P)$ be a probability space. Suppose that A, B, A_i, $i = 1, 2, \ldots$, are $\mathcal{A}$-sets and let $\mathcal{H}$ be a sub-σ-field of $\mathcal{A}$. We can show the following:
(i) $P(\emptyset|\mathcal{H}) = 0$ *a.s.*, $P(\Omega|\mathcal{H}) = 1$ *a.s.* and $0 \leq P(A|\mathcal{H}) \leq 1$ *a.s.*
(ii) If $\{A_n\}$ are disjoint then $P\left(\bigcup_n A_n|\mathcal{H}\right) = \sum_n P(A_n|\mathcal{H})$ *a.s.*
(iii) If $A \subseteq B$ then $P(B \setminus A|\mathcal{H}) = P(B|\mathcal{H}) - P(A|\mathcal{H})$ *a.s.* and $P(A|\mathcal{H}) \leq P(B|\mathcal{H})$ *a.s.*
(iv) If $A_n \uparrow A$ then $P(A_n|\mathcal{H}) \uparrow P(A|\mathcal{H})$ *a.s.* and if $A_n \downarrow A$ then $P(A_n|\mathcal{H}) \downarrow P(A|\mathcal{H})$ *a.s.*
(v) The conditional inclusion-exclusion formula $P\left(\bigcup_{i=1}^n A_i|\mathcal{H}\right) = \sum_{i=1}^n P(A_i|\mathcal{H}) - \sum_{i<j} P\left(A_i \cap A_j|\mathcal{H}\right) + \ldots$ *a.s.*, holds.
(vi) If $P(A) = 1$ then $P(A|\mathcal{H}) = 1$ *a.s.* and $P(A) = 0$ implies that $P(A|\mathcal{H}) = 0$ *a.s.*

Now we define the distribution of a random object given a σ-field.

Definition 4.15 Conditional probability distribution given a σ-field

Let $X : (\Omega, \mathcal{A}, P) \to (\mathcal{X}, \mathcal{G})$, be a random object, $\mathcal{H}$ a sub-σ-field of $\mathcal{A}$ and consider $\mathcal{Q}$ the collection of all probability measures on $(\mathcal{X}, \mathcal{G})$. The conditional probability distribution of X given $\mathcal{H}$ (or $X|\mathcal{H}$) is a function of two arguments, denoted by $Z(\omega, G) = Q_X(G|\mathcal{H})(\omega)$, such that (i) $Z(\omega, .)$ is one of the probability measures from the collection $\mathcal{Q}$, for each $\omega \in \Omega$ and (ii) $Z(., G)$ is (a version of) the conditional probability of $X^{-1}(G)$ given $\mathcal{H}$, that is, $Z(., G) = P(X \in G|\mathcal{H})(.) = P(X^{-1}(G)|\mathcal{H})(.)$, for each $G \in \mathcal{G}$. To simplify the notation, we will write $Q_X(.|\mathcal{H})$ for the conditional distribution of $X|\mathcal{H}$.

Some comments about this definition are given next.

Remark 4.13 (Conditional distribution) We note the following.

1. Note that if we let $\mathcal{H} = \sigma(Y)$, for some random object Y, then $Q_X(.|Y)(.)$ can be thought of as the conditional distribution of X given Y.

2. From property (i), $Z(\omega, .) = Q_X(.|\mathcal{H})(\omega)$ is a function that takes outcomes from Ω and gives us an element of $\mathcal{Q}$ (that is a probability measure) and therefore if we equip $\mathcal{Q}$ with a σ-field then we can talk about a random probability distribution. In fact, $Q_X(.|\mathcal{H})(\omega)$ defines a random distribution as we vary $\omega \in \Omega$, but it does not map to any probability measure in $\mathcal{Q}$, just those with the additional property that they are versions of the conditional probability $P(X^{-1}(G)|\mathcal{H})$, for each $G \in \mathcal{G}$.

Motivated by the discussion above, the definition of a random distribution is provided next.

Definition 4.16 Random distribution

Let $(\mathcal{X}, \mathcal{G})$ be a measurable space and let $\mathcal{Q}$ denote the collection of all probability measures on $(\mathcal{X}, \mathcal{G})$. The measurable space of probability measures on $(\mathcal{X}, \mathcal{G})$ is denoted by $(\mathcal{Q}, \mathcal{V})$, where $\mathcal{V}$ is the smallest σ-field such that for each $G \in \mathcal{G}$ the function $U : \mathcal{Q} \to [0, 1]$ defined by $U(Q) = Q(G)$, $Q \in \mathcal{Q}$, is $\mathcal{V}$-measurable. A random distribution D on $(\mathcal{X}, \mathcal{G})$ is a measurable function from some probability space $(\Omega, \mathcal{A}, P)$ to $(\mathcal{Q}, \mathcal{V})$, that is, $D^{-1}(V) \in \mathcal{A}$, $\forall V \in \mathcal{V}$.

Conditional probability calculations are significantly simplified using a density. The definition is similar to the unconditional case.

Definition 4.17 Conditional density given a σ-field

Let $X : (\Omega, \mathcal{A}, P) \to (\mathcal{X}, \mathcal{G})$ be a random vector ($\mathcal{X} \subseteq \mathcal{R}^p$), $\mathcal{H}$ a sub-σ-field of $\mathcal{A}$ and let $Q_X(.|\mathcal{H})$ be the conditional distribution of $X|\mathcal{H}$. A nonnegative measurable function q on $(\Omega \times \mathcal{X}, \mathcal{A} \otimes \mathcal{G})$ is called a conditional density for X with respect to μ, given $\mathcal{H}$, if

$$P(X^{-1}(G)|\mathcal{H})(\omega) = \int_G q(\omega, x)\mu(dx),$$

for all $\omega \in \Omega$ and $G \in \mathcal{G}$.

The definition holds for q that are the Radon-Nikodym derivatives of the conditional distribution of $X|\mathcal{H}$, $Q_X(.|\mathcal{H})$, with respect to the measure μ, that is, it suffices to have $Q_X(.|\mathcal{H})$ be absolutely continuous with respect to μ. Since q is a function of ω we may call it a random density.

The existence and uniqueness (in the *a.s.* sense) of conditional distributions can be established in a rigorous framework regardless of the probability space on which they are defined and regardless of the conditioning σ-field. See for example Fristedt and Gray (1997, Section 21.4) or Billingsley (2012, theorem 33.3).

Next we collect conditional densities based on two random variables.

Remark 4.14 (Conditional density) Let X_i be $(\mathcal{X}_i, \mathcal{G}_i)$-valued random variables, $i = 1, 2$, defined on a common probability space. Assume that the distribution of the random vector (X_1, X_2) has a density f with respect to a σ-finite product measure

$\mu = \mu_1 \times \mu_2$ on $\mathcal{G}_1 \otimes \mathcal{G}_2$ and let $g(x_1) = \int\limits_{X_2} f(x_1, x_2)\mu_2(dx_2)$. Then the function

$$h(\omega, x_2) = \begin{cases} f(X_1(\omega), x_2)/g(X_1(\omega)) & \text{if } g(X_1(\omega)) > 0 \\ \int\limits_{X_1} f(x_1, x_2)\mu_1(dx_1) & \text{if } g(X_1(\omega)) = 0 \end{cases}$$

is a conditional density of X_2 with respect to μ_2 given $\sigma(X_1)$.

Before we tackle conditional expectation, let us discuss why we require the two conditions in definition 4.14 of conditional probability.

Remark 4.15 (Insight on the requirements of conditional probability) We have seen that if $A_n \in \mathcal{A}$, $n = 1, 2, \ldots$, are disjoint events then

$$P\left(\bigcup_n A_n|\mathcal{H}\right) = \sum_n P(A_n|\mathcal{H}) \ a.s. \tag{4.30}$$

The question is: can we choose a version of the conditional probability so that $P(.|\mathcal{H})(\omega)$ is a measure on $(\Omega, \mathcal{A})$ for almost all $\omega \in \Omega$? The problem is that for any $A \in \mathcal{A}$ we can choose a version of the conditional probability $P(A|\mathcal{H})(\omega)$ but for fixed $\omega \in \Omega$, $P(.|\mathcal{H})(\omega)$ might not be countably additive on $(\Omega, \mathcal{A})$. The reason is that (4.30) will hold except for some $\omega \in N(A_1, A_2, \ldots)$, a null set $[P]$ that depends on the sequence $\{A_n\}$. Thus, as we vary the sequence $\{A_n\}$, the set of ωs where (4.30) might fail is

$$M = \cup\{N(A_1, A_2, \ldots) : A_n \in \mathcal{A}, \text{ disjoint}\},$$

and M can be an uncountable union of sets of measure 0 that can be of positive measure or even not in $\mathcal{A}$. The conditions take care of this situation. See Billingsley (2012, p. 471) for a counterexample that shows that conditional probabilities may not end up giving measures.

Next we collect the definition of conditional expectation.

Definition 4.18 Conditional expectation given a σ-field

Suppose that $X : (\Omega, \mathcal{A}, P) \to (\mathcal{X}, \mathcal{G})$ is an integrable random variable and let $\mathcal{H}$ be a sub-σ-field of $\mathcal{A}$. Then there exists a random variable $Y = E(X|\mathcal{H})$ $a.s.$ called the conditional expectation of X given $\mathcal{H}$, with the properties (i) Y is $\mathcal{H}$-measurable and integrable and (ii) Y is such that

$$\int\limits_H YdP = \int\limits_H XdP,$$

for all $H \in \mathcal{H}$.

Some consequences of this important definition are presented next.

Remark 4.16 (Conditional expectation) The existence of Y is easily established in a similar fashion as in theorem 4.18, that is, Y is essentially a Radon-Nikodym derivative. Any such random variable $Y = E(X|\mathcal{H})$ will be called a version of the conditional expectation of $X|\mathcal{H}$. We note the following.

1. Condition (ii) may be rewritten in terms of expectations $[P]$, that is $\forall H \in \mathcal{H}$, we have

$$E^P(YI_H) = E^P(XI_H), \tag{4.31}$$

and will be adapted from now on in proofs. We can remove the integrability requirement for X in the definition and Y in condition (i) provided that $E^P(YI_H)$ and $E^P(XI_H)$ are defined (in which case they will be equal). Consequently, in proofs we only need to verify that a candidate random variable Y is $\mathcal{H}$-measurable and that (4.31) holds in order to have $Y = E(X|\mathcal{H})$ $a.s.$

2. Conditional expectation can be defined equivalently in terms of the conditional probability distribution $Q_X(.|\mathcal{H})$ or its density $q(\omega, x)$ (when it exists) as the random variable

$$E(X|\mathcal{H})(\omega) = \int\limits_{\mathcal{X}} xQ_X(dx|\mathcal{H})(\omega) \text{ a.s.},$$

or

$$E(X|\mathcal{H})(\omega) = \int\limits_{\mathcal{X}} xq(\omega, x)\mu(dx) \text{ a.s.}$$

3. The conditional variance of X given $\mathcal{H}$ is naturally defined by

$$Var(X|\mathcal{H}) = E^{Q_X(.|\mathcal{H})}\left[(X - E(X|\mathcal{H}))^2\right] \; a.s.$$

provided that $E(X^2|\mathcal{H}) < +\infty$. We can easily show that $Var(X|\mathcal{H}) = E(X^2|\mathcal{H}) - [E(X|\mathcal{H})]^2$ $a.s.$

The proofs of the following statements are requested as exercises. To prove the conditional versions of the original theorems like the MCT and DCT one needs to work with an arbitrarily fixed ω and use the unconditional versions of the theorems.

Theorem 4.20 (Properties of conditional expectation given a σ-field) Let $X, Y, X_n : (\Omega, \mathcal{A}, P) \to (\mathcal{X}, \mathcal{G})$ be integrable random variables and let $\mathcal{H}$, $\mathcal{H}_1$, $\mathcal{H}_2$, be sub-σ-fields of $\mathcal{A}$. Then we can show the following.
(i) If $X = c$ $a.s.$ for some constant c then $E(X|\mathcal{H}) = c$ $a.s.$
(ii) $E(E(X|\mathcal{H})) = E(X)$ $a.s.$
(iii) $Var(X) = E(Var(X|\mathcal{H})) + Var(E(X|\mathcal{H}))$ $a.s.$
(iv) Linearity: $E(aX + bY|\mathcal{H}) = aE(X|\mathcal{H}) + bE(Y|\mathcal{H})$ $a.s.$ for any constants $a, b \in \mathcal{R}$.
(v) Monotonicity: If $X \le Y$ $a.s.$ then $E(X|\mathcal{H}) \le E(Y|\mathcal{H})$ $a.s.$
(vi) $|E(X|\mathcal{H})| \le E(|X||\mathcal{H})$ $a.s.$
(vii) Conditional DCT: If $X_n \overset{a.s.}{\to} X$, $|X_n| \le Y$ and Y is integrable then $E(X_n|\mathcal{H}) \overset{a.s.}{\to} E(X|\mathcal{H})$.
(viii) Conditional MCT: If $0 \le X_1 \le X_2 \le \ldots$ and $X_n \overset{a.s.}{\to} X$ then $0 \le E(X_n|\mathcal{H}) \overset{a.s.}{\to} E(X|\mathcal{H})$.
(ix) Conditional Jensen inequality: Let $g : I \to \mathcal{R}$ be a convex function. Then

$$E(g \circ X|\mathcal{H}) \ge g(E(X|\mathcal{H})) \; a.s.$$

(x) Conditional Chebyshev inequality: For any $c > 0$ and almost all $\omega \in \Omega$ for which $E(X|\mathcal{H})(\omega) < +\infty$, we have

$$P(|X - E(X|\mathcal{H})| \geq c|\mathcal{H})(\omega) \leq Var(X|\mathcal{H})(\omega)/c^2 \ a.s.$$

(xi) Successive conditioning: If $\mathcal{H}_1 \subset \mathcal{H}_2 \subset \mathcal{A}$ then

$$E(E(X|\mathcal{H}_1)|\mathcal{H}_2) = E(X|\mathcal{H}_1) \ a.s.,$$

and

$$E(E(X|\mathcal{H}_2)|\mathcal{H}_1) = E(X|\mathcal{H}_1) \ a.s.,$$

that is, in successive conditioning we always end up conditioning with respect to the smallest sub-σ-field.
(xii) Suppose that Y is $\mathcal{H}$-measurable and let $g : \mathcal{X} \times \mathcal{X} \to \overline{\mathcal{R}}$. Then

$$E(g \circ (X, Y)|\mathcal{H})(\omega) = \int_{\mathcal{X}} g(x, Y(\omega))Q_X(dx|\mathcal{H})(\omega),$$

for almost all $\omega \in \Omega$.

Example 4.11 (Conditional expectation given a σ-field) For what follows assume that $X, Y : (\Omega, \mathcal{A}, P) \to (\mathcal{X}, \mathcal{G})$ are integrable random variables and let $\mathcal{H}$ be a sub-σ-field of $\mathcal{A}$.

1. Assume that X is $\mathcal{H}$-measurable. We claim that $E(X|\mathcal{H}) = X \ a.s.$ Indeed, let the candidate be the random variable $Z = X$ so that Z is obviously $\mathcal{H}$-measurable and the condition $E(ZI_H) = E(XI_H)$, $\forall H \in \mathcal{H}$, is trivially satisfied.

2. Assume that X is $\mathcal{H}$-measurable and assume that XY is integrable. Then we claim that $E(XY|\mathcal{H}) = XE(Y|\mathcal{H}) \ a.s.$ Let $Z = XE(Y|\mathcal{H})$ denote the candidate random variable and note that since X is $\mathcal{H}$-measurable and $E(Y|\mathcal{H})$ is $\mathcal{H}$-measurable by definition, then Z is $\mathcal{H}$-measurable. We need to show that for all $H \in \mathcal{H}$, we have

$$E(ZI_H) = E((XY)I_H),$$

or that

$$E^P(XE(Y|\mathcal{H})I_H) = E^P(XYI_H). \tag{4.32}$$

By the definition of $E(Y|\mathcal{H})$ we have

$$E^P(E(Y|\mathcal{H})I_H) = E^P(YI_H). \tag{4.33}$$

Consider first the case where $X = I_A$, for some set A that is $\mathcal{H}$-measurable. Thus (4.32) is written as

$$E^P(I_A E(Y|\mathcal{H})I_H) = E^P(I_A Y I_H),$$

which reduces to (4.33) (by replacing H with $H \cap A \in \mathcal{H}$) and consequently the indicator satisfies the claim. From linearity of expectation, (4.32) is satisfied for any simple random variable X. For any nonnegative X there exists (by theorem 3.4) an increasing sequence of simple functions X_n such that $|X_n| \leq |X|$ and $X_n \uparrow X$. Since

$|X_nY| \leq |XY|$ and XY is integrable, the conditional DCT implies that $E(X_nY|\mathcal{H}) \stackrel{a.s.}{\to} E(XY|\mathcal{H})$ *a.s.* But $E(X_nY|\mathcal{H}) = X_nE(Y|\mathcal{H})$ *a.s.* since X_n is a simple random variable and therefore

$$E(XY|\mathcal{H}) = \lim_{n\to+\infty} E(X_nY|\mathcal{H}) = \lim_{n\to+\infty} X_nE(Y|\mathcal{H}) = XE(Y|\mathcal{H}) \ a.s.$$

If X is $\mathcal{R}$-valued we take $X = X^+ - X^-$ so that linearity of conditional expectation yields the general case of the claim.

4.3.3 Conditional Independence Given a σ-field

The definition of conditional independence is given first, followed by several results that illustrate the usefulness of conditioning. In particular, there are cases where conditioning on an appropriate σ-field can turn dependence into independence.

Definition 4.19 Conditional independence given a σ-field

Suppose that $(\Omega, \mathcal{A}, P)$ is a probability space and let $\mathcal{A}_1$, $\mathcal{A}_2$, $\mathcal{H}$ be sub-σ-fields of $\mathcal{A}$. The σ-fields $\mathcal{A}_1$ and $\mathcal{A}_2$ are said to be conditionally independent given $\mathcal{H}$ if

$$P(A_1 \cap A_2|\mathcal{H}) = P(A_1|\mathcal{H})P(A_2|\mathcal{H}) \ a.s.,$$

for all $A_1 \in \mathcal{A}_1$ and $A_2 \in \mathcal{A}_2$.

We prove a basic theorem on conditional independence that can be used to prove subsequent results.

Theorem 4.21 (Conditional independence of sub-σ-fields) Consider a probability space $(\Omega, \mathcal{A}, P)$ and let $\mathcal{H}$, $\mathcal{A}_1$, $\mathcal{A}_2$, be sub-σ-fields of $\mathcal{A}$. Suppose that $\mathcal{A}_2 \subseteq \mathcal{H}$. Then $\mathcal{A}_1$ and $\mathcal{A}_2$ are conditionally independent given $\mathcal{H}$.

Proof. If $A_1 \in \mathcal{A}_1$ and $A_2 \in \mathcal{A}_2 \subseteq \mathcal{H}$, then $P(A_1 \cap A_2|\mathcal{H}) = P(A_1|\mathcal{H})I_{A_2} = P(A_1|\mathcal{H})P(A_2|\mathcal{H})$ *a.s.* The first equality follows from example 4.11.2 since $P(A_1 \cap A_2|\mathcal{H}) = E(I_{A_1 \cap A_2}|\mathcal{H}) = E(I_{A_1}I_{A_2}|\mathcal{H})$ and $\mathcal{A}_2 \subseteq \mathcal{H}$ implies I_{A_2} is $\mathcal{H}$-measurable. The second equality holds from example 4.10.1. ■

Several results regarding conditional independence are presented below.

Remark 4.17 (Conditional independence) In what follows let $(\Omega, \mathcal{A}, P)$ be a probability space and let $\mathcal{H}$, $\mathcal{V}$, $\mathcal{H}_1$, $\mathcal{H}_2$, be sub-σ-fields of $\mathcal{A}$.

1. Conditional Borel-Cantelli lemma Let $\{A_n\}_{n=1}^{+\infty}$ be a sequence of events conditionally independent given $\mathcal{H}$. Then $P(\overline{\lim} A_n|\mathcal{H}) = 1$ *a.s.*, if $\sum\limits_{n=1}^{+\infty} P(A_n|\mathcal{H}) = +\infty$ *a.s.* and 0 *a.s.*, if $\sum\limits_{n=1}^{+\infty} P(A_n|\mathcal{H}) < +\infty$ *a.s.* and moreover

$$P(\overline{\lim} A_n|\mathcal{H}) = P\left(\left\{\omega \in \Omega : \sum_{n=1}^{+\infty} P(A_n|\mathcal{H})(\omega) = +\infty\right\}\right).$$

2. If $\mathcal{H}_1$ and $\mathcal{H}_2$ are conditionally independent given $\mathcal{H}$ then $\mathcal{H}_1$ and $\sigma(\mathcal{H}_2, \mathcal{H})$ are conditionally independent given $\mathcal{H}$.

3. Assume that $\mathcal{H}_1 \subseteq \mathcal{H}$, $\mathcal{H}_2 \subseteq \mathcal{V}$ and suppose that $\mathcal{H}$ and $\mathcal{V}$ are independent. Then $\mathcal{H}$ and $\mathcal{V}$ are conditionally independent given $\sigma(\mathcal{H}_1, \mathcal{H}_2)$.

4. Suppose that $\mathcal{H} \subseteq \mathcal{H}_1$. Then $\mathcal{H}_1$ and $\mathcal{H}_2$ are conditionally independent given $\mathcal{H}$ if and only if $\forall A \in \mathcal{H}_2$ we have $P(A|\mathcal{H}) = P(A|\mathcal{H}_1)$ *a.s.*

5. Let $\mathcal{H}_1$ and $\mathcal{H}_2$ be independent with $\mathcal{H} \subseteq \mathcal{H}_1$. Then $\mathcal{H}_1$ and $\sigma(\mathcal{H}, \mathcal{H}_2)$ are conditionally independent given $\mathcal{H}$.

Example 4.12 (Random walk) Consider a sequence of iid random variables $\{X_n\}_{n=1}^{+\infty}$ on a probability space $(\Omega, \mathcal{A}, P)$ and common distribution Q. Define the sequence of partial sums $S_n = \sum\limits_{i=1}^{n} X_i$, $n = 1, 2, \ldots$ and set $S_0 = 0$. The random sequence $S = \{S_n\}_{n=0}^{+\infty}$ is called a random walk with steps $\{X_n\}_{n=1}^{+\infty}$ and step distribution Q. We claim that for any n the random vector $(S_0, S_1, \ldots, S_n)$ and the random sequence $(S_n, S_{n+1}, \ldots)$ are conditionally independent given $\sigma(S_n)$, that is, the past (before time n) and the future (after time n) are conditionally independent given the present (time n). To see that this is the case, consider remark 4.17.5 and set $\mathcal{H} = \sigma(S_n)$, $\mathcal{H}_1 = \sigma(X_1, X_2, \ldots, X_n)$ and $\mathcal{H}_2 = \sigma(X_{n+1}, X_{n+2}, \ldots)$. Since $\mathcal{H} \subseteq \mathcal{H}_1$ the last remark yields that $\mathcal{H}_1$ and $\sigma(\mathcal{H}, \mathcal{H}_2)$ are conditionally independent given $\mathcal{H}$. But since $(S_0, S_1, \ldots, S_n)$ is $\mathcal{H}_1$-measurable and $(S_n, S_{n+1}, \ldots)$ is $\sigma(\mathcal{H}, \mathcal{H}_2)$-measurable the claim is established. This conditional property is essentially the idea behind the Markov property and will be treated extensively in Chapters 6 and 7 along with random walks and their applications.

4.4 Summary

In this chapter we have presented a rigorous treatment of probability theory from a mathematical perspective. Since this is an area of statistics that has great overlap with mathematics and has been developed for over a century, there is a plethora of texts on probability and measure theory that the reader can turn to for additional results and exposition. A good summary of the bibliography (books and textbooks) regarding probability theory topics can be found in Fristedt and Gray (1997, Appendix F). The greatest influence on probability theory is perhaps the classic text by Kolmogorov (1933, 1956). We will only mention here recent texts that would be the standard for researchers in statistics or mathematicians with an appetite for probability theory. Such excellent texts include Feller (1968, 1971), Rényi (1970), Chung (1974), Fristedt and Gray (1997), Kallenberg (1986, 2002), Dudley (2002), Durrett (2010), Cinlar (2010), Billingsley (2012) and Klenke (2014).

4.5 Exercises

Probability measures and random variables

Exercise 4.1 Give an example of a probability measure P on a field $\mathcal{A}_0$ of subsets of a space Ω.

Exercise 4.2 Prove theorem 4.2.

Exercise 4.3 Let P and Q be two probability measures on the measurable space $(\Omega, \mathcal{A})$. Define $u(P, Q) = dQ/(dP + dQ)$ and assume that P and Q are absolutely continuous with respect to the Lebesgue measure. Show that $P \perp Q$ if and only if $\int_\Omega u(P, Q)dP = 0$.

Exercise 4.4 Prove all the statements of lemma 4.1.

Exercise 4.5 Define the set function $\mathcal{P} : \Omega \to [0, 1]$ as $\mathcal{P}(A) = card(A)/3^n$, for all $A \subseteq \mathcal{A} = 2^\Omega$, where $\Omega = \{(\omega_1, \ldots, \omega_n) : \omega_i \in \{0, 1, 2\}, i = 1, 2, \ldots, n\}$.
(i) Show that $\mathcal{A}$ is a σ-field.
(ii) Show that $\mathcal{P}$ is a discrete probability measure.
(iii) Find the probability of the events $A_n = \{(\omega_1, \ldots, \omega_n) : \sum\limits_{i=1}^{n} \omega_i = 2n\}, n = 1, 2, \ldots$

Exercise 4.6 Assume that P is a finitely additive probability measure on a field $\mathcal{A}$ and assume that for $A_1, A_2, \cdots \in \mathcal{A}$, if $A_n \downarrow \emptyset$ then $P(A_n) \downarrow 0$. Show that P is countably additive.

Exercise 4.7 Show that $X : (\Omega, \mathcal{A}, P) \to (\mathcal{R}, \mathcal{B}_1)$ is a random variable if and only if for each $x \in \mathcal{R}$ we have $\{\omega : X(\omega) \leq x\} \in \mathcal{A}$.

Exercise 4.8 Let P and Q be two probability measures on the measurable space $(\Omega, \mathcal{A})$ and μ a σ-finite measure on the same measurable space, such that $P \ll \mu$ and $Q \ll \mu$. Let $p = \left[\frac{dP}{d\mu}\right]$ and $q = \left[\frac{dQ}{d\mu}\right]$ denote the Radon-Nikodym derivatives and define the Kullback-Leibler divergence by

$$I^{KL}(P, Q) = \int\limits_{X} \ln\left(\frac{dP}{dQ}\right) dP = \int\limits_{X} \ln\left(\frac{p(x)}{q(x)}\right) p(x)\mu(dx).$$

Show that $I^{KL}(P, Q) \geq 0$ with equality if and only if $P = Q$.

Exercise 4.9 (Bonferroni's inequality) Let $A_1, \ldots, A_n$ be events in a probability space $(\Omega, \mathcal{A}, P)$. Show that

$$P\left(\bigcup_{i=1}^{n} A_i\right) \geq \sum_{i=1}^{n} P(A_i) - \sum_{i,j=1, i<j}^{n} P(A_i \cap A_j).$$

Exercise 4.10 (Kounias' inequality) Show that for any events $A_1, \ldots, A_n$ in a probability space $(\Omega, \mathcal{A}, P)$ we have

$$P\left(\bigcup_{i=1}^{n} A_i\right) \leq \min_{j} \left\{\sum_{i=1}^{n} P(A_i) - \sum_{i=1, i\neq j}^{n} P(A_i \cap A_j)\right\}.$$

Expectation

Exercise 4.11 Assume that $X \sim Cauchy(0, 1)$. Show that $E(X^n)$ does not exist for any $n \geq 1$.

Exercise 4.12 Consider $X \sim Gamma(a, b)$ and find $E(X^c)$ along with conditions on c under which this expectation exists.

Exercise 4.13 Let $X \geq 0$ be some random variable with density f. Show that

$$E(X^c) = \int_0^{+\infty} cx^{c-1} P(X > x)dx,$$

for all $c \geq 1$ for which the expectation is finite.

Exercise 4.14 Give an example of a random variable X with $E\left(\frac{1}{X}\right) \neq \frac{1}{E(X)}$ and an example of a random variable Y with $E\left(\frac{1}{Y}\right) = \frac{1}{E(Y)}$.

Exercise 4.15 Let F be the distribution function of an $\mathcal{R}^+$-valued random variable X and let $\varphi : \mathcal{R} \to \mathcal{R}$ be monotonic and left-continuous. Then the expectation of $\varphi \circ X$ exists as a member of $\overline{\mathcal{R}}$ and

$$E(\varphi \circ X) = \varphi(0) + \int_0^{+\infty} [1 - F(x)] d\varphi(x).$$

Exercise 4.16 Show that remark 4.8.6 holds.

Exercise 4.17 Let $\{X_n\}_{n=1}^{+\infty}$ and $\{Y_n\}_{n=1}^{+\infty}$ be sequences of random variables. Show that

$$Cov\left(\sum_{i=1}^{n} X_i, \sum_{j=1}^{m} Y_j\right) = \sum_{i=1}^{n} \sum_{j=1}^{m} Cov(X_i, Y_j),$$

and therefore

$$Var\left(\sum_{i=1}^{n} X_i\right) = \sum_{i=1}^{n} \sum_{j=1}^{n} Cov(X_i, X_j).$$

Exercise 4.18 Assume that $X \sim \mathcal{N}(0, 1)$. Find the expected value of the random variable $|X|^{2n+1}$, $n = 0, 1, 2, \ldots$

Exercise 4.19 Give an example of a real-valued random variable with the property that all non-central moments are finite but the mgf is defined only at 0.

Independence

Exercise 4.20 Show that any two events A and B of positive probability cannot be disjoint and independent simultaneously.

Exercise 4.21 Use the Fubini theorem to show that for X and Y independent real-valued random variables on $(\Omega, \mathcal{A}, P)$ we have
(i) $P(X \in B - y)$ is a $\mathcal{B}_1$-measurable function of y for all $B \in \mathcal{B}_1$ and
(ii) $P(X + Y \in B) = \int_{\Omega} P(X \in B - y)P(dy)$.

Exercise 4.22 Prove all statements of remark 4.5.

Exercise 4.23 Let $\mathcal{A}_1$ and $\mathcal{A}_2$ be independent σ-fields on $(\Omega, \mathcal{A}, P)$. Show that if a set A is both in $\mathcal{A}_1$ and $\mathcal{A}_2$ then $P(A) = 0$ or 1.

Exercise 4.24 Show that for any independent random variables X and Y defined on the same space, $Var(X + Y) = Var(X) + Var(Y)$.

Exercise 4.25 Let X and Y be two random variables. Show that the integrability of

$X+Y$ does not imply that of X and Y separately. Show that it does imply integrability if X and Y are independent.

Exercise 4.26 Give an example of two dependent random variables with Pearson correlation equal to 0.

Exercise 4.27 Consider two random variables X and Y defined on the same measurable space $(\Omega, \mathcal{A})$. Give an example of two probability measures P_1 and P_2 where X is independent of Y in $(\Omega, \mathcal{A}, P_1)$ but they are not independent in $(\Omega, \mathcal{A}, P_2)$.

Exercise 4.28 (Mutual information) Let X and Y be two discrete random variables with joint pmf $f_{X,Y}$ and marginal pmfs f_X and f_Y. Show that
(i) $E(\ln f_X(x)) \geq \ln E(f_Y(x))$, and
(ii) the mutual information defined by

$$I = E\left(\ln\left(\frac{f_{X,Y}(X,Y)}{f_X(X)f_Y(Y)}\right)\right),$$

is such that $I \geq 0$ with equality if and only X and Y are independent.

Characteristic functions

Exercise 4.29 Show that the cf of a $\mathcal{N}_p(\mathbf{0}, \Sigma)$ distributed random vector is $\exp\left\{-\frac{1}{2}\mathbf{t}^T\Sigma\mathbf{t}\right\}$.

Exercise 4.30 Let X, Y be independent $\mathcal{N}(0, 1)$ random variables. Find the cf of the random vector $(X - Y, X + Y)$.

Exercise 4.31 Let X have a Laplace density $f(x|\lambda) = \frac{\lambda}{2}e^{-\lambda|x|}$, $x \in \mathcal{R}$, $\lambda > 0$. Show that the cf of X is $\varphi_X(t) = \frac{\lambda^2}{\lambda^2+t^2}$.

Exercise 4.32 Assume that $z_1, \dots, z_n, w_1, \dots, w_n$ are complex numbers of modulus $|.|$ at most 1. Show that

$$\left|\prod_{i=1}^{n} z_i - \prod_{i=1}^{n} w_i\right| \leq \sum_{i=1}^{n} |z_i - w_i|. \tag{4.34}$$

Exercise 4.33 Show that for any $x \in \mathcal{R}$ we have

$$\left|e^{itx} - \left(1 + itx - \frac{1}{2}t^2x^2\right)\right| \leq \min\{|tx|^2, |tx|^3\},$$

and as a result for a random variable X with cf $\varphi_X(t)$, $E(X) = 0$ and $\sigma^2 = E(X^2)$ the following holds

$$\left|\varphi_X(t) - \left(1 - \frac{1}{2}t^2\sigma^2\right)\right| \leq E\left[\min\{|tX|^2, |tX|^3\}\right] < \infty. \tag{4.35}$$

Exercise 4.34 If $\varphi_X(t)$ is a cf then show that $e^{c(\varphi_X(t)-1)}$ is a cf, for $c \geq 0$.

Conditional probability, expectation and independence

For this portion of the exercises assume the following: let $(\Omega, \mathcal{A}, P)$ a probability space and assume that A, A_i, $i = 1, 2, \dots$, are $\mathcal{A}$-sets and let $\mathcal{H}$ be a sub-σ-field of $\mathcal{A}$.

Exercise 4.35 Prove all the statements of theorem 4.19.

Exercise 4.36 Show that if X is $\overline{\mathcal{R}}$-valued then

$$E(X|\mathcal{H}) = E(X^+|\mathcal{H}) - E(X^-|\mathcal{H}) \text{ a.s.},$$

assuming that the equality holds *a.s.* on the set where either side is defined.

Exercise 4.37 Prove all the statements of theorem 4.20.

Exercise 4.38 For a nonnegative random variable X we can show that

$$E(X|\mathcal{H}) = \int_0^{+\infty} P(X > t|\mathcal{H})dt \ a.s.$$

Exercise 4.39 Consider the random vector (X_1, X_2, X_3) with joint density

$$P(X_1 = x_1, X_2 = x_2, , X_3 = x_3) = p^3 q^{x_3-3},$$

where $0 < p < 1$, $p + q = 1$, $x_1 = 1, \ldots, x_2 - 1$, $x_2 = 2, \ldots, x_3 - 1$, $x_3 = 3, 4, \ldots$ and show that

(i) $E(X_3|X_1, X_2) = X_2 + 1/p \ a.s.$ and

(ii) $E(a^{X_3}|X_1, X_2) = pa^{X_2+1}/(1 - qa) \ a.s.$, for $0 \le a \le 1$.

Exercise 4.40 For any integers $n, m \ge 1$ and discrete random variables $Y, X_1, X_2, \ldots$. we have

$$E[E(Y|X_1, \ldots, X_{n+m})|X_1, \ldots, X_n] = E(Y|X_1, \ldots, X_n) \ a.s.$$

Exercise 4.41 Consider the random vector $\mathbf{X} \sim \mathcal{N}_2(\mathbf{0}, I_2)$. Find the density of the random variable $X_1|X_2 > 0$ and $E(X_1|X_2 > 0)$.

Exercise 4.42 Assume that $X \sim Exp(\theta)$. Show that $E(X|X > t) = t + \frac{1}{\theta}$.

Exercise 4.43 Assume that $\begin{bmatrix} \mathbf{X}_{p\times 1} \\ \mathbf{Y}_{q\times 1} \end{bmatrix} \sim \mathcal{N}_{p+q}\left(\boldsymbol{\mu} = \begin{bmatrix} \boldsymbol{\mu}_x \\ \boldsymbol{\mu}_y \end{bmatrix}, \boldsymbol{\Sigma} = \begin{bmatrix} \boldsymbol{\Sigma}_{11} & \boldsymbol{\Sigma}_{12} \\ \boldsymbol{\Sigma}_{21} & \boldsymbol{\Sigma}_{22} \end{bmatrix}\right)$.

Show that $\mathbf{X}|\mathbf{Y} = \mathbf{y} \sim \mathcal{N}_p\left(\boldsymbol{\mu}_x + \boldsymbol{\Sigma}_{12}\boldsymbol{\Sigma}_{22}^{-1}(\mathbf{y} - \boldsymbol{\mu}_y), \boldsymbol{\Sigma}_{11} - \boldsymbol{\Sigma}_{12}\boldsymbol{\Sigma}_{22}^{-1}\boldsymbol{\Sigma}_{21}\right)$.

Exercise 4.44 Let X and Y have a bivariate normal density with zero means, variances σ_X^2, σ_Y^2 and correlation ρ. Show that

(i) $E(X|X + Y = w) = \frac{\sigma_X^2 + \rho\sigma_X\sigma_Y}{\sigma_X^2 + 2\rho\sigma_X\sigma_Y + \sigma_Y^2} w$, and

(ii) $Var(X|X + Y = w) = \frac{\sigma_X^2\sigma_Y^2(1-\rho^2)}{\sigma_X^2 + 2\rho\sigma_X\sigma_Y + \sigma_Y^2}$.

Chapter 5

Convergence of Random Objects

5.1 Introduction

A sequence of random objects $\{X_n\}_{n=1}^{+\infty}$ (also denoted by $X = (X_1, X_2, \ldots)$ or $\{X_n, n \geq 1\}$) is simply a collection of random objects X_n often defined on the same probability space $(\Omega, \mathcal{A}, P)$ and taking values in the same space $(\mathcal{X}, \mathcal{G})$ with corresponding distributions $\{Q_n\}_{n=1}^{+\infty}$, under independence, or simply Q, if X is an iid sequence, i.e., $X_n \sim Q$, for all $n = 1, 2, \ldots$. Such sequences are called random or stochastic sequences and will be investigated extensively in Chapter 6 under the assumption that the distribution of X_{n+1} observed at time $n + 1$ given the present value X_n is independent of all past values.

The iid case has been discussed extensively in previous chapters. In particular, the rudimentary definitions and results of section 2.3.4 are easily incorporated in the theory developed in Chapter 4 for any standard random object (such as random variables, vectors and matrices) and will be taken as given for random objects that live in $(\mathcal{R}^p, \mathcal{B}_p)$. In addition, showing a.s. convergence can be accomplished using Kolmogorov's 0-1 law (theorem 4.12) and the Borel-Cantelli lemmas (lemma 4.1).

In contrast, extending the weak and strong laws of large numbers is not a trivial task for more complicated random objects, like point processes and random sets. However, there exist versions of these important theorems and typically for special cases; for example, a SLLN can be obtained for random compact and convex sets using the concept of the support function. We study these methods in the *TMSO-PPRS* text.

In this chapter we study the asymptotic behavior of a sequence of real-valued random variables and its partial sums, as well as the limiting behavior of the sequence of distributions $\{Q_n\}_{n=1}^{+\infty}$.

5.2 Existence of Independent Sequences of Random Variables

The main assumption up to this point has been that $\{X_n\}_{n=1}^{+\infty}$ are independent and therefore we can define what we mean by an independent sequence of random objects (or a random sample of objects). A natural question that arises is whether or not such a sequence exists for a given sequence of probability measures and we address this first in the following existence theorem about independent sequences of random variables. Note that this result is a special case of Kolmogorov's existence theorem (see section 7.3.3) or theorem 4.8 when the probability space is the same.

Theorem 5.1 (Existence of iid sequence) If $\{Q_n\}_{n=1}^{+\infty}$ is a sequence of probability measures on $(\mathcal{R}, \mathcal{B}_1)$, then there exists an independent sequence $\{X_n\}_{n=1}^{+\infty}$ of random variables on some probability space $(\Omega, \mathcal{A}, P)$ such that X_n has distribution Q_n.

Requiring the same probability space or target space is not necessary but it is an assumption made frequently since it is painstaking (in some cases impossible) to study the behavior of the sequence without it.

5.3 Limiting Behavior of Sequences of Random Variables

Consider a sequence of random vectors $\{\mathbf{X}_n\}_{n=1}^{+\infty}$, and recall the convergence concepts introduced in Chapter 2. Having acquired the rigorous probabilistic framework of Chapter 4, we are now able to discuss convergence of random sequences in probability, in distribution and a.s. convergence, in a rigorous framework. We begin by presenting two of the most important results in asymptotic theory; the Slutsky and Cramér theorems, and their extensions. Moreover, we will show under which conditions we have consistency for an MLE estimator.

5.3.1 Slutsky and Cramér Theorems

The Slutsky theorems are the standard tool we use to show convergence in probability and in law for random sequences and their transformations. The proof is requested as an exercise.

Theorem 5.2 (Slutsky) Let $C(\mathbf{f})$ denote the set of continuity points of $\mathbf{f} : \mathcal{R}^d \to \mathcal{R}^k$.

(i) If $\mathbf{X}_n \in \mathcal{R}^d$ and $\mathbf{X}_n \overset{w}{\to} \mathbf{X}$ and if $\mathbf{f}$ is such that $P(\mathbf{X} \in C(\mathbf{f})) = 1$, then $\mathbf{f}(\mathbf{X}_n) \overset{w}{\to} \mathbf{f}(\mathbf{X})$.

(ii) If $\mathbf{X}_n \overset{w}{\to} \mathbf{X}$ and $(\mathbf{X}_n - \mathbf{Y}_n) \overset{p}{\to} \mathbf{0}$, then $\mathbf{Y}_n \overset{w}{\to} \mathbf{X}$.

(iii) If $\mathbf{X}_n \in \mathcal{R}^d$ and $\mathbf{Y}_n \in \mathcal{R}^k$, $\mathbf{X}_n \overset{w}{\to} \mathbf{X}$ and $\mathbf{Y}_n \overset{w}{\to} \mathbf{c}$, some constant, then $\begin{pmatrix} \mathbf{X}_n \\ \mathbf{Y}_n \end{pmatrix} \overset{w}{\to} \begin{pmatrix} \mathbf{X} \\ \mathbf{c} \end{pmatrix}$.

(iv) If $\mathbf{X}_n \in \mathcal{R}^d$ and $\mathbf{X}_n \overset{p}{\to} \mathbf{X}$ and if $\mathbf{f}$ is such that $P(\mathbf{X} \in C(\mathbf{f})) = 1$, then $\mathbf{f}(\mathbf{X}_n) \overset{p}{\to} \mathbf{f}(\mathbf{X})$.

(v) If $\mathbf{X}_n \xrightarrow{p} \mathbf{X}$ and $(\mathbf{X}_n - \mathbf{Y}_n) \xrightarrow{p} \mathbf{0}$, then $\mathbf{Y}_n \xrightarrow{p} \mathbf{X}$.

(vi) If $\mathbf{X}_n \in \mathcal{R}^d$ and $\mathbf{Y}_n \in \mathcal{R}^k$, $\mathbf{X}_n \xrightarrow{p} \mathbf{X}$ and $\mathbf{Y}_n \xrightarrow{p} \mathbf{Y}$, then $\begin{pmatrix}\mathbf{X}_n \\ \mathbf{Y}_n\end{pmatrix} \xrightarrow{p} \begin{pmatrix}\mathbf{X} \\ \mathbf{Y}\end{pmatrix}$.

Note that we can replace the mode of convergence $\xrightarrow{p}$ above with $\xrightarrow{a.s.}$ and the results of Slutsky theorem still hold.

Example 5.1 (Asymptotic normality of the t-test statistic) Assume that $X_i \overset{iid}{\sim} \mathcal{N}(\mu, \sigma^2)$, $i = 1, 2, \ldots, n$, $\mu \in \mathcal{R}$, $\sigma > 0$. From the WLLN and exercise 5.4, we have $\overline{X}_n = \frac{1}{n}\sum_{i=1}^{n} X_i \xrightarrow{w} \mu$, and the usual CLT (exercise 2.82) yields $\sqrt{n}\left(\overline{X}_n - \mu\right)/\sigma \xrightarrow{w} \mathcal{N}(0, 1)$. Moreover, the WLLN on the sequence $\{X_i^2\}$ yields $\frac{1}{n}\sum_{i=1}^{n} X_i^2 \xrightarrow{w} E\left(X^2\right)$. Now from part (iv) of theorem 5.2, $\overline{X}_n^2 \xrightarrow{w} \mu^2$, so that parts (iii) and (iv) yield $S_n^2 = \frac{1}{n}\sum_{i=1}^{n} X_i^2 - \overline{X}_n^2 \xrightarrow{w} E\left(X^2\right) - \mu^2 = \sigma^2$. Finally, another appeal to parts (iii) and (iv) gives $\sqrt{n}\left(\overline{X}_n - \mu\right)/S_n \xrightarrow{w} \mathcal{N}(0, 1)$, and therefore the asymptotic distribution of the t-test statistic is

$$t_{n-1} = \sqrt{n-1}\left(\overline{X}_n - \mu\right)/S_n \xrightarrow{w} \mathcal{N}(0, 1),$$

provided that $S_n \neq 0$.

A great consequence of the Slutsky theorems involves the asymptotic normality of functions of the sample moments.

Theorem 5.3 (Cramér) Let $\mathbf{g}$ be a mapping $\mathbf{g} : \mathcal{R}^d \to \mathcal{R}^k$ such that $\nabla \mathbf{g}$ is continuous in a neighborhood of $\boldsymbol{\mu} \in \mathcal{R}^d$. If $\mathbf{X}_n$ is a sequence of d-dimensional random vectors such that $\sqrt{n}(\mathbf{X}_n - \boldsymbol{\mu}) \xrightarrow{w} \mathbf{X}$, then $\sqrt{n}(\mathbf{g}(\mathbf{X}_n) - \mathbf{g}(\boldsymbol{\mu})) \xrightarrow{w} (\nabla \mathbf{g}(\boldsymbol{\mu}))\mathbf{X}$. In particular, if $\sqrt{n}(\mathbf{X}_n - \boldsymbol{\mu}) \xrightarrow{w} \mathcal{N}_d(0, \Sigma)$, with Σ the covariance matrix, then

$$\sqrt{n}(\mathbf{g}(\mathbf{X}_n) - \mathbf{g}(\boldsymbol{\mu})) \xrightarrow{w} \mathcal{N}_d(0, (\nabla \mathbf{g}(\boldsymbol{\mu}))\Sigma(\nabla \mathbf{g}(\boldsymbol{\mu}))^T).$$

The latter result (under normality) is known as the Delta method.

Example 5.2 (Delta method) Assume that $X_i \overset{iid}{\sim} \mathcal{N}(\mu, \sigma^2)$, $i = 1, 2, \ldots, n$, so that the usual CLT yields $\sqrt{n}\left(\overline{X}_n - \mu\right) \xrightarrow{w} \mathcal{N}(0, \sigma^2)$, where $\overline{X}_n = \frac{1}{n}\sum_{i=1}^{n} X_i$. Consider the asymptotic distribution of $1/\overline{X}_n$. Let $g(x) = 1/x$, and note that $g'(x) = -1/x^2$, and therefore, $g'(\mu) = -1/\mu^2$. An appeal to Cramér's theorem gives

$$\sqrt{n}\left(\frac{1}{\overline{X}_n} - \frac{1}{\mu}\right) \xrightarrow{w} \mathcal{N}\left(0, \frac{\sigma^2}{\mu^4}\right),$$

provided that $\mu \neq 0$.

5.3.2 Consistency of the MLE

Now we turn to investigating MLE consistency. The following theorem summarizes all the requirements in order to prove consistency of an MLE.

Theorem 5.4 (MLE consistency) Assume that $\mathbf{X}_i \overset{iid}{\sim} f(\mathbf{x}|\boldsymbol{\theta})$, $i = 1, 2, \ldots, n$. Let $\boldsymbol{\theta}_0$ denote the true parameter value of $\boldsymbol{\theta} \in \Theta$. Assume that the following conditions hold:
(C1) Θ is compact,
(C2) $f(\mathbf{x}|\boldsymbol{\theta})$ is upper semicontinuous in $\boldsymbol{\theta}$ for all $\mathbf{x}$,
(C3) there exists a function $K(\mathbf{x})$ such that $E_{\boldsymbol{\theta}_0}|K(\mathbf{X})| < \infty$ and

$$\ln \frac{f(\mathbf{x}|\boldsymbol{\theta})}{f(\mathbf{x}|\boldsymbol{\theta}_0)} \leq K(\mathbf{x}),$$

for all $\boldsymbol{\theta}$ and $\mathbf{x}$,
(C4) for all $\boldsymbol{\theta} \in \Theta$ and sufficiently small $\varepsilon > 0$, $\sup\limits_{|\boldsymbol{\theta}'-\boldsymbol{\theta}|<\varepsilon} f(\mathbf{x}|\boldsymbol{\theta}')$ is continuous in $\mathbf{x}$, and
(C5) $f(\mathbf{x}|\boldsymbol{\theta}) = f(\mathbf{x}|\boldsymbol{\theta}_0) \Rightarrow \boldsymbol{\theta} = \boldsymbol{\theta}_0$.
Then any sequence of MLEs $\widehat{\boldsymbol{\theta}}_n$ is strongly consistent for $\boldsymbol{\theta}$.

We present an example where the MLE is not consistent.

Example 5.3 (The MLE is not always consistent) This example is due to Neyman and Scott (1948). Assume that $X_{ij} \sim \mathcal{N}(\mu_i, \sigma^2)$, $i = 1, 2, \ldots, n$, $j = 1, 2, \ldots, d$, are samples of size d from n independent normal populations, with common unknown variance and different unknown means. Let $\mathbf{x}$ denote all the samples, so that the likelihood function is given by

$$\begin{aligned} L(\theta|\mathbf{x}) &= \prod_{i=1}^{n}\prod_{j=1}^{d}(2\pi\sigma^2)^{-1/2}\exp\left\{-\frac{1}{2\sigma^2}(x_{ij}-\mu_i)^2\right\} \\ &= (2\pi\sigma^2)^{-nd/2}\exp\left\{-\frac{1}{2\sigma^2}\sum_{i=1}^{n}\sum_{j=1}^{d}(x_{ij}-\mu_i)^2\right\}, \end{aligned}$$

and therefore, log-likelihood becomes

$$l(\theta|\mathbf{x}) = \ln L(\theta|\mathbf{x}) = -\frac{nd}{2}\ln(2\pi) - \frac{nd}{2}\ln(\sigma^2) - \frac{1}{2\sigma^2}\sum_{i=1}^{n}\sum_{j=1}^{d}(x_{ij}-\mu_i)^2.$$

It is straightforward to see that the MLE of μ_i is given by $\widehat{\mu}_i = \frac{1}{d}\sum\limits_{j=1}^{d} x_{ij}$, $i = 1, 2, \ldots, n$, whereas the MLE of σ^2 is given by $\widehat{\sigma}^2 = \frac{1}{nd}\sum\limits_{i=1}^{n}\sum\limits_{j=1}^{d}(x_{ij}-\widehat{\mu}_i)^2$. Note that $\widehat{\sigma}^2$ is not an unbiased estimator of σ^2, i.e., letting $S_i^2 = \frac{1}{d}\sum\limits_{j=1}^{d}(x_{ij}-\widehat{\mu}_i)^2$, we have $E(S_i^2) = \frac{d-1}{d}\sigma^2$. Since for all $i = 1, 2, \ldots, n, \ldots$, the sample variance S_i^2 follows the same law, the arithmetic mean $\widehat{\sigma}^2 = \frac{1}{n}\sum_{i=1}^{n} S_i^2$, tends a.s. (via an appeal to the SLLN on the sequence $\{S_i^2\}_{i=1}^n$) to the common expectation $\frac{d-1}{d}\sigma^2$, and therefore the MLE $\widehat{\sigma}^2$ is not consistent as $n \to \infty$ and d is kept fixed.

5.4 Limiting Behavior of Probability Measures

Important theorems such as the MCT and Lebesgue DCT allow us to pass the limit, as $n \to \infty$, under the integral sign when we integrate sequences of functions (random variables) X_n with respect to a probability measure P that does not depend on n. More precisely, we have obtained results to help us write

$$E(X_n) = \int X_n dP \to E(X) = \int X dP, \tag{5.1}$$

when X_n converges (in some sense, e.g., a.s. convergence) to X, as $n \to \infty$. Naturally, one wonders if there are similar results when n is affecting the probability measure over which we compute the integral and not the function we integrate, that is,

$$\int f dP_n \to \int f dP, \tag{5.2}$$

for some function f when P_n converges (in some sense, e.g., when $P_n(B) \to P(B)$, $\forall B$) to P, as $n \to \infty$. Note that since a measure P_n is uniquely determined by its corresponding distribution function F_n and in view of equation (3.17) we simplify the problem and consider the distribution function F_n instead of the probability measure P_n, that is, we investigate the limiting behavior of the integral $\int f dF_n = \int f dP_n$.

5.4.1 Integrating Probability Measures to the Limit

We begin by presenting some classic results involving equation (5.1) and weak convergence followed by an investigation of equation (5.2).

Theorem 5.5 (Helly-Bray) Assume that $X_n \xrightarrow{w} X$ or equivalently that $F_n \xrightarrow{w} F$ where F_n and F are the distribution functions of X_n and X, respectively, and let $C(F) = \{x : F \text{ is continuous at } x\}$ denote the set of continuity points of F. Then $E(g(X_n)) \to E(g(X))$, as $n \to \infty$, for all continuous functions g that vanish outside $[a, b]$.

Proof. We treat the univariate case. Since g is a continuous function vanishing outside $[a, b]$, then g is uniformly continuous (Appendix Section A.5.3), that is, $\forall \varepsilon > 0, \exists \delta > 0 : |x - y| < \delta \Longrightarrow |g(x) - g(y)| < \varepsilon$. Choose $a = x_0 < x_1 < x_2 < \cdots < x_k = b$, with $x_i \in C(F)$ and some δ (e.g., $b - a$) such that $|x_i - x_{i-1}| < \delta$ so that $|g(x) - g(x_i)| \leq \varepsilon$, for all $x \in (x_{i-1}, x_i]$, $i = 1, 2, \ldots, k$. Therefore we can write $|g(x) - g^*(x)| \leq \varepsilon$, for all $x \in [a, b]$, where $g^*(x) = \sum_{i=1}^{k} g(x_i) I_{(x_{i-1}, x_i]}(x)$ and since

$F_n \xrightarrow{w} F$ it follows that

$$\begin{aligned} E(g^*(X_n)) &= \sum_{i=1}^{k} g(x_i)E[I_{(x_{i-1},x_i]}(X_n)] = \sum_{i=1}^{k} g(x_i)[F_n(x_i) - F_n(x_{i-1})] \\ &\to \sum_{i=1}^{k} g(x_i)[F(x_i) - F(x_{i-1})] = \sum_{i=1}^{k} g(x_i)E[I_{(x_{i-1},x_i]}(X)] = E(g^*(X)), \end{aligned}$$

as $n \to \infty$. Moreover we have

$$\begin{aligned} |E(g(X_n)) - E(g(X))| &= |E[g(X_n)] - E[g^*(X_n)] + E[g^*(X_n)] - \\ &\quad E[g^*(X)] + E[g^*(X)] - E[g(X)]| \\ &\leq \left|E[g(X_n)] - E[g^*(X_n)]\right| + \left|E[g^*(X_n)] - E[g^*(X)]\right| + \\ &\quad \left|E[g^*(X)] - E[g(X)]\right| \\ &\leq \left|E[g(X_n) - g^*(X_n)]\right| + \left|E[g^*(X_n) - g^*(X)]\right| + \\ &\quad \left|E[g^*(X) - g(X)]\right| \\ &\leq E|g(X_n) - g^*(X_n)| + E|g^*(X_n) - g^*(X)| + \\ &\quad E|g^*(X) - g(X)| \\ &\leq 2\varepsilon + E|g^*(X_n) - g^*(X)| \to 2\varepsilon. \end{aligned}$$

Since this holds for all $\varepsilon > 0$ we must have $E(g(X_n)) \to E(g(X))$. ■

The converse of the latter theorem is also true and there is a more general version as we see below.

Remark 5.1 (Helly-Bray) The following conditions are equivalent:
(i) $X_n \xrightarrow{w} X$,
(ii) $E(g(X_n)) \to E(g(X))$, as $n \to \infty$, for all continuous functions g that vanish outside a compact set,
(iii) $E(g(X_n)) \to E(g(X))$, as $n \to \infty$, for all continuous bounded functions g, and
(iv) $E(g(X_n)) \to E(g(X))$, as $n \to \infty$, for all bounded measurable functions g such that $P(X \in C(g)) = 1$.

All of the directions (i) $\Longrightarrow$ (ii) or (iii) or (iv) are known as the Helly-Bray theorem. Furthermore, note that the dimension of X_n and X has no effect on the statements, although the proofs would need to be slightly modified (see for example Ferguson, 1996, theorem 3).

Example 5.4 (Counterexamples to Helly-Bray theorem) Let us discuss the necessity of some of these conditions through counterexamples.

1. Let $g(x) = x$ and take $X_n = n$, w.p. $\frac{1}{n}$ and $X_n = 0$, w.p. $\frac{n-1}{n}$. Then $X_n \xrightarrow{w} X = 0$ but $E(g(X_n)) = n\frac{1}{n} = 1 \nrightarrow E(g(0)) = 0$. Therefore one cannot remove the boundedness requirement for g in (iii) and (iv).

2. Let $g(x) = I(x > 0)$, $x \geq 0$ and assume that $P(X_n = \frac{1}{n}) = 1$. Then $X_n \xrightarrow{w} X = 0$ but $E(g(X_n)) = 1 \nrightarrow E(g(0)) = 0$ and thus the continuity requirement from (i) and (ii) cannot be removed. Similarly, in (iv) it is required that $P(X \in C(g)) = 1$.

We define weak convergence in the space of probability measures via convergence in the space of distribution functions.

Definition 5.1 Weak convergence of probability measures

Let F_n, F be the cdfs corresponding to the probability measures P_n and P. We say that the sequence of probability measures P_n converges weakly to P if and only if $F_n \overset{w}{\to} F$ and write $P_n \overset{w}{\to} P$ (recall definition 2.3.3).

The following theorem connects weak and a.s. convergence.

Theorem 5.6 (Skorohod representation) Suppose that Q_n and Q are probability measures on $(\mathcal{R}, \mathcal{B}_1)$ and that $Q_n \overset{w}{\to} Q$. Then there exist random variables X_n and X on a common probability space $(\Omega, \mathcal{A}, P)$ such that X_n has distribution Q_n, X has distribution Q and $X_n \overset{a.s.}{\to} X$ as $n \to \infty$.

Proof. The construction is related to that of theorem 4.6. Take $\Omega = (0, 1)$, $\mathcal{A} = \mathcal{B}((0, 1))$, the Borel sets of $(0, 1)$ and let $P = \mu_1$, the Lebesgue measure on $(0, 1)$ (i.e., the uniform distribution). Consider the distribution functions F_n and F corresponding to Q_n and Q and set $X_n(\omega) = \inf\{x : F_n(x) \geq \omega\}$ and $X(\omega) = \inf\{x : F(x) \geq \omega\}$. We notice that $F_n(x) \geq \omega$ if and only if $X_n(\omega) \leq x$, owing to the fact that F is nondecreasing and $\{x : F_n(x) \geq \omega\} = [X_n(\omega), +\infty)$; the infimum in the definition is there in order to handle the case where F is the same for many x. Therefore we can write

$$P(X_n \leq x) = P(\{\omega : X_n(\omega) \leq x\}) = P(\{\omega : \omega \leq F_n(x)\}) = F_n(x),$$

so that X_n has distribution function F_n and therefore distribution Q_n. Similarly, X has distribution Q and it remains to show that $X_n \overset{a.s.}{\to} X$.

Let $\varepsilon > 0$ and fix $\omega \in (0, 1)$. Select an x such that $X(\omega) - \varepsilon < x < X(\omega)$ and $Q(\{x\}) = 0$ which leads to $F(x) < \omega$. Now $F_n(x) \to F(x)$, as $n \to \infty$, implies that for n large enough, $F_n(x) < \omega$ and hence $X(\omega) - \varepsilon < x < X_n(\omega)$. Letting $n \to \infty$ and $\varepsilon \downarrow 0$ we conclude that $\liminf_n X_n(\omega) \geq X(\omega)$, for any $\omega \in (0, 1)$. Next choose ω' satisfying $0 < \omega < \omega' < 1$ and $\varepsilon > 0$. Select again a point y such that $X(\omega') < y < X(\omega') + \varepsilon$, with $Q(\{y\}) = 0$ and note that $\omega < \omega' \leq F(X(\omega')) \leq F(y)$, so that for n large enough, $\omega \leq F_n(y)$ and hence $X_n(\omega) \leq y < X(\omega') + \varepsilon$. Letting $n \to \infty$ and $\varepsilon \downarrow 0$ we have that $\limsup_n X_n(\omega) \leq X(\omega')$, for ω, ω' with $0 < \omega < \omega' < 1$. Let $\omega' \downarrow \omega$ with ω a continuity point of X so that $X(\omega') \downarrow X(\omega)$. Consequently, we have the other direction, namely, $\limsup_n X_n(\omega) \leq X(\omega)$ and hence $X_n(\omega) \to X(\omega)$, as $n \to \infty$, provided that $\omega \in (0, 1)$ is a continuity point of X.

Since X is nondecreasing on $(0, 1)$ the set $D \subseteq (0, 1)$ of discontinuities of X is countable so that $P(D) = 0$. Finally, we may write

$$P(X_n \to X, \text{ as } n \to \infty) = P(\{\omega : X_n(\omega) \to X(\omega), \text{ as } n \to \infty\}) = 1,$$

and the claim is established. ■

Next we discuss a generalization of the Skorohod's theorem to Polish spaces.

Remark 5.2 (Generalization to Polish spaces) The original version of Skorohod's theorem (Skorohod, 1956) is more general and can be applied to any complete, separable metric space (i.e., a Polish space), not just the real line $\mathcal{R}$. In addition, definition 5.1 can be reformulated as follows: let $(\mathcal{X}, \rho)$ be a Polish space and define $\{P_n\}_{n=1}^{+\infty}$ as a sequence of probability measures on $(\mathcal{X}, \mathcal{B}(\mathcal{X}))$ and let P be another probability measure on the same space. Then $P_n \stackrel{w}{\to} P$ if and only if

$$\lim_{n\to\infty} \int_{\mathcal{X}} g(x)P_n(dx) \to \int_{\mathcal{X}} g(x)P(dx), \tag{5.3}$$

for all bounded, continuous real-valued functions f on $\mathcal{X}$. Obviously, $\mathcal{X} = \mathcal{R}$ is a Polish space so that this general definition contains definition 5.1 as a special case, provided that (5.3) holds.

The next theorem shows that (5.3) holds over intervals $[a, b] \subset \mathcal{R}$.

Theorem 5.7 (Portmanteau) If $\{F_n\}_{n=1}^{+\infty}$ is a sequence of distribution functions with $F_n \stackrel{w}{\to} F$ and $a, b \in C(F)$, then for every real-valued, uniformly continuous function g on $[a, b]$ we have

$$\lim_{n\to\infty} \int_a^b g dF_n \to \int_a^b g dF.$$

Proof. For $\varepsilon > 0$ choose by uniform continuity a $\delta > 0$ so that $|f(x)-g(y)| < \varepsilon$, for $|x - y| < \delta$, $x, y \in [a, b]$. Select $x_i \in C(F)$, $i = 2, \ldots, k$, such that $a = x_1 < x_2 < \cdots < x_{k+1} = b$ and $\max\limits_{1\le i\le k}|x_{i+1} - x_i| < \delta$. Then we can write

$$H_n = \int_a^b g dF_n - \int_a^b g dF = \sum_{i=1}^{k} \left\{ \int_{x_i}^{x_{i+1}} g(x)dF_n(x) - \int_{x_i}^{x_{i+1}} g(x)dF(x) \right\},$$

and adding and subtracting $\int_{x_i}^{x_{i+1}} g(x_i)dF_n(x) + \int_{x_i}^{x_{i+1}} g(x_i)dF(x)$, above we have

$$\begin{aligned} H_n &= \sum_{i=1}^{k} \left\{ \int_{x_i}^{x_{i+1}} g(x)dF_n(x) - \int_{x_i}^{x_{i+1}} g(x_i)dF_n(x) \right\} \\ &\quad + \sum_{i=1}^{k} \left\{ \int_{x_i}^{x_{i+1}} g(x_i)dF_n(x) - \int_{x_i}^{x_{i+1}} g(x_i)dF(x) \right\} \\ &\quad + \sum_{i=1}^{k} \left\{ \int_{x_i}^{x_{i+1}} g(x_i)dF(x) - \int_{x_i}^{x_{i+1}} g(x)dF(x) \right\}, \end{aligned}$$

which leads to

$$H_n = \sum_{i=1}^{k}\left\{\int_{x_i}^{x_{i+1}}[g(x)-g(x_i)]dF_n(x)\right\} + g(x_i)[F_n(x_{i+1})-F_n(x_i)-F(x_{i+1})+F(x_i)]$$
$$+\sum_{i=1}^{k}\left\{\int_{x_i}^{x_{i+1}}[g(x_i)-g(x)]dF(x)\right\}.$$

Moreover, we can bound $|H_n|$ from above by

$$|H_n| \le \sum_{i=1}^{k}\left\{\int_{x_i}^{x_{i+1}}|g(x)-g(x_i)|\,dF_n(x)\right\} + g(x_i)\,|F_n(x_{i+1})-F_n(x_i)-F(x_{i+1})+F(x_i)|$$
$$+\sum_{i=1}^{k}\left\{\int_{x_i}^{x_{i+1}}|g(x_i)-g(x)|\,dF(x)\right\}.$$

First we notice that $|g(x)-g(x_i)|<\varepsilon$ so that we can write

$$|H_n| \le \varepsilon\sum_{i=1}^{k}\{F_n(x_{i+1})-F_n(x_i)\} + g(x_i)\,|F_n(x_{i+1})-F_n(x_i)-F(x_{i+1})+F(x_i)|$$
$$+\varepsilon\sum_{i=1}^{k}\{F(x_{i+1})-F(x_i)\},$$

and expanding the two telescoping sums yields

$$|H_n| \le \varepsilon[F_n(x_{k+1})-F_n(x_1)+F(x_{k+1})-F(x_1)]$$
$$+g(x_i)\,|F_n(x_{i+1})-F_n(x_i)-F(x_{i+1})+F(x_i)|,$$

with both summands converging to 0 as $n\to\infty$, since $F_n(x_i)\to F(x_i)$, for $x_i\in C(F)$. Thus $|H_n|\to 0$, as $n\to\infty$ and the result is established. ■

There are many versions of the Portmanteau theorem that involve different properties of the function g. For more details see Feller (1971), Fristedt and Gray (1997), and Billingsley (2012).

5.4.2 Compactness of the Space of Distribution Functions

We collect some essential results on establishing a compactness property for the space of distribution functions. Versions of these theorems and their proofs can be found in Billingsley (2012, p. 359). The first theorem can be used to prove many of the results that follow in this section.

Theorem 5.8 (Helly selection) For every sequence $\{F_n\}_{n=1}^{+\infty}$ of distribution functions there exists a subsequence $\{F_{n_k}\}_{k=1}^{+\infty}$ and a nondecreasing, right continuous function F such that $\lim\limits_{k\to\infty}F_{n_k}(x)=F(x)$ at continuity points of F.

Note that even though the $\{F_n\}_{n=1}^{+\infty}$ in the Helly selection theorem are cdfs, their limit need not be a cdf in general. For example, if F_n has a unit jump at n, then the limit is $F(x)=0$, which is not a valid distribution function. In terms of the corresponding probability measures $\{P_n\}_{n=1}^{+\infty}$ of $\{F_n\}_{n=1}^{+\infty}$, the problem is described by a P_n that has unit mass at n and the mass of P_n "escapes to infinity" as $n \to +\infty$. To avoid such cases we introduce the concept of tightness of probability measures.

Definition 5.2 Tightness

A sequence of probability measures $\{Q_n\}_{n=1}^{+\infty}$ on $(\mathcal{R}, \mathcal{B}_1)$ is said to be uniformly tight if for every $\varepsilon > 0$ there exists a finite interval $(a, b]$ such that $Q_n((a, b]) > 1 - \varepsilon$, for all n. In terms of the corresponding distribution functions F_n and F, the condition is that for all $\varepsilon > 0$ there exist x and y such that $F_n(x) < \varepsilon$ and $F_n(y) > 1 - \varepsilon$, for all n.

The following theorem connects tightness and weak convergence of subsequences of probability measures.

Theorem 5.9 (Tightness and weak convergence) Tightness is a necessary and sufficient condition that for every subsequence of probability measures $\{Q_{n_k}\}_{k=1}^{+\infty}$ there exists a sub-subsequence $\{Q_{n_{k_j}}\}_{j=1}^{+\infty}$ and a probability measure Q such that $Q_{n_{k_j}} \stackrel{w}{\to} Q$, as $j \to \infty$.

Based on the last two theorems, we can prove the following result that connects weak convergence and the classic approach to defining compactness in standard Polish spaces like $\mathcal{R}^p$, i.e., a space is compact if every convergent sequence of elements of a space has a convergent subsequence in the space. In our case we work in the space of all probability measures.

Theorem 5.10 (Weak convergence and subsequences) A sequence of probability measures $\{Q_n\}_{n=1}^{+\infty}$ converges weakly to a probability measure Q if and only if every subsequence $\{Q_{n_k}\}_{k=1}^{+\infty}$ has a subsequence $\{Q_{n_{k_j}}\}_{j=1}^{+\infty}$ converging weakly to Q.

We end this section by collecting some basic results involving weak convergence and their proofs are requested as exercises.

Remark 5.3 (Weak convergence results) Let F_n and F be distribution functions in $\mathcal{R}$ and let Q_n and Q denote the corresponding distributions.

1. $F_n \stackrel{w}{\to} F$ if and only if there is a dense subset D of $\mathcal{R}$ such that for every $x \in D$, $F_n(x) \to F(x)$, as $n \to \infty$.

2. If G is a distribution function in $\mathcal{R}$ with $F_n \stackrel{w}{\to} G$ and if $F_n \stackrel{w}{\to} F$ then $F = G$.

3. Suppose that Q_n, Q have densities f_n and f with respect to a common σ-finite measure μ. If $f_n \to f$ a.e. $[\mu]$ then $Q_n \stackrel{w}{\to} Q$.

4. If $\{Q_n\}_{n=1}^{+\infty}$ is a tight sequence of probability measures on a Polish space and if each subsequence converges weakly to Q then $Q_n \xrightarrow{w} Q$.

5. Tightness in Polish spaces Tightness can be defined for Polish spaces as follows. A distribution Q on a Polish space $\mathcal{X}$ is tight if $\forall \varepsilon > 0$, there exists a compact subset $K \subset \mathcal{X}$ such that $Q(K^c) < \varepsilon$. A family of distributions $\mathcal{Q}$ is uniformly tight if $\forall \varepsilon > 0$, there exists a compact subset $K \subset \mathcal{X}$ such that $Q_n(K^c) < \varepsilon$, for all $Q \in \mathcal{Q}$.

6. Relatively sequential compact A family $\mathcal{Q}$ of distributions defined on a Polish space is relatively sequentially compact if every sequence $\{Q_n\}_{n=1}^{+\infty}$, with $Q_n \in \mathcal{Q}$, has a convergent subsequence.

7. Let $\{Q_n\}_{n=1}^{+\infty}$ be a relatively sequentially compact sequence of distributions defined on a Polish space and assume that every convergent subsequence has the same limiting probability measure Q. Then $Q_n \xrightarrow{w} Q$.

8. Prohorov theorem A family of probability measures defined on a Polish space is relatively sequentially compact if and only if it is uniformly tight.

5.4.3 Weak Convergence via Non-Central Moments

Next we connect convergence in non-central moments with weak convergence. First we discuss when a distribution function F is uniquely determined by its moments, since knowing all non-central moments $a_k = \int\limits_{\mathcal{R}} x^k dF(x)$, $k = 1, 2, \ldots$, of a random variable with cdf F, does not necessarily imply that F is uniquely determined by $\{a_k\}_{k=1}^{+\infty}$.

Necessary and sufficient conditions for F to be uniquely determined by $\{a_k\}_{k=1}^{+\infty}$, can be found in Lin (2017), and the references therein. However, some of these conditions are not easy to verify in general. For example, Carleman's condition requires $\sum\limits_{k=1}^{+\infty} \left(\frac{1}{a_{2k}}\right)^{\frac{1}{2k}} = +\infty$, whereas, Cramér's condition requires that the mgf exists and is finite, i.e., $E(e^{tX}) < \infty$, for all $t \in [-c, c]$, for some constant $c > 0$ (see Lin, 2017, p. 4). For what follows assume that F is uniquely determined by its moments.

The usefulness of convergence in all moments is illustrated in the next theorem. Showing weak convergence by proving convergence of moments is referred to as the "method of moments," which should not be confused with the method of finding point estimators, which we saw in Chapter 2, remark 2.4.1. For a proof see Chung (1974, p. 103).

Theorem 5.11 (Fréchet-Shohat) If $\{F_n\}_{n=1}^{+\infty}$ is a sequence of distribution functions whose moments are finite and such that $q_{n,k} = \int\limits_{\mathcal{R}} x^k dF_n(x) \to a_k < +\infty$, as

$n \to \infty$, $k = 1, 2, \ldots$, where $\{a_k\}_{k=1}^{+\infty}$ are the non-central moments of a uniquely determined distribution function F, then $F_n \xrightarrow{w} F$.

5.5 Random Series

Random series appear almost exclusively in statistics, in particular, in the form of point estimators of parameters (e.g., sample averages). Therefore, it is important to consider the distribution of a partial sum (or the average) and study its behavior as the sample size increases.

5.5.1 Convolutions

Distributions of partial sums are known as convolutions. We begin with the general definition under independence.

Definition 5.3 Convolution

Let $X_1, X_2, \ldots, X_n$ be independent random objects taking values on the same space with distributions $Q_1, Q_2, \ldots, Q_n$, respectively. Then the distribution of the random object $S_n = X_1 + X_2 + \cdots + X_n$ is called a convolution and is denoted by $Q_1 * Q_2 * \cdots * Q_n$. If $Q_i = Q$, $i = 1, 2, \ldots, n$, the distribution of S_n is called the n-fold convolution of Q with itself, denoted by Q^{*n}.

Example 5.5 (Random convolutions) Let $X_1, X_2, \ldots,$ be independent random variables with a common distribution Q. Let N be a random variable independent of the X_i and taking values in $\mathcal{Z}_0^+ = \{0, 1, \ldots\}$ with positive probabilities p_i, $i \in \mathcal{Z}_0^+$. We are interested in the random variable $S_N = X_1 + X_2 + \cdots + X_N$. The conditional distribution of S_N given that $N = n$ is Q^{*n} and hence the distribution of S_N is given by the infinite mixture distribution $\sum\limits_{n=0}^{+\infty} p_n Q^{*n}$.

The following remark presents some essential results that connect probability distributions, convolutions and weak convergence.

Remark 5.4 (Convolution results) The addition symbol "+" in the definition of S_n needs to be defined appropriately and depends on the target space of the random objects. For example, in $\mathcal{R}^p$ it is standard vector addition but in the space $\mathcal{K}$ of compact subsets of $\mathcal{R}^p$ we use Minkowski addition "$\oplus$" so that

$$S_n = X_1 \oplus X_2 \oplus \cdots \oplus X_n = \{x_1 + x_2 + \cdots + x_n : x_i \in X_i, \forall i\},$$

where the random objects X_i are random sets in $\mathcal{K}$. We collect some remarks on convolutions below.

1. Consider the case of two independent random vectors $\mathbf{X}_1$ and $\mathbf{X}_2$ defined on $(\Omega, \mathcal{A}, P)$ with distributions Q_1, Q_2 and taking values in $(\mathcal{R}^j, \mathcal{B}_j)$ and $(\mathcal{R}^k, \mathcal{B}_k)$, respectively, so that the random vector $(\mathbf{X}_1, \mathbf{X}_2)$ takes values in $(\mathcal{R}^{j+k}, \mathcal{B}_{j+k})$ and has distribution $Q_1 \times Q_2$. Using the Fubini theorem we can write

$$(Q_1 \times Q_2)(B) = \int_{\mathcal{R}^j} Q_2(B_{\mathbf{x}_1}) Q_1(d\mathbf{x}_1),$$

for all Borel sets B in $\mathcal{B}_{j+k}$, with $B_{\mathbf{x}_1} = \{\mathbf{x}_2 : (\mathbf{x}_1, \mathbf{x}_2) \in B\} \in \mathcal{R}^k$, the $\mathbf{x}_1$-section of B, for all $\mathbf{x}_1 \in \mathcal{R}^j$. Replacing B with $(A \times \mathcal{R}^k) \cap B$ in the latter equation for $A \in \mathcal{B}_j$ and $B \in \mathcal{B}_{j+k}$, we have

$$(Q_1 \times Q_2)((A \times \mathcal{R}^k) \cap B) = \int_A Q_2(B_{\mathbf{x}_1}) Q_1(d\mathbf{x}_1),$$

and since

$$Q_2(B_{\mathbf{x}_1}) = P(\{\omega : \mathbf{X}_2(\omega) \in B_{\mathbf{x}_1}\}) = P(\{\omega : (\mathbf{x}_1, \mathbf{X}_2(\omega)) \in B\}) = P((\mathbf{x}_1, \mathbf{X}_2) \in B), \tag{5.4}$$

we establish that

$$P((\mathbf{X}_1, \mathbf{X}_2) \in B) = \int_{\mathcal{R}^j} P((\mathbf{x}_1, \mathbf{X}_2) \in B) Q_1(d\mathbf{x}_1),$$

and

$$P(\mathbf{X}_1 \in A, (\mathbf{X}_1, \mathbf{X}_2) \in B) = \int_A P((\mathbf{x}_1, \mathbf{X}_2) \in B) Q_1(d\mathbf{x}_1). \tag{5.5}$$

Now consider the case of two random variables X_1 and X_2 ($j = k = 1$) and let $B = \{(x_1, x_2) : x_1 + x_2 \in H\}$, $H \in \mathcal{B}_1$. Using equations (5.4) and (5.5) we can write the distribution of $X_1 + X_2$ as

$$P(X_1 + X_2 \in H) = \int_{\mathcal{R}} P(X_2 \in H - x) Q_1(dx) = \int_{\mathcal{R}} Q_2(H - x) Q_1(dx),$$

where $H - x = \{h - x : h \in H\}$. Therefore the convolution of Q_1 and Q_2 is the measure $Q_1 * Q_2$ defined by

$$(Q_1 * Q_2)(H) = \int_{\mathcal{R}} Q_2(H - x) Q_1(dx). \tag{5.6}$$

2. Convolution can be defined based on distribution functions and densities. Indeed, if F_1 and F_2 are the cdfs corresponding to Q_1 and Q_2, the distribution function corresponding to $Q_1 * Q_2$ is denoted by $F_1 * F_2$ and letting $H = (-\infty, y]$ in (5.6) we can write

$$(F_1 * F_2)(y) = \int_{\mathcal{R}} F_2(y - x) dF_1(x).$$

Moreover, if X_1 and X_2 are continuous and have densities f_1 and f_2, respectively, with respect to the Lebesgue measure μ_1, then the convolution $Q_1 * Q_2$ has density denoted by $f_1 * f_2$ and defined by

$$(f_1 * f_2)(y) = \int_{\mathcal{R}} f_2(y - x) f_1(x) d\mu_1(x).$$

3. If Q, R and Q_n and R_n, $n = 1, 2, \dots$, are probability measures on $\mathcal{R}$ with $Q_n \overset{w}{\to} Q$ and $R_n \overset{w}{\to} R$ then $Q_n * R_n \overset{w}{\to} Q * R$.

4. The distribution Q in $\mathcal{R}$ is called stable if for each n and iid random variables X, $X_i \sim Q$, $i = 1, 2, \dots, n$, there exist constants $c_n > 0$ and b_n, such that the distribution of the sum $S_n = \sum_{i=1}^{n} X_i \sim Q^{*n}$ is the same as that of the random variable $a_n X + b_n$, for all n and Q is not degenerate (concentrated in one point). If $b_n = 0$ then Q is called strictly stable.

5. If $U = Q^{*n}$, for some distribution U and positive integer n, then Q is called an nth convolution root of U and we write $Q = P^{*\frac{1}{n}}$. Distributions that have an nth convolution root for every $n = 1, 2, \ldots$, are said to be infinitely divisible. For example, the normal and Poisson distributions are infinitely divisible.

The characteristic function plays a prominent role in identifying the distribution of a convolution as shown in the following theorem.

Theorem 5.12 (Convolution distributions via characteristic functions) Let $X_1, X_2, \ldots, X_n$ be independent random variables taking values on the same space $(\mathcal{R}, \mathcal{B}_1)$ with characteristic functions $\varphi_1, \varphi_2, \ldots, \varphi_n$ and distributions $Q_1, Q_2, \ldots, Q_n$, respectively. Then the distribution (convolution) $Q_1 * Q_2 * \cdots * Q_n$ of the random variable $S_n = \sum_{i=1}^{n} X_i$ has cf $\beta = \prod_{i=1}^{n} \varphi_i$. Consequently, if $Q_i = Q$ and $\varphi_i = \varphi$, $i = 1, 2, \ldots, n$, the cf of the n-fold convolution Q^{*n} is given by φ^n.

Proof. Using theorem 4.15 and the definition of the characteristic function we can write

$$\begin{aligned} \beta(t) &= E\left(e^{itS_n}\right) = E\left(\exp\left\{it\sum_{i=1}^{n} X_i\right\}\right) = E\left(\prod_{i=1}^{n} \exp\{itX_i\}\right) \\ &= \prod_{i=1}^{n} E\left(\exp\{itX_i\}\right) = \prod_{i=1}^{n} \varphi_i(t). \end{aligned}$$

■

5.5.2 Fourier Inversion and the Continuity Theorem

The importance of characteristic functions is further exemplified in the following theorems and remarks, where we connect characteristic functions with densities, weak convergence and convolutions.

Theorem 5.13 (Fourier inversion) Assume that the random variable $X : (\Omega, \mathcal{A}, P) \to (\mathcal{R}, \mathcal{B}_1)$ has cf φ, with $|\varphi|$ Lebesgue integrable. Then the distribution $Q = PX^{-1}$ has a bounded continuous density f given by

$$f(x) = (2\pi)^{-1} \int_{-\infty}^{+\infty} e^{-itx} \varphi(t) dt. \tag{5.7}$$

In the discrete case, with X taking distinct values x_j, $j \geq 1$, we have

$$f(x_j) = \lim_{T \to \infty} \frac{1}{2T} \int_{T}^{T} e^{-itx_j} \varphi(t) dt, \ j \geq 1. \tag{5.8}$$

Proof. We prove the continuous case only and request the discrete case as an exercise. In view of remark 4.10 and the form of the convolution in remark 5.4.2, the left side of the Parseval relation

$$\int_{-\infty}^{+\infty} \frac{\sigma}{\sqrt{2\pi}} e^{-\frac{\sigma^2(x-v)^2}{2}} f(x)dx = \frac{1}{2\pi} \int_{-\infty}^{+\infty} e^{-ivy} e^{-\frac{y^2}{2\sigma^2}} \varphi(y)dy,$$

is simply the convolution $N_\sigma * Q$, of a normal random variable $Z_\sigma \sim N\left(0, \frac{1}{\sigma^2}\right)$ with density f_σ and distribution N_σ and the random variable X with distribution $Q = PX^{-1}$, where X has some density f corresponding to Q. Letting $\sigma \to \infty$ in the latter equation we see on the LHS that $N_\sigma * Q$ tends to Q, or in terms of the densities, $f_\sigma * f$ approaches f (using BCT to pass the limit under the integral sign), where this f is the bounded continuous function defined in equation (5.7) (since $e^{-\frac{y^2}{2\sigma^2}} \to 1$, as $\sigma \to \infty$, in the RHS of the latter equation and $|\varphi|$ the Lebesgue integrable). To see that this f is indeed the density of Q, we write

$$(N_\sigma * Q)(B) = \int_B (f_\sigma * f)(y)dy \to \int_B f(y)dy,$$

with $(N_\sigma * Q)(B) \to Q(B)$, as $\sigma \to \infty$, so that $Q(B) = \int_B f(y)dy$, for any $B \in \mathcal{B}_1$ and the result is established. ■

Now we can connect weak convergence with pointwise convergence of characteristic functions. The following theorem is also known as the Cramér-Wold device.

Theorem 5.14 (Continuity theorem) A sequence of probability distributions $\{Q_n\}_{n=1}^{+\infty}$ on $\mathcal{R}$ converges weakly to a distribution function Q if and only if the sequence $\{\varphi_n\}_{n=1}^{+\infty}$ of their characteristic functions converge pointwise to a complex-valued function φ which is continuous at 0 and φ is the cf of Q.

Proof. ($\Longrightarrow$) This direction follows immediately from the Helly-Bray theorem (remark 5.1), since $g_t(X) = e^{itX} = \cos(tX) + i\sin(tX)$ is bounded and continuous with respect to X so that $Q_n \stackrel{w}{\to} Q$ implies that $E(g_t(X_n)) = \varphi_n(t) \to E(g_t(X)) = \varphi(t)$, as $n \to \infty$, for all $t \in \mathcal{R}$.
($\Longleftarrow$) Assume that $\varphi_n(t) \to \varphi(t)$, as $n \to \infty$, for all $t \in \mathcal{R}$, where φ is continuous at 0. First note that the limit $Q = \lim_{n\to\infty} Q_n$ exists. Indeed, using the Helly selection theorem on the corresponding distribution functions F and F_n of Q and Q_n, there exists a subsequence $\{F_{n_k}\}_{k=1}^{+\infty}$ and a nondecreasing, right continuous function F such that $\lim_{k\to\infty} F_{n_k}(x) = F(x)$ at continuity points of F. Now apply the special Parseval relation on $\{F_{n_k}, \varphi_{n_k}\}$ to obtain

$$\int_{-\infty}^{+\infty} e^{-\frac{\sigma^2(x-v)^2}{2}} F_{n_k}(dx) = \frac{1}{\sqrt{2\pi\sigma^2}} \int_{-\infty}^{+\infty} e^{-ivy} e^{-\frac{y^2}{2\sigma^2}} \varphi_{n_k}(y)dy,$$

and letting $k \to \infty$ we have

$$\int_{-\infty}^{+\infty} e^{-\frac{\sigma^2(x-v)^2}{2}} F(dx) = \frac{1}{\sqrt{2\pi\sigma^2}} \int_{-\infty}^{+\infty} e^{-ivy} e^{-\frac{y^2}{2\sigma^2}} \varphi(y)dy.$$

Since for given φ the latter equation determines F uniquely, the limit F is the same for all convergent subsequences $\{F_{n_k}\}_{k=1}^{+\infty}$.

Now we may apply the special Parseval relation to $\{Q_n, \varphi_n\}$ so that

$$\int_{-\infty}^{+\infty} e^{-\frac{\sigma^2(x-v)^2}{2}} Q_n(dx) = \frac{1}{\sqrt{2\pi\sigma^2}} \int_{-\infty}^{+\infty} e^{-ivy} e^{-\frac{y^2}{2\sigma^2}} \varphi_n(y)dy,$$

and sending n to infinity

$$\int_{-\infty}^{+\infty} e^{-\frac{\sigma^2(x-v)^2}{2}} Q(dx) = \int_{-\infty}^{+\infty} e^{-ivy} \varphi(y) \frac{1}{\sqrt{2\pi\sigma^2}} e^{-\frac{y^2}{2\sigma^2}} dy,$$

we have that φ is the characteristic function of Q. Moreover, the RHS of the latter equation is the expectation of the bounded function $e^{-ivy}\varphi(y)$ with respect to a $N(0, \sigma^2)$ random variable. Sending σ to 0 this distribution is concentrated near the origin and so the right side tends to $\varphi(0)$ whenever φ is continuous at the origin and since $\varphi_n(0) = 1$ we have $\varphi(0) = 1$. As $\sigma \to 0$ the LHS tends to $Q(\overline{\mathcal{R}})$ so that $Q(\overline{\mathcal{R}}) = 1$ and Q is a valid probability measure. ■

Next we collect some consequences of the inversion and continuity theorems.

Remark 5.5 (Inversion and continuity for densities and cdfs) An immediate consequence of the continuity theorem is that φ is continuous everywhere and the convergence $\varphi_n \to \varphi$ is uniform in every finite interval. Based on the inversion theorem we can prove the following.

1. Continuity for densities Let φ_n and φ be integrable such that

$$\int_{-\infty}^{+\infty} |\varphi_n(t) - \varphi(t)|dt \to 0,$$

as $n \to +\infty$. By the inversion theorem the corresponding distributions Q_n and Q have densities f_n and f, respectively. Using the inversion formula we have that

$$|f_n(x) - f(x)| \leq (2\pi)^{-1} \int_{-\infty}^{+\infty} |\varphi_n(t) - \varphi(t)|dt,$$

and therefore $f_n \to f$ uniformly.

2. Inversion for cdfs Let F be the distribution function of the cf φ, where $|\varphi(t)|$ is integrable. Then for any $h > 0$

$$\frac{F(x+h) - F(x)}{h} = (2\pi)^{-1} \int_{-\infty}^{+\infty} \frac{e^{-itx} - e^{-it(x+h)}}{ith} \varphi(t)dt,$$

with $\frac{e^{-itx} - e^{-it(x+h)}}{ith} \to e^{itx}$ and $\frac{F(x+h)-F(x)}{h} \to F'(x) = f(x)$, as $h \to 0$, where f is the density of F.

5.5.3 Limiting Behavior of Partial Sums

Now we address the question of existence of a limit for the partial sum $S_n = \sum_{i=1}^{n} X_i$, as $n \to \infty$, where the $\{X_n\}_{n=1}^{+\infty}$ are independent, i.e., we investigate convergence of the random series $\sum_{n=1}^{+\infty} X_n$. Recall that from the SLLN (and the WLLN) we know that under mild conditions in the iid case, S_n/n converges a.s. (and in probability) to $E(X_1)$ (see section 2.3.4). Next we obtain necessary and sufficient conditions on the $\{X_n\}_{n=1}^{+\infty}$ and their distributions $\{Q_n\}_{n=1}^{+\infty}$ in order for S_n to converge. First we collect the following theorem, requested as an exercise.

Theorem 5.15 (Kolmogorov's maximal inequality) Assume that $X_1, X_2, \ldots, X_n$ are independent random variables with mean 0 and finite variances. Then for any $\varepsilon > 0$ we have

$$P(\max_{1\leq k\leq n} |S_k| \geq \varepsilon) \leq \frac{Var(S_n)}{\varepsilon^2}.$$

A classic convergence criterion is collected next.

Remark 5.6 (Cauchy criterion) When we need to assess convergence and we do not have a candidate for the limit we can use a stochastic version of the deterministic Cauchy criterion for convergence of real series. More precisely, the sequence $\{X_n\}_{n=1}^{+\infty}$ converges a.s. if and only if $\lim_{n,m\to+\infty} |X_n - X_m| = 0$ a.s. In order to use this criterion in practice we can show that if

$$\liminf_{n\to+\infty} \lim_{m\to+\infty} P(\sup_{1\leq k\leq m} |X_{n+k} - X_n| > \varepsilon) = 0, \tag{5.9}$$

for any $\varepsilon > 0$, then $\{X_n\}_{n=1}^{+\infty}$ converges a.s.

Using Kolmogorov's inequality we can prove the following.

Theorem 5.16 (a.s. convergence for random series) Assume that $X_1, X_2, \ldots, X_n$ are independent and have zero mean. If $\sum_{n=1}^{+\infty} Var(X_n)$ converges then $\sum_{n=1}^{+\infty} X_n$ converges a.s.

Proof. Using Kolmogorov's inequality on the sequence $\{X_{n+m}\}_{n=1}^{+\infty}$, for $m \geq 1$, we have

$$P(\max_{1\leq k\leq m} |S_{n+k} - S_n| > \varepsilon) \leq \frac{1}{\varepsilon^2} \sum_{k=n+1}^{n+m} Var(X_k),$$

for all $\varepsilon > 0$. Now assume that $\sum_{n=1}^{+\infty} Var(X_n) < \infty$. Then the RHS above goes to 0 as we let $m \to \infty$ first and then as we let $n \to \infty$. As a result, condition (5.9) of the Cauchy criterion is satisfied and therefore S_n converges a.s. ■

The last result we require before we prove our main result is the following (requested as an exercise).

Theorem 5.17 (Convergence of the series of variances) Suppose that $\{X_n\}_{n=1}^{+\infty}$ is a bounded sequence of independent random variables. If $\sum_{n=1}^{+\infty}(X_n - a_n)$ converges a.s. for some real sequence a_n then $\sum_{n=1}^{+\infty} Var(X_n)$ converges.

From example 4.4.1 and Kolmogorov's 0-1 law we know that S_n converges or diverges a.s. and the corresponding event has probability 0 or 1. The hard part here is to identify the conditions under which we get the probability to be either 0 or 1 and Kolmogorov's three-series theorem tells us exactly when this happens.

Theorem 5.18 (Kolmogorov three series) Assume that $\{X_n\}_{n=1}^{+\infty}$ is independent and consider the series

$$\sum_{n=1}^{+\infty} P(|X_n| > c), \quad \sum_{n=1}^{+\infty} E\left[X_n I(|X_n| \le c)\right], \text{ and } \sum_{n=1}^{+\infty} Var\left[X_n I(|X_n| \le c)\right]. \tag{5.10}$$

Then S_n converges a.s. if and only if all the three series converge, $\forall c > 0$.

Proof. ($\Longrightarrow$) Assume that S_n converges a.s. and fix a $c > 0$. Since $X_n \to 0$ a.s., we have that $\sum_{n=1}^{+\infty} X_n I(|X_n| \le c)$ converges a.s. and using the Borel-Cantelli lemma 4.1.3 we have that the first series converges. Theorem 5.17 implies that the third series converges, which in turn, using theorem 5.16 implies that $\sum_{n=1}^{+\infty}[X_n I(|X_n| \le c) - E(X_n I(|X_n| \le c))]$ converges a.s. As a result, the second series converges a.s. since $\sum_{n=1}^{+\infty} X_n I(|X_n| \le c)$ converges a.s.
($\Longleftarrow$) Now suppose that the series (5.10) converge. From theorem 5.16 we have that $\sum_{n=1}^{+\infty}[X_n I(|X_n| \le c) - E(X_n I(|X_n| \le c))]$ converges a.s. and since $\sum_{n=1}^{+\infty} E(X_n I(|X_n| \le c))$ converges so does $\sum_{n=1}^{+\infty} X_n I(|X_n| \le c)$. But $\sum_{n=1}^{+\infty} P(|X_n| > c) < \infty$ and the Borel-Cantelli lemma 4.1.1 imply that $P(X_n \ne X_n I(|X_n| \le c) \text{ i.o.}) = 0$ and therefore we must have that S_n converges a.s. ■

5.5.4 Central Limit Theorems

Now we turn to investigating the asymptotic distribution of transformations of partial sums, i.e., random variables of the form $a_n(S_n - b_n)$. The usual Central Limit Theorem (CLT, exercise 2.82) is quite restrictive since it requires independent and identically distributed $\{X_n\}_{n=1}^{+\infty}$ with the same mean $\mu = E(X_n)$ and finite variance $\sigma^2 = Var(X_n) < \infty$. In particular, we have $a_n(S_n - b_n) \xrightarrow{w} Z \sim \mathcal{N}(0, 1)$, with $a_n =$

$1/\sqrt{n\sigma^2}$ and $b_n = n\mu$. There are several versions of CLTs where these requirements are relaxed and we discuss the most well-known theorems, i.e., the Lindeberg and Lyapounov theorems, in this section. An example of a CLT for iid sequences with infinite variance can be found in Durrett (2010, p 131). First we discuss triangular arrays of random variables.

Remark 5.7 (Triangular arrays) Consider random variables $X_{n,m}$, $m = 1, 2, \ldots, k_n$, $n = 1, 2, \ldots$, and wlog assume that $E(X_{n,m}) = 0$ and let $\sigma_{n,m}^2 = E(X_{n,m}^2) < \infty$. For each n we assume that $X_{n,1}, \ldots, X_{n,k_n}$ are independent. A collection of random variables with these properties is known as a triangular array. Further define

$$S_n = X_{n,1} + \cdots + X_{n,k_n},$$

and let

$$\sigma_n^2 = \sigma_{n,1}^2 + \cdots + \sigma_{n,k_n}^2 = Var(S_n).$$

Note that there is no loss of generality if we assume $\sigma_n^2 = 1$ since we can replace $X_{n,m}$ by $X_{n,m}/\sigma_n$. The following conditions are required in the CLTs below.

1. Lindeberg condition For all $\varepsilon > 0$ we assume that

$$\lim_{n \to +\infty} \sum_{m=1}^{k_n} \frac{1}{\sigma_n^2} E\left[|X_{n,m}|^2 I(|X_{n,m}| \geq \varepsilon\sigma_n)\right] = 0. \tag{5.11}$$

2. Lyapounov's condition Suppose that $|X_{n,m}|^{2+\delta}$ are integrable, for some $\delta > 0$ and assume that

$$\lim_{n \to +\infty} \sum_{m=1}^{k_n} \frac{1}{\sigma_n^{2+\delta}} E\left[|X_{n,m}|^{2+\delta}\right] = 0. \tag{5.12}$$

Note that Lindeberg's condition is a special case of Lyapounov's since the sum in (5.11) is bounded from above by the sum in (5.12).

Now we relax the identically distributed condition of the usual CLT.

Theorem 5.19 (Lindeberg's CLT) Consider the setup of remark 5.7 and assume that condition (5.11) holds. Then $S_n/\sigma_n \xrightarrow{w} Z \sim \mathcal{N}(0, 1)$.

Proof. Without loss of generality assume that $\sigma_n = 1$. We will show that the characteristic function of S_n converges to the characteristic function of a $\mathcal{N}(0, 1)$. Let $\varphi_{n,m}(t)$ denote the characteristic function of $X_{n,m}$, so that owing to independence for a given n, the characteristic function of S_n is given by

$$\varphi_n(t) = \prod_{m=1}^{k_n} \varphi_{n,m}(t).$$

First using equation (4.35) we note that

$$\left|\varphi_{n,m}(t) - \left(1 - \frac{1}{2}t^2\sigma_{n,m}^2\right)\right| \leq E\left[\min\{|tX_{n,m}|^2, |tX_{n,m}|^3\}\right] < \infty,$$

and for $\varepsilon > 0$ the RHS can be bounded from above by

$$\int_{\{\omega:|X_{n,m}(\omega)|<\varepsilon\}} |tX_{n,m}|^3 dP + \int_{\{\omega:|X_{n,m}(\omega)|\geq\varepsilon\}} |tX_{n,m}|^2 dP \leq \varepsilon|t|^3\sigma_{n,m}^2 + t^2 \int_{\{\omega:|X_{n,m}(\omega)|\geq\varepsilon\}} X_{n,m}^2 dP.$$

Since $\sigma_n^2 = \sum\limits_{m=1}^{k_n} \sigma_{n,m}^2 = 1$, using Lindeberg's condition and the fact that $\varepsilon > 0$ is arbitrary we obtain

$$\begin{aligned}\sum_{m=1}^{k_n} \left|\varphi_{n,m}(t) - \left(1 - \frac{1}{2}t^2\sigma_{n,m}^2\right)\right| &\leq \sum_{m=1}^{k_n} \left[\varepsilon|t|^3\sigma_{n,m}^2 + t^2 E\left[|X_{n,m}|^2 I(|X_{n,m}| \geq \varepsilon)\right]\right] \\ &\leq \varepsilon|t|^3 + t^2 \sum_{m=1}^{k_n} E\left[|X_{n,m}|^2 I(|X_{n,m}| \geq \varepsilon)\right] \to 0,\end{aligned}$$

for each fixed t. It remains to show that

$$\varphi_n(t) = \prod_{m=1}^{k_n} \left(1 - \frac{1}{2}t^2\sigma_{n,m}^2\right) + o(1) \tag{5.13}$$

$$= \prod_{m=1}^{k_n} e^{-\frac{1}{2}t^2\sigma_{n,m}^2} + o(1) = \exp\left\{-\frac{1}{2}t^2 \sum_{m=1}^{k_n} \sigma_{n,m}^2\right\} + o(1), \tag{5.14}$$

or

$$\varphi_n(t) = e^{-\frac{1}{2}t^2} + o(1).$$

For any $\varepsilon > 0$

$$\sigma_{n,m}^2 = \int |X_{n,m}|^2 dP \leq \varepsilon^2 + \int_{\{\omega:|X_{n,m}(\omega)|\geq\varepsilon\}} |X_{n,m}|^2 dP,$$

so that the Lindeberg condition gives

$$\max_{1\leq m\leq k_n} \sigma_{n,k}^2 \to 0, \tag{5.15}$$

as $n \to \infty$. For large n, $|1 - \frac{1}{2}t^2\sigma_{n,m}^2| < 1$ so that using equation (4.34) we can write

$$\left|\prod_{m=1}^{k_n} \varphi_{n,m}(t) - \prod_{m=1}^{k_n} \left(1 - \frac{1}{2}t^2\sigma_{n,m}^2\right)\right| \leq \sum_{m=1}^{k_n} \left|\varphi_{n,m}(t) - \left(1 - \frac{1}{2}t^2\sigma_{n,m}^2\right)\right| = o(1),$$

which establishes the asymptotic equation (5.13). In addition, using equation (4.34) again we can write

$$\begin{aligned}\left|\prod_{m=1}^{k_n} e^{-\frac{1}{2}t^2\sigma_{n,m}^2} - \prod_{m=1}^{k_n} \left(1 - \frac{1}{2}t^2\sigma_{n,m}^2\right)\right| &\leq \sum_{m=1}^{k_n} \left|e^{-\frac{1}{2}t^2\sigma_{n,m}^2} - 1 - \left(-\frac{1}{2}t^2\sigma_{n,m}^2\right)\right| \\ &\leq t^4 e^{t^2} \sum_{m=1}^{k_n} \sigma_{n,m}^4,\end{aligned}$$

with the last inequality established since for any complex number z we have

$$|e^z - 1 - z| \leq |z|^2 \sum_{m=2}^{\infty} \frac{|z|^{m-2}}{k!} \leq |z|^2 e^{|z|}.$$

Finally, using (5.15) we establish the second asymptotic equality of (5.14). ■

In view of remark 5.7.2 we have the following as a result of theorem 5.19.

Theorem 5.20 (Lyapounov's CLT) Consider the setup of remark 5.7 and assume that condition (5.12) holds. Then $S_n/\sigma_n \overset{w}{\to} Z \sim \mathcal{N}(0,1)$.

We end this section by discussing how one can relax the independence assumption of the random sequence and still obtain a CLT.

Definition 5.4 ***m*-dependence and stationarity**

A sequence of random variables $X = \{X_n\}_{n=1}^{+\infty}$ is said to be m-dependent if, for every integer $k \geq 1$, the sets of random variables $\{X_1, \dots, X_k\}$ and $\{X_{m+k+1}, X_{m+k+2}, \dots\}$ are independent. The random sequence X is called stationary if for any positive integers k and l, the joint distribution of $(X_k, \dots, X_{k+l})$ does not depend on k.

Clearly, for $m = 0$, a 0-dependent sequence becomes independent. Stationarity of random sequences will be studied extensively in Chapter 6. The following result provides a CLT for m-dependent and stationary sequences. For a proof see Ferguson (1996, p. 70).

Theorem 5.21 (CLT under *m*-dependence and stationarity) Consider a sequence of random variables $X = \{X_n\}_{n=1}^{+\infty}$, and assume that X is m-dependent and stationary, with $Var(X_n) < +\infty$. Then

$$\sqrt{n}\,(S_n/n - \mu) \overset{w}{\to} Z \sim \mathcal{N}(0, \sigma^2),$$

where $\mu = E(X_n)$, and $\sigma^2 = \sigma_{00} + 2\sigma_{01} + \dots + 2\sigma_{0m}$, with $\sigma_{0i} = Cov(X_n, X_{n+i})$.

5.6 Summary

In this chapter we discussed an important aspect of classical probability theory, asymptotics. Although we typically never see the whole population, our models and the estimators of their parameters are based always on a sample of size n from the whole population. Therefore if our statistical procedure is to be useful, sending n to $+\infty$ should provide us with a good understanding of the true underlying model that realizes the data we observe, e.g., a sequence of probability measures we build based on a sample of size n should approach the true probability measure.

The theoretical results of this chapter also tell us that as n goes to $+\infty$ we have statistical procedures (Central Limit Theorems) that can assess when partial sums converge in distribution to a standard normal. Moreover, we have discussed how one may develop theory to show a compactness property of the general space of distribution functions. Recent texts on topics from asymptotics include Ferguson (1996), Fristedt and Gray (1997), Kallenberg (2002), Dudley (2002), Lehmann (2004), Durrett (2010), Çinlar (2010), Billingsley (2012) and Klenke (2014).

5.7 Exercises

Limiting behavior of iid random sequences

Exercise 5.1 Show that if $\mathbf{X}_n \stackrel{w}{\to} \mathbf{c}$, some constant, then $\mathbf{X}_n \stackrel{p}{\to} \mathbf{c}$.

Exercise 5.2 Prove theorem 5.2.

Exercise 5.3 Prove that $\mathbf{X}_n \stackrel{a.s.}{\to} \mathbf{X}$ if and only if for all $\varepsilon > 0$, $P(\|\mathbf{X}_k - \mathbf{X}\| < \varepsilon$, for all $k \geq n) \to 1$, as $n \to +\infty$.

Exercise 5.4 Show that $\mathbf{X}_n \stackrel{a.s.}{\to} \mathbf{X} \Rightarrow \mathbf{X}_n \stackrel{p}{\to} \mathbf{X} \Rightarrow \mathbf{X}_n \stackrel{w}{\to} \mathbf{X}$.

Exercise 5.5 Prove theorem 5.3.

Exercise 5.6 We say that $\mathbf{X}_n$ converges to $\mathbf{X}$ in the rth mean, $r > 0$ and write $\mathbf{X}_n \stackrel{r}{\to} \mathbf{X}$, if $E\|\mathbf{X}_n - \mathbf{X}\|^r \to 0$, as $n \to +\infty$. Show that if $\mathbf{X}_n \stackrel{r}{\to} \mathbf{X}$, for some $r > 0$, then $\mathbf{X}_n \stackrel{p}{\to} \mathbf{X}$.

Exercise 5.7 If $\mathbf{X}_n \stackrel{a.s.}{\to} \mathbf{X}$ and $|\mathbf{X}_n|^r \leq Z$ for some $r > 0$ and random variable Z with $EZ < \infty$, then $\mathbf{X}_n \stackrel{r}{\to} \mathbf{X}$.

Exercise 5.8 If $\mathbf{X}_n \stackrel{a.s.}{\to} \mathbf{X}$, $\mathbf{X}_n \geq \mathbf{0}$ and $E\mathbf{X}_n \to E\mathbf{X} < \infty$, then $\mathbf{X}_n \stackrel{r}{\to} \mathbf{X}$, for $r = 1$.

Exercise 5.9 Show that $\mathbf{X}_n \stackrel{p}{\to} \mathbf{X}$ if and only if for every subsequence $\{n_k\}$ there is a sub-subsequence $\{n_{k_j}\}$ such that $\mathbf{X}_{n_{k_j}} \stackrel{a.s.}{\to} \mathbf{X}$, as $j \to \infty$.

Exercise 5.10 Prove the Fréchet-Shohat theorem.

Exercise 5.11 Prove theorem 5.1.

Exercise 5.12 Let $\{X_n\}_{n=1}^{+\infty}$ be a sequence of iid random variables with $X_n \sim U(0,1)$, $n \geq 1$. Prove that for all $a > 1$, $\lim\limits_{n\to\infty} \dfrac{1}{n^a X_n} = 0\ a.s.$

Exercise 5.13 Prove statements 1-4, 7 and 8 of remark 5.3.

Exercise 5.14 Let $\{X_n\}_{n=1}^{+\infty}$ be a sequence of extended real-valued random variables defined on a probability space $(\Omega, \mathcal{A}, P)$. Assume that $X_n \stackrel{a.s.}{\to} X$ and $E|X_n| < \infty$ for all n. Show that $\{X_n\}_{n=1}^{+\infty}$ is uniformly integrable if and only if $\lim\limits_{n\to+\infty} E(|X_n|) = E(|X|) < \infty$.

Exercise 5.15 Prove the strong consistency of the MLE, i.e., theorem 5.4.

Exercise 5.16 (Asymptotic normality of the MLE) Let $\mathbf{X}_1, \ldots, \mathbf{X}_n$ be iid with density $f(\mathbf{x}|\boldsymbol{\theta})$ $[\mu]$, $\boldsymbol{\theta} = [\theta_1, \ldots, \theta_k] \in \Theta$, and $\boldsymbol{\theta}_0$ denote the true value of the parameter and define the log-likelihood by

$$l_n(\boldsymbol{\theta}) = \nabla_{\boldsymbol{\theta}} \ln \prod_{i=1}^{n} f(\mathbf{x}_i|\boldsymbol{\theta}) = \sum_{i=1}^{n} \nabla_{\boldsymbol{\theta}} \ln f(\mathbf{x}_i|\boldsymbol{\theta}), \tag{5.16}$$

where $\nabla_{\boldsymbol{\theta}}$ denotes the gradient with respect to $\boldsymbol{\theta}$. Assume that
(C1) Θ is an open subset of $\mathcal{R}^k$,
(C2) all second partial derivatives $\frac{\partial^2}{\partial\theta_i\partial\theta_j} f(\mathbf{x}|\boldsymbol{\theta})$, $i, j = 1, 2, \ldots, k$, exist and are continuous in $\mathbf{x}$, and may be passed under the integral sign in $\int f(\mathbf{x}|\boldsymbol{\theta}) d\mu(\mathbf{x})$,
(C3) there exists a function $K(\mathbf{x})$ such that $E_{\boldsymbol{\theta}_0}[K(\mathbf{x})] < \infty$ and each component of

the Hessian $\nabla^2 \ln f(\mathbf{x}|\boldsymbol{\theta}) = \left[\left(\frac{\partial^2}{\partial\theta_i\partial\theta_j} \ln f(\mathbf{x}|\boldsymbol{\theta})\right)\right]$ is bounded in absolute value by $K(\mathbf{x})$ uniformly in some area of $\boldsymbol{\theta}_0$,
(C4) the Fisher information matrix $I_{\mathbf{x}}^F(\boldsymbol{\theta}_0) = -E\left[\nabla^2 \ln f(\mathbf{x}|\boldsymbol{\theta}_0)\right]$ is positive definite, and
(C5) $f(\mathbf{x}|\boldsymbol{\theta}) = f(\mathbf{x}|\boldsymbol{\theta}_0)$ a.e. $[\mu] \Rightarrow \boldsymbol{\theta} = \boldsymbol{\theta}_0$.
Then there exists a strong consistent sequence $\widehat{\boldsymbol{\theta}}_n$ of roots of the likelihood equation (5.16) such that

$$\sqrt{n}\left(\widehat{\boldsymbol{\theta}}_n - \boldsymbol{\theta}_0\right) \stackrel{w}{\to} \mathcal{N}_k\left(\mathbf{0}, \left[I_{\mathbf{x}}^F(\boldsymbol{\theta}_0)\right]^{-1}\right).$$

Exercise 5.17 Let $X_1, \dots, X_n$ a random sample from a distribution with cdf $F(x)$. Define the Empirical distribution function or Empirical cdf, as $F_n(x, \omega) = \frac{1}{n}\sum\limits_{i=1}^{n} I_{(-\infty,x]}(X_i(\omega))$.
(i) Show that $F_n(x, .) \stackrel{a.s.}{\to} F(x)$, for a fixed x.
(ii) (Glivenko-Cantelli) Let $D_n(\omega) = \sup\limits_{x\in\mathcal{R}}|F_n(x, \omega) - F(x)|$. Then F_n converges to F uniformly a.s., i.e., $D_n \stackrel{a.s.}{\to} 0$.

Exercise 5.18 Prove theorem 5.21.

Exercise 5.19 (Bernstein-von Mises theorem) Let $\mathbf{X}_1, \dots, \mathbf{X}_n$ be an iid sample from $f(\mathbf{x}|\boldsymbol{\theta})$, and consider a continuous prior $\pi(\boldsymbol{\theta}) > 0$, for all $\boldsymbol{\theta} \in \Theta \subset \mathcal{R}^k$. Let $\mathbf{x} = (\mathbf{x}_1, \dots, \mathbf{x}_n)$ denote the data, $L_n(\boldsymbol{\theta}|\mathbf{x})$ the likelihood function, and $\widehat{\boldsymbol{\theta}}_n$ the MLE under the assumptions of Cramér's theorem. Furthermore, denote the posterior distribution by $\pi_n(\boldsymbol{\theta}|\mathbf{x}) = \frac{L_n(\boldsymbol{\theta}|\mathbf{x})\pi(\boldsymbol{\theta})}{m_n(\mathbf{x})}$, let $\boldsymbol{\vartheta} = \sqrt{n}(\boldsymbol{\theta} - \widehat{\boldsymbol{\theta}}_n)$, and denote the density of a $\mathcal{N}_k\left(\mathbf{0}, \left[I_{\mathbf{x}}^F(\boldsymbol{\theta})\right]^{-1}\right)$ by $\phi(\boldsymbol{\theta})$. If

$$\int\limits_{\Theta} \frac{L_n(\widehat{\boldsymbol{\theta}}_n + \boldsymbol{\vartheta}/\sqrt{n}|\mathbf{x})}{L_n(\widehat{\boldsymbol{\theta}}_n|\mathbf{x})}\pi(\widehat{\boldsymbol{\theta}}_n + \boldsymbol{\vartheta}/\sqrt{n})d\boldsymbol{\vartheta} \stackrel{a.s.}{\to} \int\limits_{\Theta} \exp\left\{-\frac{1}{2}\boldsymbol{\vartheta}^T I_{\mathbf{x}}^F(\boldsymbol{\theta}_0)\boldsymbol{\vartheta}\right\} d\boldsymbol{\vartheta}\pi(\boldsymbol{\theta}_0),$$

show that

$$\int\limits_{\Theta} |\pi_n(\boldsymbol{\vartheta}|\mathbf{x}) - \phi(\boldsymbol{\vartheta})|\, d\boldsymbol{\vartheta} \stackrel{a.s.}{\to} 0,$$

that is, the posterior distribution of $\sqrt{n}(\boldsymbol{\theta} - \widehat{\boldsymbol{\theta}}_n)|\mathbf{x}$ approaches a normal density $\mathcal{N}_k\left(\mathbf{0}, \left[I_{\mathbf{x}}^F(\boldsymbol{\theta})\right]^{-1}\right)$, as $n \to \infty$, and this limiting distribution does not depend on the selection of the prior $\pi(\boldsymbol{\theta})$.

Exercise 5.20 Show that the Bayes $\widetilde{\boldsymbol{\theta}}_n$ rule under quadratic loss is asymptotically normal, i.e., show that

$$\sqrt{n}\left(\widetilde{\boldsymbol{\theta}}_n - \boldsymbol{\theta}_0\right) \stackrel{w}{\to} \mathcal{N}_k\left(\mathbf{0}, \left[I_{\mathbf{x}}^F(\boldsymbol{\theta}_0)\right]^{-1}\right), \tag{5.17}$$

where $\boldsymbol{\theta}_0 \in \Theta \subset \mathcal{R}^k$, the true value of the parameter. Estimators that satisfy (5.17) for any $\boldsymbol{\theta}_0 \in \Theta$, are called asymptotically efficient.

Convolutions and characteristic functions

Exercise 5.21 Let $X \sim Binom(n, p)$. First show that $\varphi_X(t) = (q + pe^{it})^n$ and then use the inversion theorem to show that $f(x) = C_x^n p^x q^{n-x}$, $x = 0, 1, \ldots, n$, $q = 1 - p$, $0 \leq p \leq 1$.

Exercise 5.22 Prove equation (5.8).

Exercise 5.23 Show that if Q, R, Q_n and R_n, $n = 1, 2, \ldots$, are probability measures on $\mathcal{R}$ with $Q_n \overset{w}{\to} Q$ and $R_n \overset{w}{\to} R$ then $Q_n * R_n \overset{w}{\to} Q * R$.

Exercise 5.24 Assume that $X \sim Poisson(\lambda t)$ and $Y \sim Poisson(\mu t)$, $\lambda, \mu, t > 0$. Find the characteristic function of $X + Y$ and identify its distribution.

Exercise 5.25 Consider a $\mathcal{N}(0, 1)$ random variable X and let Q be its distribution. Find Q^{*2}.

Random Series

Exercise 5.26 Prove theorem 5.15.

Exercise 5.27 **(SLLN)** Suppose that $\{X_n\}_{n=1}^{+\infty}$ is independent and $\sum\limits_{n=1}^{+\infty} Var(X_n/b_n)$ converges for some strictly positive, increasing, real sequence $\{b_n\}$ with $b_n \to \infty$, as $n \to \infty$. Then $(S_n - E(S_n))/b_n \to 0$, a.s., where $S_n = \sum\limits_{i=1}^{n} X_i$. Further show that the result holds for convergence in $\mathcal{L}^2$ (recall remark 3.25.1).

Exercise 5.28 Prove the consequence of the Cauchy criterion, equation (5.9) of remark 5.6.

Exercise 5.29 Suppose that $\{X_n\}_{n=1}^{+\infty}$ is a bounded sequence of independent random variables. If $\sum\limits_{n=1}^{+\infty} (X_n - a_n)$ converges a.s. for some real sequence a_n then $\sum\limits_{n=1}^{+\infty} Var(X_n)$ converges.

Simulation and computation technique

Exercise 5.30 **(Importance sampling)** Consider a random variable X with density f_X and support $\mathcal{X}$. Suppose that we want to estimate the expectation

$$E(g(X)) = \int\limits_{\mathcal{X}} g(x) f_X(x) dx,$$

and it is difficult to sample from the density f_X (so that Monte Carlo integration is not an option) or $g(X)$ has very large variance. Instead of sampling from the random variable $X \sim f_X$, consider a random variable $Y \sim f_Y$, with the same support $\mathcal{X}$ as the random variable X and assume that we can easily generate values from the density f_Y, say, $Y_1, \ldots, Y_n \overset{iid}{\sim} f_Y$. The density f_Y is known as the importance density. Define

$$I_n = \frac{1}{n} \sum_{i=1}^{n} g(Y_i) \frac{f_X(Y_i)}{f_Y(Y_i)},$$

and show that

(i) $E^X(g(X)) = E^Y(I_n) = E^Y\left[g(Y)\frac{f_X(Y)}{f_Y(Y)}\right]$,

(ii) $Var(I_n) = \frac{1}{n}\left[E^Y\left(g(Y)\frac{f_X(Y)}{f_Y(Y)}\right)^2 - \left[E^X(g(X))\right]^2\right]$, and

(iii) $I_n \overset{a.s.}{\to} E^X(g(X))$ as $n \to \infty$.

Chapter 6

Random Sequences

6.1 Introduction

Every object around us evolves over time and the underlying mechanics of the changes observed on the object are known as an evolution process. Therefore, in order to capture the evolution or growth process of an object we need to define and investigate a collection of random objects, say $X = (X_t : t \in T)$, where T is called the time parameter set. Time t can be treated as discrete or continuous and the values of X_t can be discrete or continuous, which leads to four possible cases.

Although we will see examples for all cases, we present our theoretical development depending on the nature of T. In particular, if T is countable then the collection X is called a discrete parameter stochastic process (or a random sequence), otherwise it is called a continuous time parameter stochastic process (or simply a stochastic process). The former is the subject of this chapter and the latter case is treated in the next chapter. For what follows, we write $n \geq 0$ for $n \in \{0, 1, 2, \ldots\}$ and $n > 0$ for $n \in \{1, 2, \ldots\}$.

6.2 Definitions and Properties

We begin with the general definition of a random sequence.

Definition 6.1 Random sequence

A random sequence with state space Ψ is a collection of random objects $X = \{X_n : n \geq 0\}$, with X_n taking values in Ψ and defined on the same probability space $(\Omega, \mathcal{A}, P)$.

We introduce several random sequences in the following.

Example 6.1 (Random sequences) We discuss some specific examples of random sequences next.

1. Bernoulli process Suppose that the sample space is of the form $\Omega = \{(\omega_1, \omega_2, \ldots) : \omega_i \text{ is a Success or Failure}\}$ (recall the infinite coin-flip space) and let $p = P(S)$, $q = 1 - p = P(F)$, $0 \leq p, q \leq 1$. For each $\omega \in \Omega$ and $n \geq 1$, define $X_n(\omega) = 0$ or 1, if $\omega_n = F$ or $\omega_n = S$, so that $P(X_n = 1) = p$ and $P(X_n = 0) = q$. Assuming that the X_n are independent we can think of the X_n as the number of successes (0 or 1) at the nth trial. Then the collection $X = \{X_n : n \geq 1\}$ is a discrete time-parameter stochastic process known as the Bernoulli process with success probability p and state space $\Psi = \{0, 1\}$.

2. Binomial process Assume that $X = \{X_n : n \geq 1\}$ is a Bernoulli process with probability of success p. Define $N_n(\omega) = 0$, if $n = 0$ and $N_n(\omega) = X_1(\omega) + \cdots + X_n(\omega)$, $n \geq 1$, for all $\omega \in \Omega$. Then N_n is the number of successes in the first n-trials and the random sequence $N = \{N_n : n \geq 0\}$, with $N_0 = 0$, is known as the Binomial process. It is easy to see that for all $n, k \geq 0$, we have

$$N_{n+m} - N_m | N_0, N_1, N_2, \ldots, N_m \stackrel{d}{=} N_{n+m} - N_m \stackrel{d}{=} N_n \sim Binom(n, p).$$

3. Times of successes Let X be a Bernoulli process with probability of success p. For fixed ω, a given realization of X is $X_1(\omega), X_2(\omega), \ldots$, a 0-1 sequence. Denote by $T_1(\omega), T_2(\omega), \ldots$, the indices corresponding to 1s. Then T_k denotes the trial number where the kth success occurs. Clearly, $\Psi = \{1, 2, \ldots\}$ and $T = \{T_k : k \geq 1\}$ is a discrete time-parameter stochastic process. It is possible that $T_1(\omega) = +\infty$ (no success occurs) in which case $T_1(\omega) = T_2(\omega) = \cdots = +\infty$. There is a fundamental relationship between the process T and the number of successes N_n. In particular, we can show that $\{\omega : T_k(\omega) \leq n\} = \{\omega : N_n(\omega) \geq k\}$. Now if $T_k(\omega) = n$, then there are exactly $(k-1)$-successes in the first $n-1$ trials and a success at the nth trial. The converse is also true. As a result, for all $k, m \geq 1$ we can show that

$$T_{k+m} - T_m | T_1, T_2, \ldots, T_m \stackrel{d}{=} T_{k+m} - T_m \stackrel{d}{=} T_k \sim NB(k, p),$$

a negative binomial random variable.

4. Reliability analysis A component in a large system has a lifetime whose distribution is geometric with $p(m) = pq^{m-1}$, $m = 1, 2, \ldots$ and suppose that when the component fails it is replaced by an identical one. Let $T_1, T_2, \ldots$, denote the times of failure. Then $U_k = T_k - T_{k-1}$ denotes the lifetime of the kth item replaced with $U_k \stackrel{iid}{\sim} Geo(p)$, for all k. Consequently, we can think of $T = \{T_k : k \geq 1\}$ as the times of "successes" in a Bernoulli process. Now suppose that past records indicate that the first 3 failures occurred at times 3, 12 and 14 and we wish to estimate the time of the 5th failure, namely, T_5. We have

$$\begin{aligned} E(T_5 | T_1, T_2, T_3) &= E(T_5 | T_3) = E(T_3 + (T_5 - T_3) | T_3) \\ &= T_3 + E(T_5 - T_3 | T_3) = T_3 + E(T_5 - T_3) \\ &= T_3 + E(T_2) = T_3 + 2/p, \end{aligned}$$

and therefore $E(T_5 | T_1 = 3, T_2 = 12, T_3 = 14) = 14 + 2/p$.

The following properties are satisfied by many random sequences including the ones of the previous example.

Definition 6.2 Stationarity and independent increments

A random sequence $X = \{X_n : n \geq 0\}$ is said to have stationary increments if the distribution of the increment $X_{n+k} - X_n$ does not depend on n. The random sequence X is said to have independent increments if for any $k \geq 1$ and $n_0 = 0 < n_1 < n_2 < \cdots < n_k$ the increments $X_{n_1} - X_{n_0}, X_{n_2} - X_{n_1}, \ldots, X_{n_k} - X_{n_{k-1}}$ are independent.

We discuss some classic examples in order to appreciate the latter definition.

Example 6.2 (Random walks) A random sequence defined through sums of iid random variables is a random walk. In particular, let $X_n \sim Q$, $n > 0$, defined on $(\Omega, \mathcal{A}, P)$ and taking values in Ψ for some distribution Q. Define $S_n(\omega) = 0$, if $n = 0$ and $S_n(\omega) = X_1(\omega) + \cdots + X_n(\omega)$, $n > 0$, for all $\omega \in \Omega$. Then $S = \{S_n : n \geq 0\}$ is a discrete time parameter stochastic process with state space Ψ.

1. First note that $S_{n+m} - S_n = X_{n+1} + \cdots + X_{n+m} \stackrel{d}{=} X_1 + \cdots + X_m = S_m$ since the X_n are iid and hence S has stationary increments. Moreover, any increments are sums of iid random variables so that S has independent increments and therefore

$$S_{n+m} - S_n | S_1, \ldots, S_n \stackrel{d}{=} S_{n+m} - S_n \stackrel{d}{=} S_m.$$

2. Simple symmetric random walks A simple random walk S in $\mathcal{Z}^p$ is a random walk with state space $\mathcal{Z}^p$ or $\mathcal{R}^p$ whose step distribution Q is supported by the set $\mathcal{S} = \{\pm e_i, i = 1, 2, \ldots, p\}$, where e_i, $i = 1, 2, \ldots, p$, the standard basis vectors in $\mathcal{R}^p$ and $Q(\mathcal{S}) = 1$, $Q(C) < 1$, for $C \subset \mathcal{S}$. It is also called a "nearest neighbor" random walk because successive states $S_n(\omega)$ and $S_{n+1}(\omega)$ are always a unit apart. For $p = 1$, $e_1 = \pm 1$ and we have a simple random walk on $\mathcal{Z}$ with steps of size 1 in either the positive or negative direction. If $Q(\{-1\}) = Q(\{1\}) = 1/2$ the random walk is called symmetric.

Example 6.3 (Returns to 0) Consider a simple random walk S on $\mathcal{Z}$, with $S_0 = 0$ and step distribution Q with $p = Q(\{1\}) = 1 - Q(\{-1\})$, $0 < p < 1$. Denote the sequence of steps by $X = (X_1, X_2, \ldots)$ and assume that the simple random walk S is adapted to a filtration $\mathcal{F} = (\mathcal{F}_1, \mathcal{F}_2, \ldots)$. We are interested in the first return to time $\{0\}$, i.e., in the random variable $T_1(\omega) = \inf\{n > 0 : S_n(\omega) = 0\}$, where $\inf \emptyset = +\infty$. Note that $\{\omega : T_1(\omega) \leq n\} = \bigcup_{m=1}^{n} \{\omega : S_n(\omega) = 0\} \in \mathcal{F}_n$, since $\{\omega : S_n(\omega) = 0\} \in \mathcal{F}_n$ and therefore T_1 is a stopping time. Similarly, for $j > 1$ we define the jth return time to 0 of S recursively as

$$T_j(\omega) = \inf\left\{n > T_{j-1}(\omega) : S_n(\omega) = 0\right\}.$$

It is straightforward to see that T_j is a stopping time.

The following results give some insight on the distribution of T_j for random walks. The proofs are requested as exercises.

Theorem 6.1 (Return times) Let T_j denote the jth return time of some random walk to 0, $j = 1, 2, \ldots$ and let W denote the distribution of T_1. Then the distribution of the sequence $(0, T_1, T_2, \ldots)$ is that of a random walk in $\overline{\mathcal{Z}}^+$ with step distribution W.

Corollary 6.1 (Number of visits) Let Q denote the step distribution of the random walk S and define the random variable

$$V(\omega) = \inf\left\{j : T_j(\omega) = \infty\right\} = \#\left\{n \geq 0 : S_n(\omega) = 0\right\},$$

denoting the number of visits of the random walk to state 0. Then we can show that if $\sum_{n=0}^{+\infty} Q^{*n}(\{0\}) = +\infty$, then $V = +\infty$ and if $\sum_{n=0}^{+\infty} Q^{*n}(\{0\}) < +\infty$, the distribution of V is geometric with

$$P(\{\omega : V(\omega) = 1\}) = P(\{\omega : T_1(\omega) = +\infty\}) = \left(\sum_{n=0}^{+\infty} Q^{*n}(\{0\})\right)^{-1}.$$

A first classification of random walks is given next.

Definition 6.3 Random walk classification

For a simple random walk S in $\mathcal{Z}$ define $A_n = \{\omega : S_n(\omega) = 0\}$. Then S is called recurrent if $P(\limsup A_n) = 1$ and transient otherwise.

Since $\omega \in \limsup A_n \Leftrightarrow \omega \in A_n$ i.o., a recurrent random walk visits state 0 again and again while a transient random walk will visit 0 a last time (even just once since $S_0 = 0$) never to return. The corollary can then be rephrased as: $P(\limsup A_n) = 1$, if $\sum_{n=0}^{+\infty} P(A_n) = +\infty$, or 0, if $\sum_{n=0}^{+\infty} P(A_n) < +\infty$ (recall the Borel-Cantelli lemmas 4.1).

6.3 Martingales

Martingales play an important role in mathematics and statistics, and can be used as a tool to define more complicated random objects, including stochastic integrals. They owe much of their development to gambling modeling and we will explore some of these applications later in this section. We begin with the definition of a filtration and an adapted sequence of random variables to the filtration.

Definition 6.4 Filtration and adapted sequence

Let $X = (X_1, X_2, \ldots)$ be a sequence of $(\mathcal{R}, \mathcal{B}_1)$-valued random variables defined on a probability space $(\Omega, \mathcal{A}, P)$ and let $\mathcal{F} = (\mathcal{F}_1, \mathcal{F}_2, \ldots)$ be a sequence of sub-σ-fields of $\mathcal{A}$. When the collection of sub-σ-fields $\mathcal{F}$ satisfies $\mathcal{F}_n \subseteq \mathcal{F}_{n+1}$, we say that $\mathcal{F}$ is a filtration or that the $\mathcal{F}_n$ form a filtration in $\mathcal{A}$. If X_n is $\mathcal{F}_n$-measurable, for all n, then we say that the sequence X is adapted to the filtration $\mathcal{F}$.

The formal definition of a martingale is given next.

Definition 6.5 Martingale (MG)

Let $X = (X_1, X_2, \ldots)$ be a sequence of $(\mathcal{R}, \mathcal{B}_1)$-valued random variables defined on a probability space $(\Omega, \mathcal{A}, P)$ and let $\mathcal{F} = (\mathcal{F}_1, \mathcal{F}_2, \ldots)$ be a sequence of sub-σ-fields of $\mathcal{A}$. The sequence $\{(X_n, \mathcal{F}_n) : n = 1, 2, \ldots\}$ is a martingale if the following conditions hold:
(i) $\mathcal{F}_n \subseteq \mathcal{F}_{n+1}$,
(ii) X_n is $\mathcal{F}_n$-measurable,
(iii) $E|X_n| < +\infty$, and
(iv)

$$E(X_{n+1}|\mathcal{F}_n) = X_n \ a.s. \tag{6.1}$$

We write $(X, \mathcal{F})$ or $\{X_n, \mathcal{F}_n\}$ to abbreviate the notation for a MG and we say that X is a MG relative to the filtration $\mathcal{F}$.

For what follows in this section we assume the setup of this definition.

Remark 6.1 (Martingale properties and definitions) Condition (iii) is simply integrability for the random variable X_n, whereas condition (iv) expresses that given the information about the experiment up to time n, which is contained in $\mathcal{F}_n$, we do not expect the average value of the random variable to change at the next time period $n + 1$. We collect some essential definitions and results on MGs below.

1. Sub-MG If we replace condition (iv) with $E(X_{n+1}|\mathcal{F}_n) \geq X_n \ a.s.$, then $(X, \mathcal{F})$ is called a sub-martingale (sMG). Using equation (4.31) we can show that

$$E((X_{n+1} - X_n)I_F) \geq 0, \ \forall F \in \mathcal{F}_n.$$

2. Super-MG If condition (iv) becomes $E(X_{n+1}|\mathcal{F}_n) \leq X_n \ a.s.$, then $(X, \mathcal{F})$ is called a super-martingale (SMG). Similarly as above, $(X, \mathcal{F})$ is a SMG if and only if $E((X_{n+1} - X_n)I_F) \leq 0, \ \forall F \in \mathcal{F}_n$. There is nothing super about a SMG. In fact, in terms of gambling, SMG leads to unfavorable games for the player. See the gambling examples that follow in this section.

3. Minimal filtration There is always at least one filtration with respect to which X_n is adapted. Let $\mathcal{F}^X = (\mathcal{F}_1^X, \mathcal{F}_2^X, \ldots)$, where $\mathcal{F}_n^X = \sigma(X_1, X_2, \ldots, X_n) \subseteq \mathcal{A}$. Clearly $\mathcal{F}_n^X \subseteq \mathcal{F}_{n+1}^X$, for all $n = 1, 2, \ldots$, so that $\mathcal{F}^X$ is a filtration and moreover X is adapted to $\mathcal{F}^X$. The filtration $\mathcal{F}^X$ is called the minimal filtration of X. Suppose that $(X, \mathcal{F})$ is a MG. Then

$$E(X_{n+1}|\mathcal{F}_n^X) = E(E(X_{n+1}|\mathcal{F}_n)|\mathcal{F}_n^X) = E(X_n|\mathcal{F}_n^X) = X_n \ a.s.,$$

and thus $(X, \mathcal{F}^X)$ is also a MG, where $\mathcal{F}_n^X \subseteq \mathcal{F}_n$, since $\mathcal{F}_n^X$ is the smallest σ-field from $\mathcal{F}^X$ or $\mathcal{F}$ with respect to which X_n is measurable. Therefore $\mathcal{F}^X$ is the smallest filtration with respect to which X is a MG. When we say that X is a MG without indicating a filtration we simply mean relative to $\mathcal{F}^X$. Finally, note that if one augments a filtration with additional information, an adapted sequence remains adapted.

4. $\{X_n, \mathcal{F}_n\}$ is a SMG if and only if $\{-X_n, \mathcal{F}_n\}$ is a sMG. This result allows us to prove claims for a sMG only and then immediately obtain the claim for a SMG.

5. $(X, \mathcal{F})$ is a MG if and only if it is both a sMG and a SMG.

6. If $(X, \mathcal{F})$ is a MG then $E(X_{n+k}|\mathcal{F}_n) = X_n$, for all $n, k = 1, 2, \ldots$. To see this write

$$E(X_{n+2}|\mathcal{F}_n) = E(E(X_{n+2}|\mathcal{F}_{n+1})|\mathcal{F}_n) = E(X_{n+1}|\mathcal{F}_n) = X_n \; a.s.,$$

and use induction.

7. If X is a MG then $E(X_{n+1}|X_1, \ldots, X_n) = X_n \; a.s.$

8. Condition (iv) may be rewritten in many forms. Starting with the integral form of the definition of conditional probability $E(X_{n+1}|\mathcal{F}_n) = X_n \; a.s.$ if and only if

$$\int_F X_{n+1} dP = \int_F X_n dP,$$

$\forall F \in \mathcal{F}_n$. Letting $F = \Omega$, we obtain $E(X_1) = E(X_2) = \ldots$. Now letting $\Delta_1 = X_1$, $\Delta_n = X_n - X_{n-1}$, $n = 2, 3, \ldots$, linearity allows condition (iv) to be written equivalently as $E(\Delta_{n+1}|\mathcal{F}_n) = 0 \; a.s.$ Note that since $X_n = \Delta_1 + \Delta_2 + \cdots + \Delta_n$, the random vectors $(X_1, \ldots, X_n)$ and $(\Delta_1, \ldots, \Delta_n)$ generate the same σ-field: $\sigma(X_1, \ldots, X_n) = \sigma(\Delta_1, \ldots, \Delta_n)$.

9. If $\{X_n, \mathcal{F}_n\}$ and $\{Y_n, \mathcal{F}_n\}$ are MGs then so are $\{X_n \pm Y_n, \mathcal{F}_n\}$ and $\{\max(X_n, Y_n), \mathcal{F}_n\}$.

10. If X is an iid sequence of integrable random variables with $E(X_n) = 0$, for all $n = 1, 2, \ldots$, then $S_n = \sum_{i=1}^{n} X_i$ is a MG. To see this, first note that $\sigma(X_1, \ldots, X_n) = \sigma(S_1, \ldots, S_n)$. Therefore we can write

$$E(S_{n+1}|\sigma(S_1, \ldots, S_n)) = E(S_{n+1}|\sigma(X_1, \ldots, X_n)) = E(S_n + X_{n+1}|X_1, \ldots, X_n) = S_n,$$

$a.s.$, since $E(X_{n+1}|\sigma(X_1, \ldots, X_n)) = 0$.

11. If Y is an integrable random variable defined on $(\Omega, \mathcal{A}, P)$ and $\mathcal{F}$ is any filtration in $\mathcal{A}$ then the sequence X with $X_n = E(Y|\mathcal{F}_n)$ is a MG relative to $\mathcal{F}$.

12. Let g be a convex, increasing, $\mathcal{R}$-valued function and $(X, \mathcal{F})$ a sMG such that $g(X_n)$ is integrable. Then $\{g(X_n), \mathcal{F}_n\}$ is a sMG.

13. Let $(X, \mathcal{F})$ be a MG and g be a convex, $\mathcal{R}$-valued function, such that $g(X_n)$ is integrable. Then $\{g(X_n), \mathcal{F}_n\}$ is a sMG.

14. Etemadi inequality Let X be a sMG. Then for $c > 0 : P\left(\max\limits_{0 \le k \le n} X_k > c\right) \le E(|X_n|)/c$.

15. Kolmogorov inequality Let X be a MG. Then for all $c > 0$ and $p \ge 1$: $P\left(\max\limits_{0 \le k \le n} |X_k| > c\right) \le E(|X_n|^p)/c^p$.

Example 6.4 (Martingale transformations) If X is a MG then so is $\{X_n^+\}$. If $|X_n|^n$ is integrable then $\{|X_n|^n\}$ is a sMG.

Example 6.5 (Random walks and martingales) Recall example 4.12.

1. In view of remark 6.1.10, if the steps have zero mean, i.e., $E(X_n) = 0$, then the random walk S is a MG.

2. Let S be an $\mathcal{R}$-valued random walk, starting at 0, with steps from the random sequence $X = (X_1, X_2, \dots)$ having finite mean. Define $\mathcal{H}_n = \sigma(S_n/n, S_{n+1}/(n+1), \dots)$ and note that $\mathcal{H}_n = \sigma(S_n, X_{n+1}, X_{n+2}, \dots)$. Then

$$E\left(\frac{1}{n}S_n|\mathcal{H}_{n+1}\right) = \frac{1}{n}\sum_{i=1}^{n} E(X_i|\mathcal{H}_{n+1}) = \frac{1}{n}\sum_{i=1}^{n} S_{n+1}/(n+1) = S_{n+1}/(n+1)$$

$a.s.$, since $E(X_i|X_1 + \dots + X_n) = (X_1 + \dots + X_n)/n, i = 1, 2, \dots, n$. Now let $Z_n = S_n/n$ so that $E(Z_n|\mathcal{H}_{n+1}) = Z_{n+1}$ $a.s.$, leads to a reverse version of the definition of a MG.

3. Let $S = (S_0, S_1, S_2, \dots)$ be a random walk with steps $\{X_n\}$ that are independent with $E(X_n) = 0$ and $\sigma_n^2 = E(X_n^2) < +\infty$. Then S is a MG from remark 6.1.10 and remark 6.1.13 implies that $Y_n = S_n^2$, is a sMG.

Random sequences may work backward in time as we saw in the latter example.

Definition 6.6 Reverse martingale

A random sequence $Z = (Z_1, Z_2, \dots)$ is called a reverse MG relative to a reverse filtration $\mathcal{F} = (\mathcal{F}_1, \mathcal{F}_2, \dots)$ $(\mathcal{F}_{n+1} \subseteq \mathcal{F}_n)$ if for all $n = 1, 2, \dots$, $E(Z_n|\mathcal{F}_{n+1}) = Z_n$ $a.s.$, $E(Z_n) < +\infty$ and Z is adapted to $\mathcal{F}$. It is called a reverse SMG if $E(Z_n|\mathcal{F}_{n+1}) \leq Z_n$ $a.s.$ and a reverse sMG if $E(Z_n|\mathcal{F}_{n+1}) \geq Z_n$ $a.s.$

Example 6.6 (Gambling) Let $X_1, X_2, \dots$, be iid random variables on a probability space $(\Omega, \mathcal{A}, P)$ such that $P(X_n = 1) = 1 - P(X_n = -1) = p$, that is, we have an infinite coin flip space with the assignment Heads$\leftrightarrow$ 1 and Tails$\leftrightarrow$ -1. Assume that at the nth toss of the coin we bet \$$b$ and we win \$$b$ if and only if $X_n = 1$ or lose \$$b$ if and only if $X_n = -1$. A gambling strategy is a sequence of functions $b_n : \{-1, 1\}^n \to [0, +\infty)$, such that $b_n(X_1, \dots, X_n)$ denotes the amount we bet in the nth stage where $\{-1, 1\}^n$ is the space of all sequences of -1 and 1 of length n. Let S_0 be our initial fortune and denote by S_n the fortune after the nth stage. Then $S_{n+1} = S_n + X_{n+1}b_n(X_1, \dots, X_n)$ is the fortune after stage $n + 1$. In our example $b_n(X_1, \dots, X_n) = b$, the amount of dollars we bet at each stage.

Now the game is fair if $p = 1/2$, favorable if $p > 1/2$ and unfavorable $p < 1/2$. Since $\sigma(S_1, \dots, S_n) = \sigma(X_1, \dots, X_n)$ we may write

$$E(S_{n+1}|S_1, \dots, S_n) = E(S_{n+1}|X_1, \dots, X_n) = E(S_n + X_{n+1}b_n(X_1, \dots, X_n)|X_1, \dots, X_n),$$

where b_n and S_n are $\sigma(X_1, \dots, X_n)$-measurable and X_{n+1} is independent of

$X_1, \ldots, X_n$ so that

$$\begin{aligned} E(S_{n+1}|S_1, \ldots, S_n) &= S_n + E(X_{n+1})b_n(X_1, \ldots, X_n) \\ &= S_n + (2p-1)b_n(X_1, \ldots, X_n), \end{aligned}$$

since $E(X_{n+1}) = 2p - 1$. Note that $E(X_{n+1}) = 0$, if $p = 1/2$, so that $S = (S_1, S_2 \ldots)$ is a MG, negative, if $p < 1/2$ and hence S is a SMG and positive if $p > 1/2$, so that S is a sMG. Moreover

$$E(S_{n+1}) = E(E(S_{n+1}|\sigma(S_1, \ldots, S_n))) = E(S_n) + E(X_{n+1})E(b_n(X_1, \ldots, X_n)),$$

and therefore the average fortune after the $n+1$ stage, $E(S_{n+1})$, is the same as $E(S_n)$ if S is a MG $(p = 1/2)$, at most $E(S_n)$ if S is a SMG $(p < 1/2)$, and at least $E(S_n)$ if S is a sMG $(p > 1/2)$. Now assume that Y_n denotes the gambler's fortune after n stages where at some of the stages the gambler is allowed to skip playing. The next theorem shows that if the game is fair (that is, S is a MG) or favorable (S is a sMG) the game remains of the same nature and the skipping strategy does not increase the expected earnings.

The starting point in studying MG behavior is the optimal skipping theorem.

Theorem 6.2 (Optimal skipping by Halmos) Let $\{X_n, \mathcal{F}_n\}$ be a MG (or a sMG). Define indicator random variables $\varepsilon_k = I((X_1, \ldots, X_k) \in B_k)$, where $B_k \in \mathcal{B}_k$. Let $Y_1 = X_1$, $Y_2 = X_1 + \varepsilon_1(X_2 - X_1), \ldots, Y_n = Y_{n-1} + \varepsilon_{n-1}(X_n - X_{n-1})$. Then $\{Y_n, \mathcal{F}_n\}$ is a MG (or a sMG) and $E(Y_n) = E(X_n)$ (or $E(Y_n) \leq E(X_n)$), for all $n = 1, 2, \ldots$

Proof. We have

$$\begin{aligned} E(Y_{n+1}|\mathcal{F}_n) &= E(Y_n + \varepsilon_n(X_{n+1} - X_n)|\mathcal{F}_n) = Y_n + \varepsilon_n E(X_{n+1} - X_n|\mathcal{F}_n) \\ &= Y_n + \varepsilon_n(E(X_{n+1}|\mathcal{F}_n) - X_n) \ a.s. \end{aligned}$$

Consequently, if $\{X_n, \mathcal{F}_n\}$ is a MG, $E(X_{n+1}|\mathcal{F}_n) = X_n$, so that $E(Y_{n+1}|\mathcal{F}_n) = Y_n$ and $\{Y_n, \mathcal{F}_n\}$ is a MG. Moreover, if $E(X_{n+1}|\mathcal{F}_n) \geq X_n$ we have $E(Y_{n+1}|\mathcal{F}_n) \geq Y_n$ so that $\{Y_n, \mathcal{F}_n\}$ is a sMG. We prove $E(Y_n) = E(X_n)$, $n = 1, 2, ...$, using induction. Clearly $E(Y_1) = E(X_1)$ by definition. Assume that it holds for $n = k : E(X_k - Y_k) = 0$ (MG) or ≥ 0 (sMG). We show that the claim holds for $n = k + 1$. First we write

$$X_{k+1} - Y_{k+1} = X_{k+1} - Y_k - \varepsilon_k(X_{k+1} - X_k) = (1 - \varepsilon_k)(X_{k+1} - X_k) + (X_k - Y_k),$$

and taking expectation conditionally on $\mathcal{F}_k$ above we obtain

$$E(X_{k+1} - Y_{k+1}|\mathcal{F}_k) = (1 - \varepsilon_k)E(X_{k+1} - X_k|\mathcal{F}_k) + X_k - Y_k.$$

Since $E(X_{k+1} - X_k|\mathcal{F}_k) = 0$ (MG) or ≥ 0 (sMG), we have

$$E(X_{k+1} - Y_{k+1}|\mathcal{F}_k) \begin{cases} = X_k - Y_k \ (MG) \\ \geq X_k - Y_k \ (sMG) \end{cases}.$$

Taking expectation in the latter equations we have

$$E[E(X_{k+1} - Y_{k+1}|\mathcal{F}_k)] = E(X_{k+1} - Y_{k+1}) \begin{cases} = 0 \ (MG) \\ \geq 0 \ (sMG) \end{cases},$$

and the result is established. ■

6.3.1 Filtrations and Stopping Times

A filtration represents the information obtained by observing an experiment up to time n where time here refers to the index of the sequence $X_1, X_2, \ldots, X_n, \ldots$, so that at time n we observe the value X_n from the sequence. A stopping time N can be thought of as the time at which the observations of the experiment are to be stopped. The definition of a stopping time requires that the decision to stop observing at a certain time n be based on the information available up to time n. Recall that $\overline{\mathcal{Z}}^+ = \{1, 2, \ldots, +\infty\}$ and for a filtration $\mathcal{F}$ define $\mathcal{F}_\infty$ to be the σ-field $\sigma(\mathcal{F}_1, \mathcal{F}_2, \ldots) = \sigma\left(\bigcup_{n=1}^{+\infty} \mathcal{F}_n\right)$. The latter will be denoted by $\mathcal{F}_n \uparrow \mathcal{F}_\infty$.

Definition 6.7 Stopping time

Let $(\Omega, \mathcal{A})$ be a measurable space and let $\mathcal{F} = \{\mathcal{F}_n : n = 1, 2, \ldots\}$ denote a filtration in $\mathcal{A}$. A $\overline{\mathcal{Z}}^+$-valued measurable function N defined on $(\Omega, \mathcal{A})$ is called a stopping time with respect to (or relative to) the filtration $\mathcal{F}$ if $\{\omega \in \Omega : N(\omega) \leq n\} \in \mathcal{F}_n$, for all $n \in \overline{\mathcal{Z}}^+$. We say that an event $A \in \mathcal{A}$ is prior to N if and only if $A \cap \{\omega \in \Omega : N(\omega) \leq n\} \in \mathcal{F}_n$, for all $n \in \overline{\mathcal{Z}}^+$. The collection of all prior events A is denoted by $\mathcal{F}_N$.

If X is a random sequence then a stopping time for this sequence is by definition a stopping time relative to the minimal filtration $\mathcal{F}^X$. Next we collect some additional results on stopping times.

Remark 6.2 (Stopping times) The following are straightforward to show.

1. A $\overline{\mathcal{Z}}^+$-valued random variable N is a stopping time relative to the filtration $\mathcal{F}$ if and only if $\{\omega \in \Omega : N(\omega) = n\} \in \mathcal{F}_n, n \in \overline{\mathcal{Z}}^+$.

2. The prior events $\mathcal{F}_N$ form a σ-field. In terms of interpretation, consider that when we observe a random sequence $X = (X_1, X_2, \ldots)$ up to a stopping time N we obtain a certain amount of information. If X is adapted to a filtration $\mathcal{F}$ and N is a stopping time relative to $\mathcal{F}$ then $\mathcal{F}_N$ as defined contains exactly this amount of information.

3. $A \in \mathcal{F}_N$ if and only if $A \cap \{\omega \in \Omega : N(\omega) = n\} \in \mathcal{F}_n$, for all $n \in \overline{\mathcal{Z}}^+$.

4. If $M \leq N$ are two stopping times relative to the same filtration $\mathcal{F}$ then $\mathcal{F}_M \subseteq \mathcal{F}_N$.

Several examples on filtrations and stopping times are collected next.

Example 6.7 (Filtrations and stopping times) Assume the setup of definition 6.7 and let X be a random sequence defined on $(\Omega, \mathcal{A}, P)$.

1. Consider a random walk S with $S_n = \sum_{i=1}^{n} X_i$ and assume that the step distribution is $\mathcal{N}(\mu, \sigma^2)$. Define $N(\omega) = \inf\{n \geq 1 : S_n(\omega) \in I_n\}$, where $I_n = (a_n, b_n)$, an open interval. If the filtration $\mathcal{F}$ is taken to be the minimal filtration $\mathcal{F}^X$ then N is a stopping time.

2. Not every discrete random variable is also a stopping time. Indeed, consider N to be the last time the sequence X is at some value x (we will refer to this as "X visits x at time N" later in this chapter), that is, $X_N = x$ and $X_{N+k} \neq x$, for all $k = 1, 2, \ldots,$ thereafter. Then N is not a stopping time for X.

3. If $N = c$ is some constant then N is a stopping time.

4. If N is a stopping time and $\{X_n\}$ a MG then $\{X_{N \wedge n}\}$ is a SMG.

Example 6.8 (Gambler's ruin) Recall the setup of example 6.6. One usually assumes that $b_n(X_1, \ldots, X_n) \leq S_n$, that is, we bet less than our total fortune at stage n. Let $\tau = \inf\{n \geq 1 : S_n = 0\}$ denote the first stage n after which our fortune is gone, also known as gambler's ruin. Clearly, τ is a stopping time since $(\tau = n)$ depends on $X_1, X_2, \ldots, X_n$ only. The interesting questions in this case include: whether gambler's ruin occurs at a finite stage, i.e., $P(\tau < \infty)$; how many stages on the average until we lose our fortune, i.e., $E(\tau)$, and the probability that our game will last at least $n-1$ stages, i.e., $P(\tau \geq n)$. These questions can be easily answered using concepts presented later in this chapter.

6.3.2 Convergence Theorems

The concept of MGs is due to Lévy but it was Doob who developed the theory. In particular, studying the asymptotic behavior of MGs is due to Doob and his convergence theorem, which provides a great tool to assess the existence of a limit distribution for a random sequence. The proof of the limit theorem requires the concept of upcrossings by the random sequence and the corresponding upcrossings theorem.

Definition 6.8 Upcrossings

Let $X_1, X_2, \ldots, X_n$ be random variables on a probability space $(\Omega, \mathcal{A}, P)$ and let $a < b$, be real constants. We define the number of upcrossings U_{ab} of the interval (a, b) by $X_1, \ldots, X_n$ as follows: let

$$\begin{aligned} T_1 &= \min_{1 \leq k \leq n}\{k : X_k \leq a\},\ T_2 = \min_{T_1 < k \leq n}\{k : X_k \geq b\}, \\ T_3 &= \min_{T_2 < k \leq n}\{k : X_k \leq a\},\ T_4 = \min_{T_3 < k \leq n}\{k : X_k \geq b\}, \ldots, \end{aligned}$$

and set $T_i = +\infty$, if the condition specified cannot be met (e.g., if for all $i = 4, 5, \ldots,$ $X_i < b$, then $T_4 = +\infty$). Then we define N as the number of finite T_i's and

$$U_{ab} = \begin{cases} N/2 & N \text{ is even} \\ (N-1)/2 & N \text{ is odd} \end{cases}.$$

Next we collect the upcrossings theorem.

Theorem 6.3 (Upcrossings theorem by Doob) Let $(X, \mathcal{F})$ be a sMG. Then
$$E(U_{ab}) \leq E[(X_n - a)^+]/(b - a).$$

Proof. Assume for now that $a = 0$ and $X_j \geq 0$, for all j, so that we need to show $E(U_{0b}) \leq \frac{1}{b}E(X_n)$. Let T_i and ε_i be defined as in the optimal skipping theorem. Note that with the assumption $a = 0$ we have $X_{T_i} \leq a \iff X_{T_i} = 0$. Define $Y_n = X_1 + \sum_{i=1}^{n-1} \varepsilon_i(X_{i+1} - X_i)$, that is, Y_n denotes the total increase during upcrossing plus X_1 plus the contribution at the end and consequently $Y_n \geq bU_{0b}$, the reason being that in order to reach b the sequence has to hit 0 and then go up at least b units to get one upcrossing. This implies $U_{0b} \leq \frac{Y_n}{b}$ and taking expectation we have
$$E(U_{0b}) \leq \frac{E(Y_n)}{b} \leq \frac{E(X_n)}{b},$$
with the second inequality following by the optimal skipping theorem. Therefore the claim holds for $X_j \geq 0$ and $a = 0$.

Now in the general case, $\{(X_k - a)^+, \mathcal{A}_k\}$ is a sMG (from Jensen's inequality) and the number of upcrossings U_{ab} by the sequence $\{X_k\}$ is the same as the number of upcrossings U_{0b-a} by the sequence $\{(X_k - a)^+\}$ since $X_k \leq a \iff X_k - a \leq 0$ and $(X_k - a)^+ = 0$ and moreover $X_k \geq b \iff X_k - a \geq b - a$ so that $(X_k - a)^+ \geq b - a$. Hence the general result follows from the proof above for U_{0b}. ■

We are now ready to present a first MG convergence theorem.

Theorem 6.4 (Doob's convergence theorem) Let $\{X_n, \mathcal{F}_n\}$ be a sMG. If $\sup_n E(X_n^+) < +\infty$ then there exists an integrable random variable X_∞ such that $X_n \overset{a.s.}{\to} X_\infty$.

Proof. We are interested in the probability of the event $A = \{\omega \in \Omega : X_n(\omega)$ does not converge to a finite limit $X_\infty(\omega)$ or is $\pm\infty\}$ which can be written equivalently as
$$A = \bigcup_{\substack{a<b \\ a,b \text{ rational}}} \{\omega \in \Omega : \liminf_n X_n(\omega) < a < b < \limsup_n X_n(\omega)\}.$$

If for some $a < b$ we have $P(\{\omega \in \Omega : \liminf_n X_n(\omega) < a < b < \limsup_n X_n(\omega)\}) > 0$ then this implies that $\{X_n\}$ has an infinite number of upcrossings of (a, b) on a set of positive probability, that is, $E(U_{ab}) = +\infty$. Define $U_{a,b;n}$ to be the number of upcrossing of (a, b) by $X_1, ..., X_n$. Then $U_{a,b;n} \uparrow U_{ab}$ and by the MCT: $EU_{a,b;n} \uparrow$

$EU_{ab} = +\infty$. But by the upcrossing theorem

$$EU_{a,b;n} \leq \frac{1}{b-a}E((X_n - a)^+) \leq \frac{1}{b-a}\sup_n E(X_n + a) < \infty,$$

so that the DCT yields $EU_{ab} < +\infty$, a contradiction. Therefore $X_n \stackrel{a.s.}{\rightarrow} X_\infty$ and it remains to show that the limit is integrable.

Recall that $|X_n| = X_n^+ + X_n^- = 2X_n^+ - X_n$ and note that

$$E(|X_n|) \leq 2E(X_n^+) - E(X_n) \leq 2\sup_n E(X_n^+) - E(X_1) = M < \infty,$$

since $EX_n \geq EX_1$ for a sMG. Applying Fatou's lemma on $|X_n| \geq 0$ we obtain

$$\liminf_n E|X_n| \geq E[\liminf_n |X_n|] = E|X_\infty|,$$

so that

$$E|X_\infty| \leq \liminf_n E|X_n| \leq M < \infty,$$

and the integrability of the limit is established. ■

A special case of Doob's convergence theorem is as follows; if $X_n \geq 0$, $n = 1, 2, \ldots$, is a SMG then $X_n \stackrel{a.s.}{\rightarrow} X$ with $E(X) \leq E(X_1)$. The following example should provide the motivation on the necessity of replacing a fixed time of the sequence X_n with a stopping time T, and study X_T instead of X_n.

Example 6.9 (Gambling) Let EX_T denote the average fortune when we stop playing the game where T is some random variable. If $1 = T_1 < T_2 < T_3 < \cdots < T_n < \ldots$, is a sequence of stopping times *a.s.* finite ($P(T_i < \infty) = 1$) and assuming that $\{X_{T_n}\}$ is a MG then under certain conditions we can show that $E(X_{T_n}) = E(X_1)$ (see theorem 6.5 below). Consequently, for a MG it does not matter when we choose to stop playing since on the average we will have the same fortune as that after the first stage. This is a result of a general result we collect next.

Let us collect some conditions for well-behaved stopping times.

Definition 6.9 Sampling integrability conditions

Let $X = (X_1, X_2, \ldots)$ be a random sequence of $\mathcal{R}$-valued random variables with finite mean. A $\overline{\mathcal{Z}}^+$-valued random variable T that is *a.s.* finite is said to satisfy the sampling integrability conditions for X if the following two conditions hold

$$(A): E(|X_T|) < +\infty, \tag{6.2}$$

and

$$(B): \liminf_n \int_{\{T>n\}} |X_n| dP = 0. \tag{6.3}$$

The main result that allows us to replace fixed times n with a sequence of stopping times, while maintaining the MG property of the sequence, is the optional sampling theorem.

Theorem 6.5 (Optional sampling) Let $\{X_n\}$ be a sMG and $\{T_n\}$ an increasing sequence of *a.s.* finite stopping times relative to the same filtration (the minimal filtration $\mathcal{F}^X$). Define $Y_n(\omega) = X_{T_n(\omega)}(\omega)$, $\mathcal{F}_n = \mathcal{F}^X_{T_n}$, $n = 1, 2, \ldots$ and assume that each T_n satisfies the sampling integrability conditions for X. Then $\{Y_n, \mathcal{F}_n\}$ is a sMG and if $\{X_n\}$ is a MG then so is $\{Y_n, \mathcal{F}_n\}$.

Proof. First we show that $\mathcal{F} = \{\mathcal{F}_1, \mathcal{F}_2, ...\}$ is a filtration. Let $A \in \mathcal{F}_n = \mathcal{F}_{T_n}$ so that $A \cap \{T_n = i\} \in \mathcal{F}_i$ for all i. Then

$$A \cap \{T_{n+1} \leq k\} = \bigcup_{i=1}^{k} [A \cap \{T_n = i\}] \cap \{T_{n+1} \leq k\}.$$

and since

$$A \cap \{T_n = i\} \in \mathcal{F}_i \subseteq \mathcal{F}_k,$$

for $i \leq k$ and $\{T_{n+1} \leq k\} \in \mathcal{F}_k$ we have that $A \in \mathcal{F}_{T_{n+1}}$ and therefore $\mathcal{F}$ is a filtration. Furthermore, note that $\{Y_n\}$ is adapted to $\mathcal{F}$.

Now by condition (A) of the sampling integrability conditions we only need to show that $E(Y_{n+1}|\mathcal{F}_n) \overset{(=\text{ if } X \text{ a MG})}{\geq} Y_n$ or equivalently

$$\int_A Y_{n+1} dP \overset{(=\text{ if } X \text{ a MG})}{\geq} \int_A Y_n dP, \tag{6.4}$$

for all $A \in \mathcal{F}_n$. Fix $n > 0$ and $A \in \mathcal{F}_n$. Since T_n is *a.s.* finite we can write $A = \bigcup_{j=1}^{+\infty} [A \cap \{T_n = j\}]$. Then it suffices to show that (6.4) holds for sets $D_j = A \cap \{T_n = j\} \in \mathcal{F}_j$. For fixed $k > j$ we have

$$\begin{aligned}
\int_{D_j} Y_{n+1} dP &= \sum_{i=j}^{k} \int_{D_j \cap \{T_{n+1}=i\}} Y_{n+1} dP + \int_{D_j \cap \{T_{n+1}>k\}} Y_{n+1} dP \\
&= \sum_{i=j}^{k} \int_{D_j \cap \{T_{n+1}=i\}} X_i dP + \int_{D_j \cap \{T_{n+1}>k\}} X_k dP - \int_{D_j \cap \{T_{n+1}>k\}} (X_k - Y_{n+1}) dP,
\end{aligned}$$

where we used the fact that $Y_n = X_{T_n}$ in the first integral of the RHS. Combine the last term in the sum and the middle integral in order to write

$$\int_{D_j \cap \{T_{n+1}=k\}} X_k dP + \int_{D_j \cap \{T_{n+1}>k\}} X_k dP = \int_{D_j \cap \{T_{n+1}\geq k\}} X_k dP.$$

Since $\{T_{n+1} \geq k\} = \Omega \smallsetminus \{T_{n+1} \leq k - 1\} \in \mathcal{F}_{k-1}$, $D_j \in \mathcal{F}_j \underset{j<k}{\subset} \mathcal{F}_{k-1}$ and $\{X_n\}$ is a sMG it follows that

$$\int_{D_j \cap \{T_{n+1}>k\}} X_k dP \overset{(=\text{ if } X \text{ a MG})}{\geq} \int_{D_j \cap \{T_{n+1}\geq k\}} X_{k-1} dP.$$

Now combine this term with the $i = k - 1$ from the sum and obtain again an inequality due to the sMG structure of $\{X_n\}$. Continue this process by induction to get

$$\int_{D_j} Y_{n+1} dP \overset{(=\text{ if } X \text{ a MG})}{\geq} \int_{D_j \cap \{T_{n+1}\geq j\}} X_j dP - \int_{D_j \cap \{T_{n+1}>k\}} (X_k - Y_{n+1}) dP.$$

By condition (B) of the sampling integrability conditions, we have

$$\int_{D_j\cap\{T_{n+1}>k\}} X_k dP \to 0,$$

as $k \to \infty$ and

$$\int_{D_j\cap\{T_{n+1}\geq k\}} Y_{n+1} dP \to 0,$$

since $\{T_{n+1} \geq k\} \downarrow \emptyset$ as $k \to \infty$. Therefore

$$\int_{D_j} Y_{n+1} dP \overset{(=\text{ if } X \text{ a MG})}{\geq} \int_{D_j\cap\{T_{n+1}>j\}} X_j dP,$$

and since $D_j \subset \{T_n = j\} \subset \{T_{n+1} \geq j\}$, we have $D_j \cap \{T_{n+1} > j\} = D_j$ and thus

$$\int_{D_j} Y_{n+1} dP \overset{(=\text{ if } X \text{ a MG})}{\geq} \int_{D_j} Y_n dP,$$

which completes the proof. ■

The sampling integrability conditions can be hard to show. The following theorem discusses some special cases when conditions (A) and (B) hold.

Theorem 6.6 (Proving the sampling integrability conditions) Let $\{X_n\}$ be a sMG and $\{T_n\}$ an increasing sequence of *a.s.* finite stopping times relative to the minimal filtration. The sampling integrability conditions (A) and (B) are valid if one of the following conditions holds: (i) for any $n \geq 1$ there exists an integer $k_n > 0$ such that $T_n \leq k_n$ *a.s.*, or (ii) $E\left[\sup_n |X_n|\right] < +\infty$. In particular this is true if $\{X_n\}$ are uniformly bounded.

Proof. To prove $(i) \Longrightarrow (A)$ we write

$$\int |X_{T_n}| dP = \sum_{i=1}^{k_n} \int_{\{T_n=i\}} |X_i| dP \leq \sum_{i=1}^{k_n} E|X_i| < +\infty.$$

To show $(i) \Longrightarrow (B)$ note that $P(T_n > k) = 0$, for all $k > k_n$.

Now let $Z = \sup_n |X_n|$. Then $\int |X_{T_n}| dP \leq E(Z) < +\infty$ so that $(ii) \Longrightarrow (A)$. To show that $(ii) \Longrightarrow (B)$ we have

$$\int_{\{T_n>k\}} |X_k| dP \leq \int_{\{T_n>k\}} Z dP \to 0,$$

as $k \to \infty$ since Z is integrable. ■

The following remark connects MGs, uniform integrability and the sampling integrability conditions.

Remark 6.3 (Martingales and uniform integrability) Recall definition 3.17 and remark 3.17 involving uniform integrability. In terms of a random sequence $X = (X_1, X_2, \dots)$ we can write the definition as

$$\lim_{a\to\infty} \sup_n \int_{|X_n|\geq a} |X_n| dP = 0,$$

which is a stronger requirement than condition (B) of the sampling integrability conditions. We briefly discuss some uniform integrability results for random sequences.

1. By definition, $\{X_n\}$ is uniformly integrable if and only if $\forall \varepsilon > 0, \exists a \in \mathcal{R}$ such that if $c \geq a$ then $\int\limits_{\{|X_n|>c\}} |X_n| dP < \varepsilon$, for all $n = 1, 2, \dots$, which is also equivalent to $\lim\limits_{c\to\infty} \overline{\lim\limits_{n}} \int\limits_{|X_n|\geq c} |X_n| dP = 0$.

2. If $\{X_n\}$ is uniformly integrable then $\{X_n\}$ are integrable. Indeed, we can write

$$\int |X_n| dP = \int\limits_{\{|X_n|>c\}} |X_n| dP + \int\limits_{\{|X_n|\leq c\}} |X_n| dP \leq \varepsilon + c < +\infty.$$

3. Assume that $\{X_n\}$ is uniformly integrable. Then $\sup\limits_{n} \int |X_n| dP < +\infty$.

4. If $|X_n| \leq Y$ and Y is integrable then $\{X_n\}$ is uniformly integrable. To see this we write $\int\limits_{\{|X_n|\geq c\}} |X_n| dP \leq \int\limits_{\{Y\geq c\}} Y dP \to 0$, as $c \to \infty$.

5. Suppose that $|X_n| \leq M < +\infty$. Then $\{X_n\}$ is uniformly integrable.

6. Let $\{X_n\}$ be uniformly integrable. Then we can show that (a) $\int \underline{\lim\limits_{n}} X_n dP \leq \overline{\lim\limits_{n}} \int X_n dP$ and (b) if $X_n \overset{a.s.}{\to} X$ or $X_n \overset{p}{\to} X$ then X is integrable and $\int X_n dP \to \int X dP$, as $n \to \infty$.

7. The random sequence $\{X_n\}$ is uniformly integrable if and only if both of the following hold (a) $\int |X_n| dP$ are uniformly bounded and (b) $\lambda_n(A) = \int\limits_{A} |X_n| dP$ is a uniformly continuous set function, i.e., $\lambda_n(A) \to 0$, as $P(A) \to 0$, uniformly in P.

8. If the random sequence X is a reverse MG then it is uniformly integrable.

9. If $S \leq T$ are stopping times that satisfy the sampling integrability conditions for a sMG X, then $E(X_S) \leq E(X_T)$, since for any sMG $E(X_n) \leq E(X_{n+m})$, $n, m = 1, 2, \dots$.

Recall remark 3.25.1. The following remark connects $\mathcal{L}^p$-convergence (Section 3.7.5) with uniform integrability and MGs.

Remark 6.4 (Martingales and $\mathcal{L}^p$-convergence) We can show the following.

1. If $X_n \overset{p}{\to} X$ and $\{|X_n|^q\}$, $q > 0$, are uniformly integrable then $X_n \overset{\mathcal{L}^q}{\to} X$. Note that the other direction is always true, namely, $X_n \overset{\mathcal{L}^q}{\to} X \Longrightarrow X_n \overset{p}{\to} X$ and no additional requirements are needed.

2. If $\{X_n\}$ are integrable and $X_n \overset{\mathcal{L}^1}{\to} X$ then $E[X_n I_A] \to E[X I_A]$, for any $A \in \mathcal{A}$.

3. If $\{X_n\}$ is a MG and $X_n \stackrel{\mathcal{L}^1}{\to} X$ then $X_n = E(X|\mathcal{F}_n^X)$.

4. For a sMG the following are equivalent: (a) it is uniformly integrable, (b) it converges *a.s.* and in $\mathcal{L}^1$ and (c) it converges in $\mathcal{L}^1$.

5. For a MG the following are equivalent: (a) it is uniformly integrable, (b) it converges *a.s.* and in $\mathcal{L}^1$, (c) it converges in $\mathcal{L}^1$, (d) there exists an integrable random variable Y such that $X_n = E(Y|\mathcal{F}_n^X)$ *a.s.*

The equivalent statements (a) and (d) in remark 6.4.5 state that a MG $\{X, \mathcal{F}\}$ has its random sequence X uniformly integrable if and only if X has the representation $X_n = E(Y|\mathcal{F}_n)$, for some random variable Y. The next lemma (in view of remark 6.1.11) collects the direction $(d) \Longrightarrow (a)$. It can be used to prove the subsequent convergence theorem.

Lemma 6.1 (Uniformly integrability and filtrations) Let Y be an integrable random variable on $(\Omega, \mathcal{A}, P)$ and assume that $\mathcal{F} = (\mathcal{F}_1, \mathcal{F}_2, \dots)$ is a filtration in $\mathcal{A}$. Then the random variables $\{X_i = E(Y|\mathcal{F}_i)\ a.s.\}_{i \in I}$, for some index set I are uniformly integrable.

Proof. We have

$$\int_{\{|X_i| \geq c\}} |X_i| dP \leq \int_{\{|X_i| \geq c\}} E[|Y|\,|\mathcal{F}_i] dP = \int_{\{|X_i| \geq c\}} E[|Y|] dP,$$

by the definition of conditional expectation and since $\{|X_i| \geq c\} \in \mathcal{F}_i$. By Markov inequality we can write

$$P(|X_i| \geq c) \leq \frac{E(|X_i|)}{c} \leq \frac{E(E[|Y|\,|\mathcal{F}_i])}{c} = \frac{E(|Y|)}{c} \to 0,$$

as $c \to \infty$, which completes the proof since P over the set $\{|X_i| \geq c\}$ goes to zero. ■

A second MG convergence is collected next.

Theorem 6.7 (Martingale convergence) Let $(\Omega, \mathcal{A}, P)$ be a probability space and $\mathcal{F} = (\mathcal{F}_1, \mathcal{F}_2, \dots)$ be a filtration of $\mathcal{A}$. Let $\mathcal{F}_\infty = \sigma\left(\bigcup\limits_{n=1}^{+\infty} \mathcal{F}_n\right)$ and assume that Y is an integrable random variable on $(\Omega, \mathcal{A}, P)$. If we set $X_n = E(Y|\mathcal{F}_n)$ *a.s.*, $n = 1, 2, \dots$, then $X_n \stackrel{a.s.}{\to} E(Y|\mathcal{F}_\infty)$ and $X_n \stackrel{\mathcal{L}^1}{\to} E(Y|\mathcal{F}_\infty)$.

Proof. From remark 6.1.11 $\{X_n, \mathcal{A}_n\}$ is a MG and by the previous lemma X is uniformly integrable so that remark 6.4.5 $(a) \Longrightarrow (b)$, yields the convergence statements $X_n \stackrel{a.s.}{\to} X_\infty$ and $X_n \stackrel{\mathcal{L}^1}{\to} X_\infty$. We need to show that this "last element" X_∞ of the sequence X is of the form $X_\infty = E(Y|\mathcal{F}_\infty)$. The definition of X_n and remark 6.4.3 imply $E(Y|\mathcal{F}_n) = X_n = E(X_\infty|\mathcal{F}_n)$ *a.s.* so that

$$\int_F Y dP = \int_F X_\infty dP$$

for all $F \in \mathcal{F}_n$. Since Y and X_∞ are integrable and $\bigcup\limits_{n=1}^{+\infty} \mathcal{F}_n$ is a π-system, the $\pi - \lambda$ theorem implies that the latter equation holds for all $F \in \mathcal{F}_\infty$. Finally, since X_∞ is $\mathcal{F}_\infty$-measurable it follows that $X_\infty = E(Y|\mathcal{F}_\infty)$. ■

The following is an immediate consequence of the latter theorem.

Remark 6.5 (Lévy's 0-1 law) An important consequence of theorem 6.7 is that $E(Y|X_1, \ldots, X_n)$ converges *a.s.* and in $\mathcal{L}^1$ to the random variable $E(Y|X_1, \ldots, X_n, \ldots)$. In particular, Lévy's 0-1 law is a special case, namely, if $\mathcal{F}_n \uparrow \mathcal{F}_\infty$ (i.e., $\mathcal{F}_n$ and $\mathcal{F}_\infty$ as in theorem 6.7), then $E(I_F|\mathcal{F}_n) \to I_F$ *a.s.* for any $F \in \mathcal{F}_\infty$.

Both convergence theorems we have seen can be used to assess whether a sequence of random variables X converges *a.s.* since if we show it is a MG or sMG then theorems 6.4 and 6.7 can be applied.

In our treatment of MGs thus far we have assumed that the index of $X = (X_1, X_2, \ldots)$ and $\mathcal{F} = (\mathcal{F}_1, \mathcal{F}_2, \ldots)$ is an integer $n = 1, 2, \ldots$ and this n is treated as the "discrete time" parameter of the random sequence. The case where $X = (X_t : t \geq 0)$ and $\mathcal{F} = (\mathcal{F}_t : t \geq 0)$, for "continuous time" $t \geq 0$ requires similar definitions for stopping times and filtrations and will be treated extensively in Chapter 7 including the all-important MG problem. Clearly, the discrete parameter case involves a countable number of random objects and is much easier to handle, while the continuous time parameter case assumes an uncountable collection of random variables, making the study of such processes more complicated. We revisit this discussion in Chapter 7 with the formal definitions of stochastic sequences and processes.

The next theorem, due to Doob, suggests that in order to study convergence of sub and super MGs, all we need to work with are MGs. For a proof see Durrett (2010, p. 237).

Theorem 6.8 (Doob decomposition) Let $\{X_n, \mathcal{F}_n\}$ be a sMG. There exist random variables Y_n and Z_n with $X_n = Y_n + Z_n$, for all n, where $\{Y_n, \mathcal{F}_n\}$ is a MG, $Z_1 = 0$, Z_n is $\mathcal{F}_{n-1}$-measurable for $n \geq 2$ and $Z_n \leq Z_{n+1}$ *a.e.*, for all n. The Y_n and Z_n are uniquely determined.

6.3.3 Applications

We end this section with some examples and applications of MGs.

Example 6.10 (Polya urns) An urn contains r red and b black balls and after a single ball is drawn the ball is returned and c more balls of the color drawn are added to the urn. Let $X_0 = r/(b + r)$ and X_n, $n \geq 1$, denote the proportion of red balls after the nth draw. First we verify that $\{X_n\}$ is a MG. Suppose that there are r_n red balls and b_n black balls at time n. Then we draw a red ball with probability $r_n/(r_n + b_n)$ in which case we need to add $c+1$ red balls back to the urn (including the one drawn) and hence the fraction of red balls after the $(n+1)$ draw is $X_{n+1} = (r_n + c)/(r_n + b_n + c)$. Similarly, we draw a black ball with probability $b_n/(r_n + b_n)$ in which case we need to add $c+1$ black balls back to the urn and hence the fraction of red balls after the $(n+1)$ draw is $X_{n+1} = r_n/(r_n + b_n + c)$. Therefore, we have

$$X_{n+1} = \begin{cases} (r_n + c)/(r_n + b_n + c) & \text{w.p. } r_n/(r_n + b_n) = X_n \\ r_n/(r_n + b_n + c) & \text{w.p. } b_n/(r_n + b_n) = 1 - X_n \end{cases},$$

so that

$$E(X_{n+1}|X_n) = \frac{r_n + c}{r_n + b_n + c}\frac{r_n}{r_n + b_n} + \frac{r_n}{r_n + b_n + c}\frac{b_n}{r_n + b_n} = \frac{r_n}{r_n + b_n} = X_n,$$

and hence $\{X_n\}$ is a MG and Doob's convergence theorem guarantees the existence of a limit distribution, i.e., $X_n \stackrel{a.s.}{\to} X_\infty$, for some random variable X_∞. The reader can verify that the distribution of the limit X_∞ is a *Beta* $(r/c, b/c)$ with density

$$f_{X_\infty}(x) = \frac{x^{r/c-1}(1-x)^{b/c-1}}{B(r/c, b/c)},$$

for $0 \leq x \leq 1$.

Example 6.11 (Radon-Nikodym derivatives) Let $(\Omega, \mathcal{A}, P)$ be a probability space, $\mathcal{F} = (\mathcal{F}_1, \mathcal{F}_2, \dots)$ a filtration of $\mathcal{A}$ such that $\mathcal{F}_n \uparrow \mathcal{F}_\infty$ and suppose that μ is a finite measure on $(\Omega, \mathcal{A})$. Denote the restrictions of P and μ on $(\Omega, \mathcal{F}_n)$ by P_n and μ_n, respectively. Suppose that $\mu_n \ll P_n$ and define $X_n = \left[\frac{d\mu_n}{dP_n}\right]$. First note that X_n is integrable $[P]$ and $\mathcal{F}_n$-measurable by definition so that we can write $\int\limits_F X_n dP = \mu(F)$, for all $F \in \mathcal{F}_n$. But $\mathcal{F}_n \subset \mathcal{F}_{n+1}$ and therefore $F \in \mathcal{F}_{n+1}$ so that $\int\limits_F X_{n+1} dP = \mu(F)$. Since the two integrals are equal we have

$$\int\limits_F X_n dP = \int\limits_F X_{n+1} dP,$$

thus establishing that $\{X_n, \mathcal{F}_n\}$ is a MG.

Example 6.12 (Likelihood ratios) Let $X = (X_1, X_2, \dots)$ be a random sequence and suppose that the joint densities (models) of $(X_1, \dots, X_n)$ are either p_n or q_n. To help us choose one of the two models we introduce the random variables

$$Y_n = Y_n(X_1, \dots, X_n) = \frac{q_n(X_1, \dots, X_n)}{p_n(X_1, \dots, X_n)},$$

for $n = 1, 2, \dots$ and we expect Y_n to be small or large depending on whether the true density is p_n or q_n, respectively. Therefore, we are interested in the asymptotic behavior of such ratios. First note that $\sigma(Y_1, \dots, Y_n) = \sigma(X_1, \dots, X_n)$. Wlog assume that p_n is positive. If p_n is the true density then the conditional density of X_{n+1} given $X_1, \dots, X_n$, is the ratio p_{n+1}/p_n and hence we can show that

$$E(Y_{n+1}|X_1 = x_1, \dots, X_n = x_n) = Y_n(x_1, \dots, x_n),$$

for any $x_1 \in \mathcal{R}, \dots, x_n \in \mathcal{R}$, that is, $\{Y_n\}$ form a MG.

Example 6.13 (Gambling: double or nothing) Let X be a random sequence for which the conditional distribution of X_{n+1} given $\sigma(X_0, X_1, X_2, \dots)$ is the discrete uniform distribution on the two-point set $\{0, 2X_n\}$, where we have taken $X_0 = 1$. This sequence represents the fortune of a gambler that starts with \$1 and bets his entire fortune at each stage of a fair game. After each bet the fortune either doubles or vanishes, with each event occurring with probability 1/2. Clearly, $\{X_n\}$ is a MG since

$$E(X_{n+1}|X_0, \dots, X_n) = 2X_n 0.5 = X_n \ a.s.,$$

and the remaining conditions are trivially satisfied by the sequence X. Define $T(\omega) = \inf\{n > 0 : X_n(\omega) = 0\}$, the first hitting time of 0. Then T is *a.s.* finite and since $E(X_T) = E(0) = 0 < 1 = E(X_0)$, T does not satisfy the sampling integrability conditions for $\{X_n\}$. In particular, condition (A) is satisfied but (B) is not.

6.3.4 Wald Identities and Random Walks

We connect random walks with the sampling integrability conditions (p. 218) and MGs next.

Theorem 6.9 (Wald identities) Let $S = (S_0 = 0, S_1, S_2, \dots)$ be a random walk on $\mathcal{R}$ with steps $X = (X_1, X_2, \dots)$ and $E(|X_i|) < +\infty$. If τ is a stopping time that satisfies the sampling integrability conditions then

$$E(S_\tau) = E(S_1)E(\tau), \tag{6.5}$$

and

$$Var(S_\tau) = Var(S_1)E(\tau), \tag{6.6}$$

where we agree that $0 * \infty = 0$.

Proof. Let $\mu = E(S_1)$ so that $E(S_n) = n\mu$ and define $Y_n = S_n - n\mu$. Then $\{Y_n : n \geq 0\}$ is a random walk whose step distribution has mean zero and hence it is a MG. Now apply the Optional Sampling Theorem with $T_n = \min\{T, n\}$ to obtain

$$0 = E(Y_{T_n}) = E(S_{T_n}) - \mu E(T_n).$$

By the MCT we have

$$\mu E(T) = \mu \lim_{n\to+\infty} E(T_n) = \lim_{n\to+\infty} E(S_{T_n}), \tag{6.7}$$

and the first sampling integrability condition $E(|S_T|) < +\infty$. Thus the second sampling integrability condition and the DCT imply that

$$\begin{aligned}\liminf_{n\to+\infty} E(|S_{T_n} - S_T|) &\leq \liminf_{n\to+\infty} E\left[(|S_{T_n}| + |S_T|)I(T > n)\right] \\ &\leq \liminf_{n\to+\infty} E\left[|S_{T_n}|I(T > n)\right] + \limsup_{n\to+\infty} E\left[|S_T|I(T > n)\right] = 0.\end{aligned}$$

The latter equation implies that $S_{T_n} \xrightarrow{\mathcal{L}^1} S_T$ and since the steps are integrable, remark 6.2.2 with $A = \Omega$ yields $E[S_{T_n}] \to E[S_T]$, as $n \to +\infty$. Combine this with (6.7) and we have (6.5).

Now to prove (6.6), first note that from (6.5) we have $E(S_T) = 0$ and hence $Var(S_T) = E(S_T^2)$. Define for all $n \geq 0$, $Z_n = S_n^2 - n\sigma^2$ and note that

$$E(Z_{n+1}|S_0, S_1, ..., S_n) = Z_n,$$

so that $\{Z_n\}$ is a MG with respect to the minimal filtration of S. We follow similar steps as above and establish (6.6). ■

We finish this first treatment of random walks with the following theorem. The proof is requested as an exercise.

Theorem 6.10 (Conditional independence) Let $(S_0 = 0, S_1, S_2, \dots)$ a random walk. Then for any $n > 0$ we can show that $(S_0, S_1, S_2, \dots, S_n)$ and $(S_n, S_{n+1}, \dots)$ are conditionally independent given $\sigma(S_n)$.

6.4 Renewal Sequences

We are interested in an important class of random sequences which is defined in terms of random walks.

Definition 6.10 Renewal sequences via random walks

Consider a random walk $T = \{T_m : m = 0, 1, 2, \dots\}$ in $\overline{\mathcal{Z}}^+$ satisfying $T_{m+1}(\omega) \geq 1 + T_m(\omega)$, and define the random sequence $X = (X_1, X_2, \dots)$ by $X_n(\omega) = 1$, if $T_m(\omega) = n$, for some $m \geq 0$ and 0, otherwise.

1. Renewal sequence X is called the renewal sequence corresponding to T. We may write $X_n(\omega) = \sum\limits_{m=0}^{+\infty} I(T_m(\omega) = n)$. The random walk T can be obtained from X by setting $T_0(\omega) = 0$ and $T_m = \inf\{n > T_{m-1} : X_n = 1\}$, $m > 0$, where $\inf \emptyset = +\infty$.

2. Times of renewals The integers $n > 0$ for which $X_n = 1$ are called renewal times. If $T_m < \infty$ then T_m is called the mth renewal time.

3. Waiting times The time difference $T_m - T_{m-1}$ is called the mth waiting time. The distribution of the first renewal time T_1 is known as the waiting distribution W (it is the same as the step distribution of the random walk T).

Suppose that X is a 0-1 valued random sequence. The next theorem helps us assess when X is a renewal sequence (or if T is a random walk). The proof is requested as an exercise.

Theorem 6.11 (Renewal criterion) A random 0-1 sequence $X = (X_0, X_1, X_2, \dots$

) is a renewal sequence if and only if $X_0 = 1$ *a.s.* and

$$P(X_n = x_n, 0 < n \leq r + s) = P(X_n = x_n, 0 < n \leq r)P(X_{n-r} = x_n, r < n \leq r + s), \tag{6.8}$$

for all $r, s > 1$ and all 0-1 sequences $(x_1, \ldots, x_{r+s})$ such that $x_r = 1$.

The latter theorem states that renewal processes are 0-1 sequences that start over independently after each visit to state 1, i.e., they "regenerate" themselves. Note that if $x_r = 0$, equation (6.8) does not necessarily hold for a renewal sequence $(x_0, x_1, \ldots)$. In view of this the random set $\Sigma_X = \{n : X_n = 1\}$ is called a regenerative set in $\mathcal{Z}^+$.

Two of the important quantities we need in order to study renewal processes are the renewal and potential measures.

Definition 6.11 Renewal and potential measures

Let $X = (X_0, X_1, X_2, \ldots)$ denote a renewal sequence.

1. Renewal measure The random σ-finite measure R given by $R(B) = \sum\limits_{n \in B} X_n$ is called the renewal measure of X, where $R(B)$ describes the number of renewals occurring during a time set B.

2. Potential measure The function $\Pi(B) = E(R(B)) = E\left(\sum\limits_{n \in B} X_n\right)$, is a (non-random) σ-finite measure and is called the potential measure of the renewal sequence X.

3. Potential sequence The density of the potential measure Π of X with respect to the counting measure is the potential sequence of X. It satisfies $\pi_n = \Pi(\{n\}) = P(X_n = 1)$, where $(\pi_0 = 1, \pi_1, \pi_2, \ldots)$ is the potential sequence of X.

We collect some results regarding the renewal and potential measures below.

Remark 6.6 (Renewal and potential measures) Let W be a waiting time distribution for a renewal sequence X corresponding to a random walk T in $\overline{\mathcal{Z}}^+$. For any set $B \subseteq \mathcal{Z}^+$ and any non-negative integer m, it follows that $W^{*m}(B) = P(T_m \in B)$, where W^{*0} is the delta distribution at 0 and since $T_m = (T_m - T_{m-1}) + (T_{m-1} - T_{m-2}) + \ldots (T_1 - T_0)$, with each step $T_m - T_{m-1}$ of the random walk T being iid distributed according to W.

1. Relating Π and W We can show that the potential measure Π and the waiting time distribution W are related via

$$\Pi(B) = \sum_{m=0}^{+\infty} W^{*m}(B), \tag{6.9}$$

for all sets $B \subseteq \mathcal{Z}^+$. To see this, first note that $X_n = \sum\limits_{m=0}^{+\infty} I(T_m = n)$, with $W^{*m}(\{n\}) =$

$E(I(T_m = n))$ and use the Fubini theorem to obtain

$$\begin{aligned}\Pi(B) &= \sum_{n\in B} E(X_n) = \sum_{n\in B} E\left(\sum_{m=0}^{+\infty} I(T_m = n)\right) = \sum_{m=0}^{+\infty}\sum_{n\in B} E(I(T_m = n)) \\ &= \sum_{m=0}^{+\infty}\sum_{n\in B} W^{*m}(\{n\}) = \sum_{m=0}^{+\infty} W^{*m}(B).\end{aligned}$$

2. Measure generating function The pgf of a distribution Q on $\mathcal{Z}_0^+$, is defined by $\rho(s) = \sum_{n=0}^{+\infty} Q(\{n\})s^n$, $0 \le s \le 1$. It is useful to extend this definition to measures other than probability measures and we accomplish this by removing 1 from the domain. In particular, the measure generating function of a measure Π on $\mathcal{Z}_0^+$ that has a bounded density $(\pi_0, \pi_1, \dots)$ with respect to the counting measure is the function

$$\phi(s) = \sum_{n=0}^{+\infty} \pi_n s^n, \tag{6.10}$$

for $0 \le s < 1$. It is straightforward to show that two measures on $\mathcal{Z}_0^+$ with bounded densities are equal if their corresponding measure generating functions are the same.

3. Consider a renewal sequence with W and Π the waiting time distribution and potential measure, respectively, and let ϕ_W denote the measure generating function of W.
(i) Then we can show that the measure generating function of Π is $\phi_\Pi(s) = (1 - \phi_W(s))^{-1}$, for $0 \le s < 1$.
(ii) Furthermore, we have $\Pi = (\Pi * W) + \delta_0$, where δ_0 is the delta distribution at 0.

4. Let W_i, Π_i, ϕ_i, $i = 1, 2$, the waiting time distribution, potential measure and measure generating function of W_i, respectively, for two renewal sequences. The following are equivalent: (i) $W_1 = W_2$, (ii) $\Pi_1 = \Pi_2$, (iii) $\phi_1 = \phi_2$.

Example 6.14 (Potential sequence) Consider the sequence $u = (1, 0, 1/4, 1/4, \dots)$. We show that it is a potential sequence. Note that the measure generating function of u is given by $\psi(s) = 1 + s^2/(4(1 - s))$. Setting this to $(1 - \varphi)^{-1}$ yields $\varphi(s) = s^2/(2 - s)^2$, the entertained measure generating function of Π. To show that the given sequence is a potential sequence, we only need to show that φ is the measure generating function of some probability distribution on $\mathcal{Z}^+$. For that we need to obtain its expansion as a power series with coefficients that are positive, the coefficient of s^0 is 0 and $\phi(1-) < 1$. At the same time we will get a formula for the waiting time distribution W. Using the Binomial theorem we have

$$\varphi(s) = s^2(2 - s)^{-2} = (s^2/4)\sum_{n=0}^{+\infty} C_n^{-2}(-s/2)^n = \sum_{n=2}^{+\infty} C_{n-2}^{-2}(-s/2)^n = \sum_{n=1}^{+\infty}(n - 1)2^{-n}s^n,$$

and therefore $W(\{n\}) = (n - 1)2^{-n}$, $n > 0$.

Next we obtain some results involving the renewal measure R of X. The first is requested as an exercise.

Theorem 6.12 (Total number of renewals) Let X denote a renewal sequence with potential measure Π and waiting time distribution W. If $\Pi(\mathcal{Z}_0^+) < \infty$ then the distribution of $R(\mathcal{Z}_0^+)$ (the total number of renewals) is geometric with mean $\Pi(\mathcal{Z}_0^+)$. If $\Pi(\mathcal{Z}_0^+) = \infty$ then $R(\mathcal{Z}_0^+) = \infty$ *a.s.* and in any case $\Pi(\mathcal{Z}_0^+) = 1/W(\{\infty\})$ (with $1/0 = \infty$).

Many asymptotic results for a random sequence X can be proven easily via asymptotic results on an underlying renewal sequence used to describe X. The following SLLN will be utilized later in this chapter in order to study the limiting behavior of Markov chains.

Theorem 6.13 (SLLN for renewal sequences) Let R denote the renewal measure, Π the potential measure, X the renewal sequence and $\mu = \sum\limits_{n=1}^{+\infty} nW(\{n\})$, $\mu \in [1, \infty]$, the mean of the waiting time distribution W. Then we can show that

$$\lim_{n\to+\infty} \Pi(\{0, 1, \ldots, n\})/n = \lim_{n\to+\infty} R(\{0, 1, \ldots, n\})/n = \lim_{n\to+\infty} \sum_{k=0}^{n} X_k/n = 1/\mu \text{ a.s.} \tag{6.11}$$

Proof. Recall that $\Pi(\{0, 1, ..., n\}) = E(R(\{0, 1, ..., n\})) = E\left(\sum\limits_{k=0}^{n} X_k\right)$ and note that $R_n = R(\{0, 1, ..., n\}) \leq n$. Then using the BCT on the sequence R_n/n we establish the first and second equalities in (6.11).

To prove the third equality, let T denote the random walk corresponding to X and assume first that $\mu < +\infty$. Note that for a fixed ω, $n \in [T_m(\omega), T_{m+1}(\omega)]$ implies that $R_n(\omega) = m$ so that

$$T_{R_n(\omega)-1}(\omega) \leq n \leq T_{R_n(\omega)}(\omega) \Rightarrow \frac{T_{R_n(\omega)-1}(\omega)}{R_n(\omega)-1}\frac{R_n(\omega)-1}{R_n(\omega)} \leq \frac{n}{R_n(\omega)} \leq \frac{T_{R_n(\omega)}(\omega)}{R_n(\omega)}. \tag{6.12}$$

Now by the SLLN $T_m/m \to \mu$ a.s. as $m \to +\infty$, since $T_m = T_1 + (T_2 - T_1) + ... + (T_m - T_{m-1})$ is a sum of iid random variables with distribution W. Furthermore, $\lim\limits_{n\to+\infty} R_n = +\infty$ since if $\lim\limits_{n\to+\infty} R_n = R(\mathcal{Z}^+) = M < \infty$, for some finite M then by the definition of R_n we would have $T_{M+1} = +\infty$, a contradiction, because T_{M+1} has finite mean ($E(T_{M+1}) = (M+1)\mu$). Consequently,

$$\lim_{n\to+\infty} \frac{T_{R_n(\omega)}(\omega)}{R_n(\omega)} = \lim_{m\to+\infty} \frac{T_m(\omega)}{m} = \mu \text{ a.s.},$$

so that taking the limit as $n \to +\infty$ in (6.12) yields $\lim\limits_{n\to+\infty} R_n/n = \frac{1}{\mu}$ a.s., as desired.

The case $\mu = +\infty$ follows similar steps. First note that

$$T_{R_n(\omega)-1}(\omega) \leq n \Rightarrow \frac{n}{R_n(\omega)} \geq \frac{T_{R_n(\omega)-1}(\omega)}{R_n(\omega)-1}\frac{R_n(\omega)-1}{R_n(\omega)},$$

so that if $\lim\limits_{n\to+\infty} R_n = +\infty$ we have $\lim\limits_{n\to+\infty} \frac{n}{R_n} \geq \mu = \infty$ or $\lim\limits_{n\to+\infty} \frac{R_n}{n} = 0$. When $\lim\limits_{n\to+\infty} R_n < +\infty$ we have right away that $\lim\limits_{n\to+\infty} \frac{R_n}{n} = 0$. ■

We are now ready to state a classification result for renewal sequences. As we will see classification of the states of any random sequence can be accomplished via classification of an appropriate renewal sequence.

Definition 6.12 Renewal sequence classification

A renewal sequence and the corresponding regenerative set are called transient if the renewal measure is finite *a.s.* and they are called recurrent otherwise. A recurrent renewal sequence is called recurrent null if the mean waiting time is infinite; otherwise, it is called positive recurrent (or recurrent non-null).

A simple modification of a renewal sequence is required, if we do not have a renewal at the initial state of the sequence.

Remark 6.7 (Delayed renewal sequence) Let $S = (S_0 = 0, S_1, S_2, \ldots)$ be a random walk in $\mathcal{Z}$ and for a fixed integer x consider the random sequence $X = (X_0, X_1, \ldots)$ defined by

$$X_n = I(S_n = x) = \begin{cases} 1 & S_n = x \\ 0 & S_n \neq x \end{cases},$$

so that X marks the visits made by S to x. If $x \neq 0$ then X is not a renewal sequence. However, once the point x is reached by S then thereafter X behaves like a renewal sequence.

Motivated by the latter example we have the formal definition of a delayed renewal sequence.

Definition 6.13 Delayed renewal sequence

Let X be an arbitrary 0-1 sequence and let $T = \inf\{n \geq 0 : X_n = 1\}$. Then X is called a delayed renewal sequence if either $P(T = \infty) = 1$ or $P(T < \infty) < 1$ and conditional on the event $\{T < \infty\}$ the sequence $Y = (X_{T+n} : n = 0, 1, \ldots)$ is a renewal sequence.

The random variable T is called the delay time and the distribution of T is called the delay distribution of X. The waiting time distribution of Y is the same as that of X. Note that X reduces to a renewal sequence if and only if $P(X_0 = 1) = 1$.

The last classification type involves the concept of periodicity.

Definition 6.14 Period of a renewal sequence

The period γ of a renewal sequence is the greatest common divisor (GCD) of the support of the corresponding potential measure with the GCD taken to be ∞ if the support is $\{0\}$. Note that the period is also the GCD of the support of the waiting time distribution.

Example 6.15 (Renewal sequence period) Let Π be the potential measure and assume that its support is $\mathcal{S} = \{1, 2, 3, 4, \dots\}$. Then the period of the corresponding renewal sequence is 1 and if $\mathcal{S} = \{2, 4, 6, 8, \dots\}$ then the period is 2.

The following theorem can be useful in studying the asymptotic behavior of random sequences. The proof is requested as an exercise.

Theorem 6.14 (Renewal theorem) Let $\{\pi_n : n \geq 0\}$ denote the potential sequence and $\mu = \sum\limits_{n=1}^{+\infty} nW(\{n\})$, $\mu \in [1, \infty]$, the mean of the waiting time distribution W, of a renewal sequence with finite period γ. Then we can show that

$$\lim_{n \to +\infty} \pi_{n\gamma} = \frac{\gamma}{\mu}. \tag{6.13}$$

Next we give an alternative classification criterion based on the potential sequence.

Corollary 6.2 (Classification) Let $\{\pi_n : n \geq 0\}$ denote the potential sequence of a renewal sequence X. Then X is recurrent if and only if $\sum\limits_{n=0}^{+\infty} \pi_n = +\infty$, in which case X is null recurrent if and only if $\lim\limits_{n \to +\infty} \pi_n = 0$.

Proof. Follows immediately from theorem (6.12) and theorem (6.14). ∎

The following remark summarizes several results connecting random walks and renewal sequences.

Remark 6.8 (Random walks and renewal sequences) Let $S = (S_0 = 0, S_1, S_2, \dots)$ be a random walk in $\mathcal{Z}$ with step distribution Q and consider the renewal sequence X defined by $X_n = I(S_n = 0)$, $n \geq 0$.

1. Classification The random walk is called transient, positive recurrent or null recurrent depending on the classification of the renewal sequence X.

2. Potential sequence The potential sequence $\{\pi_n : n \geq 0\}$ of X is related to the step distribution as follows

$$\pi_n = P(X_n = 1) = P(S_n = 0) = Q^{*n}(\{0\}). \tag{6.14}$$

3. Step and waiting time distributions Let β denote the cf of Q, i.e., $\beta(u) = \int e^{iux} Q(dx)$. From equation (6.14) and the inversion theorem (p. 196) we have

$$\pi_n = Q^{*n}(\{0\}) = (2\pi)^{-1} \int\limits_{-\pi}^{\pi} \beta^n(u) du,$$

where β^n is the cf of Q^{*n}. Multiply by s^n and sum over all $n \geq 0$ to obtain

$$\left(1 - \phi_{T_1}(s)\right)^{-1} = \sum_{n=0}^{+\infty} \pi_n s^n = (2\pi)^{-1} \int_{-\pi}^{\pi} (1 - s\beta(v))^{-1} dv,$$

where ϕ_{T_1} is the measure generating function of the waiting time distribution W (or first return time $T_1 = \inf\{n > 0 : S_n = 0\}$) of X. Note that we used remark 6.6.3, part (i), for the LHS above. As a consequence, we have

$$\phi_{T_1}(s) = 1 - \left[(2\pi)^{-1} \int_{-\pi}^{\pi} (1 - s\beta(v))^{-1} dv\right]^{-1},$$

and we have related the first return time T_1 to 0 by S, with the cf of the step distribution Q.

4. Every random walk on $\mathcal{Z}$ is either transient or null recurrent (except for the random walk with $S_n = 0$, for all n).

Example 6.16 (Potential sequence for a simple random walk) Consider the simple random walk S on $\mathcal{Z}$ with step distribution $Q(\{1\}) = p = 1 - Q(\{-1\})$ and define $X_n = I_{\{0\}} \circ S_n$, $n \geq 0$. The cf of Q is

$$\beta(u) = E(e^{iuS_1}) = pe^{iu} + (1 - p)e^{-iu},$$

$u \in \mathcal{R}$, so that the cf of $S_n \sim Q^{*n}$ is given by

$$\beta^n(u) = \left(pe^{iu} + (1 - p)e^{-iu}\right)^n = E(e^{iuS_n}) = \sum_{k=0}^{+\infty} e^{iuk} P(S_n = k).$$

By the inversion theorem the density corresponding to β^n (i.e., $\pi_n = Q^{*n}(\{0\})$) is given by

$$\pi_n = (2\pi)^{-1} \int_{-\pi}^{\pi} \beta^n(u) du = (2\pi)^{-1} \sum_{k=0}^{n} C_k^n p^k (1 - p)^{n-k} \int_{-\pi}^{\pi} e^{(2k-n)iu} du.$$

and since

$$\int_{-\pi}^{\pi} e^{(2k-n)iu} du = 2\sin((2k - n)\pi)/(2k - n),$$

we can write

$$\pi_n = (2\pi)^{-1} \sum_{k=0}^{n} C_k^n p^k (1 - p)^{n-k} 2\sin((2k - n)\pi)/(2k - n).$$

Note that as $k \to n/2$ we have $\lim_{n \to +\infty} \sin((2k-n)\pi)/(2k-n) = \pi$, while if $k \neq n/2$ we have $\sin((2k-n)\pi)/(2k-n) = 0$. Consequently, we can write the potential sequence as $\pi_n = (2\pi)^{-1} C_{n/2}^n p^{n/2} (1 - p)^{n-n/2} 2\pi$ and thus

$$\pi_n = C_{n/2,n/2}^n p^{n/2} (1 - p)^{n/2},$$

for n even. For odd n, we have $\pi_n = 0$, since renewals (visits to 0) can only happen at even times.

6.5 Markov Chains

We have seen many sequences of random variables where the future states of the process are independent of the past given the present, and this is the idea behind the Markov property. We begin with the general definition of a Markov chain.

6.5.1 Definitions and Properties

Definition 6.15 Markov chain

The random sequence $X = \{X_n : n \geq 0\}$ that is adapted to a filtration $\mathcal{F} = \{\mathcal{F}_n : n \geq 0\}$ is called a Markov chain with respect to the filtration $\mathcal{F}$ if $\forall n > 0$, $\mathcal{F}_n$ and $\sigma(\{X_m : m \geq n\})$ are conditionally independent given $\sigma(X_n)$.

1. Transition function The transition functions R_n of a Markov chain X are defined as the conditional distributions of X_{n+1} given $\sigma(X_n)$, $n \geq 0$. The distribution R_0 is called the initial distribution and R_n, $n > 0$, is called the nth transition function (or kernel).

2. Time homogeneity If R_n is the same for all $n > 0$ then X is called time-homogeneous ($R_1 = R_2 = \ldots$).

For any Markov chain X we may write

$$R(X_n, B) = P(X_{n+1} \in B|\sigma(X_0, X_1, \ldots, X_n)) = P(X_{n+1} \in B|\sigma(X_n)),$$

for the nth transition function and read it as "the probability that the Markov chain is in states B at time $n+1$ given that it was at state X_n at time n."

The study of Markov chains in discrete time, and Markov processes (Chapter 7) in continuous time is unified via the concept of a transition operator.

Definition 6.16 Transition operator

Let $\mathbb{M} = \{\mu_x : x \in \Psi\}$, denote a collection of distributions where Ψ is a Borel space (see appendix remark A.4) and assume that for each measurable set $B \subseteq \Psi$ the function $x \mapsto \mu_x(B)$ is measurable (fixed B).

1. Left transition operator The left transition operator T corresponding to the collection $\mathbb{M}$ of transition distributions operates on the space of all bounded, $\mathcal{R}$-valued measurable functions on Ψ and is given by

$$(T(f))(x) = \int_{\Psi} f(y)\mu_x(dy). \tag{6.15}$$

To simplify the notation we write $Tf(x)$.

1. Right transition operator The right transition operator corresponding to $\mathbb{M}$, also denoted by T, operates on the space of probability measures on Ψ and is given by

$$(Q(T))(B) = \int_{\Psi} \mu_x(B)Q(dx). \tag{6.16}$$

We simplify the notation by writing $QT(B)$.

We connect transition functions and operators below.

Remark 6.9 (Transition functions and operators) Transition functions and operators are related as follows. Given a transition function R the transition distributions are given by $\mu_x(.) = R(x, .)$, $x \in \Psi$. The collection of transition distributions $\mathbb{M} = \{\mu_x : x \in \Psi\}$ uniquely determines the left and right transition operators T so that a transition function R uniquely determines a corresponding transition operator T. On the other hand since

$$TI_B(x) = \int_{\Psi} I_B(y)\mu_x(dy) = \int_{B} \mu_x(dy) = \mu_x(B),$$

for all $x \in \Psi$ and measurable sets B then a left transition operator T uniquely determines a corresponding collection $\mathbb{M}$, which in turn determines a transition function R. Similarly, a right transition operator T determines the collection $\mathbb{M}$ and the transition function since

$$\delta_x T(B) = \int_{\Psi} \mu_y(B)\delta_x(dy) = \mu_x(B),$$

where $\delta_x(y) = \begin{cases} 1 & y = x \\ 0 & y \neq x \end{cases}$, the delta distribution at x. Thus there is a one-to-one correspondence between transition operators (left or right) and transition functions.

Next, we reformulate definition 6.15 in terms of transition operators as follows.

Definition 6.17 Time-homogeneous Markov chain

Let T denote a transition operator defined in terms of a collection of probability measures $\mathbb{M} = \{\mu_x : x \in \Psi\}$, where Ψ is a Borel space. A random sequence $X = \{X_n : n \geq 0\}$ of Ψ-valued random variables that is adapted to a filtration $\mathcal{F} = \{\mathcal{F}_n : n \geq 0\}$, is called a time-homogeneous Markov chain with respect to the filtration $\mathcal{F}$ with state space Ψ and transition operator T if $\forall n \geq 0$ the conditional distribution of X_{n+1} given $\mathcal{F}_n$, is μ_{X_n}.

Unless otherwise stated, for what follows we assume that the Markov chain X is time-homogeneous and therefore μ_{X_n} does not depend on n.

Example 6.17 (Transition operators of a Markov chain) Let X be a Markov chain with respect to a filtration $\mathcal{F}$ with transition operator T and let Q_n denote the distribution of X_n. For a bounded measurable function f on Ψ we have

$$\begin{aligned} Q_{n+1}(B) &= P(X_{n+1} \in B) = E(P(X_{n+1} \in B|\mathcal{F}_n)) = E(\mu_{X_n}(B)) \\ &= \int_{\Psi} \mu_x(B)Q_n(dx) = (Q_nT)(B), \end{aligned}$$

so that $Q_{n+1} = Q_n T$. Moreover we have

$$E(f(X_{n+1})|\mathcal{F}_n) = \int_{\Psi} f(y)\mu_{X_n}(dy) = (Tf)(X_n).$$

The following remark presents several results on Markov chains.

Remark 6.10 (Markov chains) Let $X = \{X_n : n \geq 0\}$ be a Markov chain on Ψ adapted to a filtration $\mathcal{F} = \{\mathcal{F}_n : n \geq 0\}$.

1. A time-homogeneous Markov chain consists of 1) a state space Ψ, 2) a filtration (typically the minimal), 3) a transition operator T, and 4) the distribution of the first term in the sequence, Q_0.

2. The distribution of the Markov chain X is uniquely determined if we know T and Q_0. The converse is not true. Indeed, let $\Psi = \{0, 1\}$ and consider the sequence $X = (0, 0, 0, \ldots)$ (deterministic). Then we can find two different operators T_1 and T_2 such that X is a Markov chain with respect to T_1 and a Markov chain with respect to T_2.

3. Let v be a probability measure on Ψ and $f : \Psi \to \mathcal{R}$ a bounded function and then define $vf = \int_{\Psi} f dv$. We show that $(Q_0 T)(f) = Q_0(Tf)$. Indeed

$$(Q_0 T)(f) = \int_{\Psi} f(y)(Q_0 T)(dy) = \int_{\Psi} f(y) \int_{\Psi} \mu_x(dy) Q_0(dx) = \int_{\Psi} \left[\int_{\Psi} f(y)\mu_x(dy) \right] Q_0(dx),$$

so that

$$(Q_0 T)(f) = \int_{\Psi} (Tf)(x) Q_0(dx) = Q_0(Tf).$$

4. Let T^k denote the k-fold composition of T with itself defined recursively for any $k \geq 0$, (for $k = 0$ it is defined as the identity operator $T^0 = I$) based on

$$T^2 f(x) = T(Tf(x)) = \int_{\Psi} Tf(y)\mu_{Tf(x)}(dy).$$

Then we can show that $Q_n T^k = Q_{n+k}$,

$$E(f(X_{n+k})|\mathcal{F}_n) = (T^k f)(X_n) \ a.s.,$$

and

$$E(f(X_k)|X_0 = x) = (T^k f)(x),$$

for $k \geq 0$.

Most properties of a Markov chain can be studied via generators and we collect the definition next.

Definition 6.18 Discrete generator

Let X be a Markov chain with transition operator T. The discrete generator of X is the operator $G = T - I$, where I is the identity operator.

It is very often the case that we view time as a random variable. The following definition suggests that if we replace a fixed time n with an *a.s.* finite stopping time τ, the Markov property is preserved and moreover, results obtained for fixed time n can be extended similarly for the stopping time τ (e.g., see exercises 6.28 and 6.30).

Definition 6.19 Strong Markov property

Let X be a Markov chain with respect to a filtration $\mathcal{F}$ with transition operator T and assume that τ is any *a.s.* finite stopping time relative to $\mathcal{F}$. Then the random sequence $Y_n = X_{\tau+n}$, $n \geq 0$, is a Markov chain with transition operator T and it is adapted to the filtration $\{\mathcal{F}_{\tau+n} : n \geq 0\}$. In this case we say that X has the strong Markov property.

For discrete stopping times τ the extension from fixed n to random times τ is straightforward and the strong Markov property coincides with the standard Markov property. This is not the case for continuous stopping times (see definition 7.8).

6.5.2 Discrete State Space

Calculations simplify significantly when the state space Ψ is discrete (finite or countable). Assume that $\Psi = \{1, 2, \ldots, m\}$ (the case $m = +\infty$ can be developed similarly). A transition operator T in this case is thought of as an $m \times m$ matrix $T = [(T(x, y)]$, where
1) $T(x, y) \geq 0$, for all $1 \leq x, y \leq m$, and
2) each row sums up to 1, i.e., $\sum\limits_{y=1}^{m} T(x, y) = 1$, $x = 1, 2, \ldots, m$, since $\mu_x(\{y\}) = T(x, y)$, where $\mathbb{M} = \{\mu_x : x \in \Psi\}$ are the transition distributions.

A transition matrix with these two properties is known as a Markov matrix over Ψ. Note that

$$P(X_{n+1} = y) = \sum_{x=1}^{m} P(X_{n+1} = y|X_n = x)P(X_n = x),$$

so that $T(x, y) = P(X_{n+1} = y|X_n = x)$ does not depend on n and $Q_{n+1} = Q_n T$. Now if f is a vector, $f_{m\times 1} = (f(1), \ldots, f(m))'$, then the matrix T operates on f by multiplication on the left so that the notation Tf is consistent with matrix multiplication, that is

$$(Tf)(x) = \int_{\Psi} f(y)\mu_x(dy) = T_{m\times m} f_{m\times 1},$$

where $\mu_x(dy) = T(x, dy)$. Similarly, as a transition operator from the right T operates on probability measures that are row vectors, $Q_{1\times m} = (Q(\{1\}), \ldots, Q(\{m\}))$. Then the probability measure QT is identified with the vector $Q_{1\times m} T_{m\times m}$.

We present some results connecting transition probabilities with transition matrices below.

Remark 6.11 (Transition matrices) Let X be a Markov chain with transition matrix T and state space Ψ.

1. For any $n \geq 0$, $k > 0$ and $x_0, x_1, \ldots, x_k \in \Psi$, we have

$$P(X_{n+1} = x_1, \ldots, X_{n+k} = x_k | X_n = x_0) = T(x_0, x_1) \ldots T(x_{k-1}, x_k),$$

so that

$$P(X_0 = x_0, \ldots, X_k = x_k) = Q_0(\{x_0\}) T(x_0, x_1) \ldots T(x_{k-1}, x_k),$$

where Q_0 is the initial distribution of X.

2. For any $k \geq 0$ and $x, y \in \Psi$ we have

$$P(X_{n+k} = y | X_n = x) = T^k(x, y),$$

with $T^0 = I$, the identity matrix, where $T^k(x, y)$ denotes the probability that the Markov chain moves from state x to state y in k-steps and is obtained as the $(x, y)^{th}$ element of the kth power of the transition matrix T. The probability distribution $Q_k (= Q_0 T^k)$ of X_k is easily found using

$$P(X_k = y) = \sum_{x=1}^{m} P(X_k = y | X_0 = x) P(X_0 = x) = \sum_{x=1}^{m} T^k(x, y) P(X_0 = x).$$

3. Chapman-Kolmogorov equations For any $n, k \geq 0$, we have $T^{k+n} = T^k T^n$ or in terms of the elements of the matrices

$$T^{k+n}(x, y) = \sum_{z \in \Psi} T^k(x, z) T^n(z, y),$$

for any $x, y \in \Psi$.

We collect the matrix development of transition operators we have discussed thus far in the following theorem.

Theorem 6.15 (Matrix representation) Assume that T is a transition operator for a Markov chain X with countable state space Ψ. Then T can be represented as a matrix, denoted by T, with

$$T(x, y) = \delta_x T f_y = P(X_1 = y | X_0 = x),$$

$x, y \in \Psi$, where δ_x is the delta function at x and f_y is the indicator function of the singleton $\{y\}$. The representation has the property that for any distribution Q and bounded function f on Ψ, the distribution QT and the function Tf are represented by the corresponding matrix products QT and Tf and for any $k \geq 0$ the operator T^k is represented by the matrix product T^k.

Example 6.18 (Gambler's ruin) Let $\Psi = \{0, 1, 2, \ldots, 10\}$ and let $X_0 = 5$, the starting state of the sequence. Consider the transition matrix

$$T_{11\times 11} = \begin{bmatrix} 1 & 0 & 0 & 0 & 0 \\ 3/5 & 0 & 2/5 & 0 & 0 \\ \ldots & \ldots & \ldots & \ldots & \ldots \\ 0 & 0 & 3/5 & 0 & 2/5 \\ 0 & 0 & 0 & 0 & 1 \end{bmatrix},$$

so that $T(x, y) = P(X_{n+1} = y|X_n = x)$, is the transition operator of a Markov chain that describes the gambler's ruin problem, in which the gambler starts with \$5 and wins \$1 with probability 2/5 and loses \$1 with probability 3/5. The game ends if he reaches 0 or 10 dollars. The gambler's ruin at time 3 is a random variable with probability distribution given by the row vector $(0,0,0,0,0,1,0,0,0,0,0)T^3 = (0,0,0.216,0,0.432,0,0.288,0,0.064,0,0) = Q_3 = (P(X_3 = 0), P(X_3 = 1), \ldots, P(X_3 = 10))$, since $Q_0 = (0,0,0,0,0,1,0,0,0,0,0)$.

After 20 games the distribution is $Q_{20} = Q_0T^{20} = (0.5748, 0.0678, 0, 0.1183, 0, 0.0975, 0, 0.0526, 0, 0.0134, 0.0757)$, with expected fortune after 20 games given by \$2.1555. In terms of the transition probability measures μ_x, we require that $\mu_0(\{0\}) = 1 = \mu_{10}(\{10\})$ and $\mu_x(\{x-1\}) = 3/5, \mu_x(\{x+1\}) = 2/5$, $x = 1, \ldots, 9$, otherwise $\mu_x(\{y\}) = 0$, for $y \in \{x-1, x+1\}$.

Now if the gambler keeps playing to win as much as he can (no upper bound) then the state space is $\Psi = \{0, 1, 2, \ldots\}$ and the transition operator assumes the form

$$T = \begin{bmatrix} 1 & 0 & 0 & 0 & \ldots \\ 3/5 & 0 & 2/5 & 0 & \ldots \\ 0 & 3/5 & 0 & 2/5 & \ldots \\ 0 & 0 & 3/5 & 0 & \ldots \\ \ldots & \ldots & \ldots & \ldots & \ldots \end{bmatrix},$$

which is nothing but a random walk in $\mathcal{Z}_0^+$ with step distribution $Q(\{1\}) = 2/5 = 1 - Q(\{-1\})$. In terms of the transition probability measures μ_x this Markov chain requires $\mu_0(\{0\}) = 1$.

Example 6.19 (Birth-Death sequences) A Birth-Death sequence is a Markov chain $X = \{X_n : n \geq 0\}$ with state space $\mathcal{Z}_0^+$ and transition probability measures μ_x that satisfy $\mu_x(\{x-1, x, x+1\}) = 1$, $x > 0$ and $\mu_0(\{0, 1\}) = 1$. We can think of X_n as if it represents the size of a population at time n. In terms of the transition matrix T, $T(x, y) = 0$, if $|x - y| > 1$ and therefore $|X_{n+1} - X_n| \leq 1$ $a.s.$ for all $n \geq 0$. If $X_{n+1} = X_n + 1$ we say that a birth occurs at time n, whereas, if $X_{n+1} = X_n - 1$ we say that a death occurs at time n. If all the diagonal elements of T are 0 then all transitions are births or deaths. If for some $x > 0$, $T(x, x) > 0$, then the population is the same at time $n + 1$, i.e., $X_{n+1} = X_n$. Clearly, X as defined is a time-homogeneous Markov chain.

Example 6.20 (Random walks) Let $S = (S_0 = 0, S_1, \dots)$ denote a random walk on $\mathcal{Z}_0^+$ with step distribution $\{p_k : k \in \mathcal{Z}_0^+\}$ and steps $X = \{X_n : n \in \mathcal{Z}^+\}$. Since $S_{n+1} = S_n + X_{n+1}$ we have

$$P(S_{n+1} = y|S_0, \dots, S_n) = P(X_{n+1} = y - S_n|S_0, \dots, S_n) = p_{y-X_n} \text{ a.s.},$$

by the independence of X_{n+1} and $S_0, \dots, S_n$. Thus S is a Markov chain whose transition probabilities are $T(x, y) = P(S_{n+1} = y|S_n = x) = p_{y-x}$, $y \ge x$, that is, the transition matrix is of the form

$$T = \begin{bmatrix} p_0 & p_1 & p_2 & p_3 & \dots \\ 0 & p_0 & p_1 & p_2 & \dots \\ 0 & 0 & p_0 & p_1 & \dots \\ \dots & \dots & \dots & \dots & \dots \end{bmatrix}.$$

Example 6.21 (Binomial process) Let N_n denote the number of successes in n Bernoulli trials with $p = P(Success)$ (recall example 6.1.2). Since $N_{n+m} - N_m | N_0, N_1, N_2, \dots, N_m \overset{d}{=} N_{n+m} - N_m \overset{d}{=} N_n \sim Binom(n, p)$, the Binomial process $N = \{N_n : n \ge 0\}$ is a Markov chain with state space $\Psi = \mathcal{Z}_0^+$, initial distribution $Q_0(\{0\}) = 1, Q_0(\{x\}) = 0, x \in \mathcal{Z}^+$ and transition probabilities

$$T(x, y) = P(N_{n+1} = y|N_n = x) = \begin{cases} p & y = x + 1 \\ q & y = x \\ 0 & \text{otherwise} \end{cases},$$

where $q = 1 - p$, so that the transition matrix is

$$T = \begin{bmatrix} q & p & 0 & 0 & \dots \\ 0 & q & p & 0 & \dots \\ 0 & 0 & q & p & \dots \\ \dots & \dots & \dots & \dots & \dots \end{bmatrix}$$

The m-step transition probabilities are easily computed as

$$T^m(x, y) = P(N_{n+m} = y|N_n = x) = P(N_m = y - x) = C_{y-x}^m p^{y-x} q^{m-y+x},$$

$y = x, \dots, m + x$.

Example 6.22 (Times of successes) Let T_k denote the time that the kth success occurs in a Bernoulli process (recall example 6.1.3). Since $T_{k+m} - T_m | T_1, T_2, \dots, T_m \overset{d}{=} T_{k+m} - T_m \overset{d}{=} T_k \sim NB(k, p)$, $T = \{T_k : k \in \mathcal{Z}_0^+\}$ is a Markov chain with state space $\Psi = \{0, 1, 2, \dots\}$. Since $T_0 = 0$, the initial distribution is $Q_0(\{0\}) = 1, Q_0(\{x\}) = 0, x \in \mathcal{Z}^+$. The transition probabilities are easily computed as

$$T(x, y) = P(T_{k+1} = y|T_k = x) = P(T_1 = y - x) = \begin{cases} pq^{y-x-1} & y \ge x + 1 \\ 0 & \text{otherwise} \end{cases},$$

so that the transition matrix is of the form

$$T = \begin{bmatrix} 0 & p & pq & pq^2 & \dots \\ 0 & 0 & p & pq & \dots \\ 0 & 0 & 0 & p & \dots \\ \dots & \dots & \dots & \dots & \dots \end{bmatrix}.$$

For the m-step transition probabilities we have

$$T^m(x, y) = P(T_{k+m} = y|T_k = x) = P(T_m = y - x) = C_{m-1}^{y-x-1} p^m q^{y-x-m},$$

for $y \geq x + m$.

Example 6.23 (Branching processes) Let μ be a probability distribution on $\mathcal{Z}_0^+$. A branching process with branching distribution μ (also known as the offspring distribution) is a Markov chain on $\mathcal{Z}_0^+$ having transition probability measures $\mu_x = \mu^{*x}$. Branching processes can be interpreted as population models (like Birth-Death processes). More precisely, if the process is at state x we think of the population as having x members each reproducing independently. During a "birth" the member dies and produces a random number of offspring with $\mu(\{k\})$ denoting the probability that there are k offspring from a particular individual. Thus, the convolution μ^{*x} is the distribution of the upcoming population given that the population is currently at state x.

Example 6.24 (Renewal sequences) Assume that μ is a probability distribution on $\overline{\mathcal{Z}}^+$ and define

$$\mu_x = \begin{cases} \mu & x = 1 \\ \delta_{x-1} & 1 < x < \infty \\ \delta_\infty & x = \infty \end{cases}.$$

Let $X = \{X_n : n \geq 0\}$ be a Markov chain with initial state 1 and transition distributions μ_x, $x \in \overline{\mathcal{Z}}^+$ and define $Y_n = I(X_n = 1)$. Then $Y = \{Y_n : n \geq 0\}$ is a renewal sequence with waiting time distribution μ. If we replace the initial distribution δ_1 of X by an arbitrary distribution Q_0 then the sequence Y is a delayed renewal sequence.

6.5.3 The Martingale Problem

In this section we connect MGs and Markov chains. The discrete version of the MG problem is given next.

Definition 6.20 The martingale problem

Let $\mathcal{B}_\Psi$ denote all the bounded measurable functions on some Borel space Ψ. Assume that $G : \mathcal{B}_\Psi \to \mathcal{B}_\Psi$ is some operator on $\mathcal{B}_\Psi$ and let $\mathcal{F} = \{F_n : n \geq 0\}$ a filtration in a probability space $(\Omega, \mathcal{A}, P)$. A sequence X of Ψ-valued random variables on $(\Omega, \mathcal{A}, P)$ is said to be a solution to the MG problem for G and $\mathcal{F}$ if for all bounded measurable functions $f : \Psi \to \mathcal{R}$ the sequence Y defined by

$$Y_n = f(X_n) - \sum_{k=0}^{n-1} (Gf)(X_k),$$

is a martingale with respect to the filtration $\mathcal{F}$.

The following theorem shows the usefulness of the MG problem.

Theorem 6.16 (Markov chains via the martingale problem) Assume the notation of the previous definition. A sequence X of Ψ-valued random variables is a Markov chain with respect to $\mathcal{F}$ and X has generator G if and only if X is a solution to the MG problem for G and $\mathcal{F}$.

This theorem is useful in two ways; first, it gives us a large collection of MGs that can be used to analyze a given Markov chain. Second, it provides us with a way of assessing that a random sequence is Markov. The latter is more useful in the continuous time-parameter case and will be revisited later.

6.5.4 Visits to Fixed States: General State Space

We turn now to investigating the behavior of a Markov chain with respect to a fixed set of states and in particular, the random variables and probabilities of hitting times and return times. For what follows let $X = \{X_n : n \geq 0\}$ be a Markov chain with denumerable state space Ψ and transition operator T. For a fixed state $y \in \Psi$ or Borel set $B \subset \Psi$, define the probability that, starting at x, the Markov chain X ever visits (hits) state y or the set of states B by

$$H(x, y) = P(X_n = y, \text{ for some } n \geq 0 | X_0 = x), \tag{6.17}$$

and

$$H(x, B) = P(X_n \in B, \text{ for some } n \geq 0 | X_0 = x). \tag{6.18}$$

Note that $H(x, B) = 1$, $x \in B$. Similarly, define the probability that X ever returns to y or B starting from y or a state $x \in B$ by

$$H^+(y, y) = P(X_n = y, \text{ for some } n > 0 | X_0 = y), \tag{6.19}$$

and

$$H^+(x, B) = P(X_n \in B, \text{ for some } n > 0 | X_0 = x). \tag{6.20}$$

Clearly, $H(x, B) = H^+(x, B)$, for $x \in B^c$. Note that for any transition operator T we have $H^+(x, B) = (TH)(x, B)$, for fixed B.

Definition 6.21 Harmonic functions

Let G be a discrete generator on a state space Ψ and let B be a Borel subset of Ψ. A bounded measurable function $f : \Psi \to \mathcal{R}$ is G-subharmonic on B if $Gf(x) \geq 0$, $\forall x \in B$, whereas, it is called G-superharmonic on B if $-f$ is G-subharmonic on B.

It is called G-harmonic on B if it is both G-subharmonic and G-superharmonic on B, i.e., $Gf(x) \geq 0$ and $G(-f)(x) \geq 0$. If $B = \Psi$ then f is simply called G-harmonic. Note that harmonic functions f for generators G with $G(-f)(x) = -Gf(x)$, must satisfy $Gf(x) = 0$.

The following theorem gives us a way of calculating $H(x, B)$. The proof is requested as an exercise.

Theorem 6.17 (Finding hitting probabilities) Let G be the discrete generator of a Markov chain X with state space Ψ and let B be a Borel subset of Ψ. Then the function $x \mapsto H(x, B)$ is the minimal bounded function h with the properties:
(i) $h(x) = 1$, for $x \in B$,
(ii) h is G-harmonic on B^c,
(iii) $h(x) \geq 0$, for all $x \in \Psi$,
where $H(x, B)$ is minimal means that for any other function h that satisfies properties (i)-(iii), we have $H(x, B) \leq h(x)$, $\forall x \in \Psi$.

Example 6.25 (Uncountable state space Ψ) Assume that X is a Markov chain with state space $\mathcal{R}^+$ and transition distributions $\mu_x = \frac{1}{3}\delta_{x/2} + \frac{2}{3}\delta_{4x}$, where δ_y is the delta function at y. We fix $B = (0, 1]$ and calculate $H(x, B)$ by applying theorem 6.17. First note that the discrete generator G of X is given by

$$Gh(x) = (T - I)h(x) = \frac{1}{3}h(x/2) + \frac{2}{3}h(4x) - h(x),$$

since $(Th)(x) = \int\limits_{\Psi} h(y)\mu_x(dy) = \frac{1}{3}h(x/2) + \frac{2}{3}h(4x)$ and $(Ih)(x) = h(x)$. The second requirement of theorem 6.17 requires the minimal function to be harmonic for $x > 1$, that is,

$$Gh(x) = \frac{1}{3}h(x/2) + \frac{2}{3}h(4x) - h(x) = 0,$$

and setting $g(y) = h(2^y)$, $y \in \mathcal{R}$, we can write the requirement as

$$\frac{1}{3}g(y-1) - g(y) + \frac{2}{3}g(y+2) = 0, \tag{6.21}$$

for $y > 0$. If we restrict to $y \in \mathcal{Z}^+$, equation (6.21) is a third-order homogeneous linear difference equation. We try solutions of the form $g(y) = k^y$ and obtain the following condition on k

$$\frac{1}{3} - k + \frac{2}{3}k^3 = 0.$$

This polynomial has three solutions: $k_1 = -1.366$, $k_2 = 1$ and $k_3 = 0.366$. Consequently, every solution of (6.21) is of the form

$$g(y) = ak_1^y + b + ck_3^y.$$

By condition (iii) of the theorem, applied for large odd y if $a \geq 0$ and large even y if $a \leq 0$, we conclude that $a = 0$. From condition (i) we obtain $b + c = 1$. Finally, to get a minimal solution without violating condition (iii) we take $b = 0$ so that

$$g(y) = (0.366)^y,$$

for $y \in \mathcal{Z}^+$. But $g(y) = g(\lceil y \rceil)$ for $y > 0$ (or $x > 1$) and $g(y) = 1$, for $y \leq 0$ (or $x \in (0, 1]$), where $\lceil y \rceil$ denotes the ceiling of the real number y. Therefore, the hitting probability is given by

$$H(x, (0, 1]) = \begin{cases} 1 & x \in (0, 1] \\ (0.366)^m & 2^{m-1} < x \leq 2^m \end{cases}$$

$m \geq 0$.

Example 6.26 (Countable state space Ψ) Recall the Birth-Death Markov chain of example 6.19, but now assume that $\mu_0(\{0\}) = 1$, $\mu_x(\{x-1\}) = \frac{1}{x+1}$, $x > 0$ and $\mu_x(\{x\}) = \frac{x}{x+1}$, $x > 0$. We calculate $H(x, 0)$, the probability that the population dies out (at any step). First note that for $x > 0$, the discrete generator G of X is given by

$$Gh(x) = (T - I)h(x) = \frac{1}{x+1}h(x-1) + \frac{x}{x+1}h(x) - h(x).$$

The second requirement of theorem 6.17 requires the minimal function to be harmonic for $x > 0$, that is,

$$\frac{1}{x+1}h(x-1) - \frac{1}{x+1}h(x) = 0,$$

or $h(x) = h(x-1)$ and trying the solution $h(x) = k^x$, for some constant k, we get $h(x) = k$. From condition (i) $h(0) = 1$, so that $k = 1$ and condition (iii) is satisfied. Therefore $H(x, 0) = 1$, $x \geq 0$ and the population will surely die out.

6.5.5 Visits to Fixed States: Discrete State Space

We consider now the discrete state space case. Throughout this section assume that $X = \{X_n : n \geq 0\}$ is a time-homogeneous Markov chain with discrete state space Ψ and transition matrix T. Define $P_x(A) = P(A|X_0 = x)$ and $E_x(Y) = E(Y|X_0 = x)$ so that for fixed $x \in \Psi$, $P_x(A)$ is a probability measure on the measurable space $(\Omega, \mathcal{A})$ where the X_n are defined.

Remark 6.12 (Visits to fixed states) We fix a state $y \in \Psi$ and define the following important quantities used in investigating visits to fixed states.

1. Times of successive visits In general, we denote by $T_1(\omega), T_2(\omega), \ldots$, the successive indices $n \geq 1$ for which $X_n(\omega) = y$ or, in the case that there is no such n, we set $T_1(\omega) = T_2(\omega) - T_1(\omega) = \cdots = +\infty$. If y appears only a finite number of times m on the sequence X then we let $T_1(\omega), T_2(\omega), \ldots$, denote the successive indices $n \geq 1$ for which $X_n(\omega) = y$ and let $T_{m+1}(\omega) - T_m(\omega) = T_{m+2}(\omega) - T_{m+1}(\omega) = \cdots = +\infty$. Note that $T_m(\omega) \leq n$ if and only if y appears in $\{X_1(\omega), \ldots, X_n(\omega)\}$ at least m-times. Therefore, every T_m is a stopping time and the strong Markov property holds at T_m. Moreover, $X_{T_m} = y$ when $\{T_m < \infty\}$ occurs. As a consequence, we can show that

$$P_x(T_{m+1} - T_m = k|T_1, \ldots, T_m) = \begin{cases} 0 & T_m = \infty \\ P_y(T_1 = k) & T_m < \infty \end{cases} \text{ a.s.,} \tag{6.22}$$

for any $x \in \Psi$.

2. Visits in a fixed number of steps Define the probability that the chain starting at x moves to state y for the first time in exactly k-steps as

$$H_k(x, y) = P_x(T_1 = k),$$

$k \geq 1$, where $x \in \Psi$ and $T_1 = \inf\{n \geq 1 : X_n = y\}$ the first time X visits state y (first return time if $x = y$). Note the difference between the probability $H_k(x, y)$ of visiting y for the first time starting from x and $T^k(x, y) = P_x(X_k = y) = \delta_x T^k f_y$, the probability of visiting y (perhaps many times during the k-steps) starting from x, both in exactly k-steps. Now for $k = 1$ we have

$$H_1(x, y) = P_x(T_1 = 1) = P_x(X_1 = y) = T(x, y),$$

and for $k \geq 2$ we can easily see that

$$H_k(x, y) = \sum_{z \in \Psi \smallsetminus \{y\}} P_x(X_1 = z) P_z(X_1 \neq y, \ldots, X_{k-1} \neq y, X_k = y),$$

and therefore

$$H_k(x, y) = \begin{cases} T(x, y) & k = 1 \\ \sum\limits_{z \in \Psi \smallsetminus \{y\}} T(x, z) H_{k-1}(z, y) & k \geq 2 \end{cases}, \tag{6.23}$$

which is a recursive formula that can be used to compute $H_k(x, y)$. Clearly, we can write

$$H(x, y) = P_x(T_1 < \infty) = \sum_{k=1}^{+\infty} H_k(x, y), \tag{6.24}$$

for the probability that, starting from x, the Markov chain X ever visits y. Summing over all k in (6.23) we have

$$H(x, y) = T(x, y) + \sum_{z \in \Psi \smallsetminus \{y\}} T(x, z) H(z, y),$$

for all $x, y \in \Psi$, which is a system of linear equations in $H(x, y)$ that can be used to solve for $H(x, y)$, $x \in \Psi$.

Finally, assume that $\Psi = \{x_1, x_2, \ldots\}$ and let $H_k = (H_k(x_1, y), H_k(x_2, y), \ldots)$. Note that we omit state y from the calculations in (6.23) as if we have removed the yth column of T and therefore define T_{-y} to be the matrix obtained from T by replacing its yth column with zeros. As a consequence, H_1 is the yth column of T and we can write (6.23) as

$$H_k = T_{-y} H_{k-1},$$

$k \geq 2$.

3. Total number of visits For a realization of the Markov chain $X(\omega)$, $\omega \in \Omega$, let $N_y(\omega) = \sum\limits_{n=0}^{+\infty} I(X_n(\omega) = y)$ denote the total number of visits by the chain X to state y, with $N_y(\omega) = m$ if and only if $T_1(\omega) < \infty, \ldots, T_m(\omega) < \infty$ and $T_{m+1}(\omega) = \infty$. Using equation (6.22), the events $\{T_1 < \infty\}$, $\{T_2 - T_1 < \infty\}, \ldots, \{T_m - T_{m-1} < \infty\}$, $\{T_{m+1} - T_m = +\infty\}$ are independent and their probabilities are, starting at x, $H(x, y)$, $H(y, y), \ldots, H(y, y)$, $1 - H(y, y)$, respectively. Therefore, we have

$$P_y(N_y = m) = H(y, y)^{m-1}(1 - H(y, y)), \tag{6.25}$$

$m = 1, 2, \ldots$, so that $N_y | X_0 = y \sim Geo(1 - H(y, y))$ and if we start at $x \in \Psi$ we have

$$P_x(N_y = m) = \begin{cases} 1 - H(x, y) & m = 0 \\ H(x, y) H(y, y)^{m-1} (1 - H(y, y)) & m \geq 1 \end{cases}. \quad (6.26)$$

Summing over all m in (6.25) we find the probability that the total number of returns to y is finite, that is, $P_y(N_y < \infty) = \sum_{m=1}^{+\infty} P_y(N_y = m)$.

4. Expected number of visits Now if $H(y, y) = 1$ then $P_y(N_y < \infty) = 0$ (or $N_y = +\infty$ *a.s.*) so that the expected total number of visits to state y, starting from y, is $E_y(N_y) = +\infty$. In contrast, if $H(x, y) < 1$ we have the geometric distribution of (6.25) with $P_y(N_y < \infty) = 1$ and $E_y(N_y) = (1 - H(y, y))^{-1}$. Similarly, we obtain $E_x(N_y) = H(x, y)(1 - H(y, y))^{-1}$, for $x \neq y$. Putting these means in the form of a matrix $R = [(R(x, y))]$ with

$$R(x, y) = E_x(N_y) = \begin{cases} (1 - H(y, y))^{-1} & x = y \\ H(x, y) R(y, y) & x \neq y \end{cases}, \quad (6.27)$$

we have what is known as the potential matrix of X with the conventions $1/0 = \infty$ and $0 * \infty = 0$.

5. Relating T and R Using the definition of N_y and the MCT we can write

$$R(x, y) = E_x \left[\sum_{n=0}^{+\infty} I(X_n = y) \right] = \sum_{n=0}^{+\infty} E_x [I(X_n = y)] = \sum_{n=0}^{+\infty} P_x(X_n = y) = \sum_{n=0}^{+\infty} T^n(x, y),$$

or in terms of the matrices R and T, we have

$$R = I + T + T^2 + \ldots,$$

where I is the identity matrix and therefore we can compute R if we have all the powers of T. It is useful in calculations to note that

$$RT = T + T^2 + T^3 + \ldots,$$

so that

$$R(I - T) = (I - T)R = I,$$

and if $(I - T)^{-1}$ exists then we have right away $R = (I - T)^{-1}$. In general, once we solve for R we can invert equation (6.27) and compute $H(x, y)$ as

$$H(x, y) = \begin{cases} (R(y, y) - 1) R(y, y)^{-1} & x = y \\ R(x, y) R(y, y)^{-1} & x \neq y \end{cases}.$$

Appendix remark A.6 presents additional results on the calculation of H and R.

Example 6.27 (Hitting probabilities) Consider a Markov chain X with state space $\Psi = \{a, b, c\}$ and transition matrix

$$T = \begin{bmatrix} 1 & 0 & 0 \\ 1/4 & 1/4 & 1/2 \\ 1/3 & 1/3 & 1/3 \end{bmatrix},$$

fix $y = c$ and define $H_k = (H_k(a,c), H_k(b,c), H_k(c,c))'$, the k-step probabilities that we visit state c for the first time, starting from a or b or c. Writing

$$T_{-c} = \begin{bmatrix} 1 & 0 & 0 \\ 1/2 & 1/4 & 0 \\ 1/3 & 1/3 & 0 \end{bmatrix},$$

and using equation $H_k = T_{-c}H_{k-1}$ we have for $k = 1$: $H_1 = (T(a,c), T(b,c), T(c,c))'$ $= (0, 1/2, 1/3)'$ (the third column of T) and for $k \geq 2$ we can easily see that $H_2 = \left(0, \frac{1}{2}\frac{1}{4}, \frac{1}{2}\frac{1}{3}\right)'$, $H_3 = \left(0, \frac{1}{2}\left(\frac{1}{4}\right)^2, \frac{1}{2}\left(\frac{1}{3}\right)^2\right)'$, $H_4 = \left(0, \frac{1}{2}\left(\frac{1}{4}\right)^3, \frac{1}{2}\left(\frac{1}{3}\right)^3\right)'$ and therefore

$$\begin{aligned} H_k(a,c) &= 0,\ k \geq 1, \\ H_k(b,c) &= 2^{-1}(1/4)^{k-1},\ k \geq 1, \\ H_k(c,c) &= 1/3,\ k = 1, \text{ and} \\ H_k(c,c) &= 2^{-1}(1/3)^{k-1},\ k \geq 2. \end{aligned}$$

Thus starting at a the process never visits c, that is, $P_a(T_1 = +\infty) = 1$ (or $H(a,c) = 0$). Starting at b the probability that c is never visited is computed using equation (6.24) as

$$P_b(T_1 = +\infty) = 1 - P_b(T_1 < +\infty) = 1 - \sum_{k=1}^{+\infty} H_k(b,c) = 1 - \sum_{k=1}^{+\infty} 2^{-1}(1/4)^{k-1} = 1/3. \tag{6.28}$$

Finally, starting at c the probability that X never returns to c is given by

$$P_c(T_1 = +\infty) = 1 - \sum_{k=1}^{+\infty} H_k(b,c) = 1 - 1/3 - \sum_{k=2}^{+\infty} 2^{-1}(1/3)^{k-1} = 5/12. \tag{6.29}$$

The following remark shows us how we can use the renewal theory results in order to study the behavior of a Markov chain, including a first approach on how to conduct classification of states for the Markov chain.

Remark 6.13 (Renewal theory and visits to a fixed state) We can use renewal theory as an alternative way to obtain the probability $H(x,y)$ as a limit via the renewal theorem (p. 231). We restrict our development to the discrete state space case Ψ, although some of the results can be applied in general. In particular, define a random sequence $Y = \{Y_n : n \geq 0\}$, with $Y_n = I(X_n = y)$, so that Y indicates the times $n \geq 0$ for which the Markov chain X visits state $y \in \Psi$. Assuming that $X_0 = x \neq y$ and using the strong Markov property, we have that Y is a delayed renewal sequence (or simply a renewal sequence if $X_0 = y$) on the probability space $(\Omega, \mathcal{A}, P_x)$, with $P_x(.) = P(.|X_0 = x)$, for each $x \in \Psi$. Note that $P_x(Y_n = 1) = T^n(x,y) = P_x(X_n = y)$, for fixed $x, y \in \Psi$, so that $\{T^n(y,y) : n \geq 0\}$ denotes the potential sequence $\{\pi_n^y : n \geq 0\}$ for the renewal sequence corresponding to $x = y$. Therefore, we can use the renewal theory results to obtain results for $\{T^n(x,y) : n \geq 0\}$.

1. $H(x,y)$ as a limit Assume that γ_y is the period and m_y the mean waiting time of the renewal sequence Y corresponding to y. For the Markov chain X, m_y can be thought of as the mean return time to state y. Since $\pi_n^y = T^n(y,y)$, using the renewal theorem, we can show that

$$\lim_{n\to+\infty} \gamma_y^{-1} \sum_{k=0}^{\gamma_y-1} T^{n+k}(x,y) = \frac{H(x,y)}{m_y}, \tag{6.30}$$

and therefore

$$\lim_{n\to+\infty} \sum_{k=1}^{n} T^k(x,y)/n = \frac{H(x,y)}{m_y}. \tag{6.31}$$

2. Renewal measure and total number of visits Assume that $X_0 = y$. In view of the definition of N_y in remark 6.12.3 we can write the renewal measure R_y of the renewal sequence $Y_n = I(X_n = y)$ in terms of the total number of visits N_y as

$$N_y = R_y(\mathcal{Z}_0^+) = \sum_{n=0}^{+\infty} I(X_n = y).$$

An immediate consequence is a relationship between the potential measure and the potential matrix as follows

$$R(y,y) = E_y(N_y) = \sum_{n=0}^{+\infty} T^n(y,y) = \sum_{n=0}^{+\infty} \pi_n^y = \Pi_y(\mathcal{Z}_0^+),$$

where Π_y is the potential measure corresponding to the renewal sequence Y for each $y \in \Psi$. Further note that the waiting time distribution W_y describes the distribution of the time until the first renewal, or in terms of the Markov chain X, the time X visits state y for the first time, that is,

$$W_y(\{k\}) = H_k(y,y) = P_y(T_1 = k),\ k \geq 1.$$

From theorem 6.12 $\Pi_y(\mathcal{Z}_0^+) = 1/W_y(\{+\infty\})$ or in terms of potential matrix

$$W_y(\{+\infty\}) = 1/\Pi_y(\mathcal{Z}_0^+) = 1/R(y,y).$$

Consequently, we can write the mean return time to y (mean of the waiting time distribution W_y) as

$$m_y = E_y(T_1) = \sum_{k=0}^{+\infty} kH_k(y,y).$$

3. State classification The states of the Markov chain X can be classified according to the corresponding classification of the renewal sequences $Y_n = I(X_n = y)$ (definition 6.12), with $X_0 = y$, for each $y \in \Psi$. Since classification for renewal sequences works for both denumerable and countable state spaces Ψ we discuss the general case briefly here. In particular, a state $y \in \Psi$ is called recurrent if and only if $\sum_{n=0}^{+\infty} \pi_n^y = +\infty$ in which case it is called null recurrent if and only if $\lim_{n\to+\infty} \pi_n^y = 0$. The state is called transient if $\sum_{n=0}^{+\infty} \pi_n^y < +\infty$ and periodic with period γ_y, if $\gamma_y > 1$, or aperiodic if $\gamma_y = 1$.

Although the latter can be used to classify the states of a Markov chain, we provide an extensive and more systematic way of classification next.

6.5.6 State Classification

In what follows we restrict the development and definitions to the discrete state space Ψ. For denumerable Ψ one can use remark 6.13.3. For what follows let $X = \{X_n : n \geq 0\}$ be a Markov chain with state space Ψ and transition matrix T.

Definition 6.22 State classification

Fix a state $y \in \Psi$ and define $T_1 = \inf\{n \geq 1 : X_n = y\}$ as the time of first visit to y.

1. Recurrent or transient state The state y is called recurrent if $P_y(T_1 < \infty) = 1$, otherwise if $P_y(T_1 = +\infty) > 0$ then y is called transient.

2. Recurrent null or non-null state A recurrent state y is called recurrent null if $E_y(T_1) = +\infty$, otherwise it is called recurrent non-null.

3. Periodic state A recurrent state y is said to be periodic with period γ if $\gamma \geq 2$ is the largest integer for which $P_y(T_1 = n\gamma \text{ for some } n \geq 1) = 1$, otherwise if there is no such $\gamma \geq 2$, y is called aperiodic.

Equivalent forms of the classification criteria above can be obtained based on hitting probabilities and the potential matrix, as we see below.

Remark 6.14 (State classification) Let N_y denote the total number of visits to a fixed state $y \in \Psi$. In view of remark 6.13.4 it is straightforward to see that y is recurrent if and only if

$$H(y, y) = 1 \Leftrightarrow R(y, y) = +\infty \Leftrightarrow P_y(N_y = +\infty) = 1.$$

Similarly, y is transient if and only if

$$H(y, y) < 1 \Leftrightarrow R(y, y) < +\infty \Leftrightarrow P_y(N_y < +\infty) = 1.$$

Now if y is periodic with period γ then the Markov chain X can return to y only at times γ, 2γ, $3\gamma, \ldots$ and the same is true for the second return time, the third and so forth. Therefore, starting at y, $T^n(y, y) = P_y(X_n = y) > 0$, only if $n \in \{0, \gamma, 2\gamma, 3\gamma, \ldots\}$. To check if $T^n(y, y) > 0$, in view of the Chapman-Kolmogorov equations, we only need to find states $y_1, y_2, \ldots, y_{n-1} \in \Psi$ such that $T(y, y_1) > 0$, $T(y_1, y_2) > 0, \ldots, T(y_{n-1}, y) > 0$. This leads to the following criterion for periodicity.

Criterion 6.1 (Periodicity) Let γ be the GCD of the set of all $n \geq 1$ such that $T^n(y, y) > 0$. If $\gamma = 1$ then y is aperiodic, otherwise, if $\gamma \geq 2$ then y is periodic with period γ.

Next we discuss some useful properties of a Markov chain, based on its states properties.

Definition 6.23 Markov chain properties

Let $X = \{X_n : n \geq 0\}$ be a Markov chain with discrete state space Ψ.

1. For any $x, y \in \Psi$ we say that y can be reached by x if $H(x, y) > 0$ and write $x \to y$.

2. A set of states $B \subset \Psi$ is said to be closed if the only states that can be reached by states of B are those of the set B.

3. Absorbing state A state that forms a closed set is called absorbing.

4. The closed set B is irreducible if no proper subset of B is a closed set.

5. Irreducible A Markov chain is called irreducible if its only closed set is the state space Ψ.

An immediate consequence of the latter definition are the following criteria.

Criterion 6.2 (Absorbing state) A state y is absorbing if $T(y, y) = 1$.

Criterion 6.3 (Irreducible) A Markov chain is irreducible if and only if all states can be reached from each other.

Several results and definitions regarding the states of a Markov chain and the chain itself, are collected below.

Remark 6.15 (State and Markov chain properties) For a discrete state space Ψ we can show the following.

1. Let $C = \{x_1, x_2, \dots\} \subset \Psi$ be a closed set and define $Q(x_1, x_2) = T(x_1, x_2)$, $x_1, x_2 \in \Psi$. Then Q is a Markov matrix.

2. If $x \to y$ and $y \to z$ then $x \to z$ for any $x, y, z \in \Psi$.

3. If x is recurrent and $x \to y$ then $y \to x$, $H(x, y) = 1$ and y is recurrent.

4. For each recurrent state y there exists an irreducible closed set to which y belongs.

5. For any Markov chain X the recurrent states can be arranged in a unique manner into irreducible closed sets. As a consequence, any results obtained for irreducible Markov chains can be applied to irreducible closed sets.

6. From a recurrent state, only recurrent states can be reached.

7. If an irreducible closed set has finitely many states, then it has no transient states.

8. If Ψ is finite then no state is recurrent null and not all states are transient.

9. If x and y are recurrent states then either $H(x, y) = H(y, x) = 1$ or $H(x, y) = H(y, x) = 0$.

10. If x is recurrent and y is transient then $H(x, y) = 0$.

11. If $x, y \in \Psi$ with $x \to y$ and $y \to x$ then they both have the same period and have the same classification, i.e., both are recurrent null, or recurrent non-null or transient.

12. If $x, y \in \Psi$ are in the same closed set then for any state $z \in \Psi$ we have $H(z, x) = H(z, y)$.

13. Using corollary 6.2 and remark 6.13.3 we have the following in terms of the Markov chain X with transition matrix T. The state $y \in \Psi$ is recurrent if and only if $\sum\limits_{n=0}^{+\infty} T^n(y, y) = +\infty$ in which case it is null recurrent if and only if $\lim\limits_{n \to +\infty} T^n(y, y) = 0$. Moreover, the state y is transient if $\sum\limits_{n=0}^{+\infty} T^n(y, y) < +\infty$ which implies that $\lim\limits_{n \to +\infty} T^n(y, y) = 0$. The only case that remains is for recurrent non-null states, in which case $\lim\limits_{n \to +\infty} T^n(x, y)$ exists under the assumption of aperiodicity (see theorem 6.19). From remark 6.13.1 the limit of equation (6.31) always exists.

The following theorem allows us to perform classification in an efficient way for closed sets of states.

Theorem 6.18 (Irreducible Markov chain and classification) Let $X = \{X_n : n \geq 0\}$ be an irreducible Markov chain. Then either all states are transient or all are recurrent null or all are recurrent non-null. Either all states are aperiodic, otherwise if one state has period γ then all states are periodic with the same period γ.

Proof. Since X is irreducible for any two states $x, y \in \Psi$ we must have $x \to y$ and $y \to x$, which implies that there are some integers j and k such that $T^j(x, y) > 0$ and $T^k(y, x) > 0$. For the smallest such j and k let $a = T^j(x, y)T^k(y, x) > 0$.
(i) If y is recurrent then by remark 6.15.3 x is recurrent. If y is transient then x must also be transient (otherwise remark 6.15.6 would make y recurrent, a contradiction).
(ii) Suppose that y is recurrent null. Then by remark 6.15.13 $T^m(y, y) \to 0$ as $m \to \infty$. Since $T^{n+j+k}(y, y) \geq aT^n(x, x)$ we also have $T^n(x, x) \to 0$ as $n \to \infty$ and since x is recurrent this can happen only if it is recurrent null (remark 6.15.13). Replacing y with x and x with y in the above formulation, we have the converse, namely, if x is recurrent null then so is y.
(iii) Suppose that y is periodic with period γ. Since $T^{j+k}(y, y) \geq a > 0$ we must have $j + k$ a multiple of γ (remark 6.14) and therefore $T^{n+j+k}(y, y) = 0$ whenever n is not a multiple of γ so that $T^{n+j+k}(y, y) \geq aT^n(x, x)$ which implies that $T^n(x, x) = 0$ if n is not a multiple of γ. Consequently, x must be periodic with period $\gamma' \geq \gamma$ by criterion 6.1. Replacing the role of y and x, we obtain $\gamma' \leq \gamma$ so that x and y have the same period. ■

The following theorem connects state classification with limiting behavior for Markov chains.

Theorem 6.19 (State classification and limiting behavior) Fix a state $y \in \Psi$.
(i) If y is transient or recurrent null then for any $x \in \Psi$

$$\lim_{n\to+\infty} T^n(x,y) = 0. \tag{6.32}$$

(ii) If y is recurrent non-null aperiodic then

$$\pi(y) = \lim_{n\to+\infty} T^n(y,y) > 0, \tag{6.33}$$

and for any $x \in \Psi$

$$\lim_{n\to+\infty} T^n(x,y) = H(x,y)\pi(y). \tag{6.34}$$

Proof. Let $\{\pi_n^y : n \geq 0\}$ denote the potential sequence and $m_y = \sum\limits_{n=1}^{+\infty} nW_y(\{n\})$, $m_y \in [1,\infty]$, the mean of the waiting time distribution W_y of the (possibly delayed) renewal sequence defined by $Y_n = I(X_n = y)$ with finite period $\gamma_y < \infty$ (the case $\gamma_y = +\infty$ is trivial since state y is visited only once and $m_y = +\infty$). Note that by definition we have $\pi_{n\gamma_y}^y = T^{n\gamma_y}(y,y)$, $n \geq 0$.
(i) Assume that $X_0 = y$, so that $Y = \{Y_n : n \geq 0\}$ is a renewal sequence. If y is transient or recurrent null then $m_y = +\infty$ by definition in both cases and an appeal to the renewal theorem yields

$$\lim_{n\to+\infty} T^{n\gamma_y}(y,y) = \frac{\gamma_y}{m_y} = 0,$$

with $T^k(y,y) = 0$, if k is not a multiple of γ_y, so that $\lim\limits_{n\to+\infty} T^n(y,y) = 0$, in general. If $X_0 = x \neq y$, let $\tau = \inf\{n \geq 0 : X_n = y\}$ and use the strong Markov property at τ to obtain

$$\begin{aligned} T^{n+i}(x,y) &= \sum_{k=0}^{n+i} P_x(X_{n+i} = y, \tau = k) = \sum_{k=0}^{n+i} P_y(X_{n+i-k} = y)P_x(\tau = k) \\ &= \sum_{k=0}^{+\infty} I(k \leq n+i)T^{n+i-k}(y,y)P_x(\tau = k). \end{aligned}$$

Noting that $H(x,y) = \sum\limits_{k=0}^{+\infty} P_x(\tau = k)$ we use the DCT for sums to pass the limit under the infinite sum and use the result for the case $y = x$ already proven to establish

$$\begin{aligned} \lim_{n\to+\infty} T^n(x,y) &= \lim_{n\to+\infty} \sum_{k=0}^{+\infty} I(k \leq n+i)T^{n+i-k}(y,y)P_x(\tau = k) \\ &= \sum_{k=0}^{+\infty} \lim_{n\to+\infty} T^{n+i-k}(y,y)P_x(\tau = k) \\ &= \left[\lim_{n\to+\infty} T^n(y,y)\right] \sum_{k=0}^{+\infty} P_x(\tau = k) = \frac{\gamma_y}{m_y} H(x,y) = 0. \end{aligned}$$

(ii) If y is recurrent non-null aperiodic then $\gamma_y = 1$, $m_y < +\infty$ and $H(y,y) = 1$, so that applying equation (6.30) we obtain

$$\lim_{n\to+\infty} T^n(x,y) = \frac{H(x,y)}{m_y},$$

with $\lim_{n\to+\infty} T^n(y,y) = \frac{1}{m_y}$. Finally, we simply set $\pi(y) = \frac{1}{m_y} > 0$ to establish (6.33) and (6.34). ■

The next two theorems help with classification, in particular, when Ψ is countable.

Theorem 6.20 (Recurrent non-null) Let X be an irreducible Markov chain with transition matrix T and assume that $\Psi = \{x_1, x_2, \ldots\}$. Then all states are recurrent non-null if and only if there exists a solution to the system of equations

$$\pi(y) = \sum_{x\in\Psi} \pi(x)T(x,y),\ \forall y \in \Psi, \tag{6.35}$$

with

$$\sum_{x\in\Psi} \pi(x) = 1, \tag{6.36}$$

or in matrix form, the system $\boldsymbol{\pi} = \boldsymbol{\pi}T$ subject to $\boldsymbol{\pi}\mathbf{1} = 1$, where $\boldsymbol{\pi} = (\pi(x_1), \pi(x_2), \ldots)$ and $\mathbf{1} = (1, 1, \ldots)^T$. If a solution exists it is unique and

$$\pi(y) = \lim_{n\to+\infty} T^n(x,y) > 0, \tag{6.37}$$

$\forall x, y \in \Psi$.

Proof. Assume that all states are recurrent non-null. Then $\forall x, y \in \Psi$, we have $H(x,y) = 1$ and by theorem 6.19 the limits

$$\lim_{n\to+\infty} T^n(x,y) = g(y), \tag{6.38}$$

exist and they satisfy

$$g(y) > 0 \text{ and } \sum_{x\in\Psi} g(x) = 1. \tag{6.39}$$

Now if $A \subset \Psi$ with A finite then the Chapman-Kolmogorov equation yields

$$T^{n+1}(x,y) = \sum_{z\in\Psi} T^n(x,z)T(z,y) \geq \sum_{z\in A} T^n(x,z)T(z,y),$$

and taking limit as $n \to +\infty$, in view of (6.38) we obtain

$$g(y) \geq \sum_{z\in A} g(z)T(z,y),\ y \in \Psi. \tag{6.40}$$

Since A is arbitrary we may take a sequence of such subsets that increases to Ψ so that we have

$$g(y) \geq \sum_{z\in\Psi} g(z)T(z,y),\ y \in \Psi.$$

Now if the strict inequality above was true for some y then summing both sides over y we have

$$\sum_{z\in\Psi} g(y) > \sum_{z\in\Psi} g(z),$$

since $\sum_{y\in\Psi} T(z,y) = 1$, a contradiction, and therefore (6.40) is an equality. Equations (6.39) and (6.40) show that g is a solution to the system (6.35)-(6.36). To show that the solution is unique let h be another solution to (6.35)-(6.36). Then writing in matrix form we have $h = hT$ and using this iteratively we have $h = hT = hT^2 = ... = hT^n$, that is,

$$h(y) = \sum_{z\in\Psi} h(z)T^n(z, y). \quad (6.41)$$

Taking the limit as $n \to +\infty$ and applying BCT we can write

$$h(y) = \sum_{z\in\Psi} h(z) \lim_{n\to+\infty} T^n(z, y) = \sum_{z\in\Psi} h(z)g(y) = g(y), \quad (6.42)$$

$\forall y \in \Psi$ and therefore there is only one solution (6.37) is satisfied.

For the converse we need to show that existence of a solution to (6.35)-(6.36) implies that all the states are recurrent non-null. Assume that they are not. Then (6.41) would still hold and (6.42) becomes

$$h(y) = \sum_{z\in\Psi} h(z) \lim_{n\to+\infty} T^n(z, y) = \sum_{z\in\Psi} h(z)0 = 0,$$

$\forall y \in \Psi$ which contradicts $\sum_{z\in\Psi} h(z) = 1$. ■

The proof of the following theorem is requested as an exercise.

Theorem 6.21 (Recurrent null or transient) Let X be an irreducible Markov chain with state space $\Psi = \{x_1, x_2, \dots\}$, transition matrix T and let $\mathbf{T}_{-k}$ be the matrix obtained from T by deleting the kth row and the kth column for some $x_k \in \Psi$. Then all states are recurrent if and only if the only solution to the system of equations

$$u(y) = \sum_{x\in\Psi\smallsetminus\{x_k\}} u(x)\mathbf{T}_{-k}(y, x), \ \forall y \in \Psi, \quad (6.43)$$

with $0 \le u(y) \le 1$, is the solution $u(y) = 0$, $\forall y \in \Psi \smallsetminus \{x_k\}$.

We summarize state classification below.

Remark 6.16 (Performing state classification) The following steps should be used in order to perform state or Markov chain classification for a Markov chain X with states space Ψ.

1. Finite Ψ Assume that Ψ is a finite state space.
(1) Find all the irreducible closed sets. Using remark 6.15 and theorem 6.18 we conclude that all states belonging to an irreducible closed set are recurrent non-null.
(2) The remaining states are all transient.
(3) Periodicity is assessed using criterion 6.1 for each irreducible closed set and then using theorem 6.18.

2. Countable Ψ When Ψ is countable it is possible to have irreducible closed sets with countably many states that are transient or recurrent null.
(1) Find all the irreducible closed sets and apply theorem 6.20 to check if the set contains recurrent null or non-null states.
(2) If the states in a closed set are classified as recurrent non-null, apply theorem

6.21 to determine if they are transient or not.
(3) Periodicity is assessed as in step (3) above.

Example 6.28 (State classification for a simple random walk) Consider a simple random walk X on $\Psi = \{0, 1, 2, \ldots\}$ with step distribution $Q(\{1\}) = p$, $Q(\{-1\}) = q$, $0 < p < 1$, $q = 1 - p$ and a barrier at 0, that is, the transition matrix is given by

$$T = \begin{bmatrix} 0 & 1 & 0 & 0 & \ldots \\ q & 0 & p & 0 & \ldots \\ 0 & q & 0 & p & \ldots \\ \ldots & \ldots & \ldots & \ldots & \ldots \end{bmatrix}.$$

Since all states can be reached from each other the chain is irreducible. Starting at 0, in order to come back to 0 the chain must take as many steps forward as backward and therefore a return to 0 can occur only at steps numbered $2, 4, 6, \ldots$, so that state 0 is periodic with period $\gamma = 2$. From theorem 6.18, since the chain is irreducible, all states are periodic with period $\gamma = 2$. Furthermore, all states are transient, recurrent null or recurrent non-null. To assess if the states are recurrent non-null we use theorem 6.20. The system (6.35)-(6.36) can be written as

$$\begin{aligned} \pi(0) &= q\pi(1),\ \pi(1) = \pi(0) + q\pi(2), \\ \pi(2) &= p\pi(1) + q\pi(3),\ \pi(3) = p\pi(2) + q\pi(4), \end{aligned}$$

and so forth, which can be rewritten as

$$\begin{aligned} \pi(1) &= \pi(0)/q, \\ \pi(2) &= (\pi(0)/q - \pi(0))/q = \pi(0)p/q^2, \\ \pi(3) &= \left(p/q^2 - p/q\right)\pi(0)/q = \pi(0)p^2/q^3, \end{aligned}$$

and so on. Consequently, any solution is of the form

$$\pi(x) = \pi(0)q^{-1}(p/q)^{x-1}, \tag{6.44}$$

$x = 1, 2, \ldots$, for some constant $\pi(0)$ that can be found from (6.44), namely,

$$\sum_{x=0}^{+\infty} \pi(x) = 1 \Leftrightarrow \pi(0) + q^{-1}\pi(0) \sum_{x=1}^{+\infty} (p/q)^{x-1} = 1. \tag{6.45}$$

Now if $p < q$ then $p/q < 1$ and the sum in (6.45) converges so that we can write

$$\sum_{x=0}^{+\infty} \pi(x) = \pi(0)+q^{-1}\pi(0) \sum_{x=1}^{+\infty} (p/q)^{x-1} = \pi(0)+q^{-1}\pi(0)(1 - p/q)^{-1} = \pi(0)2q(q-p)^{-1},$$

and choosing $\pi(0) = (q - p)(2q)^{-1} = (1 - p/q)/2$, we can satisfy $\sum_{x=0}^{+\infty} \pi(x) = 1$. Therefore, if $p < q$

$$\pi(x) = \begin{cases} (1 - p/q)/2 & x = 0 \\ (1 - p/q)(2q)^{-1}(p/q)^{x-1} & x \geq 1 \end{cases},$$

is a solution to the system (6.35)-(6.36) and all states are recurrent non-null.

When $p \geq q$, the solution to the system has $\sum_{x=0}^{+\infty} \pi(x)$ that is either 0 (if $\pi(0) = 0$) or infinite (if $\pi(0) \neq 0$), so that there is no way to satisfy both (6.35)-(6.36) and therefore all states are not recurrent non-null. We use theorem 6.21 to help us decide if the states are transient or recurrent null. Excluding the row and column corresponding to state 0 from T we obtain the matrix $\mathbf{T}_{-0}$ and write the system of equations (6.43) as

$$u(1) = pu(2),\ u(2) = qu(1) + pu(3),\ u(3) = qu(2) + pu(4), \ldots,$$

so that in general, we have a second-order difference equation

$$u(x) = qu(x-1) + pu(x+1),$$

$x \geq 2$ and we solve it as follows. First write

$$p(u(x+1) - u(x)) = q(u(x) - u(x-1)),$$

$x \geq 2$ and from the first equation we have $p(u(2) - u(1)) = qu(1)$. Iterating on x we obtain

$$u(x+1) - u(x) = (q/p)^{x-1} (u(2) - u(1)) = (q/p)^x u(1),$$

so that for any $x \geq 1$ we have

$$\begin{aligned} u(x+1) &= (u(x+1) - u(x)) + (u(x) - u(x-1)) + \cdots + (u(2) - u(1)) + u(1) \\ &= \left[(q/p)^x + (q/p)^{x-1} + \cdots + q/p + 1\right] u(1). \end{aligned}$$

Consequently, we have obtained a solution to (6.43) of the form

$$u(x) = c\left[(q/p)^{x-1} + (q/p)^{x-2} + \cdots + q/p + 1\right],$$

$x \geq 1$ where c is some constant. If $p = q$ then $u(x) = cx$, $x \geq 1$ is a solution to (6.43) and the only way to have $0 \leq u(x) \leq 1$ is for $c = 0$, that is, if $p = q$ the only solution to (6.43) is $u(x) = 0$, $x \geq 1$ and all states are recurrent. Since we already know that they are not recurrent non-null they must be recurrent null.

For $p > q$ choosing $c = 1 - q/p$ we have that $u(x) = 1 - (q/p)^x$, $x \geq 1$, which satisfies $0 \leq u(x) \leq 1$, for all $x \geq 1$. Therefore, all states are transient for $p > q$.

6.5.7 Limiting Behavior

In theorem 6.19 we derived the limiting distribution for a Markov chain based on state classification and using renewal theory arguments. We formally collect the general definition of a limiting distribution in the following.

Definition 6.24 Limiting behavior of a Markov chain

Let $X = \{X_n : n \geq 0\}$ be a Markov chain with state space Ψ, transition operator T and corresponding transition distributions $\{\mu_x : x \in \Psi\}$. A measure π is called a stationary (or invariant) measure for X (or T) if it satisfies

$$\pi(B) = \pi T(B) = \int_{\Psi} \mu_x(B)\pi(dx), \tag{6.46}$$

for all Borel sets $B \subset \Psi$. If π is a probability measure then it is called the equilibrium (or the stationary or the limiting or the invariant) distribution and the Markov chain X is called stationary.

Several results based on the latter definition are collected below.

Remark 6.17 (Limiting behavior) From theorem 6.20 an irreducible recurrent non-null Markov chain has a finite stationary distribution always. We note the following.

1. When $\Psi = \{x_1, x_2, \ldots\}$ (countable) we can write (6.46) as

$$\pi(y) = \sum_{x \in \Psi} \pi(x) T(x, y), \ \forall y \in \Psi, \tag{6.47}$$

where $\boldsymbol{\pi}$ assumes the form $\boldsymbol{\pi} = (\pi(x_1), \pi(x_2), \ldots)$.

2. If $\boldsymbol{\pi} = Q_0$ is the initial distribution of the Markov chain then (6.46) states that $Q_1 = Q_0 T = \boldsymbol{\pi} T = \boldsymbol{\pi} = Q_0$, that is, using the Markov property and induction we have

$$P(X_n \in B) = Q_n(B) = Q_0(B) = P(X_0 \in B),$$

for all $k, n \geq 0$. Consequently, $\boldsymbol{\pi}$ represents a possible equilibrium for the chain after an infinite number of steps, i.e., the probability distribution of where the chain settles eventually.

3. Solving (6.46) for $\boldsymbol{\pi}$ (subject to $\boldsymbol{\pi}(\Psi) = 1$) can be a painstaking task in the general case. However, for discrete Ψ, theorems 6.20 and 6.21 give us a way of obtaining the solution $\boldsymbol{\pi}$ as the limit of the n-step transition probabilities $T^n(x, y)$. For a general state space $\Psi \subset \mathcal{R}^p$ there exist similar results (as the ones presented in this section) for a collection of Markov chains called Harris chains. See Robert and Casella (2004, Chapter 6) or Durrett (2010, Section 6.8), for a discussion and proofs of some of the important results.

4. Applying (6.46) repeatedly we have $\boldsymbol{\pi} = \boldsymbol{\pi} T = \boldsymbol{\pi} T^2 = \cdots = \boldsymbol{\pi} T^n$, for any $n \geq 0$.

5. Stationary measures may exist for transient chains such as random walks in $p \geq 3$ dimensions. However, if there is a stationary distribution π for a Markov chain X then all states x for which $\pi(x) > 0$ are recurrent.

6. If X is irreducible and has stationary distribution π then

$$\pi(x) = \frac{1}{E_x(\tau)},$$

where $\tau = \inf\{n > 0 : X_n = x\}$ is the first visit (or return if $X_0 = x$) time to state x.

7. If X is irreducible then the following are equivalent: (i) some x is positive recurrent, (ii) there is a stationary distribution, and (iii) all states are positive recurrent.

The two concepts that make showing that there is a stationary distribution easier are collected next.

Definition 6.25 Reversibility and detailed balance

Let $X = \{X_n : n \geq 0\}$ be a Markov chain on a state space Ψ with transition matrix T.
(i) If X is stationary it is called (time) reversible when

$$X_{n+1}|(X_{n+2} = x) \stackrel{d}{=} X_{n+1}|(X_n = x),$$

for any $x \in \Psi$.
(ii) The Markov chain X is said to satisfy the detailed balance condition if there exists a measure π such that

$$\pi(x)T(x,y) = \pi(y)T(y,x), \tag{6.48}$$

for all $x, y \in \Psi$.

Note the following about proving detailed balance.

Remark 6.18 (Showing detailed balance) Time reversibility and detailed balance are equivalent when π exists and is a density (in which case it is the equilibrium). Detailed balance guarantees that a stationary measure exists (it is a sufficient condition) but it is a stronger condition than $\pi = \pi T$. Indeed, summing (6.48) over x we obtain

$$\sum_{x\in\Psi} \pi(x)T(x,y) = \pi(y) \sum_{x\in\Psi} T(y,x) = \pi(y),$$

since $\sum\limits_{x\in\Psi} T(y,x) = 1$, which is equation (6.47), so that (6.48) gives a solution to $\pi = \pi T$. Most Markov chain Monte Carlo (MCMC) algorithms utilize detailed balance in order to show that a Markov chain X with transition operator T has limiting distribution π. In terms of definition (6.24) and equation (6.46) we can write equation (6.48) of detailed balance as

$$\int_A \pi(x)T(x,B)dx = \int_B \pi(y)T(y,A)dy,$$

or equivalently

$$\int_A \pi(x) \int_B T(x,y)dydx = \int_B \pi(y) \int_A T(x,y)dxdy,$$

and therefore

$$\int_A \int_B \pi(x)T(x,y)dydx = \int_A \int_B \pi(y)T(x,y)dydx, \tag{6.49}$$

for all Borel sets $A, B \in \mathcal{B}(\Psi)$ so that (6.49) holds if and only if (6.48) holds.

Example 6.29 (Birth-Death sequence) Recall the Birth-Death chain of example 6.19 with $\Psi = \mathcal{Z}_0^+$ and assume that $\mu_x(\{x-1\}) = q$, $\mu_x(\{x\}) = r$ and $\mu_x(\{x+1\}) = p$, with $\mu_0(\{0\}) = 0$ and $\mu_0(\{1\}) = 1$, $p + q + r = 1$. Then the measure defined by $\pi(x) = \prod\limits_{i=1}^{x}(p_{i-1}/q_i)$, $x \in \mathcal{Z}_0^+$, satisfies detailed balance. Indeed, first note that

$$\pi(x)T(x,x+1) = p_x \prod_{i=1}^{x}(p_{i-1}/q_i) = \pi(x+1)T(x+1,x),$$

and since $T(x,y) = 0$, for $|x-y| > 1$, we have

$$\pi(x)T(x,y) = \pi(y)T(y,x),$$

for all $x, y \in \mathcal{Z}_0^+$. As a consequence, Birth-Death Markov chains have an equilibrium distribution if and only if $\sum\limits_{x=0}^{+\infty}\prod\limits_{i=1}^{x}(p_{i-1}/q_i) < +\infty$ since in this case we can make π a valid density.

6.6 Stationary Sequences and Ergodicity

Stationary sequences are random sequences that model the behavior of some system that is in equilibrium. For what follows assume that $X = \{X_n : n \geq 0\}$ is a random sequence with $X_n : (\Omega, \mathcal{A}, P) \to (\Psi, \mathcal{G})$, where $(\Omega, \mathcal{A}, P)$ is a probability space and $(\Psi, \mathcal{G})$ is some Borel space. The sequence X is a random vector in the measurable space $(\Psi^\infty, \mathcal{G}^\infty)$. We begin with some basic definitions.

Definition 6.26 Stationary random sequence

Let $\tau : \Psi^\infty \to \Psi^\infty$ be a measurable map (shift transformation) defined by $\tau(X) = \tau((X_0, X_1, \ldots)) = (X_1, X_2, \ldots)$.

1. Stationarity The random sequence X is called stationary if X and the shifted sequence $\tau(X)$ have the same distribution, i.e., $X \stackrel{d}{=} \tau(X)$.

2. Shift-invariant measure If Q is the distribution of a stationary random sequence X then τ is a measure preserving transformation (or shift-invariant) on $(\Psi^\infty, \mathcal{G}^\infty, Q)$, that is, $Q(\tau^{-1}(A)) = Q(A)$, $\forall A \in \mathcal{G}^\infty$.

This following remark summarizes some of the consequences of the latter definition.

Remark 6.19 (Stationarity and ergodic theory) The theory of measure preserving transformations is called ergodic theory. Denote by $\tau^k = \tau \circ \cdots \circ \tau$, the k-fold composition of τ with itself so that $\tau^k(X) = (X_k, X_{k+1}, \ldots)$, $k \geq 0$.

1. It is easy to see that the random sequence X is stationary if and only if $X \stackrel{d}{=} \tau^k(X)$, for all $k \geq 0$. In other words, for all $k, m \geq 0$, the vectors $(X_0, X_1, \ldots, X_m)$ and $(X_k, X_{k+1}, \ldots, X_{k+m})$ have the same distribution.

2. Even if τ^{-1} does not exist (not one to one), we can still define τ^{-k}, the k-fold composition of τ^{-1} with itself. For any $k \geq 1$, define $\mathcal{T}_k = \{A \in \mathcal{G}^\infty : A = \tau^{-k}(A)\}$, which can be shown to be a sub-σ-field in $\sigma(X_k, X_{k+1}, \ldots)$ (by iterating τ^{-1} on $A \in \sigma(X_1, X_2, \ldots)$). Then a tail σ-field is defined by $\mathcal{H} = \bigcap\limits_{k=1}^{+\infty}\mathcal{T}_n$, with $\mathcal{H} \subset \mathcal{T}$, where $\mathcal{T} = \bigcap\limits_{n=1}^{+\infty}\sigma(X_n, X_{n+1}, \ldots)$, the tail σ-field of the sequence X (see definition

4.9). For $k = 1$ define

$$\mathcal{I} = \{A \in \mathcal{G}^\infty : A = \tau^{-1}(A)\}, \tag{6.50}$$

the collection of shift-invariant sets.

3. For any probability measure Q^* on $(\Psi^\infty, \mathcal{G}^\infty)$, τ need not be shift-invariant (it is if $Q^* = Q$ the distribution of a stationary random sequence X). We denote by $\mathbb{M}_\tau$ the collection of all probability measures on $(\Psi^\infty, \mathcal{G}^\infty)$ that are shift-invariant with respect to τ. A member of $\mathbb{M}_\tau$ may or may not be the distribution of some stationary sequence.

4. A random sequence $(\ldots, X_{-1}, X_0, X_1, \ldots)$ is called (two-sided) stationary if the distribution of $(X_k, X_{k+1}, \ldots)$ does not depend on $k \in \mathcal{Z}$. Definition 6.26 of stationarity is referred to as one-sided stationarity.

5. A measurable map $f : (\Omega, \mathcal{A}) \to (\mathcal{R}, \mathcal{B}_1)$ is $\mathcal{I}$-measurable if and only if $f \circ \tau = f$.

We are now ready to collect the formal definition of an ergodic sequence.

Definition 6.27 Ergodicity

A stationary random sequence X with distribution Q is ergodic if the σ-field $\mathcal{I}$ is 0-1 trivial with respect to Q. In this case the distribution Q is also called ergodic.

The main result about stationary sequences is called the ergodic theorem.

Theorem 6.22 (Birkhoff's ergodic theorem) Let $X = (X_0, X_1, \ldots)$ be an $\mathcal{R}$-valued stationary sequence and set $S_n = X_0 + \cdots + X_{n-1}$. Assume that $E|X_0| < +\infty$. Then

$$\lim_{n\to+\infty} S_n/n = E(X_0|X^{-1}(\mathcal{I}))\ a.s. \tag{6.51}$$

and

$$\lim_{n\to+\infty} E\left(\left|S_n/n - E(X_0|X^{-1}(\mathcal{I}))\right|\right) = 0, \tag{6.52}$$

that is, $S_n/n \to E(X_0|X^{-1}(\mathcal{I}))$ a.s. and in $\mathcal{L}^1$. If in addition X is an ergodic sequence then

$$\lim_{n\to+\infty} S_n/n = E(X_0)\ a.s. \tag{6.53}$$

Proof. We prove (6.51) in detail and request (6.52) as an exercise. Equation (6.53) is an immediate consequence of (6.51) under ergodicity (see remark 6.20.1 below).
Without loss of generality we may assume that $E(X_0|X^{-1}(\mathcal{I})) = 0$ a.s. Indeed, since $E|X_0| < +\infty$, $E(X_0|X^{-1}(\mathcal{I}))$ is finite a.s. and therefore we can work with the sequence $X_n - E(X_0|X^{-1}(\mathcal{I}))$ (which is stationary) once we prove the theorem for the

special case $E(X_0|X^{-1}(\mathcal{I})) = 0$. Therefore, our goal is to prove that $S_n/n \to 0$ a.s. as $n \to \infty$.

Fix $\varepsilon > 0$ and define the set

$$A_\varepsilon = \left\{\omega : \limsup_{n\to\infty} \frac{S_n(\omega)}{n} > \varepsilon\right\}.$$

We need to prove that $P(A_\varepsilon) = 0$, which implies $P\left(\limsup\limits_{n\to\infty} \frac{S_n(\omega)}{n} > 0\right) = 0$. Then applying the same argument to the sequence $-X$ we have a.s. convergence since replacing X by $-X$ leads to

$$\liminf_{n\to\infty} \frac{S_n(\omega)}{n} \geq 0 \text{ a.s.},$$

and therefore $\lim\limits_{n\to\infty} \frac{S_n(\omega)}{n} = 0$ a.s. $[P]$.

First we note that $A_\varepsilon \in X^{-1}(\mathcal{I})$ since the random variable $Z = \limsup\limits_{n\to\infty} \frac{S_n(\omega)}{n}$ is such that $Z \circ \tau = Z$ and remark 6.19.5 applies. Consequently, $E(X_0 I_{A_\varepsilon}) = E(E(X_0|X^{-1}(\mathcal{I})I_{A_\varepsilon})) = 0$ and thus $E\left[(X_0 - \varepsilon)I_{A_\varepsilon}\right] = -\varepsilon P(A_\varepsilon)$. Therefore, in order to show $P(A_\varepsilon) = 0$ it suffices to show that

$$E\left[(X_0 - \varepsilon)I_{A_\varepsilon}\right] \geq 0. \tag{6.54}$$

Define the sequence of shifted partial sums by $S_n^0 = S_n - X_0$, $n = 1, 2, \ldots$ and let

$$\begin{aligned} M_n &= \max\{S_1 - \varepsilon, S_2 - 2\varepsilon, \ldots, S_n - n\varepsilon\}, \\ M_n^0 &= \max\{S_1^0 - \varepsilon, S_2^0 - 2\varepsilon, \ldots, S_n^0 - n\varepsilon\}. \end{aligned}$$

By the definition of S_n^0 we have

$$X_0 - \varepsilon + \max(0, M_n^0) \geq M_n,$$

for $n = 0, 1, 2, \ldots$ and since $X_0 = S_1$ we have

$$X_0 - \varepsilon \geq M_n - \max(0, M_n^0).$$

Therefore

$$\begin{aligned} E\left[(X_0 - \varepsilon)I_{A_\varepsilon}\right] &\geq E\left[(M_n - \max(0, M_n^0))I_{A_\varepsilon}\right] \\ &= E\left[(\max(0, M_n) - \max(0, M_n^0))I_{A_\varepsilon}\right] + E\left[(M_n - \max(0, M_n))I_{A_\varepsilon}\right], \end{aligned}$$

for $n = 0, 1, 2, \ldots$ and since X is stationary and $A_\varepsilon \in X^{-1}(\mathcal{I})$, the random variables $\max(0, M_n)I_{A_\varepsilon}$ and $\max(0, M_n^0)I_{A_\varepsilon}$ have the same distribution so that $E\left[(\max(0, M_n) - \max(0, M_n^0))I_{A_\varepsilon}\right] = 0$. As a result,

$$E\left[(X_0 - \varepsilon)I_{A_\varepsilon}\right] \geq E\left[(M_n - \max(0, M_n))I_{A_\varepsilon}\right],$$

for $n = 0, 1, 2, \ldots$ and using the definitions of M_n and A_ε we have $\lim\limits_{n\to\infty} M_n(\omega) \geq 0$, for all $\omega \in A_\varepsilon$ so that

$$\lim_{n\to\infty} \left[(M_n - \max(0, M_n))I_{A_\varepsilon}\right] = 0,$$

a.s. $[P]$. But by the definition of M_n, $X_0 - \varepsilon \leq M_n$ so that

$$|X_0 - \varepsilon| \geq |M_n - \max(0, M_n^0)|,$$

and since X_0 is integrable, we can apply the DCT to swap the limit and the expectation sign to obtain

$$E[(X_0 - \varepsilon)I_{A_\varepsilon}] \geq \lim_{n\to\infty} E[(M_n - \max(0, M_n))I_{A_\varepsilon}]$$
$$= E\left[\lim_{n\to\infty}(M_n - \max(0, M_n))I_{A_\varepsilon}\right] = 0,$$

and equation (6.54) is proven. ■

Note the use of $X^{-1}(\mathcal{I})$ instead of $\mathcal{I}$ in Birkhoff's theorem. Since in our formulation the σ-field $\mathcal{I}$ is a sub-σ-field in $\mathcal{G}^\infty$, we have to condition on the inverse image $X^{-1}(\mathcal{I})$, which is a sub-σ-field in $\mathcal{A}$, in order to properly define the conditional expectation. Some consequences of the latter theorem follow.

Remark 6.20 (Consequences of Birkhoff's theorem) We can show the following.

1. If X is an ergodic sequence then the sub-σ-field $X^{-1}(\mathcal{I}) \subset \mathcal{A}$, is 0-1 trivial with respect to P. As a consequence, $E(X_0|X^{-1}(\mathcal{I})) = E(X_0)$ a.s. (a constant).

2. Hopf's maximal ergodic lemma For X and S_n as in theorem 6.22, with $E|X_0| < +\infty$ and $M_n = \max\{S_1, \ldots, S_n\}$ we have that $E(X_0 I(M_n > 0)) \geq 0$.

3. If X is not ergodic then $\mathcal{I}$ is not 0-1 trivial meaning that the space can be split into two sets A and A^c of positive probability such that $A, A^c \in \mathcal{I}$, i.e., $\tau(A) = A$ and $\tau(A^c) = A^c$, with $P(A) \notin \{0, 1\}$.

Example 6.30 (Stationary and ergodic sequences) Many of the sequences we have seen are stationary and ergodic sequences.

1. If the X_n are iid then X is a stationary sequence. Moreover, Kolmogorov's 0-1 law (theorem 4.12) applies to give that $\mathcal{T}$ is 0-1 trivial and therefore $\mathcal{I} \subset \mathcal{T}$ is 0-1 trivial so that X is ergodic. Since $\mathcal{I}$ is trivial, the ergodic theorem yields $\lim\limits_{n\to+\infty} S_n/n = E(X_0)$ *a.s.* (a constant) and therefore we have the SLLN as a special case.

2. Exchangeable sequences A finite or infinite sequence is called exchangeable if its distribution is invariant under permutations of its arguments. Using remark 6.19.1 we can show that these sequences are stationary. It can be shown that exchangeable sequences are ergodic if and only if they are iid.

3. Markov chain at equilibrium Assume that X is a Markov chain with countable state space Ψ and initial distribution Q_0 that is also its stationary distribution (i.e., $Q_0 = \pi$). Then X is a stationary sequence. Furthermore, it can be shown that if the Markov chain is irreducible then the sequence is ergodic. Now if X is irreducible and f a measurable function with $E^\pi(f(X)) = \sum\limits_{x\in\Psi} f(x)\pi(x) < \infty$ then the ergodic theorem applied on $f(X_n)$ gives

$$\lim_{n\to+\infty} (f(X_0) + \cdots + f(X_{n-1}))/n = \sum_{x\in\Psi} f(x)\pi(x) \ a.s.$$

4. Moving averages If $X = (X_0, X_1, \ldots)$ is a stationary sequence of $\mathcal{R}$-valued random variables then $Y = (Y_0, Y_1, \ldots)$ with $Y_n = (X_n + \cdots + X_{n+k-1})/k$, $n \geq 0$, for some $k > 0$ defines a stationary sequence.

5. Stationary non-ergodic sequence There are many stationary sequences that are not ergodic. For example, take $X_0 \sim Unif(-1, 1)$ and set $X_n = (-1)^n X_0$, $n \geq 1$. Then the sequence $X = (X_0, X_1, \ldots)$ is stationary but not ergodic.

When the σ-field $\mathcal{I}$ is not easy to work with we can use the distribution Q of X and the following approach in order to show ergodicity. For a proof of the ergodicity criterion see Fristedt and Gray (1997, p. 562).

Definition 6.28 Extremal measure

Let C be a convex set of distributions. A measure $Q \in C$ is extremal in C if Q cannot be written as a mixture of two distributions from C, that is, Q cannot be written as the mixture distribution $Q = pQ_1 + (1 - p)Q_2$, for some $0 < p < 1$ and distinct $Q_1, Q_2 \in C$.

Theorem 6.23 (Ergodicity criterion) Let $\mathbb{M}_\tau$ be the collection of all shift-invariant distributions with respect to τ on the measurable space $(\Psi^\infty, \mathcal{G}^\infty)$. A measure $Q \in \mathbb{M}_\tau$ is ergodic if and only if it is extremal in $\mathbb{M}_\tau$.

The following remark shows us how to assess ergodicity.

Remark 6.21 (Extremal measure and ergodicity) Definition 6.28 can be rephrased as: Q is extremal in C if $Q = pQ_1 + (1 - p)Q_2$ implies $Q = Q_1 = Q_2$, when $0 < p < 1$ and $Q_1, Q_2 \in C$. We note the following.

1. If $Q_1, Q_2 \in C$ are distinct ergodic measures then $Q_1 \perp Q_2$

2. Let X be a stationary Markov chain and denote by $\mathcal{E}$ the collection of all equilibrium distributions of X. Then X is ergodic if and only if its initial distribution is extremal in $\mathcal{E}$.

6.7 Applications to Markov Chain Monte Carlo Methods

Next we present a few applications of Markov chains to computational methods and in particular Bayesian computation and MCMC. We only discuss the discrete case for our illustrations below but the algorithms are given for the general case. Suppose that we wish to generate a random variable with pmf $\pi(j) = h(j)/c$, where $h(j)$, $j \in \Psi = \{1, 2, \ldots, m\}$ are positive integers, m is very large and the normalizing constant $c = \sum_{j=1}^{m} h(j)$ is hard to compute.

If we can somehow create an irreducible, aperiodic and time-reversible Markov chain with limiting distribution π then we could run the chain for n-steps, where n is large and approximate a realization from the discrete π using X_n. Moreover, we can use realizations we obtain after a large step k that can be thought of as approximate generated values from the target distribution to estimate for example $E(g(X)) = \sum_{i=1}^{m} g(i)\pi(i)$ using Monte Carlo integration by

$$E(g(X)) \simeq \frac{1}{n-k} \sum_{i=k+1}^{n} g(X_i).$$

The combination of Markov chains and Monte Carlo gave rise to the widely used approach called Markov Chain Monte Carlo (MCMC), where one constructs a Markov chain $X_0, X_1, \ldots$, that has equilibrium a given distribution π. The sampled X_i are only approximately distributed according to π and furthermore they are dependent.

The problem with the early states of the chain (the first k for instance as above) is that they are heavily influenced by the initial value $X_0 = j$ and hence we remove them from subsequent calculations. One can use what is known as the burn-in period, namely, use the first k generated states to estimate a starting state for the chain (e.g., take the average of the first k states and start the chain again with X_0 being this average). There is no theoretical justification for the burn-in, it is rather something intuitive. If the chain is indeed an irreducible aperiodic Markov chain then where we start does not matter. MCMC is best suited for high-dimensional multivariate distributions where it is difficult to find other methods of generating samples such as rejection samplers with high acceptance probabilities.

6.7.1 Metropolis-Hastings Algorithm

The first method we discuss is known in the literature as the Metropolis-Hastings (M-H) algorithm (Metropolis et al., 1953, and Hastings, 1970). Suppose that we create an irreducible Markov chain with transition matrix $\mathbf{Q} = [(q(i, j))]$, for $i, j = 1, 2, \ldots, m$. We define now a Markov chain $X = \{X_n : n \geq 0\}$ where if $X_n = i$ we generate a random variable Y with pmf $\{q(i, j)\}$, $j = 1, 2, \ldots, m$ (the i-th row from the transition matrix, known as the proposal kernel or distribution). Then if $Y = j$ we choose the next state as

$$X_{n+1} = \begin{cases} j & \text{w.p. } a(i, j) \\ i & \text{w.p. } 1 - a(i, j) \end{cases},$$

where the $a(i, j)$, will be determined.

Now the transition probabilities for the Markov chain $\{X_n\}$ become

$$\begin{aligned} p(i, j) &= q(i, j)a(i, j), \text{ if } j \neq i, \text{ and} \\ p(i, i) &= q(i, i) + \sum_{k=1, k\neq i}^{m} q(i, k)(1 - a(i, k)), \end{aligned}$$

and the chain will be time reversible with stationary probabilities $\{\pi(j)\}$ if the detailed balance condition holds, that is, $\pi(i)p(i, j) = \pi(j)p(j, i)$, $j \neq i$, or

$$\pi(i)q(i, j)a(i, j) = \pi(j)q(j, i)a(j, i), \quad j \neq i.$$

The Metropolis et al. (1953) method selects the $a(i, j)$ as

$$a(i, j) = \min\left\{\frac{\pi(j)q(j, i)}{\pi(i)q(i, j)}, 1\right\} = \min\left\{\frac{h(j)q(j, i)}{h(i)q(i, j)}, 1\right\},$$

in order to produce transition probabilities $p(i, j)$ of a time-reversible Markov chain as desired. Note that calculation of the normalizing constant is no longer an issue. The method proposed by Hastings (1970) is a special case when symmetry holds for q, i.e., $q(i, j) = q(j, i)$. The general algorithm for the discrete or continuous case is as follows.

Algorithm 6.1 (Metropolis-Hastings) Assume that the target distribution is $\pi(\mathbf{x}) \propto h(\mathbf{x})$, $\mathbf{x} \in \Psi$, with $h(\mathbf{x})$ known.

Step 1: Select a proposal distribution $q(\mathbf{y}|\mathbf{x})$ that is a density with respect to $\mathbf{y}$ for any $\mathbf{x} \in \Psi$. The collection $\{q(.|\mathbf{x}) : \mathbf{x} \in \Psi\}$ form the transition probability measures of the source Markov chain Y and must be such that Y is irreducible with state space Ψ and easy to sample from. Start the target Markov chain with some $\mathbf{X}_0 = \mathbf{x}_0$.

Step 2: Given that the chain is at state $\mathbf{X}_n$ at iteration (time) n generate $\mathbf{Y}$ from $q(\mathbf{y}|\mathbf{X}_n)$ and generate $U \sim Unif(0, 1)$.

Step 3: If

$$U < \frac{h(\mathbf{Y})q(\mathbf{X}_n|\mathbf{Y})}{h(\mathbf{X}_n)q(\mathbf{Y}|\mathbf{X}_n)},$$

then set $\mathbf{X}_{n+1} = \mathbf{Y}$, otherwise remain at the same state, i.e., $\mathbf{X}_{n+1} = \mathbf{X}_n$. Go to the previous step and continue to generate the next state.

6.7.2 Gibbs Sampling

The Gibbs sampler is an important case of the M-H algorithm and is widely used in Bayesian statistics. In general, we wish to generate a random vector $\mathbf{X} = (X_1, \ldots, X_p)$ with joint distribution $f(\mathbf{x})$ known up to a constant, i.e., $f(\mathbf{x}) = cg(\mathbf{x})$, where c is typically an intractable normalizing constant with the form of $g(\mathbf{x})$ known. The Gibbs sampler assumes that we can generate a random variable X with pmf

$$P(X = x) = P(X_i = x|X_j = x_j,\ j \neq i),$$

for any $i = 1, 2, \ldots, p$ and $x \in \Psi$. This distribution is also known as the full conditional of X_i given the rest of the Xs.

Now the implementation of the M-H algorithm is as follows: assume that the present state is $\mathbf{x} = (x_1, x_2, \ldots, x_p)$. We choose a coordinate equally likely, i.e., with probability $1/p$, say the ith coordinate. Then generate $X = x$ from $P(X = x)$ and consider the candidate for the next state as $\mathbf{y} = (x_1, \ldots, x_{i-1}, x, x_{i+1} \ldots, x_p)$. That is, the Gibbs sampler uses the M-H algorithm with proposal kernel

$$q(\mathbf{y}|\mathbf{x}) = p^{-1}P(X_i = x|X_j = x_j,\ j \neq i) = \frac{f(\mathbf{y})}{pP(X_j = x_j,\ j \neq i)},$$

and then accept the vector $\mathbf{y}$ as the next state with probability

$$a(\mathbf{x}, \mathbf{y}) = \min\left\{\frac{f(\mathbf{y})q(\mathbf{x}|\mathbf{y})}{f(\mathbf{x})q(\mathbf{y}|\mathbf{x})}, 1\right\} = \min\left\{\frac{f(\mathbf{y})f(\mathbf{x})}{f(\mathbf{x})f(\mathbf{y})}, 1\right\} = 1,$$

and therefore the Gibbs sampler always accepts the generated value! The Gibbs sampler is particularly useful in Bayesian computation and in frameworks where we work with a Gibbs distribution, e.g., Gibbs point process models or Markov random fields.

We present the algorithm for a Bayesian context. Consider modeling an experiment using a distribution for the data $f(\mathbf{x}|\boldsymbol{\theta})$, $\mathbf{x} \in \mathcal{X}$, i.e., we observe iid vectors $\mathbf{x}_1, \ldots, \mathbf{x}_n$ from f and we are interested in a full posterior analysis of the parameter vector $\boldsymbol{\theta} = (\theta_1, \ldots, \theta_p)$. Notice that each θ_i has its own marginal prior distribution $\pi_i(\theta_i|\mathbf{a}_i)$, where $\mathbf{a}_i$ is a vector of hyper-parameters and often the joint prior is $\pi(\boldsymbol{\theta}) = \prod_{i=1}^{p} \pi_i(\theta_i|\mathbf{a}_i)$. To obtain the full conditionals we write first the joint posterior distribution as

$$\pi(\boldsymbol{\theta}|\mathbf{x}_1, \ldots, \mathbf{x}_n) \propto f(\mathbf{x}_1, \ldots, \mathbf{x}_n|\boldsymbol{\theta})\pi(\boldsymbol{\theta}) = f(\mathbf{x}_1, \ldots, \mathbf{x}_n|\boldsymbol{\theta}) \prod_{i=1}^{p} \pi_i(\theta_i|\mathbf{a}_i),$$

and from here we isolate θ_i to find the full conditional for θ_i, i.e., $\pi_i(\theta_i|\mathbf{x}_{1:n}, \boldsymbol{\theta}_{-i}, \mathbf{a}_i)$.

Algorithm 6.2 (Gibbs sampler) We generate a random vector $\boldsymbol{\theta} = (\theta_1, \ldots, \theta_p)$ whose distribution is known up to a normalizing constant.
Step 1: Let $\boldsymbol{\theta}_{-i} = (\theta_1, \ldots, \theta_{i-1}, \theta_{i+1}, \ldots, \theta_p)$, the vector without the ith coordinate and obtain the forms (up to normalizing constants) of the full conditional distributions $\pi_i(\theta_i|\mathbf{x}_{1:n}, \boldsymbol{\theta}_{-i}, \mathbf{a}_i)$, $i = 1, 2, \ldots, p$.
Step 2: Suppose that the current state of the chain is $\boldsymbol{\theta}^{(r)} = (\theta_1^{(r)}, \ldots, \theta_p^{(r)})$. The Gibbs sampler has the (typically fixed for each step) updating order

draw $\theta_1^{(r+1)}$ from $\pi_1(\theta_1|\theta_2^{(r)}, \theta_3^{(r)}, \ldots, \theta_p^{(r)}, \mathbf{x}_{1:n}, \mathbf{a}_1)$ and update θ_1,
draw $\theta_2^{(r+1)}$ from $\pi_2(\theta_2|\theta_1^{(r+1)}, \theta_3^{(r)}, \ldots, \theta_p^{(r)}, \mathbf{x}_{1:n}, \mathbf{a}_2)$ and update θ_2,
...
draw $\theta_i^{(r+1)}$ from $\pi_i(\theta_i|\theta_1^{(r+1)}, \theta_2^{(r+1)}, \ldots, \theta_{i-1}^{(r+1)}, \theta_{i+1}^{(r)}, \ldots \theta_p^{(r)}, \mathbf{x}_{1:n}, \mathbf{a}_i)$
and update θ_i,
...
draw $\theta_p^{(r+1)}$ from $\pi_p(\theta_p|\theta_1^{(r+1)}, \theta_2^{(r+1)}, \ldots, \theta_{p-1}^{(r+1)}, \mathbf{x}_{1:n}, \mathbf{a}_p)$.

The new state now becomes $\boldsymbol{\theta}^{(r+1)} = (\theta_1^{(r+1)}, \ldots, \theta_p^{(r+1)})$.

6.7.3 Reversible Jump Markov Chain Monte Carlo

Sampling from varying dimension models can be a painstaking task since at each iteration the dimension of the (parameter) space can change, e.g., when we sample the parameters of a mixture of normals with a random number of components k. The problem is formulated best in terms of Bayesian model choice.

Remark 6.22 (Bayesian model choice) A Bayesian variable dimension model is defined as a collection of models

$$\mathcal{M}_k = \{f_k(.|\boldsymbol{\theta}_k) : \boldsymbol{\theta}_k \in \Theta_k\},$$

associated with a collection of priors $\pi_k(\boldsymbol{\theta}_k)$ on the parameters of these models and a prior on the indices of these models $\rho(k)$, $k \in \mathcal{K}$. The index set $\mathcal{K}$ is typically taken to be a finite set, e.g., $\mathcal{K} = \{1, 2, \ldots, k_{\max}\}$. The joint prior is given by a density

$$\pi(k, \boldsymbol{\theta}_k) = \rho(k)\pi_k(\boldsymbol{\theta}_k),$$

with respect to the standard product measure of the counting measure over $\mathcal{K}$ and (typically) the Lebesgue measure over Θ_k, where $\rho(k)$ is the density of k with respect to the counting measure over $\mathcal{K}$ and $\pi_k(\boldsymbol{\theta}_k)$ the density with respect to the Lebesgue measure over Θ_k with

$$(k, \boldsymbol{\theta}_k) \in \Theta = \bigcup_{m \in \mathcal{K}} \{m\} \times \Theta_m,$$

where $\Theta_m \subset \mathcal{R}^{d_m}$ for some $d_m \geq 1$. Now given the data $\mathbf{x}$, the model selected is the one with the largest posterior probability

$$\pi(\mathcal{M}_k|\mathbf{x}) = \frac{\rho(k) \int\limits_{\Theta_k} \pi_k(\boldsymbol{\theta}_k) f_k(\mathbf{x}|\boldsymbol{\theta}_k) d\boldsymbol{\theta}_k}{\sum\limits_{i \in \mathcal{K}} \rho(i) \int\limits_{\Theta_i} \pi_i(\boldsymbol{\theta}_i) f_k(\mathbf{x}|\boldsymbol{\theta}_i) d\boldsymbol{\theta}_i},$$

or one can use model averaging, i.e., obtain the posterior predictive distribution

$$f(\mathbf{y}|\mathbf{x}) = \sum_{k \in \mathcal{K}} \rho(k) \int\limits_{\Theta_k} \pi_k(\boldsymbol{\theta}_k|\mathbf{x}) f_k(\mathbf{y}|\boldsymbol{\theta}_k) d\boldsymbol{\theta}_k.$$

The conditioning in $\pi(k, \boldsymbol{\theta}_k)$ suggests that we condition on k before we sample $\boldsymbol{\theta}_k$. In addition, note that the standard Gibbs sampler for $(k, \boldsymbol{\theta}_k)$ will not provide moves between model spaces $\{k\} \times \Theta_k$ and $\{m\} \times \Theta_m$, $k \neq m$, since if we condition on k we can sample $\boldsymbol{\theta}_k$ but if we condition on $\boldsymbol{\theta}_k$ then k cannot move. The solution is given by a M-H type of algorithm known as Reversible Jump MCMC (RJMCMC), defined independently by Geyer and Møller (1994) and Green (1995).

Now we turn to the development of the RJMCMC algorithm.

Remark 6.23 (RJMCMC) Letting $x = (k, \boldsymbol{\theta}_k)$ the RJMCMC utilizes the M-H setup by defining a reversible kernel K that satisfies the detailed balance condition

$$\int\limits_A \int\limits_B \pi(x) K(x, dy) dx = \int\limits_A \int\limits_B \pi(y) K(dx, y) dy,$$

$\forall A, B \in \Theta$, where $\pi(x)$ is the target distribution (typically a posterior $\pi(k, \boldsymbol{\theta}_k|\mathbf{x})$). The transition kernel is decomposed in terms of the proposed jump move; that is, if we are to move to $\mathcal{M}_m$, let $q_m(x, y)$ denote the corresponding transition distribution from state x to state y and $a_m(x, y)$ the probability of accepting this move, the RJMCMC algorithm has transition kernel $K(x, B)$ given by

$$K(x, B) = \sum_{m \in \mathcal{K}} \int\limits_B q_m(x, dy) a_m(x, y) + s(x) I_B(x)$$

where

$$s(x) = \sum_{m\in\mathcal{K}} \int_{\Theta} q_m(x, dy)(1 - a_m(x, y)) + 1 - \sum_{m\in\mathcal{K}} q_m(x, \Theta),$$

the probability of no move. The detailed balance is satisfied when

$$\sum_{m\in\mathcal{K}} \int_A \pi(dx) \int_B q_m(x, dy) a_m(x, y) + \int_{A\cap B} \pi(dx) s(x) =$$
$$\sum_{m\in\mathcal{K}} \int_B \pi(dy) \int_A q_m(y, dx) a_m(y, x) + \int_{B\cap A} \pi(dy) s(y),$$

and therefore it suffices to have

$$\int_A \pi(dx) \int_B q_m(x, dy) a_m(x, y) = \int_B \pi(dy) \int_A q_m(y, dx) a_m(y, x).$$

Green's clever assumption at this stage is to require the product measure $\pi(dx)q_m(x, dy)$ to have a finite density $f_m(x, y)$ with respect to a symmetric measure ξ_m on $\Theta \times \Theta$. Then under this assumption we have

$$\begin{aligned}
\int_A \pi(dx) \int_B q_m(x, dy) a_m(x, y) &= \int_A \int_B \xi_m(dx, dy) f_m(x, y) a_m(x, y) \\
&= \int_B \int_A \xi_m(dy, dx) f_m(y, x) a_m(y, x) \\
&= \int_B \pi(dy) \int_A q_m(y, dx) a_m(y, x).
\end{aligned}$$

provided that

$$f_m(x, y) a_m(x, y) = f_m(y, x) a_m(y, x)$$

and therefore detailed balance holds by choosing

$$a_m(x, y) = \min\left\{1, \frac{f_m(y, x)}{f_m(x, y)}\right\} = \min\left\{1, \frac{\pi(dy) q_m(y, dx)}{\pi(dx) q_m(x, dy)}\right\}.$$

As elegant as it is mathematically, RJMCMC can be difficult to apply in practice, in particular, with respect to the construction of the symmetric measure ξ_m and the corresponding density $f_m(x, y)$, given the symmetry constraint. In particular, suppose that we entertain a move from $\mathcal{M}_{k_1}$ to $\mathcal{M}_{k_2}$ and wlog assume that $k_2 < k_1$. Green's idea was to supplement the parameter space Θ_{k_2} with artificial spaces (additional dimensions) in order to create a bijection between Θ_{k_1} and Θ_{k_2}. In this case the move from Θ_{k_1} to Θ_{k_2} can be represented by a deterministic transformation $\theta^{(k_2)} = T\left(\theta^{(k_1)}\right)$ and in order to achieve symmetry the opposite move from Θ_{k_2} to Θ_{k_1} is concentrated on the curve $\{\theta^{(k_1)} : \theta^{(k_2)} = T\left(\theta^{(k_1)}\right)\}$.

For the general case, if $\theta^{(k_1)}$ is augmented by a simulation $z_1 \sim f_1(z)$ and $\theta^{(k_2)}$ is augmented by a simulation $z_2 \sim f_2(z)$ so that the mapping between $\left(\theta^{(k_1)}, z_1\right)$ and $\left(\theta^{(k_2)}, z_2\right)$ is a bijection

$$\left(\theta^{(k_2)}, z_2\right) = T\left(\theta^{(k_1)}, z_1\right),$$

then the acceptance probability for the move from $\mathcal{M}_{k_1}$ to $\mathcal{M}_{k_2}$ is given by

$$a(k_1, k_2) = \min\left\{\frac{\pi\left(k_2, \theta^{(k_2)}\right)}{\pi\left(k_1, \theta^{(k_1)}\right)} \frac{\pi_{21} f_2(z_2)}{\pi_{12} f_1(z_1)} \left|\frac{\partial T\left(\theta^{(k_1)}, z_1\right)}{\partial\left(\theta^{(k_1)}, z_1\right)}\right|, 1\right\},$$

where π_{ij} denotes the probability of choosing a jump from $\mathcal{M}_{k_i}$ to $\mathcal{M}_{k_j}$ and $J = \left|\frac{\partial T(\theta^{(k_1)}, z_1)}{\partial(\theta^{(k_1)}, z_1)}\right|$ the Jacobian of the transformation. This proposal satisfies detailed balance and the symmetry condition. The transformation T we choose and the corresponding Jacobian J is what causes this method to be hard to apply for complicated models since J can be hard to calculate. The general algorithm is as follows.

Algorithm 6.3 (RJMCMC) Assume that we wish to generate from the posterior distribution $\pi(k, \boldsymbol{\theta}_k|\mathbf{x})$ which is known up to a normalizing constant. At iteration t suppose that the chain is at state $x^{(t)} = (k, \boldsymbol{\theta}_k^{(t)})$.
Step 1: Choose the model $\mathcal{M}_m$ to jump into w.p. π_{km}.
Step 2: Generate $\mathbf{z}_{km} \sim f_{km}(\mathbf{z})$ and $\mathbf{z}_{mk} \sim f_{mk}(\mathbf{z})$.
Step 3: Set $(\boldsymbol{\theta}_m, \mathbf{z}_{mk}) = T\left(\boldsymbol{\theta}_k^{(t)}, \mathbf{z}_{km}\right)$ and generate $U \sim Unif(0, 1)$.
Step 4: Accept the move and set $x^{(t+1)} = (m, \boldsymbol{\theta}_m)$ if $U < a(k, m)$, where

$$a(k, m) = \min\left\{\frac{\pi(m, \boldsymbol{\theta}_m)}{\pi\left(k, \boldsymbol{\theta}_k^{(t)}\right)} \frac{\pi_{mk} f_{mk}(\mathbf{z}_{mk})}{\pi_{km} f_{km}(\mathbf{z}_{km})} \left|\frac{\partial T_{km}\left(\boldsymbol{\theta}_k^{(t)}, \mathbf{z}_{km}\right)}{\partial\left(\boldsymbol{\theta}_k^{(t)}, \mathbf{z}_{km}\right)}\right|, 1\right\},$$

otherwise remain at the same state, i.e., set $x^{(t+1)} = x^{(t)}$. Return to step 1.

The choice of transformation and its Jacobian is one of the major difficulties in applying the RJMCMC. An elegant solution to the problem was given by Stephens (2000) with the definition of the Birth-Death MCMC (BDMCMC). In particular, BDMCMC is based on sampling methods from point process theory in order to sample from a mixture model of varying dimension. The major advantage is that it does not require calculation of a Jacobian, however, it is restrictive since it allows either a single addition of a component or a single deletion of a component. This algorithm is illustrated and utilized extensively in the *TMSO-PPRS* text.

6.8 Summary

The theory developed in this chapter provides the foundations of statistical modeling for some of the most important random sequences such as random walks and Markov chains. Standard and specialized texts in these topics include Feller (1968, 1971), Karatzas and Shreve (1991), Borodin and Salminen (1996), Durrett (1996, 2004, 2010), Fristedt and Gray (1997), Dudley (2002), Çinlar (1975, 2010), Lawler and Limic (2010), Bass (2011), Billingsley (2012), Lindgren (2013) and Klenke (2014). Building on the theory of Markov chains will allow us to model random objects over discrete time and in particular, as we see in the *TMSO-PPRS* text, point processes and random sets as they evolve over time. We collect some complementary ideas on topics from this chapter below.

Martingales

The term martingale was introduced in probability theory by Ville in 1939 but the concept had already been defined by Lévy in 1934. Lévy's 0-1 law is the first MG convergence theorem. Martingales owe their development to Doob (1940) when he began to formulate a rigorous mathematical theory. For the historical account and references see Dudley (2002, p. 382). A great number of examples and applications of MGs, in particular as they relate to gambling, can be found in Fristedt and Gray (1997), Durrett (2010), Billingsley (2012) and Klenke (2014).

Stochastic processes

It is amazing to think how many real-life processes exhibit a Markovian property, that is, the future state of the random object is independent of past states given its present state. The theoretical foundations of this idea were first developed in the early 1900s by Andrei Andreyevich Markov and for the finite state space case, where he proved the existence of the limit of $T^n(x, y)$ for an aperiodic chain with one recurrent class. In the 1930s, Andrei Nikolaevich Kolmogorov extended the methods to the infinite state space case and introduced the concepts of Markov kernels and transition functions.

The modern theory begins with Kiyoshi Itô in the 1940s where motion dynamics are described using stochastic integral equations (see Section 7.5.1) with transition functions and generators of the process being special cases. We studied the first-order Markov property here, however, there are models that allow the future state of a random object to be independent of the past, given a number of past realizations of the random object (e.g., time series models, see below). The study of random sequences becomes much harder once we depart from certain assumptions such as time-homogeneity and the Markov property.

Time series models for spatio-temporal random objects

A time series is simply a stochastic process $X = \{\mathbf{X}_t : t \in T\}$, with $T \subset \mathcal{R}$ typically discrete, $T = \{0, 1, 2, \dots\}$, and state space $\Psi \subset \mathcal{R}^p$, where the current state of the random object $\mathbf{X}_t$ is assumed to be a function of one or more past states of the object, e.g., $\mathbf{X}_t = f(\mathbf{X}_{t-1}, \mathbf{X}_{t-2}, \dots, \mathbf{X}_{t-k}) + \boldsymbol{\varepsilon}_t$, $t = k, k+1, \dots$, where the past is assumed to be fixed and the random noise is typically normal, i.e., $\boldsymbol{\varepsilon}_t \overset{iid}{\sim} \mathcal{N}_p(\mathbf{0}, \boldsymbol{\Sigma})$. Obviously, a linear function is easier to handle and it leads to the vector autoregressive time series model of order k ($VAR(k)$), given by

$$\mathbf{X}_t = \boldsymbol{\Phi}_1\mathbf{X}_{t-1} + \boldsymbol{\Phi}_2\mathbf{X}_{t-2} + \cdots + \boldsymbol{\Phi}_{t-k}\mathbf{X}_{t-k} + \boldsymbol{\varepsilon}_t, \tag{6.55}$$

$t = k, k+1, \dots$. Random walks in $\mathcal{R}^d$ are simply $VAR(1)$ models, i.e., $\mathbf{X}_t = \mathbf{X}_{t-1} + \boldsymbol{\varepsilon}_t$, $\boldsymbol{\Phi}_1 = \mathbf{I}_p$. Time series are examples of stochastic process models for data collected over time (temporal data). For more details on time series models see Cressie and Wikle (2011, Chapter 3).

6.9 Exercises

Bernoulli and related processes

Exercise 6.1 Let $X = \{X_n : n \geq 1\}$ be a Bernoulli process with probability of

success p, $N = \{N_n : n \geq 0\}$ the corresponding Binomial process with $N_0 = 0$, and let $T = \{T_k : k \geq 1\}$ be the process of success times.
(i) Show that $P(N_{n+1} = k) = pP(N_n = k - 1) + qP(N_n = k)$, $\forall n, k \geq 0$ and use this result to find $P(N_n = k)$, $k = 1, 2, \ldots, n$, $n \geq 0$.
(ii) Find $P(N_{n+m} - N_m = k), \forall m, n, k \geq 0$ and show that $N_{n+m} - N_m | N_0, N_1, N_2, \ldots, N_m \stackrel{d}{=} N_{n+m} - N_m \stackrel{d}{=} N_n$.
(iii) If $n \geq k$ and $k \geq 1$ then show that $\{\omega : T_k(\omega) \leq n\} = \{\omega : N_n(\omega) \geq k\}$ and $\{\omega : T_k(\omega) = n\} = \{\omega : N_{n-1}(\omega) = k - 1\} \cap \{\omega : X_n(\omega) = 1\}$.
(iv) Show that $T_k \sim NB(k, p)$.
(v) Prove that $T_{k+m} - T_m | T_1, T_2, \ldots, T_m \stackrel{d}{=} T_{k+m} - T_m \stackrel{d}{=} T_k$.
(vi) Show that the increments $T_1, T_2 - T_1, T_3 - T_2, \ldots,$ are iid according to a geometric $Geo(p)$.

Random walks

Exercise 6.2 Prove theorem 6.1.

Exercise 6.3 For the setup of example 6.3 find the distribution of T_1 the first return time to 0.

Exercise 6.4 Prove corollary 6.1.

Exercise 6.5 Let $X = (X_1, X_2, \ldots)$ be an iid sequence of random variables, $\mathcal{F}^X$ the corresponding minimal filtration and τ be an *a.s.* finite stopping time with respect to $\mathcal{F}^X$. Define $Y_n(\omega) = X_{\tau(\omega)+n}(\omega)$ for $n \geq 1$ and prove that the sequence $Y = (Y_1, Y_2, \ldots)$ has the same distribution as X and is independent of $\mathcal{F}_\tau$.

Exercise 6.6 Let $\tau_1 \leq \tau_2 \leq \ldots,$ be a sequence of stopping times with respect to the minimal filtration of a random walk S in $\mathcal{R}^p$. Show that the variables $S_{\tau_1}, S_{\tau_2} - S_{\tau_1}, S_{\tau_3} - S_{\tau_2}, \ldots,$ are independent.

Exercise 6.7 Prove theorem 6.10.

Exercise 6.8 Let S denote a simple random walk on $\mathcal{Z}^d$ and recall the setup of example 6.3. Show that the following are equivalent: (i) $P(T_1 < \infty) = 1$, (ii) $P(S_n = 0\ i.o) = 1$, and (iii) $\sum\limits_{n=0}^{+\infty} P(S_n = 0) = +\infty$.

Exercise 6.9 Let S denote a simple random walk on $\mathcal{Z}$ with $S_0 = 0$ and show that $X_n = |S_n|$ is a Markov chain and find the transition matrix.

Martingales, filtrations and stopping times

Exercise 6.10 Assume that $\mathcal{F} = (\mathcal{A}_1, \mathcal{A}_2, ...)$ is a filtration. Show that $\bigcup\limits_{n=1}^{+\infty} \mathcal{A}_n$ is a field.

Exercise 6.11 Prove all statements of remark 6.2.

Exercise 6.12 A branching process is a sequence $\{Z_n; n \geq 0\}$ of nonnegative integer-valued random variables with $Z_0 = 1$ and such that the conditional distribution of Z_{n+1} given $(Z_0, Z_1, \ldots, Z_n)$ is that of a sum of Z_n iid random variables each having the same distribution as Z_1. If $E(Z_1) = m \in (0, \infty)$, show that $\{W_n = Z_n/m^n; n \geq 1\}$ is a MG.

Exercise 6.13 Prove statements 5-9 of remark 6.3.

Exercise 6.14 Let $X = \{X_n; n \geq 1\}$ be a sequence of real-valued random variables defined on a probability space $(\Omega, \mathcal{A}, P)$ and having finite mean. Prove that X is a sMG with respect to a filtration $\{\mathcal{F}_n, n \geq 1\}$ to which it is adapted if and only if for all $m, n \geq 0$,

$$E(X_{n+m}|\mathcal{F}_n) \geq X_n \ a.s.$$

Exercise 6.15 Prove statements 1-3 of remark 6.4.

Exercise 6.16 Consider a sequence of iid random variables $\{X_n\}_{n=1}^{+\infty}$ on a probability space $(\Omega, \mathcal{A}, P)$. Define the sequence of partial sums $S_n = \sum_{i=1}^{n} X_i$.

(i) Show in detail that $\sigma(X_1, X_2, \ldots, X_n) = \sigma(S_1, S_2, \ldots, S_n)$.

(ii) Assume that $E(X_1^{-1})$ exists. Show that if $m \leq n$ then $E\left(\frac{S_m}{S_n}\right) = m/n$.

Exercise 6.17 Let $\{X_n\}_{n=1}^{+\infty}$ be a sequence of random variables such that the partial sums $S_n = \sum_{i=1}^{n} X_i$ are a MG. Show that $E(X_i X_j) = 0$, $i \neq j$.

Renewal sequences

Exercise 6.18 For $p \in (0, 1)$ let X be a renewal sequence with waiting time distribution $f(n) = p^{n-1}(1 - p)$, $n > 0$. Find the corresponding potential sequence.

Exercise 6.19 Prove the renewal sequence criterion, theorem 6.11.

Exercise 6.20 For $p \in [0, 1]$ let X be a renewal sequence with waiting time distribution W given by $W(\{2\}) = p = 1 - W(\{1\})$. Find the potential sequence of X.

Exercise 6.21 Prove the statements of remark 6.6, parts 3 and 4.

Exercise 6.22 Let W, $(\pi_0, \pi_1, \pi_2, ...)$ and R denote the waiting time distribution, the potential sequence and the renewal measure, respectively, for some renewal sequence. Show that

$$P(R(\{k+1, \ldots, k+l\}) > 0) = \sum_{n=0}^{k} W(\{k+1-n, \ldots, k+l-n\})\pi_n,$$

for all $k, l = 0, 1, 2, ...$ and

$$\sum_{n=0}^{k} R(\{k+1-n, \ldots, +\infty\})\pi_n = 1,$$

for all $k \in \mathcal{Z}^+$.

Exercise 6.23 Prove theorem 6.12.

Exercise 6.24 For which values of p is the sequence $(1, 0, p, p, p, \ldots)$ a potential sequence? For each such p find the density of the corresponding waiting time distribution.

Exercise 6.25 Prove theorem 6.14.

Exercise 6.26 Let W be a waiting time distribution with finite mean μ and let X

be a delayed renewal sequence corresponding to W and the delay distribution D defined by

$$D(\{n\}) = W(\{n+1, \ldots, \infty\})/\mu,$$

for $n \geq 0$. Show that $P(X_n = 1) = 1/\mu$, for all $n \geq 0$.

Exercise 6.27 For a simple random walk S in $\mathcal{Z}$ let X denote the delayed renewal sequence defined by $X_n = I_{\{x\}} \circ S_n$ for some $x \in \mathcal{Z}$. Find the pgf of the delay distribution of X and use it to find the probability that the delay equals ∞.

Markov chains

Exercise 6.28 For any Markov chain X with state space Ψ, show that

$$E(f(X_k, X_{k+1}, \ldots)|\sigma(\{X_n : n \leq k\})) = E(f(X_k, X_{k+1}, \ldots)|\sigma(X_k)),$$

for all $k \geq 1$ and bounded functions f on Ψ^∞. Extend this result by replacing k with any $a.s.$ finite stopping time τ relative to $\mathcal{F}^X$ and using the strong Markov property.

Exercise 6.29 Prove theorem 6.16.

Exercise 6.30 Let X be a Markov chain with discrete state space Ψ and operator T. Show that

$$P(X_{k+m} = y|\sigma(\{X_n : n \leq k\})) = T^m(X_k, y),$$

for all $k \geq 1$, $m \geq 0$ and $y \in \Psi$. Extend this result by replacing k with any $a.s.$ finite stopping time τ relative to $\mathcal{F}^X$ and using the strong Markov property.

Hitting probabilities

Exercise 6.31 Consider a Markov chain X with state space $\mathcal{Z}^+$ and transition measures $\mu_x(\{1\}) = \frac{1}{x+1}$ and $\mu_x(\{x+1\}) = \frac{x}{x+1}$, $x \in \mathcal{Z}^+$. Find $H(x, 1)$.

Exercise 6.32 Let X be a Markov chain with state space $\mathcal{Z}$ and transition measures $\mu_x(\{x-1\}) = \frac{1}{2}$, $\mu_x(\{x+2\}) = \frac{1}{2}$, $x \in \mathcal{Z}$. Find $H(x, 0)$.

Exercise 6.33 For all simple random walks X on $\mathcal{Z}$, find $H(x, 0)$ and use it to find $H^+(x, 0)$.

Exercise 6.34 Prove theorem 6.17.

Exercise 6.35 Consider a simple random walk X on the integers $\{0, 1, \ldots, k\}$, with $\mu_0(\{0\}) = 1 = \mu_k(\{k\})$ and $\mu_x(\{x-1\}) = p = 1 - \mu_x(\{x+1\})$, $0 < p < 1$. Find $H(x, 0)$ and $H(x, k)$.

Exercise 6.36 Let X be a Markov chain with state space $\mathcal{Z}^+$ and transition measures $\mu_x(\{x+1\}) = \frac{(x-1)(x+2)}{x(x+1)}$ and $\mu_x(\{1\}) = \frac{2}{x(x+1)}$, $x \geq 1$. Calculate $H(x, 1)$.

Exercise 6.37 Recall the Birth-Death Markov chain of example 6.19 but now assume that $\mu_0(\{0\}) = 1$, $\mu_x(\{x+1\}) = \frac{x}{x+1}$, $x \in \mathcal{Z}_0^+$ and $\mu_x(\{x-1\}) = \frac{1}{x+1}$, $x > 0$. Find $H(x, 0)$.

Exercise 6.38 Find $H_k(x, y)$, the probability of visiting y, starting from x, for the first time in k-steps, $\forall x, y \in \Psi$ and $k \geq 1$, for the Markov chains of examples 6.20 and 6.21.

State classification

Exercise 6.39 Prove equation (6.22) of remark 6.12.1.

Exercise 6.40 Prove the statement of remark 6.13.1.

Exercise 6.41 Consider an irreducible Markov chain with transition matrix T. Show that if $T(x, x) > 0$ for some x, then all the states are aperiodic.

Exercise 6.42 Prove theorem 6.21.

Exercise 6.43 Classify the states of the Markov chains in examples 6.19, 6.20 and 6.21.

Stationarity and ergodicity

Exercise 6.44 Show in detail that the random sequences of example 6.30 are stationary.

Exercise 6.45 Prove the statement of remark 6.19.5.

Exercise 6.46 Assume that X is a stationary Markov chain with countable state space Ψ. Show that if X is irreducible then the sequence is ergodic.

Exercise 6.47 Show that the shift-invariant sets of equation (6.50) form a σ-field.

Exercise 6.48 Consider a random sequence X with state space $\Psi = \{-1, 1\}$ and transition operator T such that $T(-1, 1) = 1$, $T(1, -1) = 1$ and initial distribution $P(X_0 = 1) = P(X_0 = -1) = 1/2$. Show that X is a stationary Markov chain.

Exercise 6.49 Give an example of a Markov chain that has two different stationary distributions.

Exercise 6.50 Let $S = (S_0, S_1, \dots)$ be a random walk on $\mathcal{Z}_0^+$ with transition matrix T given by $T(x, x+1) = p$, $T(x+1, x) = 1 - p$, $x \geq 0$, with $T(0, 0) = 1 - p$.
(i) Show that S is an irreducible Markov chain.
(ii) Prove that the invariant measure for S is given by

$$\pi(x) = c\,(p/(1-p))^x,$$

$x = 1, 2, \dots$, for some constant c. For what values of c and p is this the equilibrium distribution of S?
(iii) Find conditions in order for S to be ergodic.

Exercise 6.51 Prove Hopf's maximal ergodic lemma (remark 6.20.2).

Exercise 6.52 Give an example of a reversible Markov chain that is periodic.

Exercise 6.53 Show that if X is irreducible and has a stationary distribution π then any other stationary measure is a multiple of π.

Exercise 6.54 Prove equation (6.52).

Simulation and computation: use your favorite language to code the functions

Exercise 6.55 Consider the mixture of two beta distributions: $0.7 * Beta(4, 2) + 0.3 * Beta(1.5, 3)$.
(i) Build a Metropolis-Hastings algorithm to get a sample of size 1000 after a burn-in sample of 500. Choose your own $Beta(a, b)$ candidate distribution. What is the M-H acceptance rate?

(ii) Devise a random walk Metropolis algorithm to get a sample of size 1000 after a burn-in sample of 500. Use a normal proposal distribution but be careful because the target distribution only has support on [0, 1]. What is the standard deviation you chose for the random walk? What is the acceptance rate?

Exercise 6.56 Consider a random sample $X_1, ..., X_n$ from $\mathcal{N}(\mu, \sigma^2)$.

(i) Suppose that we model the parameter vector $\boldsymbol{\theta} = (\mu, \sigma^2)$, independently, according to the conjugate priors $\mathcal{N}(\xi, \tau^2)$ and $InvGamma(a, b)$.

(a) Show that the full conditional distributions are given by

$$\mu|\sigma^2, \mathbf{x} \sim \mathcal{N}\left(\frac{\tau^2\sigma^2}{n\tau^2+\sigma^2}\left(\frac{\xi}{\tau^2}+\frac{n\overline{X}}{\sigma^2}\right), \frac{\tau^2\sigma^2}{n\tau^2+\sigma^2}\right), \text{ and}$$

$$\sigma^2|\mu, \mathbf{x} \sim InvGamma\left(\frac{n}{2}+a, \left[\frac{1}{b}+\frac{1}{2}\sum_{i=1}^{n}(x_i-\mu)^2\right]^{-1}\right).$$

(b) Write a routine to implement this Gibbs sampler. Generate and plot the chains for 10000 realizations (after 1000 burn-in). Choose a starting value $\boldsymbol{\theta}_0 = [20, 1]^T$, and use $\xi = 10$, $\tau^2 = 1$, $a = 10$ and $b = 0.1$.

(ii) Now consider the Jeffreys independent prior.

(a) Give the marginal posterior distribution of μ and find the posterior mean of μ.

(b) Plot the marginal posterior density of μ.

(c) Find the 95% CS for μ.

(d) Give the marginal posterior distribution of σ^2 and find the posterior mean of σ^2.

(e) Plot the marginal posterior density of σ^2.

(f) Find the 95% equal tail CS for σ^2.

(iii) Now we consider a conjugate prior for (μ, σ^2): $\mu|\sigma^2 \sim \mathcal{N}(0, \sigma^2/\kappa)$ and $\sigma^2 \sim InvGamma(a, b)$, where $\kappa = 0.25$, $\alpha = 0.5$ and $b = 0.5$. Repeat parts (a)–(f) of part (ii).

Chapter 7

Stochastic Processes

7.1 Introduction

Continuous time parameter stochastic processes can be used to create and investigate random objects as they evolve continuously in time. The random objects we discuss in this chapter include real-valued random functions, random derivatives and random integrals. We begin with the general definition of a stochastic process.

Definition 7.1 Stochastic process

A stochastic process with state space Ψ is a collection of random objects $X = \{X_t : t \in T\}$ with X_t taking values in Ψ and defined on the same probability space $(\Omega, \mathcal{A}, P)$.

Example 7.1 (Stochastic processes) We discuss some specific examples of stochastic processes next.

1. Suppose that the experiment consists of observing the acceleration of a vehicle during the first 10 seconds of a race. Then each possible outcome is a real-valued, right-continuous function ω defined for $0 \le t \le 10$ and the sample space Ω is the set of all such functions. We define the acceleration at time t as the random variable $X_t(\omega) = \omega(t)$, for each $\omega \in \Omega$, so that the collection $X = \{X_t : 0 \le t \le 10\}$, is a continuous time-parameter stochastic process with state space $\Psi = \mathcal{R}$.

2. Arrival process Assume that the process $N = \{N_t : t \ge 0\}$ is such that the function $N_t(\omega)$, for fixed ω and as a function of t, namely, $t \mapsto N_t(\omega)$, is non-decreasing, right-continuous and increases by jumps only. Stochastic processes with these properties are known as arrival processes. The process N is a continuous time-parameter stochastic process.

3. Interarrival times Consider the process of arrivals of customers at a store and suppose that the experiment consists of measuring interarrival times. The sample space Ω is the set of all sequences $\omega = (\omega_1, \omega_2, \dots)$ with $\omega_i \ge 0$, for all i. For each $\omega \in \Omega$ and $t \ge 0$, we set $N_t(\omega) = k$ if and only if the integer k is such that

$\omega_1 + \cdots + \omega_k \leq t \leq \omega_1 + \cdots + \omega_{k+1}$ ($N_t(\omega) = 0$, if $t < \omega_1$). Then for the outcome ω, $N_t(\omega)$ denotes the number of arrivals in the time interval $[0, t]$ and therefore $N = \{N_t : t \geq 0\}$ is an arrival process with state space $\Psi = \{0, 1, \ldots\}$.

4. Queueing systems Consider an experiment where customers arrive at a store requiring service, e.g., airplanes arrive at an airport and need to land or patients arrive at a hospital and need to be treated. The system may have certain restrictions, including the number of servers (e.g., physicians in the emergency room of a hospital) or the number of customers allowed in the waiting room. Now if there are too many arrivals the physicians will not be able to keep up and the patients will experience delays in their treatment, i.e., they have to wait in line (or in the queue) to get to see the doctor and then have to wait an additional time period, the time they receive their treatment, until they exit the system.

Queueing theory refers to the area of probability theory and stochastic processes that studies the characteristics of such systems. There are several stochastic processes required in order to study these systems. Arrivals at the system can be modeled using a continuous time stochastic process $A = \{A_t : t \geq 0\}$, where A_t denotes the number of arrivals in the interval $[0, t]$. In contrast, a discrete time stochastic process $S = \{S_n : n = 1, 2, \ldots\}$ can be used to model the service times of each customer (typically iid) and similarly, for the waiting time of each customer we can build another discrete time stochastic process $W = \{W_n : n = 1, 2, \ldots\}$.

Finally, we are interested in the queue-size process $Y = \{Y_t : t \geq 0\}$, where $Y_t \in \Psi = \{0, 1, 2, \ldots\}$ denotes the total number of customers in the queue at time t which includes the number of customers in the waiting room and the number of customers being currently serviced. In addition, future values of the process Y_{t+s} at time $t + s$ depend not only on the current value Y_t but also on any departures or arrivals in the time interval $[t, t + s]$. We will see that we can make the problem more tractable by defining an underlying Markov chain that allows us to describe the moves of Y_t (see example 7.6), as well as study the general behavior of Y_t.

The following notation is used to summarize the important characteristics of a queueing system; we write $A/S/k/q$ queue, where A abbreviates the arrival process, S denotes the distribution of the service times, k denotes the number of servers and q the size of the waiting room of the queue. For example, $G/M/2/\infty$ denotes a queue with a general arrival process, an exponential service time, 2 servers and an infinite waiting room. If $q = \infty$ we typically write $G/M/2$ (omit the waiting room size).

Similar to the discrete time parameter case, the following definitions can be used to identify a wide collection of stochastic processes with certain desirable properties.

Definition 7.2 Stationarity and independence of increments

A stochastic process $X = \{X_t : t \in T\}$ is said to have stationary increments if the distribution of the increment $X_{t+s} - X_t$ does not depend on t. The stochastic process X is said to have independent increments if for any $n \geq 1$ and $s_0 = 0 < s_1 < s_2 < \cdots < s_n$, the increments $X_{s_1} - X_{s_0}, X_{s_2} - X_{s_1}, \ldots, X_{s_n} - X_{s_{n-1}}$ are independent.

7.2 The Poisson Process

The most important arrival process is the Poisson process.

Definition 7.3 Poisson process

An arrival process $N = \{N_t : t \geq 0\}$ is called a Poisson process if the following properties hold:
(P1) each jump of the function $t \mapsto N_t(\omega)$ is of unit magnitude,
(P2) $\forall t, s \geq 0$, $N_{t+s} - N_t$ is independent of $\{N_u : u \leq t\}$,
(P3) $\forall t, s \geq 0$, the distribution of $N_{t+s} - N_t$ is independent of t.
Each random variable N_t is interpreted as the number of arrivals (or events) of the process in the interval $[0, t]$.

Example 7.2 (Poisson process via random walks) Consider a random walk S in $\overline{\mathcal{R}}^+$ with $Exp(1)$ step distribution. Clearly, $S_n \sim Gamma(n, 1)$, $n > 0$ and for fixed $t \in \mathcal{R}^+$ let $N_t(\omega) = \#\{n : 0 < S_n(\omega) \leq t\}$. Since each step is *a.s.* positive, N_t *a.s.* equals the number of steps taken by the random walk before it reaches the interval $(t, +\infty)$. For $k \geq 0$ we have

$$\begin{aligned} P(\{\omega &: N_t(\omega) = k\}) = P(\{\omega : N_t(\omega) \geq k\}) - P(\{\omega : N_t(\omega) \geq k + 1\}) \\ &= P(\{\omega : S_k(\omega) \leq t\}) - P(\{\omega : S_{k+1}(\omega) \leq t\}) = t^k e^{-t}/k!, \end{aligned}$$

and therefore N_t is a *Poisson*(t) random variable.

Some comments are in order regarding properties (P1)-(P3) and their consequences.

Remark 7.1 (Poisson process properties) Note that once we relax the three assumptions the general theory cannot be applied, although one can still obtain specialized results (see for example Çinlar, 1975).

1. Property (P2) expresses the independence of the number of arrivals in $(t, t + s]$ from past history of the process up to time t and leads to the property of independent increments. Property (P3) simply states that the process is stationary.

2. Using definition 7.3 we can show that $P(N_t = 0) = e^{-\lambda t}, t \geq 0$, for some constant $\lambda \geq 0$, known as the intensity of the process. Moreover, the number of events N_t in the interval $[0, t]$ is distributed according to a *Poisson*(λt) and as a consequence we write $N \sim PP(\lambda)$ to refer to a stationary Poisson process. Note that the average number of arrivals in $[0, t]$ is $E(N_t) = \lambda t$.

3. Using (P1)-(P3) and part 2 of this remark we can easily see that $N_{t+s} - N_t | \{N_u : u \leq t\} \overset{d}{=} N_{t+s} - N_t \overset{d}{=} N_s$.

4. The three properties can be relaxed, leading to more complicated Poisson processes. Relaxing (P1) to allow jumps of any size leads to the compound Poisson process: (P1)′: the function $t \mapsto N_t(\omega)$ has finitely many jumps in any finite interval a.e. Properties (P1) and (P2) are highly applicable in applications but (P3) is not. Removing property (P3) from definition 7.3 defines the non-stationary Poisson process. Although harder to work with, we can still obtain some characteristics of the Poisson process even in this case.

5. First characterization A stochastic process $N = \{N_t : t \geq 0\}$ is a Poisson process with intensity λ if and only if (P1) holds and $E(N_{t+s} - N_t | N_u : u \leq t) = \lambda s$.

6. Second characterization A stochastic process $N = \{N_t : t \geq 0\}$ is a Poisson process with intensity λ if and only if

$$P(N_B = k) = e^{-\lambda b}(\lambda b)^k / k!,$$

$k \geq 0$, for all sets $B \subset \mathcal{R}^+$ that are unions of finite collections of disjoint intervals with the length of B being $\mu_1(B) = b$. Here N_B denotes the number of arrivals over the set B.

7. Conditioning Let $B = \cup_{i=1}^{n} A_i$ for some disjoint intervals $\{A_i\}$ with lengths a_i and set $b = a_1 + \cdots + a_n$. Then for $k_1 + \cdots + k_n = k$ with $k_1, \ldots, k_n = 0, 1, 2, \ldots$, we can show that

$$P(N_{A_1} = k_1, \ldots, N_{A_n} = k_n | N_B = k) = C^k_{k_1,k_2,\ldots,k_n} (a_1/b)^{k_1} \ldots (a_n/b)^{k_n}.$$

Example 7.3 (Queueing systems: Poisson arrivals) Consider a queueing system and assume that the arrival process is $N \sim PP(\lambda)$, that is, a Poisson process with some rate λ so that the number of events (arrivals of customers) N_t in the interval $[0, t]$ is such that $N_t \sim Poisson(\lambda t)$ (the notation for the queue becomes $M/./.$/. and if we further assume that service times follow an exponential distribution we write $M/M/./.$). Arrivals in this case occur independently of each other. The process of the times of arrivals $\{T_n\}_{n=1}^{+\infty}$ of the Poisson process can be particularly useful in calculating lengths of time including a customer's remaining time in the queue or the time until their service begins. See exercise 7.4 for some insight on this process. We discuss some classic examples of queueing systems below.

1. $M/M/1$ **queue** In the $M/M/1$ queue customers arrive at the times of a Poisson process and each requires an independent amount of service distributed as an $Exp(\mu)$ distribution. Furthermore, there is one server and an infinite size waiting room. Examples of this queue include patients arriving at a doctor's office and requiring treatment or customers waiting to use an automated vending machine.

2. $M/M/k$ **queue** Consider the $M/M/1$ queue and assume that there are k servers.

A bank with k tellers or a movie theater with k ticket counters are examples of such queues.

3. $M/M/\infty$ queue Now assume that there is an infinite number of servers so that the instant a customer arrives their service begins immediately. Large parking lots and telephone traffic can be approximated well by such a queueing system, e.g., assuming that the number of parking spots or phone lines is infinite, a customer will always find a server.

4. $M/M/k/m$ queue This queue is similar to the $M/M/k$ queue but now there are only m places in the waiting room. Customers arriving to find the waiting room of the queue full, leave the system immediately never to return. For example there is a limited number of seats, say m, in the waiting room of the department of motor vehicles and k servers to service the customers.

The Poisson process has been the subject of many texts (see Section 7.6). Here we have discussed the basic definition and properties of this process and we will discuss more of its uses once we collect Markov processes (see example 7.5). We further revisit this process in the *TMSO-PPRS* text where we consider a random object known as the Poisson point process. We collect general definitions and properties of stochastic processes next.

7.3 General Stochastic Processes

A general stochastic process is a collection of random objects $X = \{X_t : t \geq 0\}$ on some probability space $(\Omega, \mathcal{A}, P)$ where $X_t(\omega)$, for fixed $t \geq 0$, is assumed to be a measurable map from $(\Omega, \mathcal{A})$ into some measurable space $(\Psi, \mathcal{G})$. The space Ψ is known as the state space of X. We collect some important definitions and properties in the following.

Remark 7.2 (General definitions) Since $X_t(\omega)$ is really a function of two arguments, for technical reasons, it is convenient to define joint measurability properties.

1. Joint measurability A stochastic process X is called measurable if for every $B \in \mathcal{G}$ the set $\{(\omega, t) : X_t(\omega) \in B\}$ belongs to the product σ-field $\mathcal{A} \bigotimes \mathcal{B}(\mathcal{R}_0^+)$; in other words, if the mapping

$$(\omega, t) \mapsto X_t(\omega) : (\Omega \times \mathcal{R}_0^+, \mathcal{A} \bigotimes \mathcal{B}(\mathcal{R}_0^+)) \to (\Psi, \mathcal{G})$$

is measurable.

2. Equivalent stochastic processes Consider two $\mathcal{R}^p$-valued stochastic processes X and Y defined on the same probability space $(\Omega, \mathcal{A}, P)$. When viewed as functions of ω and t we would say that X and Y are the same if and only if $X_t(\omega) = Y_t(\omega)$, for all $\omega \in \Omega$ and $t \geq 0$. However, since we are working on a probability space we weaken this requirement in three different ways:
(a) Y is a modification of X if we have $P(X_t = Y_t) = 1, \forall t \geq 0$.

(b) X and Y have the same finite-dimensional distributions if $\forall n \geq 1$, real numbers $0 \leq t_1 < t_2 < \cdots < t_n < \infty$ and $B \in \mathcal{B}_{np}$, we have

$$P((X_{t_1}, \ldots, X_{t_n}) \in B) = P((Y_{t_1}, \ldots, Y_{t_n}) \in B).$$

(c) X and Y are called indistinguishable if $P(X_t = Y_t; \forall t \geq 0) = 1$.
Note that $(c) \Rightarrow (a) \Rightarrow (b)$.

3. Sample paths and RCLL For fixed $\omega \in \Omega$ the function $t \mapsto X_t(\omega)$, $t \geq 0$, is called the sample path (realization or trajectory) of the process X. When the paths of X are continuous maps then the stochastic process X is called continuous. We say that a stochastic process X has RCLL (right-continuous and left-hand limits) paths if $X_t(\omega)$, for fixed $\omega \in \Omega$, is right-continuous on $[0, \infty)$ with finite left-hand limits on $(0, \infty)$. Therefore, X consists of random objects that are random maps (or functions if $\Psi = \mathcal{R}^p$) and we need to build an appropriate measurable space in order to work with such objects. The main ingredient needed for such constructions is a well-behaved space Ψ for maps (e.g., Polish spaces). In particular, consider the following example.

Example 7.4 (Polish spaces) Recall that a Polish space is a complete metric space that has a countable dense subset. For example, $\mathcal{R}$ equipped with the Euclidean metric is a Polish space with the rationals constituting a countable set. In addition, $\overline{\mathcal{R}}$ is a Polish space using as metric $d(x, y) = |\arctan y - \arctan x|$ and the rationals as a dense subset. Completeness of the space depends on the choice of metric. Using $d(x, y)$ in $\mathcal{R}$ we do not have completeness in $\mathcal{R}$. Now $\mathcal{R}^p$ is also Polish using the Euclidean metric

$$\rho_p(\mathbf{x}, \mathbf{y}) = \sqrt{\sum_{i=1}^{p} (x_i - y_i)^2},$$

and a dense subset of the set of points with rational coordinates. Finally, $\mathcal{R}^\infty$ is the space of all infinite sequences $(x_1, x_2, \ldots.)$ of real numbers and it can be metrized using

$$\rho_\infty(\mathbf{x}, \mathbf{y}) = \sum_{i=1}^{+\infty} 2^{-i} \min\{1, |y_i - x_i|\},$$

where $\mathbf{x} = (x_1, x_2, \ldots.)$ and $\mathbf{y} = (y_1, y_2, \ldots.)$.

The following remark summarizes general results for Polish spaces.

Remark 7.3 (Polish spaces) We note the following.

1. Let (Ψ_j, ρ_j), $j = 1, 2, \ldots$, be Polish spaces and define $\Psi = \times_{j=1}^{+\infty} \Psi_j$ with the topology on Ψ being the product topology and define

$$\rho(\mathbf{x}, \mathbf{y}) = \sum_{j=1}^{+\infty} 2^{-j} \min\{1, \rho_j(x_j, y_j)\}.$$

Then it can be shown that (Ψ, ρ) is a Polish space.

2. Let $C^{\mathcal{R}}_{[0,1]}$ denote the space of continuous $\mathcal{R}$-valued functions defined on $[0,1]$ and define a metric on $C^{\mathcal{R}}_{[0,1]}$ by

$$d(f,g) = \max\{|f(t) - g(t)| : t \in [0,1]\}.$$

Then $(C^{\mathcal{R}}_{[0,1]}, d)$ is a Polish space and the Borel σ-field is given by

$$\mathcal{B}(C^{\mathcal{R}}_{[0,1]}) = \sigma\left(\{f \in C^{\mathcal{R}}_{[0,1]} : f(t) \in B; t \in [0,1], B \in \mathcal{B}(\mathcal{R})\}\right).$$

Similar results exist for $C^{\mathcal{R}}_{[0,\infty)}$, the space of continuous $\mathcal{R}$-valued functions defined on $[0,\infty)$.

3. A closed subset of a Polish space is a Polish space with the same metric.

4. For any Polish space Ψ there exists a function $f : \Psi \to B$ where $B \in \mathcal{B}([0,1]^\infty)$ such that f is continuous and f^{-1} exists.

7.3.1 Continuous Time Filtrations and Stopping Times

Next we present definitions regarding continuous time filtrations and stopping times, along with some of their properties.

Remark 7.4 (Continuous time filtrations and stopping times) For what follows, let $(\Omega, \mathcal{A}, P)$ be some probability space and let $\mathcal{F}_t$ be a sub-σ-field of $\mathcal{A}$ for each $t \geq 0$.

1. Filtrations The collection of sub-σ-fields $\mathcal{F} = \{\mathcal{F}_t : t \geq 0\}$ is called a continuous time filtration in $(\Omega, \mathcal{A})$ if $s < t$ implies that $\mathcal{F}_s \subseteq \mathcal{F}_t$. For any filtration $\mathcal{F}$ we set $\mathcal{F}_\infty = \sigma\left(\bigcup\limits_{t\geq 0} \mathcal{F}_t\right)$. The filtration $\mathcal{F}$ is called complete if $\mathcal{F}_0$ contains all null-sets $\mathcal{N}(\mathcal{A})$ of $\mathcal{A}$, where

$$\mathcal{N}(\mathcal{A}) = \{N \in \mathcal{A} : \inf\{P(A) : N \subset A, A \in \mathcal{A}\} = 0\}.$$

If $\mathcal{F}_t = \sigma(\{X_s : 0 \leq s \leq t\})$ then $\mathcal{F}$ is called the minimal filtration denoted by $\mathcal{F}^X = \{\mathcal{F}^X_t : t \geq 0\}$.

2. Right continuous filtrations Define $\mathcal{F}_{t+} = \bigcap\limits_{\varepsilon>0} \mathcal{F}_{t+\varepsilon}$ so that $\mathcal{F}_t \subseteq \mathcal{F}_{t+}$. If $\mathcal{F}_{t+} = \mathcal{F}_t, \forall t \geq 0$, then we say that the filtration $\mathcal{F}_+$ is right-continuous. Most filtrations we require in order to build a theory for general stochastic processes are right continuous. The filtration $\mathcal{F}$ is said to satisfy the usual conditions if it is right-continuous and complete.

3. Adapted process We say that the stochastic process X is adapted to a filtration $\mathcal{F}$ if X_t is measurable with respect to $\mathcal{F}_t$, $\forall t \geq 0$ and we can write the process and filtration as a pair $\{X_t, \mathcal{F}_t, \forall t \geq 0\}$.

4. Predictability Define the predictable σ-field $\mathcal{P}$ on $\Omega \times \mathcal{R}^+_0$ by

$$\mathcal{P} = \sigma(\{X : X \text{ is left continuous, bounded and adapted to } \mathcal{F}\}),$$

which is equivalent to the generated collection of all sets of the form

$$\{(\omega, t) \in \Omega \times \mathcal{R}^+_0 : X_t(\omega) > a\},$$

for $a \in \mathcal{R}$ and X is a left continuous, bounded stochastic process, adapted to $\mathcal{F}$. A stochastic process $X : \Omega \times \mathcal{R}_0^+ \to \mathcal{R}$ is called predictable if it is measurable with respect to $\mathcal{P}$. Obviously, if X is continuous, bounded and adapted then it is predictable.

5. Progressive measurability Assume that X is $\mathcal{R}^p$-valued and it is adapted to a filtration $\mathcal{F}$ that satisfies the usual conditions. If the function $X_s(\omega)$, for $(\omega, s) \in \Omega \times [0, t]$ is measurable with respect to $\mathcal{F}_t \bigotimes \mathcal{B}([0, t])$, $\forall t \geq 0$, then X is said to have the progressive measurability property with respect to $\mathcal{F}$; in other words, we have

$$\{(\omega, s) \in \Omega \times [0, t] : X_s(\omega) \in B\} \in \mathcal{F}_t \bigotimes \mathcal{B}([0, t]),$$

for all $t \geq 0$ and $B \in \mathcal{B}(\mathcal{R}^p)$. It can be shown that if X is adapted to $\mathcal{F}$ and has RCLL paths (or simply continuous paths) then it is progressively measurable. If X is not adapted but has RCLL paths then X is measurable (joint measurability, remark 7.2.1). Moreover, it can be shown that if X is adapted and right-continuous then X is progressively measurable. Finally, if X is predictable then it is progressively measurable.

6. Stopping times An $\overline{\mathcal{R}}^+$-valued random variable τ defined on $(\Omega, \mathcal{A})$ is a stopping time with respect to the filtration $\mathcal{F}$ if $\{\omega : \tau(\omega) \leq t\} \in \mathcal{F}_t$. The collection of prior events is denoted by $\mathcal{F}_\tau = \{A \subset \Omega : A \cap \{\omega : \tau(\omega) \leq t\} \in \mathcal{F}_t, \forall t \geq 0\}$ and it is a σ-field. Note that τ is measurable with respect to $\mathcal{F}_\tau$ and if τ and ξ are two stopping times with $\tau \leq \xi$ then $\mathcal{F}_\tau \subseteq \mathcal{F}_\xi$ and $\mathcal{F}_{\tau \wedge \xi} = \mathcal{F}_\tau \cap \mathcal{F}_\xi$. A stopping time τ is called an optional time for the filtration $\mathcal{F}$ if $\{\omega : \tau(\omega) < t\} \in \mathcal{F}_t$, $\forall t \geq 0$. It can be shown that τ is optional if and only if it is a stopping time of the right-continuous filtration $\mathcal{F}_+$.

7. Functions of stopping times If τ, ξ are stopping times with respect to $\mathcal{F}$ then so are $\tau \wedge \xi$, $\tau \vee \xi$ and $\tau + \xi$. If $\{\tau_n\}_{n=1}^{+\infty}$ are stopping times then so is $\sup\limits_{n\geq 1} \tau_n$. If $\{\tau_n\}_{n=1}^{+\infty}$ is a sequence of optional times then so are $\sup\limits_{n\geq 1} \tau_n$, $\inf\limits_{n\geq 1} \tau_n$, $\overline{\lim\limits_{n\to+\infty}} \tau_n$ and $\underline{\lim\limits_{n\to+\infty}} \tau_n$ (these hold also for stopping times if $\mathcal{F}_+ = \mathcal{F}$).

8. Hitting times Let X be an $\mathcal{R}^p$-valued stochastic process with continuous sample paths and A a closed subset of $\mathcal{B}(\mathcal{R}^p)$. Then the hitting time of A by X defined by $H_A(\omega) = \inf\{t \geq 0 : X_t(\omega) \in A\}$ is a stopping time with respect to any filtration with respect to which X is adapted.

9. Subordinated Markov chains to stochastic processes If X has RCLL paths and it is adapted to a filtration $\mathcal{F}$ that satisfies the usual conditions then there exists a sequence of stopping times $\{\tau_n\}_{n=1}^{+\infty}$ with respect to $\mathcal{F}$ which exhausts the jumps of X, i.e.,

$$\{(\omega, t) \in \Omega \times \mathcal{R}^+ : X_t(\omega) \neq X_{t-}(\omega)\} \subseteq \bigcup_{n=1}^{+\infty} \{(\omega, t) \in \Omega \times \mathcal{R}^+ : \tau_n(\omega) = t\}. \quad (7.1)$$

Note that this allows us to build an underlying stochastic process Y in discrete time, i.e., $Y_n = X_{\tau_n}$, $n \in \mathcal{Z}_0^+$ and then use established results (e.g., for Markov chains) to study the behavior of the stochastic process X. The stochastic process Y_n is said to be subordinated to the stochastic process X.

10. Increasing stochastic process An adapted stochastic process A is called increasing if a.e. $[P]$ (i) $A_0(\omega) = 0$, (ii) the function $t \mapsto A_t(\omega)$ is nondecreasing and right-continuous, and (iii) $E(A_t) < \infty, \forall t \geq 0$. An increasing stochastic process A is called integrable if $E(A_\infty) < \infty$, where $A_\infty = \lim\limits_{t \to +\infty} A_t$. We write $(A, \mathcal{F})$ when A is adapted to a filtration $\mathcal{F}$.

11. Classes of stopping times Let $\mathcal{L}_a^{\mathcal{F}}$ denote the collection of all stopping times τ with respect to $\mathcal{F}$ that satisfy $P(\tau \leq a) = 1$, for a given finite $a > 0$ and define $\mathcal{L}^{\mathcal{F}}$ as the collection of all stopping times with respect to $\mathcal{F}$ that satisfy $P(\tau < +\infty) = 1$.

12. Classes of stochastic processes The right-continuous stochastic process $\{X_t, \mathcal{F}_t, t \geq 0\}$ is said to be of class D if the collection of random variables $\{X_\tau\}_{\tau \in \mathcal{L}^{\mathcal{F}}}$ is uniformly integrable and of class DL if the collection of random variables $\{X_\tau\}_{\tau \in \mathcal{L}_a^{\mathcal{F}}}$ is uniformly integrable for all $0 \leq a < +\infty$. Note that if $X_t \geq 0$ $a.s.$, $\forall t \geq 0$, then X is of class DL.

13. Uniform integrability A stochastic process X is said to be uniformly integrable if the collection $\{X_t : t \geq 0\}$ is uniformly integrable (see definition 3.17).

14. Bounded variation A stochastic process X is said to have paths of bounded variation if its sample paths are functions of bounded variation w.p. 1 (see remark A.1.10). The stochastic process X is said to have paths of locally bounded variation if there exist stopping times $\tau_n \to \infty$ such that the process $X_{t \wedge \tau_n}$ has paths of bounded variation w.p. 1, for all n.

7.3.2 Continuous Time Martingales

Continuous time filtrations and MGs play a prominent role in the development of many stochastic processes. The MG definitions and results of Section 6.3 can be extended easily (for the most part) to the continuous time parameter case. The following remark summarizes the definitions, properties and basic results regarding continuous time MGs.

Remark 7.5 (Continuous time martingales) In view of remark 7.4 the generalization of discrete time MGs (Section 6.3) to continuous time MGs is straightforward.

1. Definition Consider a real-valued stochastic process $X = \{X_t : t \geq 0\}$ on $(\Omega, \mathcal{A}, P)$ that is adapted to a given filtration $\mathcal{F} = \{\mathcal{F}_t : t \geq 0\}$ and such that $E|X_t| < \infty, \forall t \geq 0$. The pair $(X, \mathcal{F}) = \{X_t, \mathcal{F}_t : t \geq 0\}$ is said to be a continuous time martingale (MG) if for every $0 \leq s < t < \infty$ we have $E(X_t|\mathcal{F}_s) = X_s$ $a.s.$ $[P]$. It is a submartingale (sMG) if $E(X_t|\mathcal{F}_s) \geq X_s$ $a.s.$ $[P]$ and a supermartingale (SMG) if $E(X_t|\mathcal{F}_s) \leq X_s$ $a.s.$ $[P]$. If $\mathcal{F}$ is not mentioned we assume the minimal filtration $\mathcal{F}^X$. If the sample paths are (right-) continuous functions then we call $(X, \mathcal{F})$ a (right-) continuous MG (or sMG or SMG).

2. Last element Assume that $(X, \mathcal{F})$ is a sMG. Let X_∞ be an integrable $\mathcal{F}_\infty$-measurable random variable. If $\forall t \geq 0$ we have $E(X_\infty|\mathcal{F}_t) \geq X_t$ *a.s.* $[P]$, then we say that $(X, \mathcal{F})$ is a sMG with last element X_∞. Similarly for the MG and SMG cases.

3. Convex transformation Let g be a convex, increasing, $\mathcal{R}$-valued function and $(X, \mathcal{F})$ a sMG such that $g(X_t)$ is integrable, $\forall t \geq 0$. Then $\{g(X_t), \mathcal{F}_t : t \geq 0\}$ is a sMG.

4. Convergence Assume that $(X, \mathcal{F})$ is a right-continuous sMG and assume that $\sup\limits_{t\geq 0} E(X_t^+) < \infty$. Then there exists an integrable last element X_∞ such that $\lim\limits_{t\to\infty} X_t(\omega) = X_\infty(\omega)$ a.e. $[P]$.

5. Optional sampling Let $(X, \mathcal{F})$ be a right-continuous sMG with a last element X_∞ and let $\tau \leq \xi$ be two optional times of the filtration $\mathcal{F}$. Then we have $E(X_\tau|\mathcal{F}_{\xi+}) \geq X_\xi$ *a.s.* $[P]$, where $\mathcal{F}_{\xi+} = \{A \subset \Omega : A \cap \{\omega : \xi(\omega) \leq t\} \in \mathcal{F}_{t+}, \forall t \geq 0\}$. If ξ is a stopping time then $E(X_\tau|\mathcal{F}_\xi) \geq X_\xi$ *a.s.* $[P]$. In particular, $E(X_\tau) \geq E(X_0)$ and if $(X, \mathcal{F})$ is a MG then $E(X_\tau) = E(X_0)$.

6. Regularity A sMG $(\mathcal{X}, \mathcal{F})$ is called regular if for every $a > 0$ and every non-decreasing sequence of stopping times $\{\tau_n\}_{n=1}^{+\infty} \subseteq \mathcal{L}_a^\mathcal{T}$ with $\tau = \lim\limits_{n\to\infty} \tau_n$, we have $\lim\limits_{n\to\infty} E(X_{\tau_n}) = E(X_\tau)$. The reader can verify that a continuous, nonnegative sMG is regular.

7. Square and uniform integrability A right-continuous MG $(X, \mathcal{F})$ is said to be square integrable if $EX_t^2 < +\infty$, $\forall t \geq 0$ (equivalently $\sup\limits_{t\geq 0} EX_t^2 < +\infty$). In addition, assume that $X_0 = 0$ *a.s.* and denote the collection of all such MGs by $\mathcal{M}_2$ (or $\mathcal{M}_2^c$ for continuous X). Moreover, if $(X, \mathcal{F})$ is a uniformly integrable MG then it is of class D. Define the $\mathcal{L}^2$ (pseudo) norm on $\mathcal{M}_2$ by

$$\|X\|_t = \sqrt{E\left[X_t^2\right]},$$

and set

$$\|X\| = \sum_{n=1}^{+\infty} 2^{-n} \min\{1, \|X\|_n\},$$

so that $\rho(X, Y) = \|X - Y\|$, $X, Y \in \mathcal{M}_2$, is a (pseudo) metric. It becomes a metric in $\mathcal{M}_2$ if we identify indistinguishable processes, i.e., for any $X, Y \in \mathcal{M}_2$, $\|X - Y\| = 0$, implies that X and Y are indistinguishable. As a result, $\mathcal{M}_2$ becomes a complete metric space equipped with ρ.

8. Doob-Meyer decomposition An extension of Doob's decomposition (theorem 6.8) to the continuous time parameter case is as follows; assume that $\mathcal{F}$ is a filtration that satisfies the usual conditions and let $(X, \mathcal{F})$, a right-continuous sMG, be of class DL. Then X can be decomposed as $X_t = M_t + A_t$, $\forall t \geq 0$, where $(M, \mathcal{F})$ is a right-continuous MG and $(A, \mathcal{F})$ is an increasing process. Requiring $(A, \mathcal{F})$ to be predictable makes the decomposition unique (up to indistinguishability). If X is of class D then $(M, \mathcal{F})$ is a uniformly integrable MG and A is integrable.

9. Quadratic variation If $X \in \mathcal{M}_2$ we have that $(X^2, \mathcal{F})$ is a nonnegative sMG (i.e., of class DL) and therefore X^2 has a unique Doob-Meyer decomposition

$$X_t^2 = M_t + A_t,\ t \geq 0, \tag{7.2}$$

where $(M, \mathcal{F})$ is a right-continuous MG and $(A, \mathcal{F})$ is an adapted, predictable, increasing process. It is convenient to center M in order to have $M_0 = 0\ a.s.$ and since $A_0 = 0\ a.s.\ [P]$, $X_0^2 = 0\ a.s.\ [P]$. If $X \in \mathcal{M}_2^c$ then M_t and A_t are continuous. We define the quadratic (or square) variation of the MG X to be the stochastic process

$$\langle X \rangle_t \stackrel{d}{=} A_t,$$

i.e., $\langle X \rangle$ is the unique, adapted, predictable, increasing process that satisfies $\langle X \rangle_0 = 0\ a.s.$ and $X^2 - \langle X \rangle\ (= M)$ is a MG. In addition, if $X, Y \in \mathcal{M}_2$ we define the cross-variation process $\langle X, Y \rangle$ by

$$\langle X, Y \rangle_t = \frac{1}{4}\left[\langle X + Y \rangle_t - \langle X - Y \rangle_t\right],$$

$t \geq 0$. Note that $XY - \langle X, Y \rangle$ is a MG and $\langle X, X \rangle = \langle X \rangle$.

10. Local martingales Let $(X, \mathcal{F})$ be a (continuous) stochastic process. We say that $(X, \mathcal{F})$ is a (continuous) local MG if
(i) there exists a nondecreasing sequence of stopping times $\{\tau_n\}_{n=1}^{+\infty}$ relative to $\mathcal{F}$ such that $\{X_t^{(n)} = X_{t \wedge \tau_n}, \mathcal{F}_t : t \geq 0\}$ is a MG for all $n \geq 1$ and
(ii) $P(\lim\limits_{n \to +\infty} \tau_n = \infty) = 1$.
If, in addition, $X_0 = 0\ a.s.\ [P]$ we write $X \in \mathcal{M}^{loc}$ (or $X \in \mathcal{M}^{c,loc}$ if X is continuous). Note that every MG is a local MG, whereas, a local MG of class DL is a MG.

11. Semi martingales A semi MG is a stochastic process X of the form $X_t = M_t + A_t$, where $M \in \mathcal{M}^{c,loc}$ is a local MG and A_t is a stochastic process with RCLL paths that are locally of bounded variation. As a result of the Doob-Meyer decomposition, sMGs and SMGs are semi MGs.

7.3.3 Kolmogorov Existence Theorem

The stochastic process $X = \{X_t : t \in T\}$, $T \subseteq \mathcal{R}_0^+$, $\Psi = \mathcal{R}$, of random variables defined on a probability space $(\Omega, \mathcal{A}, P)$ is typically defined in terms of the finite-dimensional distributions it induces in Euclidean spaces. More precisely, for each k-tuple $\mathbf{t} = (t_1, \ldots, t_k)$ of distinct elements of T, the random vector $(X_{t_1}, \ldots, X_{t_k})$ has distribution

$$P_{\mathbf{t}}(B) = P((X_{t_1}, \ldots, X_{t_k}) \in B) = Q_{\mathbf{t}}(\mathbf{Y}_k^{-1}(B)), \tag{7.3}$$

for all $B \in \mathcal{B}_k$ and $\mathbf{t} \in T^k$, where $\mathbf{Y}_k = (X_{t_1}, \ldots, X_{t_k})$ and $Q_{\mathbf{t}}$ denotes the induced probability measure, which is defined on the measurable space $(\mathcal{R}^k, \mathcal{B}_k)$. Note that this system of distributions does not completely determine the properties of a process X. In particular, (7.3) does not adhere to properties of the sample paths of the process, e.g., when X has RCLL or if X is continuous. See the discussion in remark 7.6 below in order to appreciate why this problem arises.

When a process is defined in terms of (7.3) then $\{P_{t_1,\ldots,t_k}\}$ is called a consistent family of distributions and it is easily seen that it satisfies two (consistency) properties:
(a) if $\mathbf{s} = (t_{i_1}, \ldots, t_{i_k})$ is any permutation of $\mathbf{t} = (t_1, \ldots, t_k)$ then $\forall B_i \in \mathcal{B}_1$, $i = 1, 2, \ldots, k$, we have

$$P_{\mathbf{t}}(B_1 \times \cdots \times B_k) = P_{\mathbf{s}}(B_{i_1} \times \cdots \times B_{i_k}), \tag{7.4}$$

(b) if $\mathbf{t} = (t_1, \ldots, t_k)$ with $k \geq 1$, $\mathbf{s} = (t_1, \ldots, t_{k-1})$ and $B \in \mathcal{B}_{k-1}$ we have

$$P_{\mathbf{t}}(B \times \mathcal{R}) = P_{\mathbf{s}}(B). \tag{7.5}$$

Kolmogorov's existence theorem goes the other direction. More precisely, if we are given a system of finite-dimensional distributions $\{P_{t_1,\ldots,t_k}\}$ that satisfy the consistency conditions then there exists a stochastic process $X = \{X_t : t \geq 0\}$ having $\{P_{t_1,\ldots,t_k}\}$ as its finite dimensional distributions.

We collect this important theorem next. The proof is given for stochastic process X with paths $X_t \in \mathcal{R}^{\mathcal{R}}_{[0,\infty)}$, where $\mathcal{R}^{\mathcal{R}}_{[0,\infty)}$ is the space of all $\mathcal{R}$-valued functions defined on $[0, \infty)$. First we consider a k-dimensional cylinder set in $\mathcal{R}^{\mathcal{R}}_{[0,\infty)}$ defined as a set of the form

$$C_k = \{X_t \in \mathcal{R}^{\mathcal{R}}_{[0,\infty)} : (X_{t_1}, \ldots, X_{t_k}) \in B_k\}, \tag{7.6}$$

with $t_i \in [0, \infty)$, $i = 1, 2, \ldots, k$ and base $B_k \in \mathcal{B}_k$ (recall remark 3.5.7). Let C denote the field of all cylinder sets defined on all cylinder sets in $\mathcal{R}^{\mathcal{R}}_{[0,\infty)}$ with $k < \infty$ and let $\mathcal{B}(\mathcal{R}^{\mathcal{R}}_{[0,\infty)})$ be the Borel σ-field on $\mathcal{R}^{\mathcal{R}}_{[0,\infty)}$ (the smallest σ-field containing C). Denote by T^* the set of finite sequences $\mathbf{t} = (t_1, \ldots, t_k)$ of distinct, nonnegative numbers, for all $k = 1, 2, \ldots$. Note that this construction loses track of the underlying probability space $(\Omega, \mathcal{A}, P)$, that is, we look for the induced probability measure on $\left(\mathcal{R}^{\mathcal{R}}_{[0,\infty)}, \mathcal{B}(\mathcal{R}^{\mathcal{R}}_{[0,\infty)})\right)$. However, as we see below, the existence theorem can be restated equivalently in terms of a stochastic process defined on $(\Omega, \mathcal{A}, P)$.

Theorem 7.1 (Kolmogorov existence: of a probability distribution) Let $\{P_{\mathbf{t}}\}$ be a consistent family of finite-dimensional distributions. Then there exists a probability measure Q on the measurable space $(\mathcal{R}^{\mathcal{R}}_{[0,\infty)}, \mathcal{B}(\mathcal{R}^{\mathcal{R}}_{[0,\infty)}))$ such that

$$P_{\mathbf{t}}(B) = Q\left(\left\{X_t \in \mathcal{R}^{\mathcal{R}}_{[0,\infty)} : (X_{t_1}, \ldots, X_{t_k}) \in B\right\}\right), \tag{7.7}$$

for all $B \in \mathcal{B}_k$ and $\mathbf{t} \in T^*$.

Proof. First we define a set function Q_C on the cylinders C and then use the Carathéodory extension theorem to show that there is a unique extension of Q_C on $\sigma(C) = \mathcal{B}(\mathcal{R}^{\mathcal{R}}_{[0,\infty)})$, the desired probability measure Q. That is, we need to show that the entertained Q_C is a probability measure on the field C. In particular, if $C_k \in C$ and $\mathbf{t} = (t_1, \ldots, t_k) \in T^*$ define

$$Q_C(C_k) = P_{\mathbf{t}}(B_k).$$

Clearly, $Q_C(C) \geq 0, \forall C \in \mathcal{C}$ and $Q_C\left(\mathcal{R}^{\mathcal{R}}_{[0,\infty)}\right) = 1$. Now let $A_1, A_2 \in \mathcal{C}$ be two disjoint k-dimensional cylinders with bases $B_1, B_2 \in \mathcal{B}_k$, respectively, and note that B_1 and B_2 are also disjoint (otherwise A_1 and A_2 would not be disjoint). Therefore we can write

$$Q_C(A_1 \cup A_2) = P_{\mathbf{t}}(B_1 \cup B_2) = P_{\mathbf{t}}(B_1) + P_{\mathbf{t}}(B_2) = Q_C(A_1) + Q_C(A_2),$$

so that Q_C is finitely additive. It remains to show that Q_C is countably additive. Let $B_1, B_2, \dots \in \mathcal{C}$ be a sequence of disjoint cylinders and assume that $B = \bigcup\limits_{i=1}^{+\infty} B_i \in \mathcal{C}$. Define $C_m = B \smallsetminus \bigcup\limits_{i=1}^{m} B_i \in \mathcal{C}$, $m = 1, 2, \dots$ and note that

$$Q_C(B) = Q_C(C_m) + \sum_{i=1}^{m} Q_C(B_i).$$

Countable additivity will follow if we show that

$$\lim_{m\to\infty} Q_C(C_m) = 0. \tag{7.8}$$

Since $Q_C(C_m) = Q_C(C_{m+1}) + Q_C(B_{m+1}) \geq Q_C(C_{m+1})$ the limit in (7.8) exists and note that by construction $C_1 \supseteq C_2 \supseteq C_3 \supseteq \dots$, with $\lim\limits_{m\to\infty} C_m = \bigcap\limits_{m=1}^{+\infty} C_m = \emptyset$, i.e., $\{C_m\}_{m=1}^{+\infty}$ is a decreasing sequence of $\mathcal{C}$-sets with $C_m \downarrow \emptyset$. From exercise 4.6 if we show that $Q_C(C_m) \downarrow 0$, then we will have that Q_C is countably additive and the result will be established.

We prove the latter by contradiction, that is, assume that $\lim\limits_{m\to\infty} Q_C(C_m) = \varepsilon > 0$ and we show that this assumption leads to a contradiction, so that ε has to be 0. More precisely, we will show that assuming $\varepsilon > 0$ leads to $\bigcap\limits_{m=1}^{+\infty} C_m \neq \emptyset$, a contradiction, since $\bigcap\limits_{m=1}^{+\infty} C_m = \emptyset$ by construction.

Based on the cylinders $\{C_m\}_{m=1}^{+\infty}$ with bases $\{A_{k_m}\}_{m=1}^{+\infty}$ (where $k_m \neq m$ in general) we build another sequence $\{D_m\}_{m=1}^{+\infty}$ of $\mathcal{C}$-sets with bases $\{A_m\}_{m=1}^{+\infty}$ as follows; assume first that

$$C_m = \{X_t \in \mathcal{R}^{\mathcal{R}}_{[0,\infty)} : (X_{t_1}, \dots, X_{t_{k_m}}) \in A_{k_m}\},$$

with $A_{k_m} \in \mathcal{B}_{k_m}$ and $\mathbf{t}_{k_m} = (t_1, \dots, t_{k_m}) \in T^*$, for some $k_m = 1, 2, \dots$ and since $C_{m+1} \subseteq C_m$ we permute and expand the base of C_{m+1} using the base representation of C_m so that $\mathbf{t}_{k_{m+1}}$ is an extension of $\mathbf{t}_{k_m}$ and $A_{k_{m+1}} \subseteq A_{k_m} \times \mathcal{R}^{k_{m+1}-k_m}$. In particular, define $D_1 = \{X_t \in \mathcal{R}^{\mathcal{R}}_{[0,\infty)} : X_{t_1} \in \mathcal{R}\}, \dots, D_{k_1-1} = \{X_t \in \mathcal{R}^{\mathcal{R}}_{[0,\infty)} : (X_{t_1}, \dots, X_{t_{k_1-1}}) \in \mathcal{R}^{k_1-1}\}$, $D_{k_1} = C_1$, $D_{k_1+1} = \{X_t \in \mathcal{R}^{\mathcal{R}}_{[0,\infty)} : (X_{t_1}, \dots, X_{t_{k_1}}, X_{t_{k_1+1}}) \in A_{k_1} \times \mathcal{R}\}, \dots$, $D_{k_2} = C_2$ and continue this way for all C_m. Note that by construction we have: $D_1 \supseteq D_2 \supseteq D_3 \supseteq \dots$, with $\lim\limits_{m\to\infty} D_m = \bigcap\limits_{m=1}^{+\infty} D_m = \bigcap\limits_{m=1}^{+\infty} C_m = \emptyset$, $\lim\limits_{m\to\infty} Q_C(D_m) = \varepsilon > 0$, $D_m \in \mathcal{C}$ and each D_m is of the form

$$D_m = \{X_t \in \mathcal{R}^{\mathcal{R}}_{[0,\infty)} : (X_{t_1}, \dots, X_{t_m}) \in A_m\},$$

with $A_m \in \mathcal{B}_m$. From exercise 3.29.1 there exists a compact (closed and bounded) set $K_m \subset A_m$ that can be used as a base to define a cylinder

$$E_m = \left\{X_t \in \mathcal{R}^{\mathcal{R}}_{[0,\infty)} : (X_{t_1}, \ldots, X_{t_m}) \in K_m\right\},$$

and is such that $E_m \subseteq D_m$, $Q_C(D_m \smallsetminus E_m) = P_{\mathbf{t}_m}(A_m \smallsetminus K_m) < \varepsilon/2^m$. Set $\widetilde{E}_m = \bigcap\limits_{i=1}^{m} E_i$ since the $\{E_k\}_{k=1}^{+\infty}$ may be nonincreasing and note that $\widetilde{E}_m \subseteq E_m \subseteq D_m$ with

$$\widetilde{E}_m = \left\{X_t \in \mathcal{R}^{\mathcal{R}}_{[0,\infty)} : (X_{t_1}, \ldots, X_{t_m}) \in \widetilde{K}_m\right\},$$

where

$$\widetilde{K}_m = \left(K_1 \times \mathcal{R}^{m-1}\right) \cap \left(K_2 \times \mathcal{R}^{m-2}\right) \cap \cdots \cap (K_{m-1} \times \mathcal{R}) \cap K_m,$$

which is a compact set. As a result, we can bound $P_{\mathbf{t}_m}\left(\widetilde{K}_m\right)$ away from zero since

$$\begin{aligned} P_{\mathbf{t}_m}\left(\widetilde{K}_m\right) &= Q_C(\widetilde{E}_m) = Q_C(D_m) - Q_C(D_m \smallsetminus \widetilde{E}_m) = Q_C(D_m) - Q_C\left(\bigcup_{i=1}^{m} (D_m \smallsetminus E_i)\right) \\ &\geq Q_C(D_m) - Q_C\left(\bigcup_{i=1}^{m} (D_i \smallsetminus E_i)\right) \geq \varepsilon - \sum_{i=1}^{m} \frac{\varepsilon}{2^i} > 0, \end{aligned}$$

because ε is assumed to be positive and by assumption $Q_C(D_m) \geq \varepsilon$. Therefore, $\widetilde{K}_m$ is nonempty for all $m = 1, 2, \ldots$ and we can choose some point $(x_1^{(m)}, \ldots, x_m^{(m)}) \in \widetilde{K}_m$. Since $\widetilde{K}_m \subseteq \widetilde{K}_1$ with $\widetilde{K}_1$ compact, the sequence of points $\{x_1^{(m)}\}_{m=1}^{+\infty}$ must have a convergent subsequence $\{x_1^{(m_k)}\}_{k=1}^{+\infty}$ with limit x_1. But $\{x_1^{(m_k)}, x_2^{(m_k)}\}_{k=2}^{+\infty}$ is contained in $\widetilde{K}_2$ and therefore it has a convergent subsequence with limit (x_1, x_2). Continue in this way in order to construct a $(x_1, x_2, \ldots) \in \mathcal{R}^\infty$ such that $(x_1, x_2, \ldots, x_m) \in \widetilde{K}_m$ for all $m = 1, 2, \ldots$ and as a result we have

$$\left\{X_t \in \mathcal{R}^{\mathcal{R}}_{[0,\infty)} : X_{t_1} = x_i, i = 1, 2, \ldots\right\} \subseteq \widetilde{E}_m \subseteq D_m,$$

for each m. This contradicts the fact that $\bigcap\limits_{m=1}^{+\infty} C_m = \bigcap\limits_{m=1}^{+\infty} D_m = \emptyset$ so that (7.8) holds. ∎

Now we state an equivalent version of the latter theorem (for more details see Billingsley, 2012, p. 517).

Theorem 7.2 (Kolmogorov existence: of a probability space) Let $\{P_{\mathbf{t}}\}$ be a consistent family of finite-dimensional distributions. Then there exists on some probability space $(\Omega, \mathcal{A}, P)$ a stochastic process $X = \{X_t : t \in T\}$ having $\{P_{\mathbf{t}}\}$ as its finite-dimensional distributions.

Unfortunately, this construction will not work for two very important processes, the Poisson process and Brownian motion (Wiener process). Next we discuss why the Kolmogorov existence theorem fails in those cases.

Remark 7.6 (Kolmogorov existence theorem shortcomings) Recall that the Poisson process has sample paths that are step functions (see definition 7.3), however, (7.3) defines a stochastic process that has the same finite dimensional distributions as a Poisson but it has sample paths that are not step functions. Therefore, we cannot use the Kolmogorov existence theorem to define the Poisson process.

Moreover, using Kolmogorov's existence theorem we can build a stochastic process on the sample space $\mathcal{R}^{\mathcal{R}}_{[0,\infty)}$, but there is no guarantee that the process is continuous, which is required for Brownian motion. This problem arises since the "event" $C^{\mathcal{R}}_{[0,+\infty)}$ is not even in the Borel σ-field $\mathcal{B}(\mathcal{R}^{\mathcal{R}}_{[0,\infty)})$, so that $P(X_t \in C^{\mathcal{R}}_{[0,+\infty)})$ is not even defined, when it should be equal to 1 (see for example exercise 2.7, Karatzas and Shreve, 1991, p. 53). Therefore, we need to construct the Wiener measure adhering to the continuity property for the paths of the process, namely, $C^{\mathcal{R}}_{[0,+\infty)}$-valued random variables (functions) and not general $\mathcal{R}$-valued random variables.

In order to build Brownian motion via Kolmogorov's theorem, a modification is required, known as the Kolmorogov-Čentsov theorem. See Karatzas and Shreve (1991, Section 2.2.B) for this construction and for additional discussion on the shortcomings of $\mathcal{R}^{\mathcal{R}}_{[0,\infty)}$ in defining Brownian motion. See appendix remark A.5.6 for the general definition of a locally Hölder-continuous function.

Theorem 7.3 (Kolmorogov-Čentsov) Let $X = \{X_t : 0 \le t \le T\}$ be a stochastic process on a probability space $(\Omega, \mathcal{A}, P)$, such that

$$E\,|X_t - X_s|^a \le c|t-s|^{1+b},\ 0 \le s,t \le T,$$

where $a, b, c > 0$, are some constants. Then there exists a continuous modification $\widetilde{X} = \{\widetilde{X}_t : 0 \le t \le T\}$ of X, which is locally Hölder-continuous of order $\gamma \in (0, \frac{b}{a})$, i.e., for every $\varepsilon > 0$ and $T < \infty$, there exists a finite constant $\delta = \delta(\varepsilon, T, a, b, c, \gamma) > 0$, such that

$$P\left(\left\{\omega : \left|\widetilde{X}_t - \widetilde{X}_s\right|^a \le \delta|t-s|^\gamma, 0 \le s,t \le T\right\}\right) \ge 1 - \varepsilon.$$

For a proof see Klenke (2014, p. 460). In view of remark 7.6, càdlàg spaces discussed next are a natural starting point that allow us to build stochastic processes such as the Poisson and Brownian motion in a unified framework.

7.4 Markov Processes

For what follows we concentrate on general stochastic processes that satisfy the Markov property. The latter is not the only approach in studying stochastic processes, and some texts may assume other general properties. For example, an excellent treatment of the theory and applications of stochastic processes under the assumption of stationarity can be found in Lindgren (2013).

7.4.1 Càdlàg Space

Throughout this subsection assume that (Ψ, ρ) is a Polish space. The exotic wording càdlàg is an acronym for "continue à droite, limites à gauche" which is French for right continuous with left limits (RCLL).

Definition 7.4 Càdlàg space

A function $\phi : [0, +\infty) \to \Psi$ is called càdlàg if it is right continuous and the limit $\lim_{s \uparrow t} \phi(s)$ exists, $\forall t \geq 0$. The space of càdlàg functions is called the càdlàg space and is denoted by $D^{\Psi}_{[0,+\infty)}$.

We collect some facts about càdlàg functions and spaces next.

Remark 7.7 (Càdlàg spaces) To simplify the notation we write ϕ_t for $\phi(t)$, $\forall t \geq 0$. Note that if the sample paths of a stochastic process X are RCLL then at time t the process is at state $X_t(\omega)$ which for fixed $\omega \in \Omega$ is a càdlàg function, i.e., for fixed ω, $X_t(\omega) \in D^{\Psi}_{[0,+\infty)}$. Moreover, if $\lim_{s \uparrow t} \phi(s) = \phi(t)$, then ϕ is continuous, and therefore the càdlàg space contains the continuous functions $C^{\Psi}_{[0,+\infty)}$ from $[0, +\infty)$ to Ψ, that is, $C^{\Psi}_{[0,+\infty)} \subset D^{\Psi}_{[0,+\infty)}$.

1. We turn the càdlàg space $D^{\Psi}_{[0,+\infty)}$ into a measurable space by introducing the σ-field generated by sets of the form $\{\phi : \phi_t \in B\}$, for $B \in \mathcal{B}(\Psi)$ and $t \geq 0$, so that the càdlàg measurable space is given by $(D^{\Psi}_{[0,+\infty)}, \mathcal{D}^{\Psi}_{[0,+\infty)})$ where

$$\mathcal{D}^{\Psi}_{[0,+\infty)} = \sigma(\{\phi : \phi(t) \in B\} : t \geq 0, B \in \mathcal{B}(\Psi)).$$

2. For all $\phi_1, \phi_2 \in D^{\mathcal{R}}_{[0,+\infty)}$ define

$$\rho_C(\phi_1, \phi_2) = \sum_{n=1}^{+\infty} 2^{-n} \max_{0 \leq t \leq n} \{\min(1, |\phi_1(t) - \phi_2(t)|)\}. \tag{7.9}$$

It can be shown that ρ_C is a metric such that $(D^{\mathcal{R}}_{[0,+\infty)}, \rho_C)$ is a Polish space and $\mathcal{D}^{\mathcal{R}}_{[0,+\infty)}$ is the σ-field of Borel sets in $D^{\mathcal{R}}_{[0,+\infty)}$. The latter allows us to study convergence of sequences of distributions on $D^{\mathcal{R}}_{[0,+\infty)}$ which is important when constructing complicated stochastic processes by taking the limit of a sequence of simpler stochastic processes. This result still holds if we replace $\mathcal{R}$ with a general Ψ and appropriate metric ρ_C.

In order to study Markov processes we require a definition similar to the transition operator for Markov chains.

Definition 7.5 Transition semigroup

A transition semigroup for Ψ is a collection of transition operators $T_{SG} = \{T_t : t \geq 0\}$ for Ψ satisfying the following properties:
(i) $T_0 f = f$, for all bounded measurable $\mathcal{R}$-valued functions defined on Ψ.
(ii) $T_s T_t = T_{s+t}$, $\forall t, s \in \mathcal{R}^+_0$.
(iii) $\lim_{t \downarrow 0} T_t f(x) = f(x)$, for all bounded, continuous, measurable $\mathcal{R}$-valued functions defined on Ψ and $x \in \Psi$.

We connect a transition semigroup with its corresponding transition distributions below.

Remark 7.8 (Transition semigroup and distributions) A transition semigroup has associated transition distributions $\{\mu_{x,t} : x \in \Psi, t \geq 0\}$ on Ψ with

$$T_t f(x) = \int_{\Psi} f(y)\mu_{x,t}(dy), \tag{7.10}$$

$\forall x \in \Psi$ and $t \geq 0$. In particular, the transition distributions for any $x, y \in \Psi$ are given by

$$\mu_{x,t}(B) = P(X_{t+s} \in B | X_s = x), \tag{7.11}$$

$\forall B \in \mathcal{B}(\Psi)$. In view of definition (7.5) and conditions (i)-(iii), the transition distributions must satisfy the following conditions:
(a) for all $x \in \Psi$, $\mu_{x,t} \to \mu_{x,0} = \delta_x$, as $t \downarrow 0$,
(b) for all $B \in \mathcal{B}(\Psi)$ and $t \geq 0$, the function $x \mapsto \mu_{x,t}(B)$ is measurable, and
(c) for all $B \in \mathcal{B}(\Psi)$, $s, t \geq 0$ and $x \in \Psi$, $\mu_{x,s+t}(B) = \int_{\Psi} \mu_{y,s}(B)\mu_{x,t}(dy)$.

We collect the formal definition of a time-homogeneous Markov process next.

Definition 7.6 Time-homogeneous Markov process

Let $T_{SG} = \{T_t : t \geq 0\}$ be a transition semi-group for Ψ with corresponding transition distributions $\{\mu_{x,t} : x \in \Psi, t \geq 0\}$. A $D^{\Psi}_{[0,+\infty)}$-valued random variable X adapted to a filtration $\mathcal{F} = \{\mathcal{F}_t : t \geq 0\}$ is called a time-homogeneous Markov process with respect to $\mathcal{F}$ having state space Ψ and transition semi-group T_{SG} if $\forall t, s \in \mathcal{R}_0^+$ the conditional distribution of X_{s+t} given $\mathcal{F}_s$ is $\mu_{X_s,t}$, that is, $P(X_{s+t} \in B|\mathcal{F}_s) = \mu_{X_s,t}(B)$, $\forall B \in \mathcal{B}(\Psi)$. The distribution of such a process is called a (time-homogeneous) Markov distribution with transition semi-group T_{SG}.

We will only consider time-homogeneous Markov processes in what follows and therefore we simply refer to these processes as Markov processes.

Example 7.5 (Construction of the Poisson process) Assume that $T_1, T_2, \ldots,$ is a sequence of independent, exponentially distributed random variables with $\lambda > 0$, that is, $f_{T_n}(t) = \lambda e^{-\lambda t}$, $t \geq 0$. Define a random walk $S = (S_0, S_1, \ldots)$ in $\mathcal{R}_0^+$ with $S_0 = 0$ and $S_n = \sum_{i=1}^{n} T_i$, $n \geq 1$. We can think of S_n as the time of arrival of the nth customer in some queue and of the random variables T_i, $i = 1, 2, \ldots,$ as the interarrival times. Now define a continuous time parameter, integer-valued RCLL process N by

$$N_t = \max\{n \geq 0 : S_n \leq t\},$$

for $t \geq 0$, which is adapted with respect to the minimal filtration $\mathcal{F}^N$. We can think of N_t as the number of customers that arrived in the queue up to time t. Clearly, $N_0 = 0$ *a.s.* and it can be shown that for all $0 \leq s < t$ we have $P(S_{N_s+1} > t|\mathcal{F}_s^N) = e^{-\lambda(t-s)}$, *a.s.* $[P]$ and $N_t - N_s$ is a Poisson random variable with rate $\lambda(t - s)$ independent of $\mathcal{F}_s^N$. Therefore, N_t as defined is a Poisson process (recall definition 7.3).

From conditions (P2) and (P3) of definition 7.3 we can write

$$P(N_{t+s} = y|N_u, u \leq s) = P(N_{t+s} = y|N_s),$$

so that N is a Markov process and in addition

$$P(N_{t+s} = y|N_s = x) = \begin{cases} 0 & \text{if } y < x, \\ \frac{e^{-\lambda t}(\lambda t)^{y-x}}{(y-x)!} & \text{if } y \geq x \end{cases}, \tag{7.12}$$

which is free of $s \geq 0$. As a result, N is a time-homogeneous Markov process with transition measures given by

$$\mu_{x,t}(\{y\}) = P(N_t = y|N_0 = x) = \frac{e^{-\lambda t}(\lambda t)^{y-x}}{(y-x)!} I(y \geq x) = g_t(y-x)I(y \geq x),$$

where g_t is the pmf of a $Poisson(\lambda t)$ random variable. Now based on (7.10), the transition semigroup for the Poisson process can be written as

$$T_t f(x) = \sum_{y \in \Psi} f(y) g_t(y-x) I(y \geq x) = (f * g_t)(x). \tag{7.13}$$

Example 7.6 (Queueing systems) Consider the queue-size process $X = \{X_t : t \geq 0\}$, where X_t denotes the number of customers in the system (waiting or being served) at time t. The future state of the process X_{t+s} at time $t+s$ is equal to the sum of the current number of customers in the queue X_t and number of arrivals $A_{(t,t+s]}$ during the time interval $(t, t+s]$ minus the number of services $S_{(t,t+s]}$ completed during $(t, t+s]$, that is, $X_{t+s} = X_t + A_{(t,t+s]} - S_{(t,t+s]}$. If the numbers $A_{(t,t+s]}$ and $S_{(t,t+s]}$ do not depend on any component of the queueing system before time t then the process X is Markov with discrete state space $\Psi = \{0, 1, 2, \ldots\}$.

For example, assume that the arrivals of customers follow a Poisson process and there is one server with an exponential service time (i.e., $M/M/1$ queue). Then the number of arrivals in $(t, t+s]$ is independent of everything else that happened in the queue before time t. Now since the exponential distribution has the memoryless property, the remaining service time of the customer being served at time t (if any) is independent of the past before time t. Therefore, the number $S_{(t,t+s]}$ of services completed during $(t, t+s]$ depends only on X_t and the number of arrivals $A_{(t,t+s]}$ during the time interval $(t, t+s]$ and nothing before time t.

Definition 7.7 Markov family

A Markov family of processes is a collection of Markov processes $\{X^x : x \in \Psi\}$ with common transition semi-group T_{SG} such that $X_0^x = x$ *a.s.*, $\forall x \in \Psi$. The corresponding collection of Markov distributions $\{Q^x : x \in \Psi\}$ is called a Markov family of distributions with transition semi-group T_{SG}.

Typically, we build a collection of distributions $\{Q^x : x \in \Psi\}$ on $D^{\Psi}_{[0,+\infty)}$ and treat it as a candidate for being a Markov family and this is often done without specifying the transition semi-group. The following result tells us when there exists a transition semi-group that turns a collection of probability measures into a Markov family.

Criterion 7.1 (Markov family) For any $t \geq 0$, let

$$C_t = \sigma(\{\phi_s : s \in [0,t], \phi \in D^{\Psi}_{[0,+\infty)}\}),$$

and define $C = \sigma(\{C_t : t \geq 0\})$. Then the family of distributions $\{Q^x : x \in \Psi\}$ on $(D^{\Psi}_{[0,+\infty)}, C)$ is a Markov family if and only if the following conditions are satisfied:
(i) $\forall x \in \Psi$, $Q^x(\{\phi : \phi_0 = x\}) = 1$.
(ii) $\forall t \geq 0$, the function $x \mapsto Q^x(\{\phi : \phi_t \in \cdot\})$ is a measurable function from Ψ to the measurable space of probability measures on Ψ.
(iii) $\forall x \in \Psi$, $s, t \geq 0$,

$$Q^x(\phi_{s+t} \in \cdot | C_t)(w) = Q^{w_s}(\phi \in \cdot),$$

for almost all $w \in D^{\Psi}_{[0,+\infty)}$ with respect to Q^x.

Under these conditions the transition semi-group of the Markov family is defined by

$$T_t f(x) = \int_{D^{\Psi}_{[0,+\infty)}} f(\phi_t) Q^x(d\phi),$$

for $x \in \Psi$, $t \geq 0$ and bounded measurable $\mathcal{R}$-valued function f defined on Ψ.

The next remark summarizes some important results on Markov processes.

Remark 7.9 (Markov process results) We note the following.

1. The initial distribution Q_0 of a Markov process X is the distribution of X_0. If $Q_0 = \delta_x$, for some $x \in \Psi$ then we call x the initial state of X. The Markov process X is uniquely determined by its initial distribution and transition semi-group.

2. Let $\{Q^x : x \in \Psi\}$ be a Markov family with transition semi-group T_{SG} and Q_0 a probability distribution on Ψ. Then

$$C \mapsto \int_{\Psi} Q^x(C) Q_0(dx), \tag{7.14}$$

for $C \in \mathcal{D}^{\Psi}_{[0,+\infty)}$ is the Markov distribution having initial distribution Q_0 and transition semi-group T_{SG}. Now the function $\phi \mapsto \phi_t$ (as a function of ϕ) from $D^{\Psi}_{[0,+\infty)}$ to Ψ is measurable. This function and the Markov distribution of (7.14) induce a distribution on Ψ, namely, the distribution of the state of the Markov process at time t when the initial distribution is Q_0. If this induced distribution is also Q_0 for all $t \geq 0$ then we say that Q_0 is an equilibrium for the Markov family and the corresponding transition semi-group.

3. Markov process criterion Let T_{SG} be a transition semi-group for a Polish space Ψ, X a $D^{\Psi}_{[0,+\infty)}$-valued random variable and $\mathcal{F}$ a filtration to which X is adapted. The following are equivalent:
(a) X is Markov with respect to $\mathcal{F}$.
(b) For all $s, t \geq 0$ and bounded measurable $f : \Psi \to \mathcal{R}$ we have

$$E(f(X_{s+t})|\mathcal{F}_s) = T_t f(X_s). \tag{7.15}$$

(c) For all $t > 0$ the random sequence $(X_0, X_t, X_{2t}, \dots)$ is Markov with respect to the

filtration $\{\mathcal{F}_{nt} : n = 0, 1, \ldots\}$ with transition operator T_t.
(d) Equation (7.15) holds for all $s, t \geq 0$ and continuous functions $f : \Psi \to \mathcal{R}$ that vanish at ∞ (i.e., $\forall \varepsilon > 0, \exists$ a compact set $C \subset \Psi$ such that $|f(x)| < \varepsilon$, for $x \in \Psi \setminus C$, such functions are necessarily bounded and uniformly continuous).

4. Chapman-Kolmogorov equation Let X be a Markov process with state space Ψ, initial distribution Q_0 and transition distributions $\mu_{x,t}$. Then for all $n > 0$, bounded measurable functions $f : \Psi^n \to \mathcal{R}$ and times $t_1, \ldots, t_n$ we have

$$E(f(X_{t_1}, X_{t_1+t_2}, \ldots, X_{t_1+t_2+\cdots+t_n})) = \int_\Psi \int_\Psi \ldots \int_\Psi f(x_1, \ldots, x_n)\mu_{x_{n-1},t_n}(dx_n) \ldots \mu_{x_0,t_1}(dx_1)Q_0(dx_0).$$

For a discrete state space Ψ let $p_t(x, y) = \mu_{x,t}(\{y\})$ denote the transition function. Then the Chapman-Kolmogorov equation can be written as

$$\sum_{z \in \Psi} p_t(x, z)p_s(z, y) = p_{t+s}(x, y). \tag{7.16}$$

In the discrete time parameter case, the Markov and Strong Markov properties are equivalent. This is not the case in the continuous time parameter case, that is, there are stochastic processes that are Markov but not strong Markov (see exercise 7.19).

Definition 7.8 Strong Markov property

Let X be a Markov process with respect to a filtration $\mathcal{F}$ with transition distributions $\{\mu_{x,t} : x \in \Psi, t \geq 0\}$. Then X is strong Markov with respect to $\mathcal{F}$ if for each *a.s.* finite stopping time τ with respect to the filtration $\mathcal{F}$ and for all $t \geq 0$, the conditional distribution of $X_{\tau+t}$ given $\mathcal{F}_\tau$ is $\mu_{X_\tau,t}$. The distribution of a strong Markov process is called a strong Markov distribution. Markov families whose members are all strong Markov are called strong Markov families.

7.4.2 Infinitesimal Generators

We can describe the behavior of a Markov process locally via infinitesimal generators and then develop methods that describe the global characteristics of a Markov process.

Definition 7.9 Infinitesimal generator

Let $T_{SG} = \{T_t : t \geq 0\}$ be a transition semi-group for a Polish space Ψ. Then an operator G defined by

$$Gf(x) = \lim_{t \downarrow 0} (T_t f(x) - f(x)) / t = \lim_{t \downarrow 0} (T_t - I) f(x)/t, \tag{7.17}$$

for all bounded measurable functions f for which the limit above exists with $|(T_t f(x) - f(x)) / t| < c < \infty$, for some constant $c > 0$ and for all $t \geq 0$, is called the infinitesimal generator of the transition semi-group T_{SG} and of the corresponding Markov processes and Markov families based on T_{SG}.

Let us discuss the relationship between the transition function and the infinitesimal generator for a Markov process.

Remark 7.10 (Transition function and infinitesimal generator) Based on (7.11) the transition probabilities for a Markov process are given by the transition functions

$$p_t(x, y) = P(X_t = y|X_0 = x) = \mu_{x,t}(\{y\}), \tag{7.18}$$

$\forall x, y \in \Psi$ and $t \geq 0$. Now for a given $y \in \Psi$, let $f(x) = \delta_y(x) = I(x = y)$ in (7.17) and use (7.10) and (7.11) to obtain

$$\begin{aligned} G\delta_y(x) &= \lim_{t \downarrow 0} \left(T_t\delta_y(x) - \delta_y(x)\right)/t = \lim_{t \downarrow 0} \frac{1}{t}\left(\int_{\Psi} \delta_y(t)\mu_{x,t}(dt) - \delta_y(x)\right) \\ &= \lim_{t \downarrow 0} \frac{1}{t}\left(\mu_{x,t}(\{y\}) - \delta_y(x)\right) = \lim_{t \downarrow 0} \frac{1}{t}\left(p_t(x, y) - I(x = y)\right). \end{aligned}$$

Now if $y \neq x$ then

$$G\delta_y(x) = \lim_{t \downarrow 0} \frac{1}{t} p_t(x, y),$$

and for $y = x$

$$G\delta_x(x) = \lim_{t \downarrow 0} \frac{1}{t}(p_t(x, x) - 1),$$

so that

$$G\delta_y(x) = \frac{d}{dt} p_t(x, y)|_{t=0},$$

provided that

$$p_0(x, y) = \lim_{t \downarrow 0} p_t(x, y) = \delta_x(y). \tag{7.19}$$

The latter continuity condition ensures that the Markov process is "well behaved" at 0. A transition function with this property is known as a standard transition function. Moreover, for stochastic processes with RCLL we have $\lim_{\varepsilon \downarrow 0} X_{t+\varepsilon}(\omega) = X_t(\omega)$, for all $\omega \in \Omega$ (right continuity) so that using the continuity of probability measure we can write

$$\begin{aligned} \lim_{\varepsilon \downarrow 0} p_{t+\varepsilon}(x, y) &= \lim_{\varepsilon \downarrow 0} P(X_{t+\varepsilon} = y|X_0 = x) = P(\lim_{\varepsilon \downarrow 0} X_{t+\varepsilon} = y|X_0 = x) \\ &= P(X_t = y|X_0 = x) = p_t(x, y), \end{aligned}$$

and therefore the transition function is right continuous and (7.19) holds.

The next theorem can be used to solve for a transition semi-group T_{SG} given an infinitesimal generator G.

Theorem 7.4 (Infinitesimal generator) Let $T_{SG} = \{T_t : t \geq 0\}$ be a transition semi-group for a Polish space Ψ and let G be the corresponding generator. Then for all f in the domain of G we have

$$T_t G f(x) = \lim_{h \to 0} \left(T_{t+h} f(x) - T_t f(x)\right)/h = G T_t f(x),$$

provided that $|(T_{t+h} f(x) - T_t f(x))/h| < c < \infty$, for some constant $c > 0$ and for all $t, h \geq 0$.

7.4.3 The Martingale Problem

We connect MGs and Markov processes next.

Definition 7.10 The martingale problem

Let Ψ be Polish, $\mathcal{BC}_\Psi^{\mathcal{R}}$ denote the collection of all bounded continuous functions $f : \Psi \to \mathcal{R}$ and G be a functional from $\mathcal{BC}_\Psi^{\mathcal{R}}$ to the space of bounded measurable functions on Ψ. A $D_{[0,+\infty)}^{\Psi}$-valued random variable X defined on $(\Omega, \mathcal{A}, P)$ is a solution to the MG problem for $(G, \mathcal{BC}_\Psi^{\mathcal{R}})$ if the function of $t \geq 0$ defined by

$$t \mapsto f(X_t) - \int_0^t G f(X_u) du, \tag{7.20}$$

is a continuous time MG with respect to the minimal filtration of X for all $f \in \mathcal{BC}_\Psi^{\mathcal{R}}$.

The usefulness of the MG problem in identifying Markov processes is illustrated below.

Remark 7.11 (Markov process and the martingale problem) The random variable X in (7.20) is $D_{[0,+\infty)}^{\Psi}$-valued since f is assumed to be continuous. Letting $\mathcal{F}$ denote the minimal filtration of X and noting that $\int_0^t G f(X_u) du$ is measurable with respect to $\mathcal{F}_t$ we can rewrite equation (7.20) along with the assumption that X is a MG equivalently as

$$E\left(f(X_{t+s}) | \mathcal{F}_s\right) - E\left(\int_0^{t+s} G f(X_u) du | \mathcal{F}_s\right) = f(X_s),$$

for all $s, t \geq 0$. We note the following.

1. Let G be the generator of a Markov process X with state space Ψ and let C_G denote the subset of the domain of G that contains only continuous functions. Then X is a solution to the MG problem for (G, C_G).

2. Under the setup of part 1, assume that for each $x \in \Psi$ there is a solution to the MG problem for (G, C_G) with initial state x. Then the collection of solutions obtained by varying x over Ψ is a strong Markov family with a generator that agrees with G on C_G.

7.4.4 Construction via Subordinated Markov Chains

This construction works for stochastic processes that have RCLL sample paths. In view of remark 7.4.9, we envision the construction of a stochastic process X with RCLL sample paths and a countable number of jumps where time is measured in an appropriate way, e.g., the stopping times $\{\tau_n\}_{n=1}^{+\infty}$ of equation (7.1).

Remark 7.12 (Construction of a Markov process via a Markov chain) Assume that T_d is a transition operator for a discrete state space Ψ with discrete generator $G_d = T_d - I$, let $x_0 \in \Psi$ and $\lambda > 0$ be some constant. We construct a Markov process X with infinitesimal generator $G = \lambda G_d$ and initial state x_0. Consider a Markov chain Y with initial state x_0 and transition operator T_d and a random walk $S = (S_0, S_1, \dots)$, $S_0 = 0$, having iid exponential distributed steps with parameter $\lambda > 0$, i.e., $E(S_1) = 1/\lambda$. Further assume that Y and S are independent. For $t \geq 0$, define $M_t = \max\{n \geq 0 : S_n \leq t\}$, a Poisson process with intensity λ (recall example 7.5), independent of Y. Finally, we define a stochastic process for any $t \geq 0$ by

$$X_t = Y_{M_t},$$

so that the discrete time-parameter Markov chain Y was turned into a continuous time-parameter stochastic process by viewing time through a Poisson process. Note that both X and Y take values on the same discrete state space Ψ and X is a $D^{\Psi}_{[0,+\infty)}$-valued random variable.

1. Transition function The transition function of the stochastic process X is written with respect to the transition operator T_d of the discrete Markov chain Y as follows:

$$\begin{aligned} p_t(x,y) &= P(X_t = y|X_0 = x) = P(Y_{M_t} = y|Y_0 = x) \\ &= \sum_{n=0}^{+\infty} P(Y_n = y|Y_0 = x)P(M_t = n|Y_0 = x) = \sum_{n=0}^{+\infty} T_d^n(x,y)\frac{(\lambda t)^n e^{-\lambda t}}{n!}, \end{aligned}$$

and therefore

$$p_t(x,y) = e^{-\lambda t}\sum_{n=0}^{+\infty}\frac{\lambda^n t^n}{n!}T_d^n(x,y), \tag{7.21}$$

where $T_d^n(x,y)$ is the n-step transition probability from x to y by the Markov chain Y.

2. Matrix exponential function For a matrix $T = [(T(x,y))]$ we define

$$e^{tT} = \sum_{n=0}^{+\infty}\frac{t^n}{n!}T^n,$$

the matrix exponential function. As a result, we can write (7.21) in matrix form as

$$e^{-\lambda t}\sum_{n=0}^{+\infty}\frac{\lambda^n t^n}{n!}T_d^n = e^{-\lambda t}\sum_{n=0}^{+\infty}\frac{(\lambda t)^n}{n!}T_d^n = e^{-\lambda t}e^{t\lambda T_d} = e^{t\lambda(T_d - I)} = e^{t\lambda G_d} = e^{tG},$$

and letting $T_t = [(p_t(x,y))]$ we have the matrix representation of the transition matrix of X based on the generator G as $T_t = e^{tG}$. We collect this result in the theorem below.

3. State classification Many of the properties and results we have seen based on the Markov chain Y carry over to the Markov process X, with slight modifications since time t is continuous. In particular, classification of states for the Markov process X when Ψ is discrete can be accomplished by classifying the states of the Markov chain Y. See Section 7.4.5.2 for more details.

Theorem 7.5 (Transition operator and discrete generator) Let $\{X^x : x \in \Psi\}$ be the family of stochastic processes as defined in remark 7.12 in terms of a transition operator T_d and discrete generator G_d. Then for each $x \in \Psi$, X^x is strong Markov. The corresponding Markov family of distributions has generator $G = \lambda G_d$ with transition semi-group $T_{SG} = \{T_t : t \geq 0\}$ related to G as follows

$$T_t = e^{tG} = e^{\lambda t G_d}, \tag{7.22}$$

for all $t \geq 0$.

The Markov process X thus created above is known as a pure-jump process with bounded rates. We collect the following regarding pure-jump processes.

Remark 7.13 (Pure-jump process with bounded rates) To justify the name, first note that the process has a countable number of jumps at the times of arrival of a Poisson process. Furthermore, letting μ_x denote the transition distributions of the Markov chain Y we write

$$Gf(x) = \lambda G_d f(x) = \lambda (T_d - I) f(x) = \lambda \int_{\Psi} (f(y) - f(x)) \mu_x(dy),$$

$\forall x \in \Psi$. Define

$$r(x) = \lambda \mu_x(\Psi \setminus \{x\}), \tag{7.23}$$

and build a probability measure on Ψ by taking

$$P_x(B) = \mu_x(B \setminus \{x\}) / r(x),$$

if $r(x) > 0$ and let $P_x = \delta_x$, if $r(x) = 0$, $\forall B \in \mathcal{B}(\Psi)$. Then we can rewrite (7.23) as

$$Gf(x) = r(x) \int_{\Psi} (f(y) - f(x)) P_x(dy),$$

$\forall x \in \Psi$ and bounded functions $f : \Psi \to \mathcal{R}$. The function $r(x)$ is known as the jump rate function and P_x as the jump distribution. By construction, the jump rate function is bounded above by λ, which rightly earns these processes the name pure-jump process with bounded rates. There are ways of constructing such processes with unbounded rates. For more details see Fristedt and Gray (1997, p. 636).

7.4.5 Discrete State Space

We now connect properties of RCLL paths of a stochastic process with sojourn times, thus expanding on the development of the last two sections. In addition, we will present an alternative subordinated Markov chain construction and perform state classification. For what follows assume that $X = \{X_t, t \geq 0\}$ is a time-homogeneous Markov process defined on a discrete (finite or countable) state space

Ψ via the transition function

$$p_t(x, y) = P(X_t = y|X_0 = x) \geq 0,$$

with $\sum_{y \in \Psi} p_t(x, y) = 1$, $\forall x \in \Psi$ and any $t \geq 0$. We write $P_t = [(p_t(x, y))]$ for the transition matrix.

For the exposition of this section we only need to assume that $p_t(x, y)$ is a standard transition function (i.e., continuous at 0, recall remark 7.10). In fact, based only on this condition, it can be shown that the transition function $t \mapsto p_t(x, y)$, for fixed $x, y \in \Psi$, is continuous everywhere.

Example 7.7 (Poisson process) Let $N = \{N_t : t \geq 0\}$ be a Poisson process with $\Psi = \{0, 1, 2, \ldots\}$ and transition function $p_t(x, y)$ given by (7.12). Then the transition matrix is given by

$$P_t = \begin{bmatrix} p_t(0) & p_t(1) & p_t(2) & \ldots \\ & p_t(0) & p_t(1) & \ldots \\ & & p_t(0) & \ldots \\ 0 & & & \ddots \end{bmatrix}$$

with

$$p_t(k) = \frac{e^{-\lambda t}(\lambda t)^k}{k!},$$

$k = 0, 1, 2, \ldots$, the Poisson distribution with parameter λt. From (7.13) we have

$$T_t f(x) = \sum_{y=0}^{+\infty} f(y) p_t(y - x) I(y \geq x) = \sum_{y=x}^{+\infty} f(y) p_t(y - x),$$

and letting $z = y - x \Rightarrow y = z + x$, we have

$$T_t f(x) = \sum_{z=0}^{+\infty} f(z + x) p_t(z).$$

Using the fact that $\sum_{z=0}^{+\infty} p_t(z) = 1$, the infinitesimal generator of the Poisson process is given by

$$\begin{aligned} Gf(x) &= \lim_{t \downarrow 0} (T_t f(x) - f(x))/t = \lim_{t \downarrow 0} \frac{1}{t} \left[\sum_{z=0}^{+\infty} f(z + x) p_t(z) - f(x) \right] \\ &= \lim_{t \downarrow 0} \frac{1}{t} \left[\sum_{z=0}^{+\infty} (f(z + x) - f(x)) p_t(z) \right] = \sum_{z=1}^{+\infty} [f(z + x) - f(x)] \lim_{t \downarrow 0} \frac{1}{t} p_t(z), \end{aligned}$$

where the term for $z = 0$ in the sum vanishes and we use the DCT to sum to the limit. Now for $z > 1$, the limits

$$\lim_{t \downarrow 0} \frac{1}{t} p_t(z) = \lim_{t \downarrow 0} \frac{1}{t} \frac{e^{-\lambda t}(\lambda t)^z}{z!} = \frac{\lambda^z}{z!} \lim_{t \downarrow 0} e^{-\lambda t} t^{z-1} = 0,$$

vanish, whereas, for $z = 1$ we have

$$\lim_{t \downarrow 0} \frac{1}{t} p_t(1) = \lambda.$$

Consequently, the infinitesimal generator for the Poisson Markov process is given by

$$Gf(x) = \lambda\left[f(x+1) - f(x)\right].$$

Now note that

$$p_t(x,y) = \frac{e^{-\lambda t}(\lambda t)^{y-x}}{(y-x)!}I(y \geq x), \tag{7.24}$$

so that $p_0(x,y) = I(y = x)$ and

$$\frac{d}{dt}p_t(x,y) = -\lambda p_t(x,y) + \lambda p_t(x+1,y).$$

Letting $f(x) = \delta_y(x)$, for some state $y \in \Psi$, we have

$$G\delta_y(x) = \lambda\left[\delta_y(x+1) - \delta_y(x)\right] = \lambda\left[I(y = x+1) - I(y = x)\right],$$

and on the other hand

$$\frac{d}{dt}p_t(x,y)|_{t=0} = -\lambda p_0(x,y) + \lambda p_0(x+1,y) = -\lambda I(y = x) + \lambda I(y = x+1), \tag{7.25}$$

so that $G\delta_y(x) = \frac{d}{dt}p_t(x,y)|_{t=0}$, as expected (recall remark 7.10). The form of the derivative at 0 suggests that given the transition function we can obtain the infinitesimal generator, but more importantly, the converse is also feasible; given an infinitesimal generator there may be a way to solve for the transition function of the Markov process. We formalize the usefulness of such derivatives in the next definition.

Definition 7.11 Jump rates and infinitesimal generator

Let X be a Markov process with discrete state space Ψ and standard transition function $p_t(x,y)$, $x, y \in \Psi$. Define

$$A(x,y) = \frac{d}{dt}p_t(x,y)|_{t=0} = \lim_{t \downarrow 0} \frac{1}{t}p_t(x,y), \tag{7.26}$$

the derivative of the transition function at 0, $\forall x, y \in \Psi$ and assume that it exists. We say that X jumps with rate $A(x,y)$ from x to y and that the matrix $A = [(A(x,y))]$ is the infinitesimal generator matrix of X.

Example 7.8 (Poisson process) For a Poisson process N with intensity λ the probability of at least two jumps by time t is given by

$$\begin{aligned} P(N_t &\geq 2) = 1 - P(N_t = 0) - P(N_t = 1) = 1 - e^{-\lambda t} - \lambda t e^{-\lambda t} \\ &= 1 - (1 + \lambda t)\left(1 - \lambda t + \frac{(\lambda t)^2}{2!} + \ldots\right) \\ &= \frac{(\lambda t)^2}{2!} + \cdots = o(t^2), \end{aligned}$$

so that

$$\lim_{t \downarrow 0} \frac{1}{t}P(N_t \geq 2) = 0,$$

and therefore the probability of two or more jumps in an infinitesimal interval by the Poisson process is zero. Using the transition function (7.24) and its derivative in equation (7.25) we have that the jump rate from state x to y is given by

$$A(x,y) = \frac{d}{dt}p_t(x,y)|_{t=0} = -\lambda I(y = x) + \lambda I(y = x+1),$$

so that we could jump from x to $x+1$ with rate λ, $\forall x \in \Psi$.

In what follows, we will use infinitesimal generators in order to construct Markov processes via subordinated Markov chains. First we collect important definitions and characteristics of the sample paths of the process that can be used to classify the states of the discrete Ψ.

7.4.5.1 Sample Paths and State Classification

Now consider the length of time S_t that the process X remains in the state occupied at the instant t, i.e., for any $\omega \in \Omega$ and $t \geq 0$, define

$$S_t(\omega) = \inf\{s > 0 : X_{t+s}(\omega) \neq X_t(\omega)\}, \tag{7.27}$$

the sojourn of X at the state $X_t(\omega)$ from time t and on. Based on the sojourn random variables we can perform state classification.

Definition 7.12 State classification

For any given state $x \in \Psi$ and $t \geq 0$, define the probability that the sojourn exceeds some time $s \geq 0$ given the current state of the process by

$$W_{x,t}(s) = P(S_t > s | X_t = x).$$

The state x is called
(i) absorbing if $W_{x,t}(s) = 1$, i.e., $S_t = +\infty$ a.s., $X_{t+s} = x$, $\forall s \geq 0$,
(ii) instantaneous if $W_{x,t}(s) = 0$, i.e., $S_t = 0$ a.s. and the process jumps out of the state x the instant it enters it, or
(iii) stable if $0 < W_{x,t}(s) < 1$, i.e., $0 < S_t < +\infty$ a.s. and the process stays at x for an additional amount of time that is positive but finite.

The following remark summarizes properties of the sample paths $t \mapsto X_t(\omega)$, for any $\omega \in \Omega$. For more details on these results see Çinlar (1975) and Durrett (2004).

Remark 7.14 (Sample path behavior) Assume that P_t is a standard transition function.

1. Stochastic continuity Then the stochastic process X is stochastically continuous, that is,

$$\lim_{\varepsilon \downarrow 0} P(X_{t-\varepsilon} \neq X_t) = \lim_{\varepsilon \downarrow 0} P(X_t \neq X_{t+\varepsilon}) = 0, \tag{7.28}$$

or equivalently $\lim_{t \to s} P(X_t = X_s) = 1$, for any fixed $s \geq 0$. This property guarantees well-behaved sample paths in the sense that there are no instantaneous states. In particular, stochastic continuity does not imply continuity of the sample paths, however, it ensures that X does not visit a state in Ψ only to jump out of it instantaneously.

2. Modification for smoothness For a discrete state space it is possible that $\lim_{t\to\infty} X_t(\omega) = X_\infty(\omega) \notin \Psi$, e.g., the limit escapes to $+\infty$ and Ψ does not contain the state $+\infty$ (which is usually the case). We can take care of this situation by considering a modification $\widetilde{X}$ of the original process X that leaves the probability law the same, with $\widetilde{X}$ defined on the augmented state space $\overline{\Psi} = \Psi \cup \{\Delta\}$, where $X_\infty(\omega) = \Delta$. In particular, the modification can have the following properties: for any $\omega \in \Omega$ and $t \geq 0$, either
(i) $\widetilde{X}_t(\omega) = \Delta$ and $\widetilde{X}_{t_n}(\omega) \to \Delta$ for any sequence of times $t_n \downarrow t$, or
(ii) $\widetilde{X}_t(\omega) = x$ for some $x \in \Psi$, there are sequences $t_n \downarrow t$ such that $\widetilde{X}_{t_n}(\omega) \to x$ and for any sequence $t_n \downarrow t$ such that $\widetilde{X}_{t_n}(\omega)$ has a limit in Ψ then we have $\lim_{t_n \downarrow t} \widetilde{X}_{t_n}(\omega) = x$.

3. The state Δ Since $\widetilde{X}$ is a modification of X we have
$$P(\widetilde{X}_t = \Delta | \widetilde{X}_0 = x) = 1 - P(X_t = \Delta | X_0 = x) = 1 - \sum_{y\in\Psi} p_t(x,y) = 0,$$
so that the addition of the artificial point Δ does not alter the probability distribution of the original stochastic process. From now on we let X denote the modified version $\widetilde{X}$.

4. RCLL Assume that there are no instantaneous states. For any $\omega \in \Omega$ and $t \geq 0$, if $t_n \downarrow t$ such that $\lim_{t_n \downarrow t} X_{t_n}(\omega) = x \in \Psi$ then $X_t(\omega) = x$, that is, X has right-continuous paths. Moreover, X has left-hand limits everywhere, i.e., X has RCLL paths.

5. Let Y be a modification of a stochastic process X and assume that both processes have RCLL paths. Then it can be shown that X and Y are indistinguishable.

6. For any $\omega \in \Omega$ and $t \geq 0$, one of the following holds:
(i) $X_t(\omega) = \Delta$, in which case $\lim_{t_n \downarrow t} X_{t_n}(\omega) = \Delta$ for any sequence of times $t_n \downarrow t$,
(ii) $X_t(\omega) = x$, where $x \in \Psi$ is either absorbing or stable. Then $\lim_{t_n \downarrow t} X_{t_n}(\omega) = x$ for any sequence of times $t_n \downarrow t$, or
(iii) $X_t(\omega) = x$, where $x \in \Psi$ is instantaneous. Then $\lim_{t_n \downarrow t} X_{t_n}(\omega)$ is either x or Δ or does not exist, for different sequences $t_n \downarrow t$.

Now we take a closer look at the sojourn times and their use in classification of states.

Remark 7.15 (Sojourn times and classification) Based on the definition of $W_{x,t}(s)$ and the sample path behavior of X we have the following.

1. Sojourn distribution For any $x \in \Psi$ and $t \geq 0$ we have
$$W_{x,t}(s) = P(S_t > s | X_t = x) = e^{-\lambda(x)s}, \tag{7.29}$$
for all $s \geq 0$, where $\lambda(x)$ is some constant in $[0, +\infty]$.

2. Clearly, $\lambda(x)$ denotes the sojourn rate, i.e., the expected length of time X spends at state x is $1/\lambda(x)$. We can classify the states in Ψ based on the corresponding $\lambda(x)$ values. In particular, if $\lambda(x) = 0$ then $W_{x,t}(s) = 1$ and x is absorbing, whereas, if

$0 < \lambda(x) < +\infty$ then $0 < W_{x,t}(s) < 1$ so that x is stable and $S_t | X_t = x \sim Exp(\lambda(x))$. Finally, if $\lambda(x) = +\infty$ then $W_{x,t}(s) = 0$ and x is instantaneous.

3. Note that the distribution of the sojourn times in (7.29) does not depend on t. Furthermore, $\lambda(x)$ can be interpreted as the rate at which the process X leaves state $x \in \Psi$, that is, if $\lambda(x) = 0$ we never leave (x is absorbing), whereas, if $\lambda(x) = +\infty$ we leave the state immediately (x is instantaneous).

4. If Ψ is finite then no state is instantaneous.

5. The sample paths are right continuous if and only if there are no instantaneous states.

6. Stable states For a stable state $x \in \Psi$ it makes sense to define and study the total amount of time spent by the process X at x, i.e., define the random set

$$W_x(\omega) = \{t : X_t(\omega) = x\}. \tag{7.30}$$

In view of (7.29) the set $W_x(\omega)$ consists of a union of intervals each of which has positive length and the lengths are iid exponential random variables $Exp(\lambda(x))$.

7.4.5.2 Construction via Jump Times

Assume for the moment that all states of a Markov process X are stable and consider $S = \{S_t : t \geq 0\}$, the process of sojourn times and $T = \{T_n : n = 0, 1, 2, \ldots\}$, the process of the times at which X performs a jump. We study the behavior of X based on the sojourn process S, the jump times process T and an underlying Markov chain $Y = \{Y_n : n = 0, 1, 2, \ldots\}$ that keeps track of the states that X jumps into.

More precisely, let $T_0 = 0$ and define

$$T_{n+1} = T_n + S_{T_n}, \tag{7.31}$$

for all $n = 1, 2, \ldots$, so that T_{n+1} denotes the time that X performs its $(n+1)^{th}$ jump, which is expressed in terms of the sum of the time T_n of the nth jump and the sojourn time $S_{T_n} = T_{n+1} - T_n$. We keep track of the states the process X jumps into by defining the Markov chain

$$Y_n = X_{T_n}, \tag{7.32}$$

for all $n = 0, 1, 2, \ldots$ and as a result, T is the process of the instants of transitions (jumps) for the process X and Y are the successive states visited by X.

Since all states are stable we have $0 < S_{T_n} < +\infty$ a.s. and every Y_n takes values in Ψ. Now for a typical $\omega \in \Omega$, suppose that $T_n(\omega)$ and $Y_n(\omega)$ are defined as in (7.31) and (7.32), respectively, for $n = 0, 1, 2, \ldots, m$ and assume that $Y_m(\omega) = x$, an absorbing state. Then we expect that $X_t(\omega) = x$ for all $t \geq T_m(\omega)$. By the definition of the sojourn time S_t we have $T_{m+1}(\omega) = +\infty$, but we cannot use (7.31) to obtain $T_{m+1}(\omega)$ since S_∞ is not defined and consequently we cannot obtain any of the $T_{m+2}(\omega)$, $T_{m+3}(\omega)$ and so forth. In addition, (7.32) yields $Y_{m+1}(\omega) = X_\infty(\omega) = \Delta$. In order to have Y_n in Ψ and be able to define $T_n(\omega)$ and $Y_n(\omega)$ beyond m, we alter the definition of the subordinated process as follows; let

$$T_0 = 0,\ Y_0 = X_0,\ S_\infty = +\infty,\ T_{n+1} = T_n + S_{T_n}, \tag{7.33}$$

and

$$Y_{n+1} = \begin{cases} X_{T_{n+1}} & \text{if } T_{n+1} < \infty \\ Y_n & \text{if } T_{n+1} = \infty \end{cases}, \tag{7.34}$$

for all $n = 0, 1, 2, \ldots$ and note that if there are no absorbing states these definitions reduce to (7.31) and (7.32).

The following theorem shows that Y as constructed based on (7.31) and (7.32) is a Markov chain. The proof is requested as an exercise.

Theorem 7.6 (Transitions and sojourn times) For any $n = 0, 1, 2, \ldots$ and stable $x \in \Psi$, $s \geq 0$, assume that $Y_n = x$. Then

$$P(Y_{n+1} = y, S_{T_n} > s | Y_0, Y_1, \ldots, Y_n, T_0, T_1, \ldots, T_n) = q(i, j)e^{-\lambda(x)s}, \tag{7.35}$$

where $Q = [(q(x, y))]$ is a Markov matrix, with $q(x, x) = 0$, $\forall x \in \Psi$.

Note that setting $s = 0$ in the latter theorem yields that the subordinated process Y is a Markov chain with transition matrix Q. We say that the Markov process X is irreducible recurrent if the underlying Markov chain Y is irreducible recurrent.

Note that if X is right continuous and T_n, Y_n are defined based on (7.33) and (7.34) then the latter theorem holds for absorbing states as well. In particular, if x is absorbing for the Markov process X then it is absorbing for Markov chain Y as well. Therefore, the transition matrix Q can be used for classification; if $q(x, x) = 1$ then x is absorbing, whereas, for $q(x, x) = 0$ we have that x is stable.

Remark 7.16 (Construction of the Markov process) Now we turn to the construction of a Markov process $X = \{X_t : t \geq 0\}$ based on $T = \{T_n : n = 0, 1, 2, \ldots\}$, the jump times process and the underlying Markov chain $Y = \{Y_n : n = 0, 1, 2, \ldots\}$. From definitions (7.33) and (7.34) we may write

$$X_t(\omega) = Y_n(\omega),\ \text{for } t \in [T_n(\omega), T_{n+1}(\omega)) \tag{7.36}$$

for a typical $\omega \in \Omega$. Therefore, if we find an n such that we can bound any time $t \geq 0$ between two successive jump times $T_n(\omega)$ and $T_{n+1}(\omega)$, then $X_t(\omega)$ is well defined based on the underlying Markov chain Y. Such n exist provided that the random variable $\psi(\omega) = \sup\limits_n |T_n(\omega)| = +\infty$ a.s., since if $\psi(\omega) < +\infty$ the process makes an infinite number of jumps in a finite amount of time. We avoid this case by considering finite jump rates, or more formally, by introducing the concept of regularity of Markov processes.

Definition 7.13 Regular Markov process

The stochastic process X is said to be regular if for almost all $\omega \in \Omega$ we have:

(i) the sample paths $t \mapsto X_t(\omega)$ are right continuous, and
(ii) $\psi(\omega) = \sup_n |T_n(\omega)| = +\infty$.

Several criteria for regularity are collected next.

Criterion 7.2 (Regularity criteria) The following criteria can used to assess regularity for a Markov process X.

1. If $\lambda(x) \leq \lambda$, for all $x \in \Psi$ for some constant $\lambda < \infty$ then X is regular.

2. If Ψ is finite then X is regular.

3. If there are no instantaneous states for X and every state is recurrent in the subordinated Markov chain Y then X is regular.

4. If for the underlying Markov chain Y the probability of staying in its transient states forever is zero and there are no instantaneous states for X then X is regular.

5. Assume that there exists a set D of transient states with respect to the subordinated Markov chain Y such that Y has positive probability of staying forever in D. Define $R = \sum_{n=0}^{+\infty} Q^n$ and assume that $\sum_{y \in D} R(x,y)/\lambda(y) < \infty$, for some $x \in D$. Then X is not regular.

7.4.5.3 Infinitesimal Generator and Transition Function

For what follows assume that X is a regular Markov process with transition function $p_t(x,y)$. In this section we connect the generator matrix A (recall definition 7.11) of the Markov process with the characteristics of the subordinated Markov chain Y and the sojourn rates $\lambda(x)$, $x \in \Psi$. The following theorem connects the transition function with Y and $\lambda(.)$ recursively.

Theorem 7.7 (Recursive transition function) For any $x, y \in \Psi$ and $t \geq 0$ we have

$$p_t(x,y) = e^{-\lambda(x)t} I(x=y) + \int_0^t \lambda(x) e^{-\lambda(x)s} \sum_{z \in \Psi} q(x,z) p_{t-s}(z,y) ds.$$

We can think of this result as follows; the probability of a jump from x to y is either the probability of remaining at state x after t time units if $y = x$, which means that the sojourn has not ended and this happens with probability $e^{-\lambda(x)t}$ or if there is a jump before time t, say at time s, the sojourn ends at s with probability (infinitesimally) $\lambda(x)e^{-\lambda(x)s}$, we jump to a state z with probability $q(x,z)$ and in the remaining time $t-s$ the Markov process transitions to y with probability $p_{t-s}(z,y)$. This theorem allows us to connect the generator matrix with Q and $\lambda(.)$.

Theorem 7.8 (Transition function and jump rates) For any $x, y \in \Psi$ the transition function $t \to p_t(x, y)$ as a function of $t \geq 0$ is differentiable and the derivative is continuous. The derivative at 0 is given by

$$A(x,y) = \frac{d}{dt}p_t(x,y)|_{t=0} = \begin{cases} -\lambda(x) & \text{if } x = y, \\ \lambda(x)q(x,y) & \text{if } x \neq y. \end{cases} \tag{7.37}$$

Furthermore, for any $t \geq 0$ we have

$$\frac{d}{dt}p_t(x,y) = \sum_{z\in\Psi} A(x,z)p_t(z,y) = \sum_{z\in\Psi} p_t(x,z)A(z,y). \tag{7.38}$$

When $p_t(x, y)$ is known we can find A and therefore solve for

$$\lambda(x) = -A(x,x),$$

and

$$q(x,y) = \begin{cases} A(i,j)/\lambda(x) & \text{if } \lambda(x) > 0, \\ 0 & \text{if } \lambda(x) = 0,\ x \neq y, \\ 0 & \text{if } \lambda(x) = 0, \text{ and } x = y \text{ stable}, \\ 1 & \text{if } \lambda(x) = 0, \text{ and } x = y \text{ absorbing}. \end{cases}$$

Now consider the converse of the problem. Given Q and $\lambda(.)$, or equivalently the generator matrix A, we wish to find the transition functions P_t. In particular, we have to solve the infinite system of differential equations given in (7.38) or in terms of matrices we solve

$$\frac{d}{dt}P_t = AP_t, \tag{7.39}$$

known as Kolmogorov's backward equations or

$$\frac{d}{dt}P_t = P_tA, \tag{7.40}$$

known as Kolmogorov's forward equations. The solution to these systems of equations is similar to theorem 7.5 and we present it next for the specific construction via the generator matrix A.

Theorem 7.9 (Transition matrix via generator matrix) For any $t \geq 0$ we have

$$P_t = e^{tA}. \tag{7.41}$$

Example 7.9 ($M/M/k$ queue) Consider an $M/M/k$ queue where customers arrive according to a Poisson process with intensity λ and the service time is exponential with rate μ. Customers that arrive in the system to find all servers busy, enter the waiting room and have to wait for their turn to be serviced. Further assume that the k-servers process customers independently. Increases in the queue size occur upon an arrival and the transition has jump rate $A(x, x + 1) = \lambda$ (recall example 7.8). To model the departures we let

$$A(x, x-1) = \begin{cases} x\mu & \text{if } x = 0, 1, \ldots, k \\ k\mu & \text{if } x > k \end{cases},$$

since when there are $x \leq k$ customers in the system then they are all being served and departures occur at rate $x\mu$, whereas, if there are more than k customers, all k servers are busy and departures occur independently at rate $k\mu$.

Example 7.10 ($M/M/k$ queue with balking) Consider the previous example with the twist that now customers choose to join the queue with some probability a_x that depends on the queue size at time t, i.e., $X_t = x$, or they choose to leave (balk). Therefore, increases happen at rate $A(x, x+1) = \lambda a_x$ and the rest of the setup for this queue remains the same.

7.4.5.4 Limiting Behavior

Suppose that X is a regular Markov process with transition matrix P_t and let Y denote the underlying Markov chain with transition matrix Q. In order to assess the limiting behavior of X we turn to established definitions and results for Y. Fix a state $y \in \Psi$ and let lengths of successive sojourn times in y be denoted by $S_1, S_2, \ldots$, with $S_i \overset{iid}{\sim} Exp(\lambda(y))$. Now the total time spent in state y is defined in equation (7.30) as

$$W_y(\omega) = \{t : X_t(\omega) = y\} = \int_0^{+\infty} \delta_y(Y_s(\omega))ds,$$

for a typical $\omega \in \Omega$. We turn to the underlying Markov chain Y and use its behavior to derive results for X. The following remark summarizes the case where y is transient.

Remark 7.17 (Transient state) Note that if y is transient then the total number of visits V_y to y is finite w.p. 1 and starting at $x \in \Psi$ its distribution is given by (6.25). Then we can write

$$W_y(\omega) = \begin{cases} 0 & \text{if } V_y(\omega) = 0 \\ S_1(\omega) + \cdots + S_{V_y(\omega)}(\omega) & \text{if } V_y(\omega) > 0 \end{cases},$$

which is finite w.p. 1. We note the following.

1. Let $F(x, y)$ denote the probability that starting from x the Markov chain Y ever reaches y. It is straightforward to show that

$$P(W_y \leq t | X_0 = x) = 1 - F(x, y) \exp\{-t\lambda(x)(1 - F(y, y))\}.$$

2. For any state $x \in \Psi$ and $y \in \Psi$ transient, define $R(x, y) = E(V_y | X_0 = x)$, the expected number of visits by the Markov chain Y to state y starting at x. Then

$$E(W_y | X_0 = x) = \frac{R(x, y)}{\lambda(y)} < \infty.$$

3. For any state $x \in \Psi$ and $y \in \Psi$ transient, we have

$$\lim_{t \to +\infty} P(X_t = y | X_0 = x) = \lim_{t \to +\infty} p_t(x, y) = 0.$$

If y is recurrent and it can be reached from x then $W_y = +\infty$ a.s., otherwise $W_y = 0$ a.s. and since the sojourn times are continuous (exponentially distributed) we do not have to worry about periodicity. For what follows we consider the case where the underlying Markov chain Y is irreducible recurrent.

Note that for a Markov chain with transition matrix P (recall definition 6.24), solving $\pi = \pi P$ can give us the stationary measure. However, for continuous time parameter processes we need to satisfy equation (6.46) for all $t \geq 0$, a much stronger requirement.

Definition 7.14 Limiting behavior of a Markov process

Let $X = \{X_t : t \geq 0\}$ be a Markov process with discrete state space Ψ and transition function $P_t = [(p_t(x, y))]$. A measure π is called a stationary (or invariant) measure for X if it satisfies

$$\pi = \pi P_t, \tag{7.42}$$

for all $t \geq 0$, where $\pi = [\pi(x)]_{x \in \Psi}$ is a strictly positive row vector denoting the stationary measure (unique up to a constant multiplication). If π is a probability measure then it is called the equilibrium (or stationary) distribution and the Markov process X is called stationary.

Solving equation (7.42) is not easily accomplished since P_t are not easy to compute. Instead, we use the following theorem to find the stationary distribution of X via its generator matrix A. For a proof see Çinlar (1975, p. 264).

Theorem 7.10 (Stationary distribution) Let $X = \{X_t : t \geq 0\}$ be an irreducible recurrent Markov process with discrete state space Ψ and generator matrix A. Then μ is a stationary measure for X if and only if μ is a solution to the system

$$\mu A = 0, \tag{7.43}$$

where $\mu = [\mu(x)]_{x \in \Psi}$ is a row vector denoting the stationary measure (unique up to a constant multiplication), with $\mu(x) > 0$, $\forall x \in \Psi$. Moreover, the equilibrium distribution is given by

$$\pi(y) = \lim_{t \to +\infty} P(X_t = y | X_0 = x) = \frac{\mu(y)}{\sum\limits_{x \in \Psi} \mu(x)} = \frac{\mathbf{v}(y)/\lambda(y)}{\sum\limits_{x \in \Psi} \mathbf{v}(x)/\lambda(x)}, \tag{7.44}$$

where $\mathbf{v}$ is the stationary distribution for the underlying Markov chain Y, i.e., $\mathbf{v} = \mathbf{v}Q$ and

$$\mu(y) = \mathbf{v}(y)/\lambda(y),$$

for all $y \in \Psi$.

7.4.5.5 Birth-Death Processes

Now we discuss Birth-Death processes in continuous time (recall example 6.19). This is a special class of Markov processes with many applications, including population models and queueing systems. For what follows suppose that $X = \{X_t : t \geq 0\}$ is a stochastic process with state space $\Psi = \{0, 1, 2, \dots\}$, where X_t denotes the size of a population at time t. Transitions occur when a birth or a death occurs in the population so that a jump either increases or decreases the population size by 1. The following remark summarizes the construction of the process.

Remark 7.18 (Birth-Death process) Let G_t and D_t denote the lengths of time from t until the next birth and the next death event, respectively. Assume that if $X_t = x$, then for some constants $\lambda(x)$ and $p(x)$ we have

$$P(G_t > s, D_t > s | X_u, u \leq t) = e^{-\lambda(x)s}, \tag{7.45}$$

and

$$P(G_t \leq D_t | X_u, u \leq t) = p(x), \tag{7.46}$$

so that under these assumptions X is a Markov process. We build the generator matrix of X based on $\lambda(x)$ and $p(x)$.

1. Generator Let

$$a_x = p(x)\lambda(x) \geq 0, \tag{7.47}$$
$$b_x = (1 - p(x))\lambda(x) \geq 0, \tag{7.48}$$

$x \in \Psi$ and note that a_x and b_x can be interpreted as the respective time rates of births and deaths when the population size is x. Clearly, $b_0 = 0$ since we cannot have deaths in a population of size 0 and we cannot have $(a_x, b_x) = (0, 0)$, $\forall x \in \Psi$. Therefore, from equation (7.37) we have

$$A = \begin{bmatrix} -a_0 & a_0 & & & 0 \\ b_1 & -a_1 - b_1 & a_1 & & \\ & b_2 & -a_2 - b_2 & a_2 & \\ 0 & & \ddots & \ddots & \ddots \end{bmatrix}, \tag{7.49}$$

where the underlying Markov chain Y has transition probabilities

$$q(x, x-1) = 1 - q(x, x+1) = b_x/(a_x + b_x), x \geq 1,$$

and $q(0, 1) = 1$.

2. Regularity Regularity of the Markov process X can be shown using the criteria of remark 7.2. In particular, if X is not regular then the population size escapes to ∞ in a finite time with positive probability. Otherwise, the population remains finite w.p. 1 but it may still reach ∞. More precisely, letting $R(y, y)$ denote the average number of visits to state y, starting from y, then if the series $\sum_{y \in \Psi} R(y, y)/\lambda(y)$ converges then X is not regular, whereas, if the series diverges then X is regular.

3. Limiting behavior From theorem 7.10 a limiting distribution π exists if and only if π satisfies $\pi A = 0$ and $\sum_{y=0}^{+\infty} \pi(y) = 1$ with A given by (7.49). First we find the stationary measure by solving the system $\mu A = 0$ with $\mu = [\mu(0), \mu(1), \mu(2), \dots]$ which yields the equations

$$\mu(y) = \frac{a_{y-1}}{b_y}\mu(y-1),$$

so that

$$\mu(y) = \frac{a_0 a_1 \dots a_{y-1}}{b_1 \dots b_y}\mu(0),$$

$y = 1, 2, \dots$ and therefore the limiting distribution exists if

$$\sum_{y=0}^{+\infty} \mu(y) = \mu(0) + \mu(0) \sum_{y=1}^{+\infty} \frac{a_0 a_1 \dots a_{y-1}}{b_1 \dots b_y} = \mu(0)c < \infty,$$

with

$$c = 1 + \sum_{y=1}^{+\infty} \frac{a_0 a_1 \dots a_{y-1}}{b_1 \dots b_y}. \tag{7.50}$$

Now if $c < \infty$ then choosing $\mu(0) = 1/c$ we have $\sum_{y=0}^{+\infty} \mu(y) = 1$ and the stationary distribution is given by

$$\pi(y) = \lim_{t \to +\infty} P(X_t = y | X_0 = x) = \begin{cases} \frac{1}{c} & \text{if } y = 0 \\ \frac{a_0 a_1 \dots a_{y-1}}{c b_1 \dots b_y} & \text{if } y > 0 \end{cases}, \tag{7.51}$$

for any $x \in \Psi$, whereas, if $c = +\infty$ then

$$\lim_{t \to +\infty} P(X_t = y | X_0 = x) = 0,$$

for all $x, y \in \Psi$.

We present several examples in order to illustrate these Birth-Death models.

Example 7.11 (Population model) Consider a population where each individual has an exponential lifetime with parameter b and each individual can generate an offspring (birth) after an exponential time with parameter a. We are interested in the Markov process X, with X_t denoting the population size at time $t \geq 0$. Clearly this is a Birth-Death process with birth rates $a_x = ax$ and death rates $b_x = bx$. Then the underlying Markov chain Y has transition probabilities $q(0, 0) = 1$, $q(x, x + 1) = p$, $q(x, x - 1) = 1 - p$, for $x \geq 1$, where $p = \frac{a}{a+b}$, so that $\lambda(0) = 0$ and $\lambda(x) = x(a + b)$, for $x \geq 1$ (which are not bounded from above). Now if $a \leq b$ the Markov chain Y is absorbing (the population dies out) so that from remark 7.2.3 the process X is regular. In addition, if $a > b$ using remark 7.18.2 we have

$$\sum_{y=1}^{+\infty} \frac{R(y, y)}{y(a + b)} \geq \frac{1}{a + b} \sum_{y=1}^{+\infty} \frac{1}{y} = +\infty,$$

so that X is regular in this case as well. Since the process X is regular, $P(X_t < \infty) = 1$, $\forall t \geq 0$, however, for $a > b$ the population tends to grow out of control leading to $\lim_{t \to +\infty} P(X_t \leq y) = 0$, $\forall y \in \Psi$ and therefore the population size becomes infinite.

Example 7.12 (Queueing systems) We utilize the results we presented in this section on Birth-Death processes in order to illustrate several queueing systems. As usual, let $X = \{X_t : t \geq 0\}$ denote the queue-size process and assume that arrivals follow a Poisson process with rate a, whereas, services follow exponential times with rate b.

1. $M/M/1$ **queue** This queue is a special case of a Birth-Death process with birth rates $a_x = a$ and death rates $b_x = b$. Let $r = a/b$ denote the ratio of the rate of arrivals to the rate of service times (known as the traffic intensity). Then the normalizing constant of equation (7.50) is given by

$$c = \sum_{y=0}^{+\infty} r^y = \begin{cases} +\infty & \text{if } a \geq b \\ \frac{1}{1-r} & \text{if } a < b \end{cases},$$

and therefore if $r < 1$ then the stationary distribution is

$$\pi(y) = (1-r)r^y,$$

$y = 0, 1, 2, \dots$, otherwise, if $r \geq 1$ then $\lim_{t\to+\infty} P(X_t = y) = 0$, $\forall y \in \Psi$. Therefore, if $r < 1$ the server can keep up with arrivals and the queue never explodes to infinity, otherwise, for $r \geq 1$ arrivals occur faster than the server can handle and the queue goes to infinity.

2. $M/M/1/m$ **queue** In this queue we need to modify the $M/M/1$ queue setup. In particular, we have birth rates $a_x = a$, $x = 0, 1, \dots, m$, $a_x = 0$, $x \geq m+1$ and death rates $b_x = b$, $x = 1, 2, \dots, m$, $b_x = 0$, $x \geq m+1$. Therefore, the set of states $\{m+1, m+2, \dots\}$ are transient and $C = \{0, 1, \dots, m\}$ is a recurrent irreducible set. The stationary distribution follows in similar steps as in the $M/M/1$ queue, however we work with the set C only. More precisely, assuming $r = a/b$ and $a \neq b$, we have

$$c = \sum_{y=0}^{m} r^y = \frac{1-r^{m+1}}{1-r},$$

so that the stationary distribution is given by

$$\pi(y) = \frac{1-r}{1-r^{m+1}} r^y,$$

$y \in C$ and $\pi(y) = 0$, $y \notin C$. Note that the state space for X is $\Psi = C$ but the representation as a Birth-Death process can be given for $\Psi = \{0, 1, 2, \dots\}$.

3. $M/M/k$ **queue** Now consider k-servers working independently. If there are $x < k$ customers in the system then x servers are busy so that (7.45) becomes

$$P(G_t > s, D_t > s | X_u, u \leq t) = e^{-as} e^{-xbs},$$

whereas, if there are $x \geq k$ customers in the queue, all servers are busy so that

$$P(G_t > s, D_t > s | X_u, u \leq t) = e^{-as} e^{-bs}.$$

As a result, the queue-size process is a Birth-Death process with birth rates $a_x = a$, $\forall x = 1, 2, \dots$ and death rates $b_x = bx$, $x = 1, 2, \dots, k$ and $b_x = kb$, $x \geq k+1$. The stationary distribution exists when $a < kb$.

4. $M/M/\infty$ **queue** In this case the birth rates are $a_x = a$, $\forall x = 1, 2, \dots$ and the death rates are $b_x = bx$, $x \geq 1$. Let $r = a/b$. It is straightforward to show that the stationary distribution is given by

$$\pi(y) = \frac{e^{-r} r^y}{y!},$$

$y = 0, 1, 2, \ldots$, which is nothing but a *Poisson*(r) density.

7.4.6 Brownian Motion

Now we turn to a classic continuous time parameter process with continuous state space, known as Brownian motion. We define first the one-dimensional Brownian motion and discuss some of its properties. The construction of the Wiener measure (the probability distribution of Brownian motion) is given in the next subsection.

Definition 7.15 One-dimensional Brownian motion

Given a filtration $\mathcal{F} = \{\mathcal{F}_t : t \geq 0\}$, a one-dimensional $\mathcal{F}$-Brownian motion (or Wiener process) in $[0, +\infty)$, denoted by $BM([0, +\infty), \mathcal{F})$, is a continuous real-valued process $B = \{B_t : t \geq 0\}$ defined on some probability space $(\Omega, \mathcal{A}, P)$, that is adapted to $\mathcal{F}$, with the following properties:
(a) $B_0 = 0$ *a.s.* $[P]$,
(b) for $0 \leq s < t$ the increment $B_t - B_s$ is independent of $\mathcal{F}_s$, and
(c) $B_t - B_s \sim \mathcal{N}(0, t - s)$, for all $0 \leq s < t$.
We may define Brownian motion on $[0, T]$, denoted by $BM([0, T], \mathcal{F})$, for some $T > 0$, by considering the stochastic process $\{B_t, \mathcal{F}_t : 0 \leq t \leq T\}$. We omit the filtration $\mathcal{F}$ from the notation of the Brownian motion when it is clear from the context.

We summarize several properties and results on Brownian motion below.

Remark 7.19 (Properties of Brownian motion) Assume that B is a $BM([0, +\infty))$.

1. Condition (b) can be restated as follows: if $0 = t_0 < t_1 < \cdots < t_n < \infty$, then the increments $\{B_{t_i} - B_{t_{i-1}}\}_{i=1}^n$ are independent and $B_{t_i} - B_{t_{i-1}} \sim \mathcal{N}(0, t_i - t_{i-1})$. In other words, B has stationary and independent increments.

2. We can easily see that B is a continuous, square integrable MG, a local MG (taking $\tau_n = n$) and the transformations $B_t^2 - t$ and $e^{aB_t - a^2 t/2}$, $a \in \mathcal{R}$, are MGs. However, B is not a uniformly integrable MG. In view of equation (7.2) we have $\langle B \rangle_t \stackrel{d}{=} t$ *a.s.* $[P]$.

3. The choice of filtration is important in the latter definition. When we do not mention the filtration but we are given only a Markov process $B = \{B_t : t \geq 0\}$ and we assume that B has stationary and independent increments and $B_t = B_t - B_0 \sim \mathcal{N}(0, t)$, then $\{B_t, \mathcal{F}_t^B : t \geq 0\}$ is a Brownian motion.

4. The following transformations of B are also Brownian motions on $[0, +\infty)$: (i)

$t \mapsto -B_t$, (ii) $t \mapsto (B_{s+t} - B_s)$, $\forall s \geq 0$, fixed, (iii) $t \mapsto \sqrt{a} B_{t/a}$, $a > 0$, fixed, and (iv) $t \mapsto tB_{1/t}$, $t > 0$ and $t \mapsto 0$, if $t = 0$.

5. We can define Brownian motion equivalently in terms of a Gaussian Process (GP, i.e., $G = \{G_t : t \geq 0\}$ is a GP if all finite-dimensional distributions of G are multivariate normal) as follows: (i) B is a GP, (ii) $EB_t = 0$ and $E(B_s B_t) = \min(s, t)$, and (iii) the sample paths $t \mapsto B_t$ are continuous $a.s.$

6. Translation invariance We can show that $\{B_t - B_0 : t \geq 0\}$ is independent of B_0 and it has the same distribution as Brownian motion with $B_0 = 0$.

7. Strong law It can be shown that

$$\lim_{t \to \infty} B_t / t = 0 \ a.s.$$

8. Non-differentiable For almost all $\omega \in \Omega$ the sample paths of a Brownian motion $B_t(\omega)$ are nowhere differentiable.

9. Strong Markov property Assume that B is a $BM([0, +\infty))$. For any $a.s.$ finite stopping time τ the stochastic process $B_t^* = B_{\tau+t} - B_\tau$, $\forall t \geq 0$, is $BM([0, +\infty))$ and is independent of $\mathcal{F}_\tau^B$. In this case we say that B has the strong Markov property at τ.

10. Laws of iterated logarithm Brownian motion satisfies

$$\limsup_{t \to \infty} |B_t| / \sqrt{2t \ln(\ln t)} = 1 \ a.s.,$$

and

$$\limsup_{t \to 0} |B_t| / \sqrt{2t \ln(\ln(1/t))} = 1 \ a.s.$$

11. Lévy's characterization Let $X \in \mathcal{M}^{c,loc}$ with $X_0 = 0$. The following are equivalent:
(i) X is Brownian motion,
(ii) $\langle B \rangle_t \stackrel{d}{=} t \ a.s. \ [P]$, for all $t \geq 0$, and
(iii) $X_t^2 - t$ is a local MG.

The extension to the multivariate case is straightforward.

Definition 7.16 Multi-dimensional Brownian motion

Consider Q_0, a probability measure on $(\mathcal{R}^p, \mathcal{B}_p)$, and let $B = \{B_t : t \geq 0\}$ be a continuous process defined on a probability space $(\Omega, \mathcal{A}, P)$ that takes values on $\Psi = \mathcal{R}^p$ and assume that B is adapted to a filtration $\mathcal{F} = \{\mathcal{F}_t : t \geq 0\}$. The process B is called a p-dimensional Brownian motion (denoted by $MBM([0, +\infty))$) with initial distribution Q_0 if
(a) $P(B_0 \in A) = Q_0(A)$, $\forall A \in \mathcal{B}_p$, and
(b) for all $0 \leq s < t$, the increment $B_t - B_s$ is independent of $\mathcal{F}_s$ and has a multivariate normal distribution $\mathcal{N}_p(\mathbf{0}, (t-s)\mathbf{I}_d)$.

Figure 7.1: Brownian motion realizations: (a) univariate, (b) bivariate.

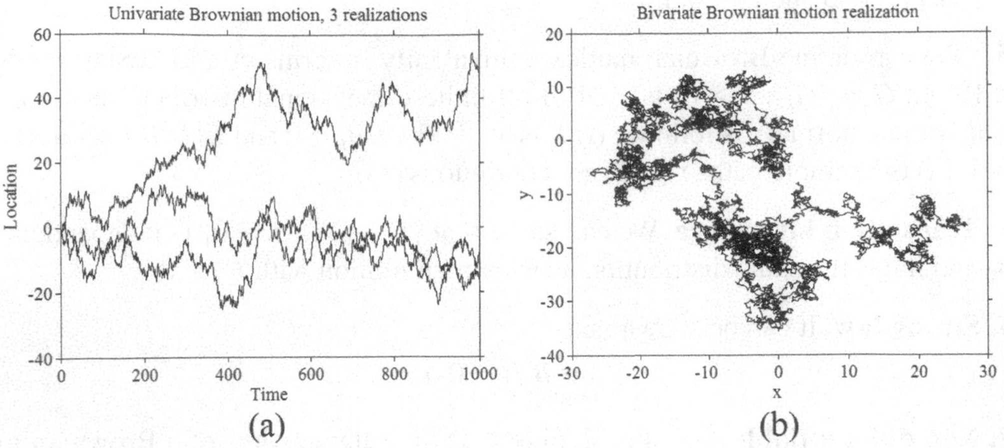

If $Q_0(\{\mathbf{x}\}) = \delta_{\mathbf{x}}$ then we say that B is a p-dimensional Brownian motion starting at $\mathbf{x} \in \mathcal{R}^p$.

Example 7.13 (Brownian motion) In Figure 7.1 (a), we simulate and display three realizations of a univariate Brownian motion, whereas in plot (b) we have a realization of a bivariate Brownian motion. See appendix section A.7 for details on the code used to generate these plots.

7.4.7 Construction of the Wiener Measure

Two questions arise when defining a Markov process with certain properties; first, does it exist (can we construct a Markov process with these properties) and second is it unique. Construction of Brownian motion has been accomplished in many different ways over the past century. The typical approach is to apply the modification of Kolmogorov's existence theorem (Kolmorogov-Čentsov, theorem 7.3) and construct a stochastic process with finite-dimensional distributions, such as those under Brownian motion assumptions. Another approach is to exploit the Gaussian property of this stochastic process and is based on Hilbert space theory.

The approach we discuss here is based on weak limits of a sequence of random walks on the space of continuous $\mathcal{R}$-valued functions $C^{\mathcal{R}}_{[0,\infty)}$. See Karatzas and Shreve (1991), Fristedt and Gray (1997), Durrett (2010), Bass (2011) and Klenke (2014) for excellent treatments of Brownian motion and examples, as well as the construction of the Wiener measure under the different settings discussed above.

Note that since $C^{\mathcal{R}}_{[0,\infty)}$ is a subset of the càdlàg space $\mathcal{D}^{\mathcal{R}}_{[0,\infty)}$, topological results we have studied for $\mathcal{D}^{\mathcal{R}}_{[0,\infty)}$ carry over immediately to the subspace $C^{\mathcal{R}}_{[0,\infty)}$, e.g., results on Polish spaces. In addition, working on the càdlàg space (RCLL paths) we are able, as we saw in example 7.5, to define the Poisson process, and as we will see next, construct Brownian motion, thus providing a unifying framework that can handle both continuity and the countable number of jumps for the sample paths of a stochastic process.

The construction that follows considers a setup similar to that of example 4.2.2, where we defined random functions on $C^{\mathcal{R}}_{[0,1]}$ and based on such functions we build the Wiener measure initially on $[0, 1]$ and then extend it to $[0, +\infty)$. Alternatively, one can define $C^{\mathcal{R}}_{[0,+\infty)}$-valued random variables directly (Karatzas and Shreve, 1991, Section 2.4).

Remark 7.20 (Sequences of random variables in $C^{\mathcal{R}}_{[0,1]}$) Consider a sequence of iid random variables $\{Z_n\}_{n=0}^{+\infty}$ with mean μ and variance σ^2, $0 < \sigma^2 < \infty$ and define the random walk $S = (S_0, S_1, \dots)$, with $S_0 = 0$ and $S_k = \sum_{i=1}^{k} Z_i$, $k \geq 1$. We define $C^{\mathcal{R}}_{[0,1]}$-valued random variables $X^{(n)}$ by defining the function $t \longmapsto X_t^{(n)}(\omega)$ as follows. First consider values for t equal to a multiple of $\frac{1}{n}$, that is, for each $n = 1, 2, \dots$, we let

$$X_{k/n}^{(n)} = \frac{1}{\sigma\sqrt{n}} \sum_{i=1}^{k} (Z_i - \mu) = \frac{S_k - k\mu}{\sigma\sqrt{n}},$$

for $k = 0, 1, 2, \dots, n$ and the random variable $X_t^{(n)}$, $0 \leq t \leq 1$, can be made continuous for $t \in [0, 1]$ by assuming linearity over each of the intervals $I_{k,n} = [(k-1)/n, k/n]$, that is, we linearly interpolate the value of $X_t^{(n)}$, for any $t \in I_{k,n}$, based on the boundary values at $X_{(k-1)/n}^{(n)}$ and $X_{k/n}^{(n)}$. Now note that the increment $X_{(k+1)/n}^{(n)} - X_{k/n}^{(n)} = (Z_{k+1} - \mu)/(\sigma\sqrt{n})$ is independent of $\mathcal{F}_{k/n}^{X^{(n)}} = \sigma(Z_1, \dots, Z_k)$ and $X_{(k+1)/n}^{(n)} - X_{k/n}^{(n)}$ has zero mean and variance $1/n = (k+1)/n - k/n$ so that apart from the normality assumption $\{X_t^{(n)} : t \geq 0\}$ is behaving like a $BM([0, 1])$.

The next theorem shows that even though the steps of the random walk are not necessarily normal, CLT guarantees that the limiting distributions of the increments of $X^{(n)}$ are multivariate normal.

Theorem 7.11 (Convergence in law) Let $\{X_t^{(n)} : t \geq 0, n \geq 1\}$ be defined as in remark 7.20.

(i) For any $0 \leq t_0 < t_1 < \dots < t_p \leq 1$, we have

$$\left(X_{t_1}^{(n)} - X_{t_0}^{(n)}, \dots, X_{t_p}^{(n)} - X_{t_{p-1}}^{(n)}\right) \overset{w}{\to} (B_{t_1 - t_0}, \dots, B_{t_p - t_{p-1}}),$$

as $n \to +\infty$, where $B_{t_k - t_{k-1}}$ are iid $\mathcal{N}(0, t_k - t_{k-1})$, $k = 1, 2, \dots, p$.

(ii) Any subsequential limit B of the sequence $\{X^{(n)} : n \geq 1\}$ has the properties:

(a) for $0 \leq t \leq 1$, $B_t \sim \mathcal{N}(0, t)$, and

(b) B has independent and stationary increments,

and as a result of (a) and (b), B is a $BM([0, 1])$.

(iii) If there exists a $B^* \neq B$ that satisfies properties (a) and (b) then B and B^* have the same distribution so that all Wiener processes on $[0, 1]$ have the same distribution and consequently all subsequential limits of $\{X^{(n)} : n \geq 1\}$ have the same distribution.

The latter part of the theorem above leads to the following definition.

Definition 7.17 Wiener measure

The common distribution Q_w of all Brownian motions on $[0, 1]$ is called the Wiener measure on $[0, 1]$.

Exercises 7.31 and 7.32 summarize the important results one needs to show in succession in order to show existence and uniqueness of the Wiener measure as well as existence of subsequential limits of $\{X^{(n)} : n \geq 1\}$. Theorem 7.12 below is also a consequence of these results.

Remark 7.21 (Existence and uniqueness) For what follows assume that Q_n is the distribution of the $C^{\mathcal{R}}_{[0,1]}$-valued random variable $X^{(n)}$, $n \geq 1$. In view of remark 5.3, parts 4-8, we require the sequence $\{Q_n\}_{n=1}^{+\infty}$ to be uniformly tight. In that case every subsequence of $\{Q_n\}_{n=1}^{+\infty}$ has a further subsequence that converges. Then from theorem 7.11, part (ii), since a convergent subsequence exists then the Wiener measure Q_w exists on $C^{\mathcal{R}}_{[0,1]}$. Uniqueness follows via theorem 7.11, part (iii), and since all limits of convergent subsequences are identical using remark 5.3.7, Q_w is defined as the weak limit of $\{Q_n\}_{n=1}^{+\infty}$, that is, $Q_n \stackrel{w}{\rightarrow} Q_w$, as $n \rightarrow +\infty$. The latter result is known as Donsker's invariance principle.

Theorem 7.12 (Donsker invariance principle) Consider a sequence of iid random variables $\{Z_n\}_{n=1}^{+\infty}$ with mean μ and variance σ^2, $0 < \sigma^2 < \infty$, defined on some probability space $(\Omega, \mathcal{A}, P)$. Let $\{X_t^{(n)} : t \geq 0\}$ be as defined in remark 7.20 and let Q_n denote the distribution of $X^{(n)}$ on the measurable space $\left(C^{\mathcal{R}}_{[0,1]}, \mathcal{B}\left(C^{\mathcal{R}}_{[0,1]}\right)\right)$. Then $\{Q_n\}_{n=1}^{+\infty}$ converges weakly to Q_w.

The generalization of the Wiener process from $[0, 1]$ to $[0, +\infty)$ is straightforward as we see below.

Remark 7.22 (Wiener measure on $[0, +\infty)$) In order to construct the Wiener measure on $C^{\mathcal{R}}_{[0,+\infty)}$ we define a $C^{\mathcal{R}}_{[0,+\infty)}$-valued random variable that has independent and stationary increments and a Wiener process when restricted to $[0, 1]$. In particular, let $\{B^{(n)} : n \geq 0\}$ be an iid sequence of $C^{\mathcal{R}}_{[0,1]}$-valued Wiener processes and for $n = 0, 1, 2, \ldots$ and $t \in [n, n + 1]$ define

$$B_t = B^{(n)}_{t-n} + \sum_{k=0}^{n-1} B^{(k)}_1,$$

so that the random function B_t is $BM([0, +\infty))$ and its distribution is the Wiener measure on $C^{\mathcal{R}}_{[0,+\infty)}$.

Example 7.14 (Measurable functionals) There are several functionals of interest that are measurable in $\left(C^{\mathcal{R}}_{[0,1]}, \mathcal{B}\left(C^{\mathcal{R}}_{[0,1]}\right)\right)$, for example

$$M(\phi) = \max\{\phi(t) : 0 \leq t \leq 1\},$$

$$I(\phi) = \int_0^1 \phi(t)dt,$$

and

$$L(\phi) = \mu_1(\{t \in [0,1] : \phi(t) > 0\}),$$

where μ_1 denotes the Lebesgue measure on $[0,1]$. The functionals $M(\phi)$ and $I(\phi)$ are continuous but $L(\phi)$ is not. It can be shown that under the Wiener measure and for any $t \geq 0$ we have

$$Q_w(\{\phi : M(\phi) \leq t\}) = \sqrt{2/\pi} \int_0^t e^{-u^2/2} du,$$

and the distribution of $I(\phi)$ is $\mathcal{N}(0, 1/3)$. Moreover, for $0 \leq t \leq 1$ we have

$$Q_w(\{\phi : L(\phi) \leq t\}) = \pi^{-1} \int_0^t (u(1-u))^{-1/2} du = 2\pi^{-1} \arcsin \sqrt{t}.$$

7.5 Building on Martingales and Brownian Motion

In this section we define two important random objects, stochastic integrals and stochastic differential (or diffusion) equations (SDEs), with applications ranging from biological sciences, physics and engineering to economics. Stochastic calculus is required when ordinary differential and integral equations are extended to involve continuous stochastic processes.

Since the most important stochastic process, Brownian motion, is not differentiable, one cannot use the same approach to defining stochastic derivatives and integrals as with ordinary calculus. Instead, we begin by defining a stochastic integral as a Riemann-Stieltjes integral with respect to a random continuous function and then define stochastic derivatives, with the latter describing the infinitesimal rate of change of a stochastic process.

We restrict the exposition to stochastic calculus with Brownian motion integrators and briefly discuss the extension to semi-MG integrators. Theoretical development, definitions and examples of stochastic integrals and differential equations can be found in Karatzas and Shreve (1991), Durrett (1996), Fristedt and Gray (1997), Durrett (2010), Bass (2011) and Klenke (2014).

7.5.1 Stochastic Integrals

For what follows let $(B, \mathcal{F}) = \{B_t, \mathcal{F}_t : t \geq 0\}$ be Brownian motion defined on a probability space $(\Omega, \mathcal{A}, P)$ with $B_0 = 0$ $a.s.$ $[P]$. Our goal in this section is to define a random object that is represented notationally by the integral

$$I_t^B(X) = \int_0^t X_s dB_s,$$

for a large class of integrands $X : \Omega\times[0, +\infty) \to \mathcal{R}$ with X_t chosen in such a way that $I_t^B(X)$ is a continuous MG with respect to $\mathcal{F}$. At first glance the stochastic integral $I_t^B(X)$ can be thought of as the stochastic version of the corresponding deterministic integral of the sample paths for a given $\omega \in \Omega$, i.e.,

$$I_t^B(X)(\omega) = \int_0^t X_s(\omega)dB_s(\omega).$$

However, since almost all paths $s \mapsto B_s(\omega)$ of Brownian motion are *a.s.* nowhere differentiable (Paley-Wiener-Zygmud theorem, Klenke, 2014, p. 467) and hence have locally infinite variation, $I_t^B(X)$ cannot be defined and understood as a Riemann-Stieltjes integral in the framework of classical integration theory. Therefore, we must use a different approach to define it and in particular, we establish the stochastic integral as an $\mathcal{L}^2$-limit. The approach of the construction is as follows: first we define $I_t^B(X)$ for simple integrands X (i.e., the sample path of X, $t \mapsto X_t(\omega)$ is a simple function), that is, $I_t^B(X)$ is a finite sum of stochastic processes. Next we extend the definition to integrands that can be approximated by simple integrands in a certain $\mathcal{L}^2$-space.

Example 7.15 (Stock prices) We can think of B_s as the price of a stock at time s and X_s as the number of shares we have (may be negative, selling short). The integral $I_t^B(X)$ represents the net profits at time t relative to our wealth at time 0. The infinitesimal rate of change of the integral $dI_t^B(X) = X_t dB_t$ represents the rate of change of the stock multiplied by the number of shares we have.

The following remarks summarize the steps required.

Remark 7.23 (Conditions on integrands) First we need to discuss the kinds of integrands X that are appropriate in defining $I_t^B(X)$.

1. Measurability condition In order for the stochastic process $I^B(X) = \{I_t^B(X) : t \geq 0\}$ to be a MG the integrands X will be required to be progressively measurable (which implies that X is adapted and measurable). The measurability requirement depends on the type of integrator we use to define the integral. See Karatzas and Shreve (1991, p. 131) for some discussion on how to develop the integral $I_t^M(X)$ under different types of integrators M (other than B) and integrands X, e.g., for $M \in \mathcal{M}_2$ we require X to be a smaller class, namely, predictable stochastic processes.

2. Integrability condition The second condition required is a certain type of integrability. For any measurable, adapted process X we define two $\mathcal{L}^2$-norms by

$$[X]_T^2 = E\left[\int_0^T X_t^2 dt\right],$$

for any $T > 0$ and let

$$[X]_\infty^2 = E\left[\int_0^{+\infty} X_t^2 dt\right].$$

Denote by $\mathcal{E}^*$ the collection of all adapted, measurable stochastic processes satisfying $[X]_\infty < \infty$. We define a metric on $\mathcal{E}^*$ by $\rho_{\mathcal{E}^*}(X, Y) = [X - Y]_\infty$, for $X, Y \in \mathcal{E}^*$ and as a result we can define limits of sequences of elements of $\mathcal{E}^*$.

Now we discuss the types of integrands used in order to define the stochastic integral.

Remark 7.24 (Simple integrands) Consider step functions (simple stochastic processes) $X : \Omega \times [0, +\infty) \to \mathcal{R}$ of the form

$$X_t(\omega) = \sum_{i=1}^{n} x_{i-1}(\omega) I_{(t_{i-1},t_i]}(t), \tag{7.52}$$

where $0 = t_0 < t_1 < \cdots < t_n$, x_i is an $\mathcal{F}_{t_i}$-measurable random variable, for $i = 1, \ldots, n$, $n \geq 0$ and $\sup_{k\geq 0}|x_k(\omega)| < +\infty$, for all $\omega \in \Omega$ and any $t \geq 0$. The collection of all such simple stochastic processes is denoted by $\mathcal{E}_0$. Note that by definition all $X \in \mathcal{E}_0$ are progressively measurable and bounded with

$$[X]_\infty^2 = E\left[\int_0^{+\infty} X_t^2 dt\right] = \sum_{i=1}^{n} E\left(x_{i-1}^2\right)(t_i - t_{i-1}) < +\infty,$$

so that $X \in \mathcal{E}_0 \subset \mathcal{E}^*$.

1. Integrating simple stochastic processes For any $X \in \mathcal{E}_0$ and $t \geq 0$ we define the stochastic integral of X with respect to B by

$$I_t^B(X) = \int_0^t X_s dB_s \stackrel{d}{=} \sum_{i=1}^{n} x_{i-1}\left(B_{t_i \wedge t} - B_{t_{i-1}\wedge t}\right),$$

which leads to a stochastic process $I^B(X) = \{I_t^B(X) : t \geq 0\}$, whereas

$$I_\infty^B(X) = \int_0^{+\infty} X_s dB_s \stackrel{d}{=} \sum_{i=1}^{n} x_{i-1}\left(B_{t_i} - B_{t_{i-1}}\right),$$

is an $\mathcal{R}$-valued random variable and both integrals are simply linear transformations of Brownian motion. Note that $E\left[I_\infty^B(X)\right]^2 = [X]_\infty^2 < \infty$.

2. Basic properties Let $X, Y \in \mathcal{E}_0$ and $0 \leq s < t < \infty$. By construction, it is easy to show that $I_t^B(X)$ is a $C_{[0,+\infty)}^{\mathcal{R}}$-valued random variable, that is a continuous time, square integrable MG with respect to $\mathcal{F}$ and moreover

$$\begin{aligned} I_0^B(X) &= 0 \; a.s. \; [P], \\ E\left[I_t^B(X)|\mathcal{F}_s\right] &= I_s^B(X) \; a.s. \; [P], \\ E\left[I_t^B(X)\right] &= [X]_t^2 = E\left[\int_0^t X_u^2 du\right] < \infty, \\ E\left[\left(I_t^B(X) - I_s^B(X)\right)^2 |\mathcal{F}_s\right] &= E\left[\int_s^t X_u^2 du|\mathcal{F}_s\right] \; a.s. \; [P], \end{aligned}$$

and

$$I_t^B(aX + bY) = aI_t^B(X) + bI_t^B(Y) \; a.s. \; [P],$$

for all $a, b \in \mathcal{R}$.

3. Approximations in E^* via simple stochastic processes If $X \in \mathcal{E}^*$ then the sequence of simple processes

$$X_t^{(n)}(\omega) = X_0(\omega)I_{\{0\}}(t) + \sum_{k=0}^{2^n-1} X_{kT2^{-n}}(\omega)I_{[kT2^{-n},(k+1)T2^{-n}]}(t),$$

$n \geq 1$, is such that $X_t^{(n)} \in \mathcal{E}_0$ and

$$\lim_{n\to\infty} \left[X_t^{(n)} - X_t\right]_T^2 = \lim_{n\to\infty} E\left[\int_0^T \left|X_t^{(n)} - X_t\right|^2 dt\right] = 0,$$

for any $T > 0$. See Karatzas and Shreve (1991, p. 132) or Klenke (2014, p. 566) for more details on the construction of the simple integrands $X^{(n)} \in \mathcal{E}_0$ and their limiting behavior. Consequently, limits of simple stochastic processes are suitable integrands in $\mathcal{E}^*$ and therefore we augment the original collection of simple integrands $\mathcal{E}_0$ with all stochastic processes that are limits of elements of $\mathcal{E}_0$, that is, we take the closure $\overline{\mathcal{E}_0}$ of the set $\mathcal{E}_0$, denoted by $\mathcal{E} = \overline{\mathcal{E}_0}$. Note that $\mathcal{E}_0 \subset \mathcal{E} \subset \mathcal{E}^*$. It can be shown that if X is progressively measurable and $[X]_\infty < \infty$ then $X \in \mathcal{E}$.

Based on the discussion of the last two remarks, the stochastic integral is now naturally introduced.

Definition 7.18 Itô integral as a random variable

For any $X \in \mathcal{E}$ consider a sequence of simple integrands $X^{(n)} \in \mathcal{E}_0$ such that

$$\lim_{n\to\infty} \left[X^{(n)} - X\right]_\infty = 0.$$

Then the stochastic integral (Itô integral random variable) of X with respect to Brownian motion $B = \{B_t, \mathcal{F}_t : t \geq 0\}$ over $[0, +\infty)$ is the square-integrable random variable $I_\infty^B(X)$ that satisfies

$$\lim_{n\to\infty} \left\| I_\infty^B\left(X^{(n)}\right) - I_\infty^B(X) \right\|_2 = 0,$$

for every such sequence $X^{(n)}$ (in other words $I_\infty^B\left(X^{(n)}\right) \xrightarrow{\mathcal{L}^2} I_\infty^B(X)$, recall remarks 3.24 and 3.25). We write

$$I_\infty^B(X) = \int_0^\infty X_s dB_s.$$

The following remark illustrates how we can build a stochastic integral over a finite interval $[0, t]$.

Remark 7.25 (Itô integral as a stochastic process) A similar approach can be used in the definition of the stochastic integral of X with respect to B over $[0, t]$. For any progressively measurable X we weaken the strong integrability condition $[X]_\infty^2 < \infty$ by requiring

$$\int_0^T X_t^2 dt < \infty, \tag{7.53}$$

for all $T > 0$. Denote by $\mathcal{E}^{loc}$ the collection of all progressively measurable stochastic processes satisfying (7.53). The definition that follows requires the following result; for any $X \in \mathcal{E}^{loc}$ there exists a sequence $\{\tau_n\}_{n=1}^{+\infty}$ of stopping times with $\tau_n \uparrow \infty$ $a.s.$ and $[X]_{\tau_n}^2 < \infty$ and hence such that $X^{(\tau_n)} \in \mathcal{E}$, where

$$X_t^{(\tau_n)} = X_t I(t \leq \tau_n),$$

for all $n \geq 1$.

Using the ideas of the last remark we can define the stochastic integral as a stochastic process.

Definition 7.19 Itô integral as a stochastic process

For any $X \in \mathcal{E}^{loc}$ let $\{\tau_n\}_{n=1}^{+\infty}$ and $X^{(\tau_n)}$ be as in remark 7.25. The Itô integral of X with respect to Brownian motion $B = \{B_t, \mathcal{F}_t : t \geq 0\}$ over $[0,t]$ is a stochastic process $I^B(X) = \{I_t^B(X), \mathcal{F}_t : t \geq 0\}$ that satisfies

$$I_t^B(X) = \int_0^t X dB = \lim_{n\to\infty} \int_0^t X_s^{(\tau_n)} dB_s \ a.s., \tag{7.54}$$

for any choice of $\{\tau_n\}_{n=1}^{+\infty}$. We write

$$\int_s^t X_u dB_u = I_t^B(X) - I_s^B(X),$$

for the integral of X over $[s,t]$, $0 \leq s \leq t \leq \infty$.

Example 7.16 (Stochastic integration) We compute $I_t^B(B) = \int_0^t B_s dB_s$. Since $B \in \mathcal{E}$ we can define the Itô integral using simple integrands $B_t^{(n)}$ and then taking the limit as follows

$$\begin{aligned} I_t^B(B) &= \int_0^t B_s dB_s = \lim_{n\to+\infty} \sum_{k=0}^{2^n-1} B_{kt2^{-n}}(\omega)\left(B_{(k+1)t2^{-n}} - B_{kt2^{-n}}\right) \\ &= \lim_{n\to+\infty} \frac{1}{2} \sum_{k=0}^{2^n-1} \left[\left(B_{(k+1)t2^{-n}}^2 - B_{kt2^{-n}}^2\right) - \left(B_{(k+1)t2^{-n}} - B_{kt2^{-n}}\right)^2\right] \\ &= B_t^2/2 - \lim_{n\to+\infty} \frac{1}{2} \sum_{k=0}^{2^n-1} \left(B_{(k+1)t2^{-n}} - B_{kt2^{-n}}\right)^2, \end{aligned}$$

where $\left(B_{(k+1)t2^{-n}} - B_{kt2^{-n}}\right)^2 \sim \mathcal{N}\left(t2^{-n}, 2t^2 2^{-2n}\right)$ and therefore the sum is $\mathcal{N}\left(t, t^2 2^{1-n}\right)$. Consequently, the sum converges in $\mathcal{L}^2$ to t so that

$$I_t^B(B) = \int_0^t B_s dB_s = B_t^2/2 - t/2.$$

The term $t/2$, although unexpected, guarantees that $I_t^B(B)$ is a continuous-time MG as required by construction. This example shows that ordinary rules of calculus do not apply to the Itô integral.

We summarize several important results about $I^B(X)$ below, including how to find solutions to stochastic integral equations.

Remark 7.26 (Properties and extensions of the stochastic integral) We note the following properties and results.

1. Assume that X is progressively measurable with $[X]_T^2 < \infty$, for all $T > 0$. The integral $I^B(X)$ as defined in (7.54) is a continuous, square integrable MG, that satisfies all the properties of remark 7.24.2. Moreover, the stochastic process $N_t = I_t^B(X) - \int_0^t X_u^2 du$, $t \geq 0$, is a continuous MG with $N_0 = 0$ $a.s.$

2. Quadratic variation Assume that $X \in \mathcal{E}^{loc}$. The integral $I^B(X)$ is a continuous local MG, with a square variation stochastic process given by $\left\langle I^B(X)\right\rangle_t = \int_0^t X_u^2 du$, $t \geq 0$.

3. Beyond Brownian motion The Itô integral need not be defined with respect to Brownian motion. In particular, starting with a simple integrand $X \in \mathcal{E}_0$ of the form (7.52) and a (local) MG $(M, \mathcal{F})$ the integral

$$I_t^M(X) \stackrel{d}{=} \sum_{i=1}^{n} x_{i-1} \left(M_{t_i \wedge t} - M_{t_{i-1} \wedge t}\right),$$

is a (local) MG with

$$\begin{aligned} E\left[\left(I_\infty^M(X)\right)^2\right] &= \sum_{i=1}^{n} E\left[x_{i-1}\left(M_{t_i \wedge t} - M_{t_{i-1} \wedge t}\right)\right] = \sum_{i=1}^{n} E\left[x_{i-1}\left(\langle M \rangle_{t_i \wedge t} - \langle M \rangle_{t_{i-1} \wedge t}\right)\right] \\ &= E\left[\int_0^\infty X_t^2 d\langle M \rangle_t\right], \end{aligned}$$

provided that the RHS is finite. We can repeat the construction procedure for B to define the integral with respect to M provided that we equip $\mathcal{E}_0$ with the norm

$$[X]_M^2 = E\left[\int_0^{+\infty} X_t^2 d\langle M \rangle_t\right],$$

and follow similar steps. Recall that $\langle B \rangle_t = t$ so that the procedure is a direct extension of the construction with respect to Brownian motion B. The extension is completed once we identify which integrands belong to the closure $\mathcal{E} = \overline{\mathcal{E}_0}$, e.g., if M is a continuous MG then X needs to be predictable in order for the integral to be a MG. In order to have progressively measurable integrands in $\mathcal{E}$ the additional requirement needed is that the square variation $\langle M \rangle$ is absolutely continuous with respect to the Lebesgue measure. See Klenke (2014, p. 572) for more details. We summarize the extension in the next part of this remark.

4. Local martingale integrators Let $(M, \mathcal{F})$ be a continuous local MG with absolutely continuous square variation $\langle M \rangle$ (with respect to the Lebesgue measure) and assume that X is a progressively measurable process that satisfies $\int_0^T X_t^2 d\langle M \rangle_t < \infty$, $a.s.$ for all $T \geq 0$. Let $\{\tau_n\}_{n=1}^{+\infty}$ be as in remark 7.25, with $\left[X^{(\tau_n)}\right]_M < \infty$ and define simple stochastic processes $X^{n,m} \in \mathcal{E}_0$, $m \geq 1$, with $\lim\limits_{m\to+\infty} \left[X^{n,m} - X^{(\tau_n)}\right]_M = 0$, $n \geq 1$.

The Itô integral of X with respect to M is defined as a limit in probability by

$$I_t^M(X) = \int_0^t XdM = \int_0^t X_u dM_u = \lim_{n\to+\infty}\lim_{m\to+\infty} I_t^M(X^{n,m}),$$

for all $t \geq 0$ and it is a continuous local MG with square variation $\left\langle I^M(X)\right\rangle_t = \int_0^t X_u^2 d\langle M\rangle_u$.

5. Arithmetic of integration Stochastic integrals are the same as ordinary integrals when it comes to linearity as we have seen in terms of linear integrands. Linearity carries over to integrators as well. In particular, if $a, b \in \mathcal{R}$, are constants, $X, Y \in \mathcal{M}^{c,loc}$ and Z, W are continuous stochastic processes, then for all $t \geq 0$ we have

$$\int_0^t Z_u d(aX_u + bY_u) = a\int_0^t Z_u dX_u + b\int_0^t Z_u dbY_u,$$

i.e., linearity with respect to the integrator. Moreover

$$\int_0^t X_u dY_u = X_t Y_t - X_0 Y_0 - \int_0^t Y_u dX_u + \langle X, Y\rangle_t, \tag{7.55}$$

is known as the product rule or integration by parts and if $Y_t = \int_0^t Z_u dX_u$ then $\int_0^t W_u dY_u = \int_0^t (W_u Z_u)\, dX_u$.

6. Itô formula Let X be a continuous local MG with paths locally of bounded variation and assume that $f : \mathcal{R} \to \mathcal{R}$ is two times continuously differentiable with bounded derivatives (denote this set of functions by $C^2(\mathcal{R}, \mathcal{R})$). Then

$$f(X_t) = f(X_0) + \int_0^t f'(X_s)dX_s + \frac{1}{2}\int_0^t f''(X_s)d\langle X\rangle_s, \tag{7.56}$$

a.s., for all $t \geq 0$. The Itô formula is also known as the change-of-variable formula and it is of extreme importance in finding solutions to stochastic integral and differential equations. For $X \in \mathcal{M}^{c,loc}$ the Itô formula shows that $f(X)$ is always a semi-MG but it is not a local MG unless $f''(x) = 0$, for all x. The multivariate extension is straightforward: assume that $X = \{\mathbf{X}_t = (X_{t1}, \ldots, X_{tp}), \mathcal{F}_t; t \geq 0\}$, is a p-variate local MG (i.e., $X_i \in \mathcal{M}^{c,loc}$), where all cross-variation stochastic processes $\left\langle X_i, X_j\right\rangle_t$ exist and are continuous for all $i, j = 1, 2, \ldots, p$ and $f \in C^2(\mathcal{R}^p, \mathcal{R})$. Then we can write

$$f(X_t) = f(X_0) + \sum_{i=1}^p \int_0^t \frac{\partial f(X_s)}{\partial x_i} dX_{si} + \frac{1}{2}\sum_{i,j=1}^p \int_0^t \frac{\partial^2 f(X_s)}{\partial x_i \partial x_j} d\left\langle X_i, X_j\right\rangle_s, \tag{7.57}$$

provided that $\int_0^t \frac{\partial f(X_s)}{\partial x_i} dX_{si}$ exist, for all $i = 1, 2, \ldots, p$.

7. Girsanov formula Girsanov's formula is another useful tool in solving stochastic integral equations. Assume that $B = \{\mathbf{B}_t = (B_{t1}, \ldots, B_{tp}), \mathcal{F}_t; t \geq 0\}$, is p-dimensional Brownian motion, where $\mathcal{F}$ satisfies the usual conditions. Let $X = \{\mathbf{X}_t = (X_{t1}, \ldots, X_{tp}), \mathcal{F}_t; t \geq 0\}$ be a vector of measurable, adapted processes satisfying $\int_0^T X_{ti}^2 dt < \infty$ *a.s.*, for all $i = 1, 2, \ldots, p$ and $T \geq 0$. Then for each i the integral $I^{B_i}(X_i) = \{I^{B_{ti}}(X_{ti}) : t \geq 0\}$ is well defined with $I^{B_i}(X_i) \in \mathcal{M}^{c,loc}$ so that we can define

$$Y_t(X) = \exp\left\{\sum_{i=1}^{p} \int_0^t X_{si} dB_{si} - \frac{1}{2} \int_0^t \mathbf{X}_s^T \mathbf{X}_s ds\right\},$$

which is a MG. Then the p-variate stochastic process $\widetilde{B} = \{\widetilde{\mathbf{B}}_t = \left(\widetilde{B}_{t1}, \ldots, \widetilde{B}_{tp}\right), \mathcal{F}_t; 0 \leq t \leq T\}$ defined by

$$\widetilde{B}_{ti} = B_{ti} - \int_0^t X_{si} ds,$$

is a p-dimensional Brownian motion for all $T \geq 0$.

Example 7.17 (Applying Itô's formula) Let X be a continuous local MG with $X_0 = 0$ and define the stochastic process $Y_t = \exp\{X_t - \langle X \rangle_t /2\}$ known as the exponential of the MG X. Applying Itô's formula to $Z_t = X_t - \langle X \rangle_t /2$ with $f(x) = e^x$ yields

$$Y_t = \exp\{X_t - \langle X \rangle_t /2\} = 1 + \int_0^t e^{Z_s} d(X_s - \langle X \rangle_s /2) + \frac{1}{2} \int_0^t e^{Z_s} d\langle X \rangle_s = 1 + \int_0^t Y_s dX_s,$$

and since Y is the stochastic integral with respect to a local MG, Y is a local MG. This stochastic integral equation can be written in terms of differentials as $dY_t = Y_t dX_t$, defining a stochastic differential equation as we see in following section.

7.5.2 Stochastic Differential Equations

We turn now to the extension of ordinary differential equations and investigate random objects obtained as solutions to stochastic differential equations (SDEs). Such random processes are known as diffusion processes and they are typically $\mathcal{R}^p$-valued, time-homogeneous strong Markov processes. We begin with the discrete case.

Remark 7.27 (Stochastic difference equations) Let B be Brownian motion, $a : \mathcal{R} \to \mathcal{R}_0^+$, $b : \mathcal{R} \to \mathcal{R}$, two measurable functions and define the difference operator $\Delta_\varepsilon f(t) = f(t + \varepsilon) - f(t)$, $t \geq 0$. Given $\varepsilon > 0$ consider the equations

$$X_{(n+1)\varepsilon} = X_{n\varepsilon} + a(X_{n\varepsilon})(B_{(n+1)\varepsilon} - B_{n\varepsilon}) + b(X_{n\varepsilon})\varepsilon, \tag{7.58}$$

$n \geq 0$. For a given starting value $X_0 = x_0$ and known a, b, these equations have a unique solution that can be shown to be a Markov chain. The solution is calculated recursively and it depends on Brownian motion. Equation (7.58) can be rewritten using the difference operator as

$$\Delta_\varepsilon X_t = a(X_t)\Delta B_t + b(X_t)\Delta_\varepsilon t, \tag{7.59}$$

for $t = n\varepsilon$, $n \geq 0$. Equation (7.59) is known as a stochastic difference equation with coefficients a and b.

A generalization of (7.59) is considered next.

Remark 7.28 (Stochastic differential equations as stochastic integrals) Assume that $B = \{\mathbf{B}_t = (B_{t1}, \ldots, B_{tp}), \mathcal{F}_t; t \geq 0\}$ is p-variate Brownian motion over $[0, +\infty)$ in a probability space $(\Omega, \mathcal{A}, P)$ with $\mathcal{F}$ chosen appropriately and let X be a p-variate stochastic process with continuous paths which is at a minimum, adapted to a filtration built using $\mathcal{F}$ and progressively measurable. Consider Borel measurable functions $a_{ij}, b_i : \mathcal{R}_0^+ \times \mathcal{R}^p \to \mathcal{R}$, $i = 1, 2, \ldots, p$, $j = 1, 2, \ldots, r$ and define the drift vector $b(t, \mathbf{x}) = (b_1(t, \mathbf{x}), \ldots, b_p(t, \mathbf{x}))^T$ and the dispersion matrix $A(t, \mathbf{x}) = [(a_{ij}(t, \mathbf{x}))]_{p \times r}$. Our goal in this section is to study SDEs of the form

$$dX_t = A(t, X_t)dB_t + b(t, X_t)dt, \tag{7.60}$$

with the starting condition $X_0 = \boldsymbol{\xi}$ where the stochastic process $X = \{\mathbf{X}_t = (X_{t1}, \ldots, X_{tp}), \mathcal{F}_t; t \geq 0\}$ assumes the role of a solution (in some sense) of the SDE, whereas, the drift vector and the dispersion matrix are treated as the coefficients of the SDE. The random variable $\boldsymbol{\xi} = (\xi_1, \ldots, \xi_p)$ has distribution Q_{ξ} and is assumed to be independent of B. Since the sample paths of B are nowhere differentiable, (7.60) has only notational usefulness. Thinking in terms of ordinary calculus one can write the SDE equivalently as

$$X_t = \boldsymbol{\xi} + \int_0^t A(s, X_s)dB_s + \int_0^t b(s, X_s)ds, \tag{7.61}$$

$t \geq 0$. The stochastic integral equation (SIE) holds $a.s.$ $[P]$ and in order to be able to define it we need any solution X_t to be adapted to the filtration $\mathcal{F}$ or an augmented filtration based on $\mathcal{F}$. Moreover, if X is a solution to (7.61) then it automatically has continuous paths by construction of the stochastic integral.

Development and examples of SDEs along with the proofs of existence and uniqueness of strong and weak solutions and their properties (e.g., strong Markov) can be found in Karatzas and Shreve (1991), Durrett (1996), Durrett (2010), Bass (2011) and Klenke (2014). The choice of filtration we use leads to different solutions. A strong solution X requires a stronger measurability condition.

A SDE can be reformulated as a MG problem and the solution is equivalent to constructing a weak solution. Pathwise uniqueness can be used to connect existence and uniqueness of weak and strong solutions (see Klenke, 2014, p. 601). Next we collect the definitions for completeness along with some illustrative examples.

Definition 7.20 Strong solution

Consider the setup of remark 7.28. Assume that the following conditions hold:

(a) $\int_0^t \left(|b_i(s, X_s)| + a_{ij}^2(s, X_s)\right) ds < \infty$ $a.s.$ $[P]$, for all $i = 1, 2, \ldots, p$, $j = 1, 2, \ldots, r$, $t \geq 0$,

(b) X is a stochastic process that satisfies (7.61) $a.s.$ or equivalently in terms of its coordinates

$$X_{ti} = \xi_i + \sum_{j=1}^{p} \int_0^t a_{ij}(s, X_s) dB_{sj} + \int_0^t b_i(s, X_s) ds, \quad (7.62)$$

holds *a.s.*, for all $i = 1, 2, \ldots, p$,
(c) $P(X_0 = \xi) = 1$, and
(d) $\mathcal{F} = \mathcal{F}^B$, the minimal filtration for B, define the augmented minimal filtration $\mathcal{H}_t = \sigma\left(\sigma(\xi) \cup \mathcal{F}_t^B\right)$ and using the null sets

$$\mathcal{N}_\Omega = \{N \subseteq \Omega : \exists H \in H_\infty, \text{ with } N \subseteq H \text{ and } P(G) = 0\},$$

further augment the filtration by defining $\mathcal{G}_t = \sigma(H_t \cup \mathcal{N}_\Omega)$, $t \geq 0$, so that $(B_t, \mathcal{H}_t)$ or $(B_t, \mathcal{G}_t)$ is Brownian motion and $\mathcal{G} = \{\mathcal{G}_t : t \geq 0\}$ satisfies the usual conditions.

Then the solution X is called a strong solution of (7.60) if it is adapted to $\mathcal{G}$. Further assume that the drift vector and the dispersion matrix are given. A strong solution X is said to satisfy strong uniqueness if for any other strong solution X_t' of the SDE we have $P(X_t = X_t', t \geq 0) = 1$.

The definition of a weak solution follows.

Definition 7.21 Weak solution

Consider the setup of remark 7.28 and assume that conditions (a) and (b) of definition 7.20 hold. In addition, assume that $\mathcal{F}$ satisfies the usual conditions so that $(B, \mathcal{F})$ is Brownian motion and $(X, \mathcal{F})$ is a continuous, adapted stochastic process. Then the triple $\{(X, B), (\Omega, \mathcal{A}, P), \mathcal{F}\}$ is called a weak solution of (7.60) and Q_ξ is called the initial distribution of the solution. We have two types of uniqueness properties.

1. Pathwise uniqueness A weak solution $\{(X, B), (\Omega, \mathcal{A}, P), \mathcal{F}\}$ is said to satisfy pathwise uniqueness if for any other weak solution $\{(X', B), (\Omega, \mathcal{A}, P), \mathcal{F}'\}$ of the SDE with $P(X_0 = X_0') = 1$ we have $P(X_t = X_t', t \geq 0) = 1$.

2. Distributional uniqueness A weak solution $\{(X, B), (\Omega, \mathcal{A}, P), \mathcal{F}\}$ is said to satisfy distributional uniqueness if for any other weak solution $\{(X', B'), (\Omega', \mathcal{A}', P'), \mathcal{F}'\}$ of the SDE with the same initial distribution $Q_{X_0} = Q_{X_0'}$, the two processes X and X' have the same distribution.

There are several approaches that we can take in trying to solve SDEs, from applying the Itô and Girsanov formulas to solving a MG problem. For an excellent detailed exposition see Karatzas and Shreve (1991, Chapter 5). We collect some comments below before we discuss some classic examples.

Remark 7.29 (Solving stochastic differential equations) We note the following.

1. Cross-variation Let $\Sigma(s, X_s) = A(s, X_s)A(s, X_s)^T = [(\Sigma_{ij}(s, X_s))]_{p \times p}$. The cross-variation process of a solution X to the SDE (7.60) is given by

$$\left\langle X_i, X_j \right\rangle_t = \sum_{k=1}^{p} \int_0^t a_{ik}(s, X_s) a_{jk}(s, X_s) ds = \int_0^t \Sigma_{ij}(s, X_s) ds. \tag{7.63}$$

2. Local martingale problem We say that a p-variate stochastic process X defined on a probability space $(\Omega, \mathcal{A}, P)$ is a solution to the local MG problem for $\Phi = \left[\left(\Phi_{ij}(s, X_s)\right)\right]_{p\times p}$ and $v = [(v_i(s, X_s))]_{p\times 1}$ (denoted by $LMG(\Phi, v)$) if the stochastic process

$$M_{ti} = X_{ti} - \int_0^t v_i(s, X_s) ds,$$

is a continuous local MG with a cross-variation process

$$\left\langle M_i, M_j \right\rangle_t = \int_0^t \Phi_{ij}(s, X_s) ds,$$

for all $i, j = 1, 2, \ldots, p$, $t \geq 0$. The solution is unique if for any other solution X' we have $Q_X = Q_{X'}$. The local MG problem is said to be well posed if there exists a solution and it is unique.

3. Weak solutions via the local martingale problem The stochastic process X is a solution of $LMG(AA^T, b)$ if and only if there exists a Brownian motion B such that $\{(X, B), (\Omega, \mathcal{A}, P), \mathcal{F}\}$ is a weak solution of (7.60) (where $\mathcal{F}$ is an appropriate filtration, see proposition 4.6 of Karatzas and Shreve, 1991). Moreover, if $LMG(AA^T, b)$ has a unique solution then there exists a unique weak solution to the SDE (7.60) that is strong Markov. Results on existence and uniqueness of the local MG problem can be found in Karatzas and Shreve (1991), Bass (2011) and Klenke (2014).

Example 7.18 (Counterexample) A strong solution is also a weak solution but the existence of a weak solution does not imply that of a strong solution. Similarly, if pathwise uniqueness holds then distributional uniqueness holds but the converse is not true. Indeed, consider the SDE

$$dX_t = sign(X_t) dB_t,$$

$t \geq 0$, where $sign(x) = 2I(x > 0) - 1$. Then it can be shown that there exists a weak solution but no strong solutions (see Karatzas and Shreve, 1991, p. 302).

Example 7.19 (Ornstein-Uhlenbeck (O-U) process) The O-U process is the solution to the SDE

$$dX_t = a dB_t - bX_t dt,$$

for $X_0 = \xi$, $a, b > 0$ and it can be used to model fluid dynamics, meaning that X_t models the velocity of a small particle suspended in a fluid. The stochastic differential adB_t represents changes in velocity as the particle moves and it hits the fluid molecules and $-bX_t dt$ models a friction effect. This SDE is tractable and it can be solved using the Itô formula. In particular, multiply both sides by e^{bt} and use the Itô formula to obtain

$$d\left(e^{bt} X_t\right) = e^{bt} dX_t + b e^{bt} X_t dt = a e^{bt} dB_t,$$

so that

$$e^{bt}X_t = \xi + a\int_0^t e^{bs}dB_s,$$

which leads to

$$X_t = e^{-bt}\xi + ae^{-bt}\int_0^t e^{bs}dB_s.$$

Now if $\xi \sim \mathcal{N}(0,\sigma^2)$ then it can be shown that X is a Gaussian process with mean 0 and covariance function

$$C(s,t) = \sigma^2 e^{-b(s+t)} + \frac{a^2}{2b}\left(e^{-b|s-t|} - e^{-b|s+t|}\right).$$

Choosing $\sigma^2 = \frac{a^2}{2b}$ we obtain $C(s,t) = \sigma^2 e^{-b|s-t|}$ which yields an isotropic covariance structure.

Example 7.20 (Brownian bridge) Consider the one-dimensional equation

$$dX_t = [(b - X_t)/(T-t)]\,dt + dB_t,$$

$0 \le t \le T$, with $X_0 = \xi$, for given $T > 0$, ξ and b constants. Using the Itô formula we can verify that

$$X_t = \xi(1 - t/T) + bt/T + (T-t)\int_0^t (T-s)^{-1}dB_s,$$

is a weak solution that satisfies pathwise uniqueness.

Ordinary partial differential equations play a pivotal role in describing phenomena deterministically, especially in physics. The following remark summarizes some of the ideas involved.

Remark 7.30 (Stochastic partial differential equations) The solutions of many classic elliptic and parabolic ordinary partial differential equations (OPDEs) can be represented as expectations of stochastic functionals in particular functionals of Brownian motion. Introducing stochastic partial differential equations (SPDEs) is accomplished via an application of the Itô formula. Recall the notation of remark 7.28 and further assume that $\Phi = \left[\left(\Phi_{ij}(\mathbf{x})\right)\right]_{p\times p}$ and $v = [(v_i(\mathbf{x}))]_{p\times 1}$ have Borel measurable elements and bounded, $\forall \mathbf{x} \in \mathcal{R}^p$. Define a second-order partial differential operator $\mathcal{O}$ for $f \in C^2(\mathcal{R}^p, \mathcal{R})$ by

$$(\mathcal{O}f)(\mathbf{x}, \Phi, v) = \frac{1}{2}\sum_{i,j=1}^{p} \Phi_{ij}(t,\mathbf{x})\frac{\partial^2 f(\mathbf{x})}{\partial x_i \partial x_j} + \sum_{i=1}^{p} v_i(\mathbf{x})\frac{\partial f(\mathbf{x})}{\partial x_i}, \tag{7.64}$$

and assume that the operator if uniformly elliptic, that is, there exists $c > 0$, such that

$$\sum_{i,j=1}^{p} y_i \Phi_{ij}(\mathbf{x}) y_j \ge c\,\|y\|_2^2,$$

for all $\mathbf{x}, \mathbf{y} \in \mathcal{R}^p$, which implies that Φ is positive definite, uniformly in $\mathbf{x} \in \mathcal{R}^p$. Uniform ellipticity has many welcomed consequences (see Bass, 2011, Chapter 40). For parabolic OPDEs an extension of (7.64) is required, namely, we define the operator O_t in a similar fashion by

$$(O_t f)(\mathbf{x}, \Phi, v) = \frac{1}{2} \sum_{i,j=1}^{p} \Phi_{ij}(t, \mathbf{x}) \frac{\partial^2 f(t, \mathbf{x})}{\partial x_i \partial x_j} + \sum_{i=1}^{p} v_i(\mathbf{x}) \frac{\partial f(t, \mathbf{x})}{\partial x_i}. \tag{7.65}$$

1. Assume that $\{(X, B), (\Omega, \mathcal{A}, P), \mathcal{F}\}$ is a weak solution of (7.60) and let $\Sigma(s, X_s) = A(s, X_s)A(s, X_s)^T = [(\Sigma_{ij}(s, X_s))]_{p \times p}$. Using the Itô formula we can show that the stochastic process M^l defined by

$$M_t^l = f(X_t) - f(X_0) - \int_0^t \left(\frac{\partial f(X_s)}{\partial s} + (Of)(\mathbf{x}, \Sigma, b) \right) ds$$

is a continuous local MG and a continuous MG if a_{ij}, b_i are bounded on the support of f. More importantly, we can verify that the stochastic process M defined by

$$M_t = f(X_t) - f(X_0) - \int_0^t (Of)(\mathbf{x}, \Phi, v) ds, \tag{7.66}$$

is a solution to $LMG(AA^T, b)$ and therefore of the SDE (7.60). Note that $(Of)(\mathbf{x}, \Sigma, b)$ depends on s via the dispersion $\Sigma(s, X_s)$ and drift $b(s, X_s)$ coefficients.

2. Building stochastic partial differential equations By construction the operator O contains partial derivatives of the second order for any $f \in C^2(\mathcal{R}^p, \mathcal{R})$ and it appears in the weak solution of the SDE (7.60). Therefore, one can describe OPDEs using the operator O on deterministic functions and then obtain their solution as an expected value with respect to the weak solution to the corresponding SPDE. Some examples are definitely in order.

Example 7.21 (Solving stochastic partial differential equations) For what follows assume that $\{(X, B), (\Omega, \mathcal{A}, P), \mathcal{F}\}$ is a weak solution of (7.60) with initial condition $X_0 = \xi$ and let Q_X^ξ be the distribution of X.

1. Poisson stochastic partial differential equations Suppose that $\lambda > 0$, $f \in C^2(\mathcal{R}^p, \mathcal{R})$ and $g \in C^1(G, \mathcal{R})$, $G \subset \mathcal{R}^p$, has compact support. The Poisson SPDE in $\mathcal{R}^p$ is given by

$$(Of)(\mathbf{x}, \Phi, v) - \lambda f(\mathbf{x}) = -g(\mathbf{x}), \tag{7.67}$$

$\mathbf{x} \in \mathcal{R}^p$, where Φ and v are as in equation (7.64). Assume that f is a solution of (7.67). By remark 7.30.1 and equation (7.66) we have

$$f(X_t) - f(X_0) = M_t + \int_0^t (Of)(X_u, \Phi, v) du,$$

where M is a MG and using integration by parts (equation 7.55, on the stochastic processes $e^{-\lambda t}$ and $f(X_t)$) we obtain

$$e^{-\lambda t} f(X_t) - f(X_0) = \int_0^t e^{-\lambda u} dM_u + \int_0^t e^{-\lambda u} (Of)(X_u, \Phi, v) du - \lambda \int_0^t e^{-\lambda u} f(X_s) ds.$$

Finally, taking expectation on both sides above with respect to Q_X^ξ and sending

$t \to \infty$ leads to

$$-f(\boldsymbol{\xi}) = E^{Q_X^{\boldsymbol{\xi}}}\left[\int_0^{+\infty} e^{-\lambda u}\left[(\mathcal{O}f)(X_u, \Phi, \nu) - \lambda f(X_s)\right] du\right],$$

and since $(\mathcal{O}f)(\mathbf{x}, \Phi, \nu) - \lambda f(\mathbf{x}) = -g(\mathbf{x})$ the solution of (7.67) can be written as

$$f(\boldsymbol{\xi}) = E^{Q_X^{\boldsymbol{\xi}}}\left[\int_0^{+\infty} e^{-\lambda u} g(X_u) du\right].$$

2. Dirichlet stochastic partial differential equations The Laplacian operator $\mathcal{O}_{\mathcal{L}}$ is obtained from equation (7.64) by setting $\Phi_{ij}(t, \mathbf{x}) = 2\delta_{ij}$, $\nu_i(\mathbf{x}) = 0$, $\delta_{ij} = I(i = j)$, for all $i, j = 1, 2, \dots, p$, i.e., $(\mathcal{O}_{\mathcal{L}}f)(\mathbf{x}) = \sum\limits_{i=1}^{p} \frac{\partial^2 f(\mathbf{x})}{\partial x_i^2}$. The Dirichlet SPDE (or Dirichlet problem) is defined using the Laplacian operator as follows; given an open ball $D = b(\mathbf{c}, r)$ in $\mathcal{R}^p$ (or disc for $p = 2$), $r > 0$ and g is a continuous function on the ball boundary $\partial b(\mathbf{c}, r)$, we want to find f such that $f \in C^2(\overline{b(\mathbf{c}, r)}, \mathcal{R})$ and

$$(\mathcal{O}_{\mathcal{L}}f)(\mathbf{x}) = 0, \tag{7.68}$$

$\mathbf{x} \in b(\mathbf{c}, r)$, subject to the initial condition $f(\mathbf{x}) = g(\mathbf{x})$, $\mathbf{x} \in \partial b(\mathbf{c}, r)$ with g bounded. When such function f exists it is called the solution to the Dirichlet problem (D, g). We can think of $f(\mathbf{x})$ as the temperature at $\mathbf{x} \in b(\mathbf{c}, r)$, when the boundary temperatures $\partial b(\mathbf{c}, r)$ are specified by g. In order to solve this equation we need to bring in a Brownian motion B explicitly, that is, let $\tau_D = \inf\{t \geq 0 : B_t \in D^c\}$ the first time that B exits the ball D and since D is a bounded open set, $\tau_D < \infty$ $a.s.$ (for a proof see Durrett, 1996, p. 95). Further define $D_n = \{\mathbf{x} \in D : \inf\limits_{\mathbf{y} \in \partial D}\{\|\mathbf{x} - \mathbf{y}\|\} > 1/n\}$. By equation (7.66) we have

$$f(X_{t \wedge D_n}) - f(X_0) = M_t + \int_0^{t \wedge D_n} (\mathcal{O}_{\mathcal{L}}f)(X_u) du.$$

Since $(\mathcal{O}_{\mathcal{L}}f)(\mathbf{x}) = 0$, for $\mathbf{x} \in b(\mathbf{c}, r)$, taking expectations with respect to $Q_X^{\boldsymbol{\xi}}$ yields

$$f(\mathbf{a}) = E^{Q_X^{\boldsymbol{\xi}}}\left[f(X_{t \wedge D_n})\right],$$

$\mathbf{a} \in D_n$ and then sending $t \to \infty$ first and then $n \to \infty$ we have that $f(X_{t \wedge D_n})$ converges to $g(X_{\tau_D})$ $a.s.$ Finally, applying the DCT we pass the limit under the expectation and the solution is given by

$$f(\boldsymbol{\xi}) = E^{Q_X^{\boldsymbol{\xi}}}\left[g(X_{\tau_D})\right], \ \boldsymbol{\xi} \in \overline{D}.$$

3. Cauchy stochastic partial differential equations This is a parabolic OPDE given by

$$\frac{\partial}{\partial t} f(t, \mathbf{x}) = (\mathcal{O}_t f)(\mathbf{x}, \Phi, \nu), \tag{7.69}$$

where now f is a function of $t \in \mathcal{R}_0^+$ and $\mathbf{x} \in \mathcal{R}^p$. The Cauchy problem involves finding an f that (i) satisfies (7.69), (ii) is bounded, (iii) f is twice continuously differentiable with bounded first and second partial derivatives with respect to $\mathbf{x} \in \mathcal{R}^p$, (iv) f is continuously differentiable with respect to $t > 0$, and (v) f satisfies the initial conditions $f(0, x) = g(x)$, $\mathbf{x} \in \mathcal{R}^p$, where g is a continuous function of $\mathbf{x} \in \mathcal{R}^p$

with compact support. Using the multivariate Itô formula of equation (7.57) we can show that the solution is given by

$$f(t,\xi) = E^{Q_{\mathbf{x}}^{\xi}}\left[g(X_t)\right], \tag{7.70}$$

$t > 0$, $\mathbf{x} \in \mathcal{R}^p$. An important special case of (7.69) is the heat equation. Consider an infinite rod extended along the x-axis of the (t, x)-plane. At $t = 0$ the temperature of the rod is $g(x)$ at location $x \in \mathcal{R}$. If $f(t, x)$ denotes the temperature of the rod at time $t \geq 0$ and location $x \in \mathcal{R}$ then f satisfies the heat equation if

$$\frac{\partial}{\partial t} f(t,\mathbf{x}) = \frac{1}{2}\frac{\partial^2}{\partial x^2} f(t,\mathbf{x}), \tag{7.71}$$

with initial condition $f(0, x) = g(x)$, $\mathbf{x} \in \mathcal{R}^p$ and it can be solved in terms of the transition density of the one-dimensional Brownian motion (see Karatzas and Shreve, 1991, Section 4.3).

7.6 Summary

We have developed theory for continuous time parameter stochastic processes, such as Markov processes, continuous-time MGs, the Poisson process and Brownian motion, and based on these processes, we collected elements of stochastic calculus. Standard and specialized texts in these topics include Feller (1968, 1971), Karatzas and Shreve (1991), Karr (1991), Borodin and Salminen (1996), Durrett (1996, 2004, 2010), Fristedt and Gray (1997), Dudley (2002), Çinlar (1975, 2010), Lawler and Limic (2010), Bass (2011), Billingsley (2012) and Klenke (2014). Building on the theory of stochastic processes will allow us to model random objects over time and in particular, as we see in the *TMSO-PPRS* text, point processes and random sets as they evolve over time. For a historical account of stochastic processes see Dellacherie and Meyer (1978). We collect some complementary ideas on topics from this chapter below.

Brownian motion

Brownian motion is named after Robert Brown, a botanist, who was the first to utilize it in the 1820s in order to model the erratic motion of particles in suspension. The rigorous foundations are due to Norbert Wiener and his work in the 1920s, where he proved the existence of the distribution of Brownian motion for the first time. A historical account and references can be found in Karatzas and Shreve (1991, pp. 126-127).

Continuous-time martingales

An extension of Doob's decomposition theorem to the continuous time parameter case (Doob-Meyer decomposition theorem) was given by Meyer in the 1960s. Applications of MGs to financial modeling can be found in Musiela and Rutkowski (2005). See Karatzas and Shreve (1991, p. 46) for a historical account and references on continuous-time MGs.

Filtering

Stochastic calculus results have found many applications in engineering and signal processing via the method of stochastic (linear) filtering. For example, the Kalman-Bucy filter arises as the solution to the SDE

$$\begin{aligned} dX_t &= A(t)dB_t^{(1)} + B(t)X_t dt, \\ dY_t &= dB_t^{(2)} + C(t)X_t dt, \end{aligned}$$

where A, B and C are $p \times p$ deterministic matrices, with elements continuous in t, $B_t^{(1)}$, and $B_t^{(2)}$ are independent Brownian motions with X_t and $B_t^{(2)}$ independent. See Øksendal (2003) and Bass (2011, Chapter 29) for more details on filtering and stochastic calculus.

Stochastic (partial) differential equations

Independent of Itô's work, Gihman developed a theory of SDEs in the late 1940s. For additional details and references on the theoretical foundations based on Itô's formulation see Karatzas and Shreve (1991, pp. 394-398), Durrett (1996) and Bass (2011). For a connection between probability theory and SPDEs see Bass (1997). Moreover, SPDEs can be used to model spatio-temporal data as illustrated in Cressie and Wikle (2011, Chapter 6 and the references therein). Additional details on SPDEs can be found in Øksendal (2003) and for PDEs the interested reader can turn to the text by Jost (2013).

Stochastic integration

The concept of a stochastic integral was first introduced by Paley, Wiener and Zygmud in the early 1930s for deterministic integrands and then by Itô in the 1940s for integrands that are stochastic processes. Doob was the first to study stochastic integrals via MG theory in the 1950s and provided a unified treatment of stochastic integration. More details and references can be found in Karatzas and Shreve (1991, pp. 236-238), Durrett (1996), Bass (2011) and Klenke (2014).

7.7 Exercises

Poisson process

Exercise 7.1 Assume that N_t is a Poisson process defined via definition 7.3. Show the following:
(i) $P(N_t = 0) = e^{-\lambda t}$, $t \geq 0$, for some constant $\lambda \geq 0$.
(ii) $\lim\limits_{t \downarrow 0} \frac{1}{t} P(N_t \geq 2) = 0$.
(iii) $\lim\limits_{t \downarrow 0} \frac{1}{t} P(N_t = 1) = \lambda$, for the same constant λ as in (i).
(iv) $N_t \sim Poisson(\lambda t)$, $t \geq 0$, for the same constant λ as in (i).

Exercise 7.2 For any Poisson process with intensity λ show that $N_{t+s} - N_t | \{N_u : u \leq t\} \overset{d}{=} N_{t+s} - N_t \overset{d}{=} N_s$.

Exercise 7.3 Prove the statements of remark 7.1, parts 5-7.

Exercise 7.4 Let $T = \{T_k : k = 1, 2, \ldots\}$ denote the stochastic process of successive instants of arrivals of a Poisson process $N \sim PP(\lambda)$.

(i) Show that the interarrival times $T_1, T_2 - T_1, T_3 - T_2, \ldots$, are iid $Exp(\lambda)$.
(ii) Find the distribution of T_k.
(iii) For any measurable function $f \geq 0$, show that

$$E\left(\sum_{n=1}^{+\infty} f(T_n)\right) = \lambda \int_0^{+\infty} f(t)dt.$$

Exercise 7.5 Consider a non-stationary Poisson process (property (P3) of definition 7.3 is removed), let $a(t) = E(N_t)$, $t \geq 0$, and define the time inverse of $a(.)$ as the function $\tau(t) = \inf\{s : a(s) > t\}$, $t \geq 0$.
(i) Show that in general $a(.)$ is non-decreasing and right continuous. For the remaining parts assume that $a(.)$ is continuous.
(ii) Define the process $M_t(\omega) = N_{\tau(t)}(\omega)$, $t \geq 0$. Show that $M = \{M_t : t \geq 0\}$ is a stationary Poisson process with intensity 1.
(iii) Show that

$$P(N_{t+s} - N_t = k) = e^{-(a(t+s)-a(t))}(a(t+s) - a(t))^k / k!, \ \forall t, s \geq 0.$$

(iv) If $T_1, T_2, \ldots$, are the successive arrival times, show that $\forall n = 1, 2, \ldots$ and $t \geq 0$, we have $P(T_{n+1} - T_n > t | T_1, \ldots, T_n) = e^{-(a(T_n+t)-a(T_n))}$.

Polish spaces and continuous time filtrations

Exercise 7.6 Recall example 7.4.
(i) Show in detail that $d(x, y)$, $\rho_p(\mathbf{x}, \mathbf{y})$ and $\rho_\infty(\mathbf{x}, \mathbf{y})$ are metrics in $\overline{\mathcal{R}}$, $\mathcal{R}^p$ and $\mathcal{R}^\infty$, respectively.
(ii) Prove that $(\overline{\mathcal{R}}, d)$, $(\mathcal{R}^p, \rho_p)$ and $(\mathcal{R}^\infty, \rho_\infty)$ are Polish spaces.

Exercise 7.7 Prove all statements of remark 7.3.

Exercise 7.8 Show that if τ and ξ are two stopping times with respect to a filtration $\mathcal{F} = \{\mathcal{F}_t, t \geq 0\}$ and $\tau \leq \xi$ then $\mathcal{F}_\tau \subseteq \mathcal{F}_\xi$.

Exercise 7.9 Let τ and ξ be two stopping times with respect to a filtration $\mathcal{F} = \{\mathcal{F}_t, t \geq 0\}$ and Y an integrable random variable in a probability space $(\Omega, \mathcal{A}, P)$. Show that
(i) $E[Y|\mathcal{F}_\tau] = E\left[Y|\mathcal{F}_{\tau\wedge\xi}\right]$ *a.s.* $[P]$ on the set $\{\omega : \tau(\omega) \leq \xi(\omega)\}$,
(ii) $E\left[E[Y|\mathcal{F}_\tau]|\mathcal{F}_\xi\right] = E\left[Y|\mathcal{F}_{\tau\wedge\xi}\right]$ *a.s.* $[P]$.

Exercise 7.10 Give an example of a stochastic process X with minimal filtration $\mathcal{F}^X$ that is not right continuous.

Exercise 7.11 Prove all the statements of remark 7.4.7.

Exercise 7.12 Show that $(D^{\mathcal{R}}_{[0,+\infty)}, \rho_C)$ is a Polish space with ρ_C defined in (7.9).

Markov processes

Exercise 7.13 Show that the Poisson process N_t (based on definition 7.3) is a Markov process that is an adapted, integer-valued RCLL process such that $N_0 = 0$ *a.s.* and for $0 \leq s < t$, $N_t - N_s$ is independent of $\mathcal{F}^N_s$ and it is Poisson distributed with mean $\lambda(t - s)$.

Exercise 7.14 Given a Poisson process N with intensity $\lambda > 0$, adapted to the minimal filtration $\mathcal{F}^N$, define the compensated Poisson process by $M_t = N_t - \lambda t$, for all $t \geq 0$ and adapted to some filtration $\mathcal{F}$. First show that $\mathcal{F}^N = \mathcal{F}$ and then prove that $(M, \mathcal{F})$ is a MG.

Exercise 7.15 Show that the Borel σ-field over $C_{[0,1]}^{\mathcal{R}}$, the space of continuous, $\mathcal{R}$-valued functions defined on [0, 1], is the same as the σ-field generated by the finite dimensional sets (cylinders) $\{\phi \in C_{[0,1]}^{\mathcal{R}} : (\phi(t_1), \ldots, \phi(t_n)) \in B\}$ for Borel sets $B \in \mathcal{B}_n$, $n \geq 1$.

Exercise 7.16 Prove theorem 7.4.

Exercise 7.17 (Dynkin's formula) Let X be a Markov process with generator G and let $\xi \leq \tau$ be *a.s.* finite stopping times with respect to $\mathcal{F}^X$. Show that

$$E(f(X_\tau) - f(X_\xi)) = E\left(\int_{\xi}^{\tau} Gf(X_u)du\right),$$

for any continuous function f in the domain of G.

Exercise 7.18 Prove theorem 7.5.

Exercise 7.19 Give an example of a Markov process that is not strong Markov. Hint: take $E = \mathcal{R}^+$ and define $X_0 = 0$, $X_t = (t - T)^+$, $t > 0$, where $T \sim Exp(c)$ denotes the sojourn at state 0 (see Çinlar, 2010, p. 449).

Exercise 7.20 Consider a Markov chain with state space $\Psi = \{1, 2\}$ and generator

$$A = \begin{bmatrix} -\mu & \mu \\ \lambda & -\lambda \end{bmatrix},$$

where $\lambda\mu > 0$.

(i) Give the form of Kolmogorov's forward equations and solve them to obtain the transition function $p_t(x, y)$, $x, y \in \Psi$.

(ii) Find A^n and use it in equation (7.41). Compare the result with that of part (i).

(iii) Find the stationary distribution π using theorem 7.10 and verify that $p_t(x, y) \to \pi(y)$, as $t \to \infty$.

Markov processes in discrete state space

Exercise 7.21 Prove the statements of remark 7.14, parts 1-2 and 4-6.

Exercise 7.22 Assume that X is a discrete-valued Markov process and with a standard transition function $p_t(x, y)$. Show that the transition function $t \mapsto p_t(x, y)$, for fixed $x, y \in \Psi$, is continuous everywhere.

Exercise 7.23 Prove the statement of remark 7.15.1.

Remark 7.31 Show that if Ψ is finite then no state is instantaneous.

Remark 7.32 Prove all the regularity criteria of remark 7.2.

Remark 7.33 Prove that the sample paths of a Markov process X over a discrete state space Ψ are right continuous if and only if there are no instantaneous states.

Exercise 7.24 Prove theorem 7.6.

Exercise 7.25 For a regular Markov process X and any $x, y \in \Psi$ and $t \geq 0$, show that

$$p_t(x, y) = e^{-\lambda(x)t} I(x = y) + \int_0^t \lambda(x) e^{-\lambda(x)s} \sum_{z \in \Psi} q(x, z) p_{t-s}(z, y) ds.$$

Exercise 7.26 Prove theorem 7.8.

Exercise 7.27 Prove all statements of remark 7.17 in succession.

Brownian motion and the Wiener measure

Exercise 7.28 Prove all statements of remark 7.19.

Exercise 7.29 Let $B \sim BM([0, +\infty))$ with $B_0 = 0$. Show that

$$P(B_s > 0, B_t > 0) = \frac{1}{4} + \frac{1}{2\pi} \arcsin\left(\sqrt{\frac{s}{t}}\right),$$

for $0 < s < t$.

Exercise 7.30 Let μ_1 denote the Lebesgue measure on $[0, 1]$. Show that

$$Q_w(\{\phi : \mu_1(\{t \in [0, 1] : \phi(t) = 0\}) = 0\}) = 1.$$

Exercise 7.31 Let $\{Q_n\}_{n=0}^{+\infty}$ be defined as in remark 7.20. Show (sequentially) that
(i) for $c > 0$ and $k = 0, 1, 2, \ldots$,

$$\lim_{m\to+\infty} \limsup_{n\to+\infty} \left(m \sup_{k \leq mt \leq k+1} \left\{ Q_n \left(\left\{ \phi : \left| \phi(t) - \phi\left(\frac{k}{m}\right) \right| > c \right\} \right) \right\} \right) = 0,$$

(ii) for $c > 0$ and $k = 0, 1, 2, \ldots$,

$$\lim_{m\to+\infty} \limsup_{n\to+\infty} \left(m Q_n \left(\left\{ \phi : \sup_{k \leq mt \leq k+1} \left\{ \left| \phi(t) - \phi\left(\frac{k}{m}\right) \right| \right\} > c \right\} \right) \right) = 0,$$

(iii) for $c > 0$

$$\lim_{m\to+\infty} \limsup_{n\to+\infty} \left(Q_n \left(\left\{ \phi : \sup_{m|t-s| \leq 1} \{ |\phi(t) - \phi(s)| \} > c \right\} \right) \right) = 0.$$

Exercise 7.32 Let $\{Q_n\}_{n=0}^{+\infty}$ be defined as in remark 7.20. Show that $\{Q_n\}_{n=0}^{+\infty}$ is uniformly tight.

Exercise 7.33 Prove theorem 7.11.

Exercise 7.34 Let $H_a = \inf\{t \geq 0 : B_t = a\}$ be the first time a Brownian motion B_t hits the state $a \in \mathcal{R}$.
(i) Show that $\{H_a : a \geq 0\}$ has independent and stationary increments.
(ii) Prove the reflection principle for Brownian motion, that is, for $a > 0$

$$P(H_a \leq t, B_t < a) = P(H_a \leq t, B_t > a).$$

(iii) Show that $P(H_a \leq t) = 2P(B_t > a)$.

Exercise 7.35 Let B_t be Brownian motion and define $H_{a,b} = \inf\{t > 0 : B_t \notin [-a, b]\}$, $a, b > 0$. Show that $P(B_{H_{a,b}} = -a) = b/(a + b) = 1 - P(B_{H_{a,b}} = b)$ and $E(H_{a,b}) = ab$.

Exercise 7.36 Let $S = (S_0, S_1, \dots)$, $S_0 = 0$, be a random walk in $\mathcal{R}$ with steps that have mean 0 and finite positive variance. Show that

$$\lim_{n \to +\infty} P\left(\frac{1}{n}\#\{k \leq n : S_k > 0\} \leq c\right) = \frac{2}{\pi} \arcsin \sqrt{c},$$

for $0 \leq c \leq 1$.

Exercise 7.37 Prove theorem 7.12.

Exercise 7.38 (Lévy process and Blumenthal 0-1 law) A Lévy process L is a generalization of a Wiener process. In particular, L is a $D^{\mathcal{R}}_{[0,+\infty)}$-valued random variable defined on $(\Omega, \mathcal{A}, P)$, such that $L_0 = 0$ *a.s.* and for any $0 = t_0 < t_1 < \cdots < t_n$ and Borel sets $A_j \in \mathcal{B}_1$, $j = 1, 2, \dots, n$, we have

$$P(L_{t_j} - L_{t_{j-1}} \in A_j,\ j = 1, 2, \dots, n) = \prod_{j=1}^{n} P(L_{t_j - t_{j-1}} \in A_j).$$

If $\{\mathcal{F}_{t+} : t \geq 0\}$ denotes the right-continuous minimal filtration of a Lévy process L in $\mathcal{R}$ then show that $\mathcal{F}_{0+}$ is 0-1 trivial.

Stochastic integration and differential equations

For the following exercises assume that $B \sim BM([0, +\infty))$.

Exercise 7.39 Prove all statements of remark 7.24.3.

Exercise 7.40 Assume that τ is a finite stopping time adapted to $\mathcal{F}^B$. Show that $\int_0^{+\infty} I_{[0,\tau]} dB_t = B_\tau$.

Exercise 7.41 (Kunita-Watanabe inequality) Assume that $X, Y \in \mathcal{M}_2^c$. Show that

$$\langle X, Y \rangle_t \leq (\langle X \rangle_t)^{1/2} (\langle Y \rangle_t)^{1/2} \ a.s.$$

Exercise 7.42 Verify that the linear SDE

$$dX_t = aX_t dB_t + bX_t dt,$$

with initial condition $X_0 = x_0$, where $a, b \in \mathcal{R}$, are constants, has the solution given by $X_t = x_0 \exp\left\{aB_t + (b - a^2/2)t\right\}$.

Exercise 7.43 Show in detail that equation (7.70) is the solution of the Cauchy problem.

Appendix A

Additional Topics and Complements

A.1 Mappings in $\mathcal{R}^p$

Remark A.1 (Maps and functions in $\mathcal{R}^d$) By writing "f is a mapping (or map) from $\mathcal{X}$ to $\mathcal{Y}$" we express the fact that $f : \mathcal{X} \to \mathcal{Y}$, i.e., f is a function that assigns elements of $\mathcal{X}$ to elements $\mathcal{Y}$ and f takes values in the (target) set $f(\mathcal{X}) = \{y \in \mathcal{Y} : \exists x \text{ such that } y = f(x)\}$. The term function is used for maps in $\mathcal{R}^d$.

1. The image of $A \subset \mathcal{X}$ through a map f is $f(A) = \{y \in \mathcal{Y} : \exists x \text{ such that } [x \in A \text{ and } y = f(x)]\}$.

2. The inverse image of B is the set $f^{-1}(B) = \{x \in \mathcal{X} : f(x) \in B\}$ and exists always even if the inverse map f^{-1} does not exist.

3. The map f is called "one to one" (1:1) when $f(x_1) = f(x_2) \Longleftrightarrow x_1 = x_2$, $\forall x_1, x_2 \in \mathcal{X}$. The map is called "onto" when $f(\mathcal{X}) = \mathcal{Y}$.

4. The map f is called bijective when it is 1:1 and onto. In this case there exists $g : \mathcal{Y} \to \mathcal{X}$, with $g(f(x)) = x$ and $f(g(y)) = y$, $x \in X$, $y \in \mathcal{Y}$. The map g is called the inverse of f and is denoted by f^{-1}.

5. If $f : \mathcal{X} \to \mathcal{Y}$ and $g : \mathcal{Y} \to \mathcal{Z}$ then define $h : \mathcal{X} \to \mathcal{Z}$, by $h(x) = g(f(x))$, $x \in \mathcal{X}$, the composition of f and g, denoted by $h = f \circ g$.

6. If $f : \mathcal{X} \to \mathcal{Y}$ then the map $g : A \to \mathcal{Y}$, defined as $g(x) = f(x)$, $\forall x \in A$, is called the restriction of f to A and is denoted by $f|A$.

7. Positive and negative parts Now consider $f : \Omega \to \overline{\mathcal{R}}$ and note that it can be written in terms of its positive and negative part as $f = f^+ - f^-$,where $f^+(\omega) = \begin{cases} f(\omega), & \text{if } 0 \leq f(\omega) \leq +\infty \\ 0, & \text{if } -\infty \leq f(\omega) \leq 0 \end{cases}$, the positive part of f and $f^-(\omega) = \begin{cases} -f(\omega), & \text{if } -\infty \leq f(\omega) \leq 0 \\ 0, & \text{if } 0 \leq f(\omega) \leq +\infty \end{cases}$, the negative part of f. Note that f^+ and f^- are non-negative and $|f| = f^+ + f^- \geq 0$.

8. Taylor expansion If $f : \mathcal{R}^p \to \mathcal{R}$ and if $\nabla^2 f = \left[\left(\frac{\partial^2 f(\mathbf{x})}{\partial x_i \partial x_j}\right)\right]$ is continuous on the sphere $\{\mathbf{x} : |\mathbf{x} - \mathbf{x}_0| < r\}$ then for $|\mathbf{t}| < r$

$$f(\mathbf{x}_0 + \mathbf{t}) = f(\mathbf{x}_0) + \nabla f(\mathbf{x}_0)\mathbf{t} + \mathbf{t}^T \left[\int_0^1 \int_0^1 v\nabla^2 f(\mathbf{x}_0 + uv\mathbf{t})dudv\right]\mathbf{t}.$$

9. Mean value If $f : \mathcal{R}^p \to \mathcal{R}^k$ and if ∇f is continuous on the sphere $\{\mathbf{x} : |\mathbf{x} - \mathbf{x}_0| < r\}$ then for $|\mathbf{t}| < r$

$$f(\mathbf{x}_0 + \mathbf{t}) = f(\mathbf{x}_0) + \left[\int_0^1 \nabla f(\mathbf{x}_0 + u\mathbf{t})du\right]\mathbf{t}.$$

10. Bounded variation Consider a partition $\Xi = \{a = \xi_0 < \xi_1 < \cdots < \xi_k = b\}$ of the bounded interval $[a, b]$ and for any $\mathcal{R}$-valued function f over $[a, b]$ define $\|f\|_\Xi = \sum_{i=1}^{k} \left|f(\xi_i) - f(\xi_{i-1})\right|$. The function f is said to be of bounded variation over $[a, b]$ if $\sup_\Xi \|f\|_\Xi < \infty$. If f is absolutely continuous then it is of bounded variation.

A.2 Topological, Measurable and Metric Spaces

Definition A.1 Topological space

A topological space $(\mathcal{X}, \mathcal{O})$, where $\mathcal{X}$ is some space and $\mathcal{O}$ is a collection of sets from $\mathcal{X}$, is a pair satisfying (i) $\emptyset \in \mathcal{T}$ and $\mathcal{X} \in \mathcal{O}$, (ii) $\mathcal{O}$ is a π−system and (iii) $\mathcal{O}$ is closed under arbitrary unions. We say that the space $\mathcal{O}$ is a topology on $\mathcal{X}$ or that $\mathcal{O}$ induces a topology on $\mathcal{X}$, with the members of $\mathcal{O}$ treated as the "open" sets, while their complements are called "closed" sets. A subfamily $\mathcal{O}_0 \subset \mathcal{O}$ is called the base of the topology if each open set $O \in \mathcal{O}$ can be represented as a union of sets from $\mathcal{O}_0$. A sub-base of a topology is a subfamily of sets $\mathcal{O}_1 \subset \mathcal{O}$ such that their finite intersections form the topology base $\mathcal{O}_0$.

Remark A.2 (Topological spaces) Assume that $(\mathcal{X}, \mathcal{O})$ is a topological space.

1. A subset A of $\mathcal{X}$ is compact if every open cover $\{O_i\}_{i=1}^{+\infty}$ of A (i.e., $A \subseteq \bigcup_{i=1}^{+\infty} O_i$, with O_i open) has a finite subcovering (i.e., $A \subseteq \bigcup_{i=1}^{n} O_{a_i}$, for some indices a_i). If every set in $2^{\mathcal{X}}$ is compact then the space $\mathcal{X}$ is compact.

2. A neighborhood of a point in a topological space is any set that contains some open set of which the point is a member.

3. Hausdorff space A topological space is Hausdorff if for any two points x and y in $\mathcal{X}$ there exist neighborhoods of x and y that have empty intersection.

4. A topological space $\mathcal{X}$ is locally compact if for any $x \in \mathcal{X}$ there exists an open set O containing x such that $\overline{O}$ is compact, where $\overline{O}$ denotes the closure of O, which is its smallest closed superset.

5. The empty set $\emptyset$ and the whole space $\mathcal{X}$ are open sets but also closed since $\emptyset^c = \mathcal{X}$ and $\mathcal{X}^c = \emptyset$.

Remark A.3 (Borel σ-fields) We collect some examples and important results on Borel and topological spaces from real analysis.

1. In any topological space, all closed sets, being the complements of open sets, are Borel sets. We could have Borel sets that are neither open nor closed. The intersection of countably many open sets is Borel but may be neither open nor closed. So what does a generic Borel set look like? Since $B \in \mathcal{B} = \sigma(\mathcal{O})$ then B can be written in terms of arbitrary unions, intersections and complements of open sets and in view of theorem 3.2, open intervals.

2. The real line, $\mathcal{R}$, is a topological space, e.g., using the topology induced from the usual metric $\rho(x, y) = |x - y|$. Standard topological spaces of interest in classical topology include the positive real line $\mathcal{R}^+ = \{x \in \mathcal{R} : x > 0\}$, the nonnegative real line $\mathcal{R}_0^+ = \{x \in \mathcal{R} : x \geq 0\}$, the p-dimensional real line $\mathcal{R}^p$, the infinite dimensional real line $\mathcal{R}^\infty$, the extended real line $\overline{\mathcal{R}} = \mathcal{R} \cup \{-\infty\} \cup \{+\infty\}$ and the positive extended real line $\overline{\mathcal{R}}^+$, using appropriately defined metrics to induce the topology.

3. All intervals in $\mathcal{R}$ are Borel sets. In particular, (a, b) is an open ball $b((a + b)/2, (b - a)/2) = \{x \in \mathcal{R} : |x - (a + b)/2| < (b - a)/2\}$, which is an open set and hence Borel. Moreover, the intervals $[a, b) = \bigcap\limits_{n=1}^{+\infty}(a - 1/n, b)$, $(a, b] = \bigcap\limits_{n=1}^{+\infty}(a, b+1/n)$, $[a, b] = \bigcap\limits_{n=1}^{+\infty}(a-1/n, b+1/n)$, $(-\infty, b) = \bigcup\limits_{n=1}^{+\infty}(-n, b)$, $(a, +\infty) = \bigcup\limits_{n=1}^{+\infty}(a, n)$, $\mathcal{R} = (-\infty, +\infty) = \bigcup\limits_{n=1}^{+\infty}(-n, n)$ and the singleton $\{a\} = \bigcap\limits_{n=1}^{+\infty}(a - 1/n, a + 1/n)$ are Borel sets as operations on countable collections of open (Borel) sets. Countable sets like the set of rational numbers $\mathcal{Q}$ are Borel sets.

4. Every open subset O of $\mathcal{R}$ can be written as a countable disjoint union of open intervals, that is, $O = \bigcup\limits_{n=1}^{+\infty}(a_n, b_n)$, with $a_n < b_n$, real numbers.

5. Consider any collection of open sets C in $\mathcal{R}$. Then there exists a countable sub-collection $\{O_n\}_{n=1}^{+\infty}$ of C, such that $\bigcup\limits_{O \in C} O = \bigcup\limits_{n=1}^{+\infty} O_n$.

6. A mapping f of a topological space $(\mathcal{X}, \mathcal{T})$ into a topological space $(\mathcal{Y}, \mathcal{O})$ is said to be continuous if the inverse image of every open set is open, that is, if $O \in \mathcal{O} \Longrightarrow f^{-1}(O) \in \mathcal{T}$. We say that the map f is continuous at a point $x_0 \in \mathcal{X}$, if given any open set $O \in \mathcal{O}$, containing $f(x_0)$, there is an open set $T \in \mathcal{T}$, containing x_0 such that $f(T) \subset O$. The map f is continuous if and only if it is continuous at each point of $\mathcal{X}$.

Remark A.4 (Measurable spaces) Note the following.

1. For any space Ω, since 2^Ω is a σ-field, $(\Omega, 2^\Omega)$ is a measurable space, in fact the largest possible we can build on Ω. So what is the point of looking for the smallest σ-field generated by some collection of sets of the space Ω, when we have already a wonderful σ-field 2^Ω to work with where every set is measurable? The σ-field 2^Ω is in general very large, so large that it can contain sets that will cause mathematical and logical problems with the Axiom of Choice (AC) and the various forms of the Continuum Hypothesis, two foundational items from modern set theory and analysis (see Vestrup, 2003, p. 17).

The AC can be described as follows: given an arbitrary family of nonempty sets we can form a new set consisting of one element from each set in the family. Such logical arguments make sense and will be taken as true since if they are not true, almost all our measure theoretic results are rendered invalid and we can build probability theory, for example, that leads to probabilities of events being 110%. These mathematical paradoxes manifest themselves when one defines measures in order to attach numbers to each of the measurable sets, for example, length of intervals in $\mathcal{R}$ is one such measure.

To further clarify this point, recall example 3.15 of a set that belongs to 2^Ω and hence is measurable in $(\Omega, 2^\Omega)$, but it is not Borel measurable in the measurable space $(\Omega, \mathcal{B}(\Omega))$ since in general $\mathcal{B}(\Omega) \subseteq 2^\Omega$. Another major reason we look for generated σ-fields has to do with the cardinality of the σ-field of the measurable space $(\Omega, \mathcal{A})$, i.e., how many sets are there in $\mathcal{A}$. For example, it is much easier to assign measures to a (possibly finite or countable) collection of sets that generate $\mathcal{A}$ rather than to all possible members of $\mathcal{A}$ directly. Clearly, it makes no sense to say that a set is "measurable" when the measure theory we develop cannot assign a measure to it! Such sets are naturally called nonmeasurable sets.

2. Two measurable spaces are called isomorphic if there exists a bijective function φ between them such that both φ and φ^{-1} are measurable.

3. A measurable space is called a Borel space if it is isomorphic to some $(A, \mathcal{B}(A))$ where A is a Borel set in $[0, 1]$. A product of a finite or countable number of Borel spaces is a Borel space. Every measurable subset A of a Borel space B is itself a Borel space.

The construction (or description) of the open sets of $\mathcal{X}$, can also be accomplished by equipping the space $\mathcal{X}$ with a metric ρ and then describe the open sets using this metric, thus, inducing a topology on $\mathcal{X}$. However, a topological space as defined is a generalization of the topological space obtained via a metric.

Definition A.2 Metric spaces

A metric space consists of a set $\mathcal{X}$ and a function $\rho : \mathcal{X} \times \mathcal{X} \to \mathcal{R}^+$, satisfying (i) $\rho(x, y) = 0$ if and only if $x = y$, (ii) $\rho(x, y) = \rho(y, x)$ (symmetry), and (iii)

$\rho(x,z) \leq \rho(x,y) + \rho(y,z)$ (triangular inequality). The function ρ is called a metric, while the metric space is denoted by $(\mathcal{X}, \rho)$.

Remark A.5 (Topology via a metric) Using the metric we can define, for all $r > 0$, the sets $b(x,r) = \{y \in \mathcal{X} : \rho(x,y) < r\}$, $\overline{b(x,r)} = \{y \in \mathcal{X} : \rho(x,y) \leq r\}$ and $\partial b(x,r) = \{y \in \mathcal{X} : \rho(x,y) = r\}$, interpreted as the open ball, the closed ball and the boundary of the ball, respectively, centered at $x \in \mathcal{X}$ of radius r.

1. An open set $B \subset \mathcal{X}$ is a set that satisfies: $\forall x \in B$, $\exists r > 0$, such that the open ball of radius r centered at x is a subset of B. A set is closed if its complement is open.

2. The interior of a set $B \subset X$, denoted by B^{Int}, is the largest open subset of B, while the closure of $C \subset \mathcal{X}$, denoted by $\overline{C}$, is its smallest closed superset. The boundary of a set B, denoted by ∂B, is defined by $\partial B = \overline{B} \backslash B^{Int}$.

3. Heine–Cantor theorem If $f : \mathcal{X} \to \mathcal{Y}$ is a continuous function between two metric spaces and $\mathcal{X}$ is compact then f is uniformly continuous, that is, $\forall \varepsilon > 0$, $\exists \delta > 0 : |x - y| < \delta \implies |f(x) - f(y)| < \varepsilon$. The difference between uniformly continuous and simply continuous f is that for uniform continuity the bound δ depends only on ε and not the points x and y.

4. A subset B of a set C in a metric space is dense in C if every ball centered at a point in C contains a member of B. A metric space is separable if it contains a countable dense subset. The metric space is complete if every Cauchy sequence converges to an element of the space.

5. A Polish space is a complete separable metric space. Every Polish space is a Borel space.

6. Let $(\mathcal{X}_1, \rho_1)$ and $(\mathcal{X}_2, \rho_2)$ be two metric spaces and consider a map $\phi : \mathcal{X}_1 \to \mathcal{X}_2$. Given a $\gamma \in (0, 1]$, we say that ϕ is locally Hölder-continuous of order γ at a point $x \in \mathcal{X}$, if there exists $\varepsilon > 0$ and $c > 0$ such that, for any $y \in \mathcal{X}$, with $\rho_1(x,y) < \varepsilon$, we have

$$\rho_2(\phi(x), \phi(y)) \leq c\rho_1(x,y)^\gamma. \tag{A.1}$$

We say that ϕ is locally Hölder-continuous of order γ if for every $x \in \mathcal{X}$, there exists $\varepsilon > 0$ and $c = c(x,\varepsilon) > 0$ such that, for all $y, z \in \mathcal{X}$ with $\rho_1(x,y) < \varepsilon$ and $\rho_1(x,z) < \varepsilon$, equation (A.1) holds. Finally, ϕ is called Hölder-continuous of order γ, if there exists a $c > 0$ such that (A.1) holds for all $x, y \in \mathcal{X}$.

Halmos (1950) developed measure theory based on rings, which are smaller classes of sets than fields.

Definition A.3 Ring

Given a set Ω, a collection of sets $\mathcal{A} \subset 2^\Omega$ is called a ring if and only if (i) $\emptyset \in \mathcal{A}$, and (ii) for all $A, B \in \mathcal{A}$ we have $A \cup B \in \mathcal{A}$ and $B \setminus A \in \mathcal{A}$. A ring is a field if and only if $\Omega \in \mathcal{A}$.

A.3 Baire Functions and Spaces

Since we are often working with measures on a space that is also a topological space then it is natural to consider conditions on the measure so that it is connected with the topological structure, thus introducing additional properties. There seem to be two classes of topological spaces where this can be accomplished: locally compact Hausdorff spaces (LCHS, see remark A.2 parts 3 and 4) and complete metric spaces ($\mathcal{R}^p$ enjoys both properties). Royden (1989, Chapters 13 and 15) and Dudley (2002, Section 7.1) present results on Baire and Borel spaces, and introduce the concept of regularity for measures and treat mappings of measure spaces. In particular, locally compact Hausdorff separable spaces (LCHSS) play a pivotal role in studying random set theory; see for example Molchanov (2005).

When defining the measurable space $(\Omega, \mathcal{A})$, the choice of the σ-field $\mathcal{A}$ determines the types of events we can describe and define probabilities for. Similarly, in defining random objects $X : (\Omega, \mathcal{A}, P) \to (\mathcal{X}, \mathcal{G})$ the mapping X and the measurable space $(\mathcal{X}, \mathcal{G})$ determine the types of random objects we can define. In order to treat the two problems we have studied Borel sets and functions throughout the book and we defined most of our random objects to be Borel measurable functions.

Certainly Borel functions make great candidates for random variables and vectors but one naturally wonders what other classes of functions would be acceptable as random variables. Baire functions allow us to define such well-behaved measurable maps and they coincide with Borel functions in $\mathcal{R}^p$, so a sceptic would argue that they are not that useful. But our purpose is to slowly move away from random vectors in $\mathcal{R}^p$ and define random objects in more complicated spaces such as the space of all Radon counting measures (point process objects) or the space of compact-convex sets (random set objects). In many of these complicated spaces what we knew about $\mathcal{R}^p$ does not work anymore and we require additional tools to define and study probability and measurability of mappings. We collect the definition of a Baire set next.

Definition A.4 Baire set

Assume that $\mathcal{X}$ is a locally compact Hausdorff space, and let $C^c(\mathcal{X})$ denote the family of all continuous, real-valued functions that vanish outside a compact subset of $\mathcal{X}$. The class of Baire sets $\mathcal{B}_{\mathcal{X}}$ is defined to be the smallest σ-field of subsets of $\mathcal{X}$ such that each function in $C^c(\mathcal{X})$ is measurable with respect to $\mathcal{B}_{\mathcal{X}}$.

Note that by definition every Baire set is also a Borel set, i.e., $\mathcal{B}_{\mathcal{X}} \subseteq \mathcal{B}(\mathcal{X})$, in general. The converse is true when $\mathcal{X}$ is a locally compact separable metric space. The definition of a Baire function follows.

Definition A.5 Baire functions

The smallest family of functions that contains the continuous functions and is closed with respect to pointwise limits is called the Baire class of functions and is denoted by $\mathcal{B}_C$. The elements of $\mathcal{B}_C$ are called Baire functions.

There are two conditions here; first, if $X \in \mathcal{B}_C$, then X is continuous, and second, if $X_n \in \mathcal{B}_C$ and the limit $\lim_{n \to +\infty} X_n(x) = X(x)$ exists for all $x \in \mathcal{X}$, then $X \in \mathcal{B}_C$. The definition depends only on continuity and no other properties of the space $\mathcal{X}$ and therefore it is applicable to any topological space. For $\mathcal{R}^p$ in particular, it can be shown that $\mathcal{B}_C$ coincides with the class of Borel measurable functions.

The importance of Baire functions as random objects in Baire spaces was seen in example 4.2, parts 2 and 3 and definition 4.16. More precisely, in defining random objects we need to build a measurable map X from $(\Omega, \mathcal{A}, P)$ into $(\mathcal{X}, \mathcal{G})$, that is, X must satisfy $X^{-1}(G) \in \mathcal{A}, \forall G \in \mathcal{G}$ which occurs if $X^{-1}(\mathcal{G}) \subseteq \mathcal{A}$. We could also define the probability space based on the smallest σ-field with respect to which X is measurable, since in general $X^{-1}(\mathcal{G}) = \sigma(X) = \{X^{-1}(G) \subset \Omega, \text{ for all } G \in \mathcal{G}\} \subseteq \mathcal{A}$, so that $X : (\Omega, \sigma(X), P) \to (\mathcal{X}, \mathcal{G})$. Baire functions X possess this structure by definition in a Baire space, namely, the Baire sets form the smallest σ-field with respect to which X is measurable. For more details on the development and applicability to defining random variables as Baire functions, the interested reader can turn to Feller (1971).

A.4 Fisher Information

Definition A.6 Fisher information

The information contained in a vector $\mathbf{X} \sim f(\mathbf{x}|\boldsymbol{\theta})$ about a parameter $\boldsymbol{\theta} \in \Theta \subseteq \mathcal{R}^p$ is defined as the matrix $I_{\mathbf{x}}^F(\boldsymbol{\theta}) = [(I_{ij})]$ where

$$I_{ij} = E\left(\frac{\partial \ln f(\mathbf{x}|\boldsymbol{\theta})}{\partial \theta_i} \frac{\partial \ln f(\mathbf{x}|\boldsymbol{\theta})}{\partial \theta_j}\right). \tag{A.2}$$

If $\frac{\partial^2 \ln f(\mathbf{x}|\boldsymbol{\theta})}{\partial \theta_i \partial \theta_j}$, $i, j = 1, 2, \dots, p$, exists and is finite then $I_{ij} = E\left(-\frac{\partial^2 \ln f(\mathbf{x}|\boldsymbol{\theta})}{\partial \theta_i \partial \theta_j}\right)$.

Regularity conditions Let $\boldsymbol{\theta} \in \Theta \subseteq \mathcal{R}^p$ and assume that the following hold.

1. Θ is an open interval in $\mathcal{R}^p$.

2. The set $\{\mathbf{x} : f(\mathbf{x}|\boldsymbol{\theta}) > 0\}$ is independent of $\boldsymbol{\theta}$.

3. For all $\boldsymbol{\theta} \in \Theta$ the derivative $\frac{\partial \ln f(\mathbf{x}|\boldsymbol{\theta})}{\partial \theta_j}$ exists almost everywhere with respect to $\mathbf{x}$.

4. The quantity $\int \cdots \int_D \prod_{i=1}^n f(x_i|\boldsymbol{\theta}) d\mathbf{x}$ can be differentiated with respect to $\boldsymbol{\theta}$ inside the integral for any set D.

5. $I_{\mathbf{x}}^F(\boldsymbol{\theta}) = \left[\left(E\left(\frac{\partial \ln f(\mathbf{x}|\boldsymbol{\theta})}{\partial \theta_i} \frac{\partial \ln f(\mathbf{x}|\boldsymbol{\theta})}{\partial \theta_j}\right)\right)\right]$ is positive definite, $\forall \boldsymbol{\theta} \in \Theta$.

6. The quantity $\int_D \cdots \int T(\mathbf{x}) \prod_{i=1}^{n} f(x_i|\boldsymbol{\theta}) d\mathbf{x}$ can be differentiated with respect to $\boldsymbol{\theta}$ inside the integral for any statistic $T(\mathbf{X})$ and set D.

Remark A.6 Note that if $X_i \stackrel{iid}{\sim} f(x|\theta)$, $i = 1, 2, \ldots, n$, then the Fisher information contained in $\mathbf{X}$ about θ is simply $I_{\mathbf{X}}^F(\theta) = \sum_{i=1}^{n} I_{X_i}^F(\theta)$.

A.5 Multivariate Analysis Topics

Decision theoretic results in a multivariate setting can be found in Anderson (2003) and Muirhead (2005). In terms of the (joint) decision problem for a mean vector and covariance matrix, we can use a generalization of the weighted quadratic loss, namely, the loss function

$$L(\delta, \theta) = (\mathbf{d} - \boldsymbol{\mu})^T J(\mathbf{d} - \boldsymbol{\mu}) + tr((D - \Sigma)G(D - \Sigma)H),$$

where the parameter is $\theta = \{\boldsymbol{\mu}, \Sigma\}$, the action is $\delta = \{\mathbf{d}, D\}$ and J, G and H are known matrices with the appropriate dimensions.

Generalized multivariate analysis involves multivariate analysis techniques under the assumption of the elliptical family of distributions, i.e., the $p \times 1$ vector $\mathbf{X}$ has density given by

$$f(\mathbf{x}|\boldsymbol{\mu}, \Sigma) = k_p \, |\Sigma|^{-1/2} \, h\left[(\mathbf{x} - \boldsymbol{\mu})^T \Sigma^{-1} (\mathbf{x} - \boldsymbol{\mu})\right],$$

where k_p is a constant that depends only on p and $h(.)$ is a real function that could depend on p, called the generator function. We write $El_p(\boldsymbol{\mu}, \Sigma; h)$ to denote the elliptical family of distributions with generator h, mean vector $\boldsymbol{\mu} = [\mu_1, \ldots, \mu_p]^T$ and covariance structure proportional to $\Sigma = [(\sigma_{ij})]$. When $\boldsymbol{\mu} = \mathbf{0}$ and $\Sigma = I_p$, the family $El_p(\mathbf{0}, I_p; h)$ is called spherical. Notice that h is such that

$$\frac{k_p \pi^{p/2}}{\Gamma(p/2)} \int_0^{+\infty} z^{p/2-1} h(z) dz = 1,$$

since $z = (\mathbf{x} - \boldsymbol{\mu})^T \Sigma^{-1} (\mathbf{x} - \boldsymbol{\mu})$ has density

$$f_z(z) = \frac{k_p \pi^{p/2}}{\Gamma(p/2)} z^{p/2-1} h(z),$$

$z > 0$. The multivariate normal is a special case for $h(z) = e^{-z/2}$.

The family can be alternatively defined via its characteristic function which is given by

$$\varphi_{\mathbf{X}}(\mathbf{t}) = e^{i\mathbf{t}^T \boldsymbol{\mu}} g(\mathbf{t}^T \Sigma \mathbf{t}),$$

for some non-negative function g (that depends on h if the density exists). For additional results one can turn to Fang and Zhang (1990) and Muirhead (2005). We recite here three important results. Assuming that $\mathbf{X} \sim El_p(\boldsymbol{\mu}, \Sigma; h)$ then we can show the following:

(a) $E(\mathbf{X}) = \boldsymbol{\mu}$ and $Var(\mathbf{X}) = a\Sigma$, where $a = -2g'(0)$.

(b) If $\mathbf{X}$ has independent coordinates then $\mathbf{X} \sim \mathcal{N}_p(\mu, \Sigma)$, where Σ is a diagonal matrix.

(c) If B is an $r \times p$ matrix of rank r ($r \leq p$) then $B\mathbf{X} \sim El_r(B\mu, B\Sigma B^T; h)$.

Consequently, elliptical and spherical families are related as follows: if $\mathbf{Y} \sim El_p(\mathbf{0}, I_p; h)$ and $\Sigma = AA^T$, for some $p \times p$ matrix A then $\mathbf{X} = A\mathbf{Y} + \mu \sim El_p(\mu, \Sigma; h)$.

A.6 State Classification

Remark A.7 (Calculating H and R) We have seen several ways of obtaining the matrix of hitting probabilities H and the potential matrix R. We complete the discussion in the following remark. Let Ψ denote the state space of a Markov chain X.

1. Recurrent state x or y Assume that y is recurrent. Then by remark 6.14 we have $H(y, y) = 1$ and therefore $R(y, y) = +\infty$. Now if $x \to y$ then by equation (6.27), $H(x, y) > 0$ and hence $R(x, y) = +\infty$. However, if $x \nrightarrow y$ then $H(x, y) = 0$ and hence $R(x, y) = 0$. Consequently, for y recurrent and any type x we have

$$R(x, y) = \begin{cases} 0, \text{ if } H(x, y) = 0 \\ +\infty, \text{ if } H(x, y) > 0 \end{cases}. \tag{A.3}$$

Now if y is transient and x recurrent then from remark 6.15.6, y cannot be reached by x and therefore $H(x, y) = 0$ and $R(x, y) = 0$. The only remaining case is when x and y are transient.

2. Transient states x and y Let $D \subset \Psi$ denote the set of transient states and assume that $x, y \in D$. Define T_D and R_D to be the matrices obtained from T and R, respectively, by deleting all rows and columns corresponding to recurrent states so that $T_D(x, y) = T(x, y)$ and $R_D(x, y) = R(x, y)$, for $x, y \in D$. Rearrange the states in Ψ so that the recurrent states precede the transient states and write the transition matrix as $T = \begin{bmatrix} T_R & 0 \\ K & T_D \end{bmatrix}$, where T_R is the transition probabilities for the recurrent states and K is the matrix of transition probabilities from the transient states to the recurrent states. Then for any $m \geq 0$ we have $T^m = \begin{bmatrix} T_R^m & 0 \\ K_m & T_D^m \end{bmatrix}$. Since $R = \sum\limits_{m=0}^{+\infty} T^m = \begin{bmatrix} \sum\limits_{m=0}^{+\infty} T_R^m & 0 \\ \sum\limits_{m=0}^{+\infty} K_m & \sum\limits_{m=0}^{+\infty} T_D^m \end{bmatrix}$ we have established that

$$R_D = \sum_{m=0}^{+\infty} T_D^m = I + T_D + T_D^2 + T_D^3 + \ldots,$$

and therefore we can easily compute R_D from the systems of equations

$$(I - T_D)R_D = I = R_D(I - T_D). \tag{A.4}$$

If D is finite then $R_D = (I - T_D)^{-1}$. In general, R_D is the minimal solution to the system of equations

$$(I - T_D)Y = I, \ Y \geq 0. \tag{A.5}$$

Indeed, R_D satisfies (A.4) so it remains to show it is minimal. Let Y be another solution of (A.5). Then

$$Y = I + T_D Y,$$

so that repeatedly replacing Y on the RHS we obtain

$$Y = I + T_D + T_D^2 + \cdots + T_D^n + T_D^{n+1} Y \geq \sum_{m=0}^{n} T_D^m, \tag{A.6}$$

since $T_D^{n+1} \geq 0$ and $Y \geq 0$. Taking limits above as $n \to +\infty$ we have $Y \geq R_D$, as claimed.

A.7 MATLAB® Code Function Calls

See the book website in order to download the MATLAB functions.

Remark A.8 (Goodness-of-fit Monte Carlo test) The function calls for example 2.18 (with the results presented in Table 2.2) are as follows:

```
x=unifrnd(0,1,10,1);MCModelAdequacyex1(100000,1,x);
x=unifrnd(0,1,50,1);MCModelAdequacyex1(100000,1,x);
x=unifrnd(0,1,100,1);MCModelAdequacyex1(100000,1,x);
x=gamrnd(10,10,10,1);MCModelAdequacyex1(100000,1,x);
x=gamrnd(10,10,50,1);MCModelAdequacyex1(100000,1,x);
x=gamrnd(10,10,100,1);MCModelAdequacyex1(100000,1,x);
x=normrnd(-10,1,10,1);MCModelAdequacyex1(100000,1,x);
x=normrnd(-10,1,50,1);MCModelAdequacyex1(100000,1,x);
x=normrnd(-10,1,100,1);MCModelAdequacyex1(100000,1,x);
```

Remark A.9 (Random object realizations) Figure 4.1 presents realizations for several random objects. The function calls are as follows:

```
%Plot a)
RandomFunctionex1([1,1,0,1,0,0,0,1,0,0,1,1,1,1,0],[2,5,7]);
%Plot b)
GenerateMarkovPP([.1],50,1,1,[0,1],[0,1],0,5000,1);
%Plot c)
HPPP(2,100,[0,1],[0,1],[0,1],1);
%seting up a mixture parameters for plot d)
trueps=[1/4,1/4,1/4,1/4];truemus=[[0,0];[0,1];[1,0];[1,1]];
truesigmas=zeros(4,2,2);truesigmas(1,:,:)=eye(2);truesigmas(2,:,:)=eye(2);
truesigmas(3,:,:)=eye(2);truesigmas(4,:,:)=eye(2);
GenPPPFromGivenMixture2d([0,1],[0,1],100,100,0,trueps,truemus,0.01*
truesigmas,1,0);
%Plot e)
RandomDiscex1(5,10,1);
```

```
%Plot f)
GRF2d([],[0,10],[0,10],[1/2,10,0,1],50,1);
```

Remark A.10 (Brownian motion) Figure 7.1 presents realizations for the one- and two-dimensional Brownian motion. The MATLAB calls are:

```
BrownianMotion(1,1000,1000,3,1);
BrownianMotion(2,1000,1000,3,1);
```

A.8 Commonly Used Distributions and Their Densities

We begin with commonly used discrete random variables.

Remark A.11 (Discrete distributions) We collect some of the important characteristics of commonly used discrete random variables.

1. Binomial We write $X \sim Binom(n, p)$ to denote a binomial random variable on n trials with probability of success $p \in [0, 1]$. The pmf is given by $f_X(x|p) = C_x^n p^x (1-p)^{n-x}$, $x = 0, 1, \ldots, n$, with $E(X) = np$, $Var(X) = np(1-p)$ and mgf $m_X(t) = (pe^t + 1 - p)^n$. If $n = 1$ this is called the Bernoulli random variable.

2. Discrete uniform We write $X \sim DUnif(N)$ to denote a discrete uniform random variable, $N = 1, 2, \ldots$. The pmf is given by $f_X(x|N) = 1/N$, $x = 1, \ldots, N$, with $E(X) = \frac{N+1}{2}$, $Var(X) = \frac{(N+1)(N-1)}{12}$ and mgf $m_X(t) = \frac{1}{N}\sum_{j=1}^{N} e^{jt}$.

3. Geometric We write $X \sim Geo(p)$ to denote a geometric random variable with probability of success $p \in [0, 1]$. The pmf is given by $f_X(x|p) = (1-p)p^{x-1}$, $x = 1, 2, \ldots$, with $E(X) = \frac{1}{p}$, $Var(X) = \frac{1-p}{p^2}$ and mgf $m_X(t) = \frac{pe^t}{1-(1-p)e^t}$, $t < -\log(1-p)$.

4. Hypergeometric We write $X \sim HyperGeo(N, M, n)$ to denote a hypergeometric random variable. The pmf is given by $f_X(x|N, M, n) = \frac{C_x^M C_{n-x}^{N-M}}{C_n^N}$, $x = 1, 2, \ldots, n$, $\max(0, M-(N-n)) \leq x \leq \min(M, n)$, with $E(X) = \frac{nM}{N}$, $Var(X) = \frac{nM}{N}\frac{(N-M)(N-n)}{N(N-1)}$. Letting $p = \frac{M}{N}$ we may write $X \sim HyperGeo(N, n, p)$.

5. Multinomial We write $\mathbf{X} \sim Multi(n, \mathbf{p})$, $\mathbf{X} = [X_1, \ldots, X_k]^T$, $\sum_{i=1}^{k} X_i = n$, to denote the multinomial random variable on n trials with probabilities of success $\mathbf{p} = [p_1, \ldots, p_k]^T$, $p_i \in [0, 1]$. The pmf is given by $f_{\mathbf{X}}(\mathbf{x}|\mathbf{p}) = C_{x_1,\ldots,x_k}^n p_1^{x_1} \ldots p_k^{x_k}$, $\sum_{i=1}^{k} x_i = n$, $x_i = 0, 1, \ldots, n$, $i = 1, 2, \ldots, k$, with $E(X_i) = np_i$, $Var(X_i) = np_i(1-p_i)$ and $Cov(X_i, X_j) = -np_ip_j$, $i, j = 1, 2\ldots, k$. Note that for $k = 2$ we get the $Binom(n, p = p_1 = 1 - p_2)$ distribution.

6. Negative binomial We write $X \sim NB(r, p)$ to denote a negative binomial random variable with probability of success $p \in [0, 1]$ and $r = 1, 2, \ldots$. The pmf is given by $f_X(x|r, p) = C_x^{r+x-1} p^r (1-p)^x$, $x = 1, 2, \ldots$, with $E(X) = \frac{r(1-p)}{p}$,

$Var(X) = \frac{r(1-p)}{p^2}$ and mgf $m_X(t) = \left(\frac{pe^t}{1-(1-p)e^t}\right)^r$, $t < -\log(1-p)$. Note that $Geo(p) \equiv NB(r = 1, p)$.

7. Poisson We write $X \sim Poisson(\lambda)$ to denote a Poisson random variable with rate $\lambda \in [0, +\infty)$. The pmf is given by $f_X(x|\lambda) = \frac{\lambda^x e^{-\lambda}}{x!}$, $x = 0, 1, \ldots$, with $E(X) = \lambda = Var(X)$ and mgf $m_X(t) = e^{\lambda(e^t-1)}$.

Remark A.12 (Continuous distributions) We now collect some of the important characteristics of commonly used continuous random variables.

1. Beta We write $X \sim Beta(a, b)$ to denote the beta random variable with parameters $a, b > 0$. The pdf is given by $f_X(x|a, b) = \frac{x^{a-1}(1-x)^{b-1}}{Be(a,b)}$, $0 \leq x \leq 1$, with $Be(a, b)$ the Beta function $Be(a, b) = \int\limits_0^1 x^{a-1}(1-x)^{b-1}dx$, $E(X) = \frac{a}{a+b}$, $Var(X) = \frac{ab}{(a+b)^2(a+b+1)}$ and mgf $m_X(t) = 1 + \sum\limits_{k=1}^{+\infty}\left(\prod\limits_{r=0}^{k-1}\frac{a+r}{a+b+r}\right)\frac{t^k}{k!}$. Note that $Be(a, b) = \frac{\Gamma(a)\Gamma(b)}{\Gamma(a+b)}$, where $\Gamma(a) = \int\limits_0^{+\infty} x^{a-1}e^{-x}dx$, the gamma function.

2. Dirichlet We write $\mathbf{X} \sim Dirichlet(\mathbf{a})$, $\mathbf{X} = [X_1, \ldots, X_k]^T$, $\sum\limits_{i=1}^k X_i = 1$, $0 \leq X_i \leq 1$, to denote the Dirichlet random vector with parameter $\mathbf{a} = [a_1, \ldots, a_k]^T$, with $a_i > 0$, $i = 1, 2 \ldots, k$. Let $a_0 = \sum\limits_{i=1}^k a_i$. The pdf is given by $f_{\mathbf{X}}(\mathbf{x}|\mathbf{a}) = \frac{\Gamma(a_0)}{\Gamma(a_1)\ldots\Gamma(a_k)}x_1^{a_1-1}\ldots x_k^{a_k-1}$, $\sum\limits_{i=1}^k x_i = 1$, $0 \leq x_i \leq 1$, $i = 1, 2, \ldots, k$, with $E(X_i) = \frac{a_i}{a_0}$, $Var(X_i) = \frac{(a_0-a_i)a_i}{a_0^2(a_0+1)}$ and $Cov(X_i, X_j) = -\frac{a_i a_j}{a_0^2(a_0+1)}$, $i, j = 1, 2 \ldots, k$. Note that for $k = 2$ we get the $Beta(a = a_1, b = a_2)$ distribution.

3. Gamma and related distributions We write $X \sim Gamma(a, b)$ to denote the gamma random variable with shape parameter $a > 0$ and scale parameter $b > 0$. The pdf is given by $f_X(x|a, b) = \frac{x^{a-1}e^{-\frac{x}{b}}}{\Gamma(a)b^a}$, $x > 0$, with $E(X) = ab$, $Var(X) = ab^2$ and mgf $m_X(t) = (1 - bt)^{-a}$, $t < 1/b$. The χ_n^2 random variable with n degrees of freedom is defined by $\chi \sim \chi_n^2 \equiv Gamma(a = \frac{n}{2}, b = 2)$, with $E(\chi) = n$ and $Var(\chi) = 2n$. Given two independent random variables $\chi_1 \sim \chi_{n_1}^2$ and $\chi_2 \sim \chi_{n_2}^2$ we define the F-distribution with n_1 and n_2 degrees of freedom via the transformation $F = \frac{\chi_1/n_1}{\chi_2/n_2} \sim F_{n_1,n_2}$. The exponential random variable with rate θ is given by $X \sim Exp(\theta) \equiv Gamma(a = 1, b = \frac{1}{\theta})$, with $E(X) = \theta$ and $Var(X) = \theta^2$. Note that $Y = 1/X$ follows the Inverse Gamma distribution, denoted by $InvGamma(a, b)$ with density $f_X(x|a, b) = \frac{e^{-\frac{1}{b\sigma^2}}}{\Gamma(a)b^a(\sigma^2)^{a+1}}$, $x > 0$, with $E(X) = \frac{1}{b(a-1)}$, if $a > 1$ and $Var(X) = \frac{1}{b^2(a-1)^2(a-2)}$, if $a > 2$.

4. Uniform We write $X \sim Unif(a, b)$ to denote the uniform random variable with parameters $a, b \in \mathcal{R}$, $a < b$. The pdf is given by $f_X(x|a,b) = \frac{1}{b-a}$, $a \le x \le b$, with $E(X) = \frac{a+b}{2}$ and $Var(X) = \frac{(b-a)^2}{12}$. Note that $Unif(0,1) \equiv Beta(a=1, b=1)$.

5. Univariate normal and related distributions We write $X \sim \mathcal{N}(\mu, \sigma^2)$ to denote the normal random variable with parameters $\mu \in \mathcal{R}$ and $\sigma > 0$. The pdf is given by $f_X(x|\mu,\sigma^2) = \frac{1}{\sqrt{2\pi\sigma^2}} e^{-\frac{1}{2\sigma^2}(x-\mu)^2}$, $x \in \mathcal{R}$, with $E(X) = \mu$, $Var(X) = \sigma^2$ and mgf $m_X(t) = e^{\mu t + \frac{t^2\sigma^2}{2}}$. Given independent random variables $Z \sim \mathcal{N}(\mu, \sigma^2)$ and $\mathcal{X} \sim \mathcal{X}_n^2$, the t-distribution with n-degrees of freedom is defined via the transformation $t = \frac{Z}{\sqrt{\frac{\mathcal{X}}{n}}}$, denoted by $t \sim t_n(\mu, \sigma^2)$, with $E(t_n) = \mu$, if $n > 1$ and $Var(t_n) = \frac{n\sigma^2}{n-2}$, if $n > 2$. The density is given by $f_t(t|\mu,\sigma^2) = \frac{\Gamma(\frac{n+1}{2})}{\sqrt{n\pi\sigma^2}\Gamma(\frac{n}{2})}\left[1 + \frac{(x-\mu)^2}{n\sigma^2}\right]^{-\frac{n+1}{2}}$, $t \in \mathcal{R}$. Note that $(\frac{t-\mu}{\sigma})^2 \sim F_{1,n}$. The Cauchy distribution is defined as a special case of the t_n for $n = 1$ by $X \sim Cauchy(\mu, \sigma^2) = t_1(\mu, \sigma^2)$. Note that the mean and the variance do not exist for the Cauchy distribution.

References

Anderson, T. W. (2003). An Introduction to Multivariate Statistical Analysis. 3rd edition, Wiley, New Jersey.

Bass, R. F. (1997). Diffusions and Elliptic Operators. New York, Springer-Verlag.

Bass, R. F. (2011). Stochastic Processes. Cambridge Series in Statistical and Probabilistic Mathematics. Cambridge University Press.

Bayarri, M. J. and Berger, J. O. (1999). Quantifying surprise in the data and model verification. Bayesian Statistics, 6, (eds. Bernardo, J. M. et al.), Oxford Science Publication, 53-82.

Berger, J. O. (1984). The robust Bayesian viewpoint (with discussion). In Robustness of Bayesian Analysis, J. Kadane (ed.), North Holland, Amsterdam.

Berger, J. O. (1985). Statistical Decision Theory and Bayesian Analysis, 2nd edition. Springer-Verlag, New York.

Berger, J.O. (1994). An overview of robust Bayesian analysis. Test, 3, 5-124 (with discussion).

Berger, J. O., Bernardo, J. M., and Sun, D. (2009). The formal definition of reference priors. The Annals of Statistics, 37, 2, 905–938.

Berger J. O., and Delampady, M. (1987). Testing precise hypothesis, (with discussion). Statistical Science, 2, 317-352.

Berliner, L. (1996). Hierarchical Bayesian time-series models. Fundamental Theories of Physics 79, 15-22.

Bernardo, J. M., and Ramon, J. M. (1998). An introduction to Bayesian reference analysis: inference on the ratio of multinomial parameters. The Statistician, 47, 101-35.

Bernardo, J. M., and Smith, A. F. M. (2000). Bayesian Theory. Wiley.

Besag, J. E. (1974). Spatial interaction and the statistical analysis of lattice systems. Journal of the Royal Statistical Society, Series B, 36, 192-225.

Billingsley, P. (2012). Probability and Measure. Anniversary Edition. Wiley.

Borodin, A. N. and Salminen, P. (1996). Handbook of Brownian Motion: Facts and Formulae. Birkhäuser, Basel.

Box, G. E. P. (1980). Sampling and Bayes inference in scientific modelling and robustness. Journal of the Royal Statistical Society, Series A, 143, 383-430.

Brown, L. D. (1975). Estimation with incompletely specified loss functions. Journal of the American Statistical Association, 70, 417-427.

Cappé, O., Robert, C.P., and Rydén, T. (2003). Reversible jump, birth-and-death and more general continuous time Markov chain Monte Carlo samplers. Journal of the Royal Statistical Society, B, 65, 679–700.

Carathéodory, C. (1918). Vorlesungen über reelle Funktionen. Teubner, Leipzig. 2d edition, 1927.

Carter, D. S. and Prenter, P. M. (1972). Exponential spaces and counting processes. Zeitschrift für Wahrscheinlichkeitstheorie und verwandte Gebiete, 21, 1-19.

Casella G. and Berger R. L. (1987). Reconciling Bayesian and Frequentist evidence in the one-sided testing problem. Journal of the American Statistical Association, 82, 106-111.

Casella, G. and Berger, R. L. (2002). Statistical Inference. 2nd edition, Duxbury Advanced Series.

Chung, K.L. (1974). A Course in Probability Theory, 2nd edition. Academic Press, New York.

Çinlar, E. (1975). Introduction to Stochastic Processes. Prentice Hall, Inc., New Jersey.

Çinlar, E. (2010). Probability and Stochastics. Springer, New York.

Cox, D. R. (1977). The role of significance tests, (with discussion). Scandinavian Journal of Statistics, 4, 49-70.

Cressie, N., and Pardo, L. (2000). Minimum φ-divergence estimator and hierarchical testing in loglinear models. Statistica Sinica, 867-884.

Cressie, N., and Wikle, C. K. (2011). Statistics for Spatio-Temporal Data. Wiley, Hoboken, NJ.

De la Horra, J. (2005). Reconciling Classical and Prior Predictive P-Values in the Two-Sided Location Parameter Testing Problem. Communications in Statistics: Theory and Methods, 34, 575 - 583.

De la Horra, J., and Rodrıguez-Bernal, M.T. (1997). Asymptotic behavior of the posterior predictive p-value. Communications in Statistics: Theory and Methods. 26, 2689-2699.

De la Horra, J., and Rodrıguez-Bernal, M.T. (1999). The posterior predictive p-value for the problem of goodness of fit. Test, 8, 117-128.

De la Horra, J., and Rodrıguez-Bernal, M.T. (2000). Optimality of the posterior predictive p-value based on the posterior odds. Communications in Statistics-Theory and Methods, 29, 181-192.

De la Horra, J., and Rodrıguez-Bernal, M.T. (2001). Posterior predictive p-values: what they are and what they are not. Test, 10, 75-86.

Dellacherie, C. and Meyer, P. A. (1978). Probability and Potential. Amsterdam, North-Holland.

Dellaportas, P., and Papageorgiou, I. (2006). Multivariate mixtures of normals with unknown number of components. Statistics and Computing, 16, 57–68.

Dempster, A. P., Laird, N. M., and Rubin, D. B. (1977). Maximum likelihood from incomplete data via the EM algorithm. Journal of the Royal Statistical Society, B, 39, 1-38.

Dey, D. K., Lu, K. and Bose, S. (1998). A Bayesian approach to loss robustness. Statistics and Decisions, 16, 65-87.

Dey, D. K. and Micheas, A. C. (2000). Ranges of posterior expected losses and epsilon-robust actions. In: Robust Bayesian Analysis (Eds. D. Rios Insua and F. Ruggeri; Springer-Verlag). Lecture Notes Monograph Series, Series #152, 145-160.

Dey, R., and Micheas, A. C. (2014). Modeling the growth of objects through a stochastic process of random sets. Journal of Statistical Planning and Inference, 151–152, 17–36.

Diaconis, P., and Ylvisaker, D. (1979). Conjugate Priors for Exponential Families. The Annals of Statistics, 7, 269-281.

Diebolt, J., and Robert, C. P. (1994). Estimation of finite mixture distributions through Bayesian sampling. Journal of the Royal Statistical Society B, 56, 2, 363-375.

Doob, J. L. (1940). Regularity properties of certain families of chance variables. Trans. Amer. Math. Soc., 47, 455–486.

Dudley, R. M. (2002). Real Analysis and Probability, 2nd edition. Cambridge University Press.

Durrett, R. (1996). Stochastic Calculus: A Practical Introduction. CRC Press.

Durrett, R. (2004). Essentials of Stochastic Processes. Springer.

Durrett, R. (2010). Probability: Theory and Examples, 4rth edition. Cambridge University Press.

Escobar, M. and West, M. (1995). Bayesian density estimation and inference using mixtures. Journal of the American Statistical Association, 90, 577–588.

Everitt, B.S., and Hand, D. J. (1981). Finite Mixture Distributions. Chapman and Hall, London.

Fang, K. T. and Zhang, Y. T. (1990). Generalized Multivariate Analysis. Springer-Verlag, Berlin Heidelberg.

Feller, W. (1968). An Introduction to Probability Theory and Its Applications, Vol. 1, 3rd edition. Wiley.

Feller, W. (1971). An Introduction to Probability Theory and Its Applications, Vol. 2, 2nd edition, Wiley.

Ferguson, T.S. (1996). A Course in Large Sample Theory. Chapman and Hall/CRC.

Fristedt, B., and Gray, L. (1997). A Modern Approach to Probability Theory. Birkhäuser.

Frühwirth-Schnatter, S. (2006). Finite Mixture and Markov Switching Models. Springer.

Fu, Q., and Banerjee, A. (2008). Multiplicative mixture models for overlapping clustering, in: International Conference on Data Mining, IEEE, Pisa, Italy, 791–796.

Gelfand, A. E. and Smith, A. (1990). Sampling-based approaches to calculating marginal densities. Journal of the American Statistical Association, 85, 410, DOI: 10.1080/01621459.1990.10476213.

Gelman, A., Carlin, J.B., Stern, H.S., and Rubin, D.B. (2004). Bayesian Data Analysis, 2nd edition. Chapman and Hall.

Gelman, A., Meng, X.L., and Stern, H. (1996). Posterior predictive assessment of model fitness via realized discrepancies. Statistica Sinica, 6, 733-807 (with discussion).

Geman, S., and Geman, D. (1984). Stochastic relaxation, Gibbs distributions, and the Bayesian restoration of images. IEEE Transactions on Pattern Analysis and Machine Intelligence, 6, 6, 721–741.

Geyer, C. J., and Thompson, E. A. (1992). Constrained Monte Carlo maximum likelihood for dependent data. Journal of the Royal Statistical Society. Series B, 657-699.

Geyer, C. J. and Møller, J. (1994). Simulation procedures and likelihood inference for spatial point patterns. Scandinavian Journal of Statistics, 21, 359-373.

Green, P. J. (1995). Reversible jump Markov chain Monte Carlo computation and Bayesian model determination. Biometrika 82, 711-732.

Halmos, P.R. (1950). Measure Theory. Van Nostrand, New York.

Hastings, W. (1970). Monte Carlo sampling methods using Markov chains and their application. Biometrika, 57, 97-109.

Heller, K.A., and Ghahramani, Z. (2007). A nonparametric Bayesian approach to modeling overlapping clusters, in: International Conference on Artificial Intelligence and Statistics, 187–194.

Hwang, J. T., Casella, G., Robert, C., Wells, M. T. and Farrell, R. H. (1992). Estimation of accuracy in testing. Annals of Statistics, 20, 490-509.

James, W., and Stein, C. (1961). Estimation with quadratic loss. In Proc. Fourth Berkeley Symp. Math. Statist. Probab., 1, 361-380, University of California Press.

Jasra, A., Holmes, C.C. and Stephens, D. A. (2005). Markov chain Monte Carlo methods and the label switching problem in Bayesian mixture. Statistical Science, 20, 50-67.

Jeffreys, H. (1946). An invariant form for the prior probability in estimation problems. Proceedings of the Royal Statistical Society of London (Ser. A), 186, 453-461.

Jeffreys, H. (1961). Theory of Probability, 3rd edition. Oxford University Press, London.

Ji, C., Merly, D., Keplerz, T.B., and West, M. (2009). Spatial mixture modelling for unobserved point processes: Examples in immunofluorescence histology. Bayesian Analysis, 4, 297-316.

Johnson, M. E. (1987). Multivariate Statistical Simulation. Wiley.

Johnson, R. A., and Wichern, D. W. (2007). Applied Multivariate Statistical Analysis, 6th ed., Pearson.

Jost, J. (2013). Partial Differential Equations. 3rd ed., Graduate Texts in Mathematics, vol 214, Springer, New York.

Jozani, M. J., Marchand, É., and Parsian, A. (2012). Bayesian and Robust Bayesian analysis under a general class of balanced loss functions. Statistical Papers, 53, 1, 51-60.

Kadane, J. B., and Chuang, D. T. (1978). Stable decision problems. The Annals of Statistics, 1095-1110.

Kallenberg, O. (1986). Random Measures. 4th ed., Akademie, Berlin.

Kallenberg, O. (2002). Foundations of Modern Probability, 2nd ed., Probability and Its Applications, Springer, New York.

Karatzas, I., and Shreve S. E. (1991). Brownian Motion and Stochastic Calculus. 2nd ed., Springer, New York.

Karr, A. (1991). Point Processes and Their Statistical Inference. 2nd ed., CRC / Marcel Dekker, Inc.

Kass, R. E., and Wasserman, L. (1996). The selection of prior distributions by formal rules. Journal of the American Statistical Association, 91, 1343-70.

Klenke, A. (2014). Probability Theory: A Comprehensive Course, 2nd edition. Springer.

Kolmogorov, A. N. (1933 and 1956). Grundbegriffe derWahrscheinlichkeitsrechnung, Ergebnisse der Math., Springer, Berlin. English translation: Foundations of the Theory of Probability, 2nd edition. Chelsea, New York, 1956.

Kottas, A. and Sanso, B. (2007). Bayesian mixture modeling for spatial Poisson process intensities, with applications to extreme value analysis. Journal of Statistical Planning and Inference (Special Issue on Bayesian Inference for Stochastic Processes), 137, 3151-3163.

Lawler, G. F. and Limic V. (2010). Random Walk. Cambridge Univ. Press, Cambridge.

Lebesgue, H. (1902). Intégrale, longueur, aire (thèse, Univ. Paris). Annali Mat. purae appl. (Ser. 3) 7: 231–359.

Lehmann, E. L. (1986). Testing Statistical Hypothesis, 2nd edition. Wiley, New York, USA.

Lehmann, E. L. (2004). Elements of Large-Sample Theory. Springer.

Lehmann, E. L. and Casella, G. (1998). Theory of Point Estimation, 2nd edition. Springer-Verlag, New York, USA.

Lin, G. D. (2017). Recent Developments on the Moment Problem. arXiv preprint arXiv:1703.01027.

Lindgren, G. (2013). Stationary Stochastic Processes: Theory and Applications. Chapman & Hall/CRC, Boca Raton.

Lindsay, B. G. (1995). Mixture Models: Theory, Geometry and Applications, volume 5 of NSF-CBMS Regional Conference Series in Probability and Statistics, Institute of Mathematical Statistics and the American Statistical Association.

Makov, U. E. (1994). Some aspects of Bayesian loss robustness. Journal of Statistical Planning and Inference, 38, 359-370.

Martin, J.-M., Insua, D. R. and Ruggeri, F. (1998). Issues in Bayesian loss robustness. Sankhya A, 60, 405-417.

Matérn, B. (1986). Spatial variation. Lecture Notes in Statistics, 36, Springer-Verlag, Berlin.

McLachlan, G. J., and Peel, D. (2000). Finite Mixture Models. Wiley-Interscience.

Meng, X. L. (1994) Posterior predictive p-values. Annals of Statistics, 22, 1142-1160.

Metropolis, N., Rosenbluth, A.W., Rosenbluth, M.N., Teller, A.H., Teller, E. (1953). Equation of state calculations by fast computing machines. Journal of Chemical Physics, 21, 1087-1092.

Micheas, A. C. Textbook website: https://www.crcpress.com/9781466515208.

Micheas, A. C. (2006). A unified approach to Prior and Loss Robustness. Communications in Statistics: Theory and Methods, 35, 309-323.

Micheas, A. C. (2014). Hierarchical Bayesian modeling of marked non-homogeneous Poisson processes with finite mixtures and inclusion of covariate information. Journal of Applied Statistics, 41, 12, 2596-2615.

Micheas, A. C. and Dey, D. K. (2003). Prior and posterior predictive p-values in the one-sided location parameter testing problem. Sankhya, A, 65, 158-178.

Micheas, A.C. and Dey, D.K. (2004). On measuring loss robustness using maximum a posteriori estimate. Communications in Statistics: Theory and Methods, 33, 1069-1085.

Micheas, A.C., and Dey, D.K. (2007). Reconciling Bayesian and frequentist evidence in the one-sided scale parameter testing problem. Communications in Statistics: Theory and Methods, 36, 6, 1123-1138.

Micheas, A.C., Wikle, C.K., and Larsen, D.R. (2012). Random set modelling of three dimensional objects in a hierarchical Bayesian context. Journal of Statistical Computation and Simulation, DOI:10.1080/00949655.2012.696647.

Micheas, A.C., and Zografos, K. (2006). Measuring stochastic dependence using φ-divergence. Journal of Multivariate Analysis 97.3, 765-784.

Mengersen, K., Robert, C. P., and Titterington, M. (2011). Mixtures: Estimation and Applications. Wiley.

Molchanov, I. (2005). Theory of Random Sets. Springer-Verlag, London.

Muirhead, R. J. (2005). Aspects of Multivariate Statistical Theory, 2nd edition. Wiley.

Musiela, M. and Rutkowski, M. (2005). Martingale Methods in Financial Modelling. 2nd ed., Stochastic modelling and applied probability, vol 36, Springer, Berlin.

Neyman, J., and Scott, E.L. (1948). Consistent estimates based on partially consistent observations. Econometrica: Journal of the Econometric Society, 1-32.

Nguyen, H. T. (2006). An Introduction to Random Sets. Chapman & Hall/CRC, Boca Raton.

Øksendal, B. (2003). Stochastic Differential Equations: An Introduction with Applications. 6th ed., Springer-Verlag, Berlin.

Pearson, K. (1894). Contributions to the mathematical theory of evolution. Philosophical Transactions of the Royal Society of London A, 185, 71–110.

Press, J.S. (2003). Subjective and Objective Bayesian Statistics, 2nd edition. Wiley.

Ramsay, J. O. and Novick, M. R. (1980). PLU robust Bayesian decision. Theory: point estimation. Journal of the American Statistical Association, 75, 901-907.

Rényi, A. (1970). Foundations of Probability. Holden Day, San Francisco, 1970.

Richardson, S., and Green, P.J. (1997). On Bayesian analysis of mixtures with an unknown number of components (with discussion). Journal of the Royal Statistical Society, B, 59, 731-792.

Rios Insua, D. and Ruggeri, F. (2000). Robust Bayesian Analysis. Lecture Notes Monograph Series, Series #152, Springer-Verlag.

Robert, C. P. (2007). The Bayesian Choice, 2nd edition. Springer.

Robert, C. P., and Casella, G. (2004). Monte Carlo Statistical Methods, 2nd edition. Springer.

Robert, C. P., and Casella, G. (2010). Introducing Monte Carlo Methods with R. Springer-Verlag.

Royden, H.L. (1989). Real Analysis, 3rd edition. Prentice Hall.

Rubin, D. B. (1984). Bayesianly justifiable and relevant frequency calculations for the applied statistician. Annals of Statistics, 12, 1151-1172.

Scricciolo, C. (2006). Convergence rates for Bayesian density estimation of infinite-dimensional exponential families. The Annals of Statistics, 34, 2897–2920.

Sethuraman, J. (1994). A constructive definition of Dirichlet priors. Statistica Sinica, 4, 639–650.

Shafer, G. (1982). Lindley's Paradox, (with discussion). Journal of the American Statistical Association, 77, 325-351.

Skorohod, A. V. (1956). Limit theorems for stochastic processes. Theory of Probability and Its Applications, SIAM, 261-290.

Stephens, M. (2000). Bayesian analysis of mixture models with an unknown number of components: An alternative to reversible jump methods. Annals of Statistics, 28, 40–74.

Stummer, W., and Vajda, I. (2010). On divergences of finite measures and their applicability in statistics and information theory. Statistics 44.2, 169-187.

Tanner, M. and Wong, W. (1987). The calculation of posterior distributions by data augmentation. Journal of the American Statistical Association, 82, 528-550.

Thomopoulos, N. T. (2013). Essentials of Monte Carlo Simulation: Statistical Methods for Building Simulation Models. Springer.

Titterington, D. M., Smith, A. F. M., and Makov, U. E. (1985). Statistical Analysis of Finite Mixture Distributions. Wiley, New York.

Vajda, I. (1989). Theory of Statistical Inference and Information, Kluwer Academic Publishers.

Varian, H.R. (1975). A Bayesian approach to real estate assessment. In Studies in Bayesian Econometrics and Statistics in Honor of Leonard J. Savage, (S.E. Feinberg and A. Zellner eds.), 195-208, Amsterdam: North Holland.

Vestrup, E. M., (2003). The Theory of Measure and Integration. John Wiley & Sons, Inc., Hoboken, New Jersey.

Wasserman, L., (2009). All of Nonparametric Statistics. Springer.

Widom, B. and Rowlinson, J. S. (1970). A new model for the study of liquid-vapor phase transitions. Journal of Chemichal Physics, 52, 1670-1684.

Wikle, C.K., Berliner, L. and Cressie, N. (1998). Hierarchical Bayesian space-time models. Environmental and Ecological Statistics 5, 2, 117-154.

Wolpert, R. and Ickstadt, K. (1998). Poisson/Gamma random field models for spatial statistics. Biometrika, 85, 251-267.

Zellner, A. (1986). Bayesian estimation and prediction using asymmetric loss functions. Journal of the American Statistical Association, 81, 446-451.

Zellner, A. and Geisel, M.S. (1968). Sensitivity of control to the form of the criterion function. In: The Future of Statistics, D.G. Watts, ed., Academic Press, 269-289. (With comments by J.C. Kiefer and H.O. Hartley and the authors' responses).

Index

Printed in the United States
by Baker & Taylor Publisher Services

Printed in the United States
by Baker & Taylor Publisher Services